T0178577

# Lecture Notes in Artificial Intelligence    11828

## Subseries of Lecture Notes in Computer Science

Series Editors

Randy Goebel
*University of Alberta, Edmonton, Canada*
Yuzuru Tanaka
*Hokkaido University, Sapporo, Japan*
Wolfgang Wahlster
*DFKI and Saarland University, Saarbrücken, Germany*

Founding Editor

Jörg Siekmann
*DFKI and Saarland University, Saarbrücken, Germany*

More information about this series at http://www.springer.com/series/1244

Petra Kralj Novak · Tomislav Šmuc ·
Sašo Džeroski (Eds.)

# Discovery Science

22nd International Conference, DS 2019
Split, Croatia, October 28–30, 2019
Proceedings

 Springer

*Editors*
Petra Kralj Novak 🆔
Jožef Stefan Institute
Ljubljana, Slovenia

Tomislav Šmuc 🆔
Rudjer Bošković Institute
Zagreb, Croatia

Sašo Džeroski
Jožef Stefan Institute
Ljubljana, Slovenia

ISSN 0302-9743        ISSN 1611-3349    (electronic)
Lecture Notes in Artificial Intelligence
ISBN 978-3-030-33777-3      ISBN 978-3-030-33778-0    (eBook)
https://doi.org/10.1007/978-3-030-33778-0

LNCS Sublibrary: SL7 – Artificial Intelligence

© Springer Nature Switzerland AG 2019
The chapter "Sparse Robust Regression for Explaining Classifiers" is Open Access. This chapter is licensed under the terms of the Creative Commons Attribution 4.0 International License (http://creativecommons.org/licenses/by/4.0/). For further details see license information in the chapter.
This work is subject to copyright. All rights are reserved by the Publisher, whether the whole or part of the material is concerned, specifically the rights of translation, reprinting, reuse of illustrations, recitation, broadcasting, reproduction on microfilms or in any other physical way, and transmission or information storage and retrieval, electronic adaptation, computer software, or by similar or dissimilar methodology now known or hereafter developed.
The use of general descriptive names, registered names, trademarks, service marks, etc. in this publication does not imply, even in the absence of a specific statement, that such names are exempt from the relevant protective laws and regulations and therefore free for general use.
The publisher, the authors and the editors are safe to assume that the advice and information in this book are believed to be true and accurate at the date of publication. Neither the publisher nor the authors or the editors give a warranty, expressed or implied, with respect to the material contained herein or for any errors or omissions that may have been made. The publisher remains neutral with regard to jurisdictional claims in published maps and institutional affiliations.

This Springer imprint is published by the registered company Springer Nature Switzerland AG
The registered company address is: Gewerbestrasse 11, 6330 Cham, Switzerland

# Preface

The Discovery Science conference presents a unique combination of latest advances in the development and analysis of methods for discovering scientific knowledge, coming from machine learning, data mining, and intelligent data analysis, with their application in various scientific domains.

The 22nd International Conference on Discovery Science (DS 2019) was held in Split, Croatia, during October 28–30, 2019. This was the first time the conference was organized as a stand-alone event. For its first 20 editions, DS was co-located with the International Conference on Algorithmic Learning Theory (ALT). In 2018 it was co-located with the 24th International Symposium on Methodologies for Intelligent Systems (ISMIS 2018).

DS 2019 received 63 international submissions. Each submission was reviewed by at least three Program Committee (PC) members. The PC decided to accept 21 regular papers and 19 short papers. This resulted in an acceptance rate of 33% for regular papers.

The conference included three keynote talks. Marinka Žitnik (Stanford University) contributed a talk titled "Representation Learning as a New Approach to Biomedical Research," Guido Caldarelli (IMT Lucca and ECLT Venice) gave a presentation titled "The Structure of Financial Networks," and Dino Pedreschi (University of Pisa), contributed a talk titled "Data and Algorithmic Bias: Explaining the Network Effect in Opinion Dynamics and the Training Data Bias in Machine Learning." Abstracts of the invited talks with short biographies of the invited speakers are included in these proceedings.

Besides the presentation of regular and short papers in the main program, the conference offered two new sessions. The "PhD Symposium" gave an opportunity to PhD students at an early stage of their studies to participate in the conference by presenting the topics of and early results from their research and discuss their work and experiences with peers, senior researchers and leading experts working on similar problems. The session titled "Late Breaking Contributions" featured poster and spotlight presentations of very recent research results on topics related to Discovery Science.

We are grateful to Springer for their long-term support, which got even stronger this year. Springer publishes the conference proceedings, as well as a regular special issue of the Machine Learning journal on Discovery Science. The latter offers authors a chance of publishing in this prestigious journal significantly extended and reworked versions of their DS conference papers, while being open to all submissions on DS conference topics.

This year, Springer (LNCS and Machine Learning journal), supported the best student paper awards. For DS 2019, the awardees are Anton Björklund, Andreas Henelius, Emilia Oikarinen, Kimmo Kallonen and Kai Puolamäki (for the paper "Sparse Robust Regression for Explaining Classifiers") and Yannik Klein, Michael

Rapp and Eneldo Loza Mencía (for the paper "Efficient Discovery of Expressive Multi-label Rules Using Relaxed Pruning.") We would like to thank the Best Paper Award committee composed of Dragan Gamberger and Toon Calders for their precious and timely evaluations.

On the program side, we would like to thank all the authors of submitted papers, the PC members and the additional reviewers for their efforts in evaluating the submitted papers, as well as the keynote speakers. On the organization side, we would like to thank all the members of the Organizing Committee: Tomislav Lipić, Ana Vidoš, Matija Piškorec and Ratko Mileta, for the smooth preparation and organization of all conference associated activities. We are also grateful to the people behind EasyChair for developing the conference organization system that proved to be an essential tool in the paper submission and evaluation process, as well as in the preparation of the Springer proceedings.

The DS 2019 conference was organized under the auspices of the Rudjer Bošković Institute in Zagreb. The event was also supported by the Project of the Croatian Center for Excellence in Data Science and Advanced Cooperative Systems. Significant support, especially through human resources, was also provided by the Jožef Stefan Institute from Ljubljana. Finally, we are indebted to all conference participants, who contributed to making this exciting event a worthwhile endeavor for all involved.

October 2019

Petra Kralj Novak  
Tomislav Šmuc  
Sašo Džeroski

# Organization

## General Chair

Sašo Džeroski        Jožef Stefan Institute, Slovenia

## Program Committee Chairs

Petra Kralj Novak       Jožef Stefan Institute, Slovenia
Tomislav Šmuc        Rudjer Bošković Institute, Croatia

## PhD Symposium Chair

Tomislav Lipić        Rudjer Bošković Institute, Croatia

## Proceedings Chair

Matija Piškorec       Rudjer Bošković Institute, Croatia

## Web and Social Media Chairs

Ratko Mileta        Rudjer Bošković Institute, Croatia
Matija Piškorec      Rudjer Bošković Institute, Croatia

## Local Arrangements Chair

Ana Vidoš         Rudjer Bošković Institute, Croatia

## Program Committee

Annalisa Appice       University of Bari Aldo Moro, Italy
Martin Atzmueller     Tilburg University, The Netherlands
Viktor Bengs        Paderborn University, Germany
Concha Bielza Lozoya   Universidad Politécnica de Madrid, Spain
Albert Bifet         LTCI, Telecom ParisTech, France
Alberto Cano        Virginia Commonwealth University, USA
Michelangelo Ceci     University of Bari Aldo Moro, Italy
Bruno Cremilleux     University of Caen Normandy, France
Claudia d'Amato      University of Bari Aldo Moro, Italy
Nicola Di Mauro      University of Bari Aldo Moro, Italy
Ivica Dimitrovski     Ss. Cyril and Methodius University in Skopje,
                       North Macedonia

| | |
|---|---|
| Wouter Duivesteijn | Eindhoven University of Technology, The Netherlands |
| Lina Fahed | IMT Atlantique, France |
| Hadi Fanaee | University of Oslo, Norway |
| Nicola Fanizzi | University of Bari Aldo Moro, Italy |
| Stefano Ferilli | University of Bari Aldo Moro, Italy |
| Johannes Fürnkranz | Technische Universität Darmstadt, Germany |
| Mohamed Gaber | Birmingham City University, UK |
| João Gama | University of Porto, Portugal |
| Dragan Gamberger | Rudjer Bošković Institute, Croatia |
| Makoto Haraguchi | Hokkaido University, Japan |
| Kouichi Hirata | Kyushu Institute of Technology, Japan |
| Jaakko Hollmén | Aalto University, Finland |
| Eyke Huellermeier | Paderborn University, Germany |
| Alípio Jorge | University of Porto, Portugal |
| Masahiro Kimura | Ryukoku University, Japan |
| Dragi Kocev | Jožef Stefan Institute, Slovenia |
| Stefan Kramer | Johannes Gutenberg University Mainz, Germany |
| Vincenzo Lagani | Ilia State University, Georgia |
| Pedro Larranaga | University of Madrid, Spain |
| Nada Lavrač | Jožef Stefan Institute, Slovenia |
| Jurica Levatić | Institute for Research in Biomedicine, Spain |
| Tomislav Lipić | Rudjer Bošković Institute, Croatia |
| Francesca Alessandra Lisi | University of Bari Aldo Moro, Italy |
| Gjorgji Madjarov | Ss. Cyril and Methodius University in Skopje, North Macedonia |
| Giuseppe Manco | Institute for High Performance Computing and Networking, Italy |
| Sanda Martinčić-Ipšić | University of Rijeka, Croatia |
| Elio Masciari | Institute for High Performance Computing and Networking, Italy |
| Anna Monreale | University of Pisa, Italy |
| Siegfried Nijssen | Université Catholique de Louvain, Belgium |
| Rita P. Ribeiro | University of Porto, Portugal |
| Panče Panov | Jožef Stefan Institute, Slovenia |
| Ruggero G. Pensa | University of Torino, Italy |
| Bernhard Pfahringer | University of Waikato, New Zealand |
| Gianvito Pio | University of Bari Aldo Moro, Italy |
| Pascal Poncelet | LIRMM Montpellier, France |
| Jan Ramon | French Research Institute for Digital Sciences, France |
| Chedy Raïssi | French Research Institute for Digital Sciences, France |
| Marko Robnik-Šikonja | University of Ljubljana, Slovenia |
| Kazumi Saito | University of Shizuoka, Japan |
| Marina Sokolova | University of Ottawa and Institute for Big Data Analytics, Canada |
| Jerzy Stefanowski | Poznan University of Technology, Poland |

Ljupčo Todorovski            University of Ljubljana, Slovenia
Luis Torgo                   Dalhousie University, Canada
Herna Viktor                 University of Ottawa, Canada
Albrecht Zimmermann          Université Caen Normandie, France
Blaž Zupan                   University of Ljubljana, Slovenia

## Additional Reviewers

Ahmadi, Mohsen                       Oliveira, Mariana
Barracchia, Emanuele Pio             Pasquadibisceglie, Vincenzo
Cancela, Brais                       Pisani, Francesco S.
Chambers, Lorraine                   Stepišnik, Tomaž
Fernandes, Sofia                     Tabassum, Shazia
Ghomeshi, Hossein                    Tornede, Tanja
Guarascio, Massimo                   Wever, Marcel
Koptelov, Maksim                     Zopf, Markus
Kulikovskikh, Ilona

# Keynote Talks

# The Structure of Financial Networks

Guido Caldarelli

IMT School for Advanced Studies,
Lucca and European Centre for Living Technology, Venice

**Abstract.** Financial inter-linkages play an important role in the emergence of financial instabilities and the formulation of systemic risk can greatly benefit from a network approach. In this talk, we focus on the role of linkages along the two dimensions of contagion and liquidity, and we discuss some insights that have recently emerged from network models. With respect to the issue of the determination of the optimal architecture of the financial system, models suggest that regulators have to look at the interplay of network topology, capital requirements, and market liquidity. With respect to the issue of the determination of systemically important financial institutions, the findings indicate that both from the point of view of contagion and from the point of view of liquidity provision, there is more to systemic importance than just size. In particular for contagion, the position of institutions in the network matters and their impact can be computed through stress tests even when there are no defaults in the system.

We present an overview of the use of networks in Finance and Economics. We show how this approach enables us to address important questions as, for example, the stability of financial systems and the systemic risk associated with the functioning of the interbank market. For example with DebtRank, a novel measure of systemic impact inspired by feedback-centrality we are able to measure the nodes that become systemically important at the peak of the crisis. Moreover, a systemic default could have been triggered even by small dispersed shocks. The results suggest that the debate on too-big-to-fail institutions should include the even more serious issue of too-central-to-fail. All these results are new in the field and allow for a better understanding and modelling of different Financial systems.

**Keywords:** Financial networks · Systemic risk · Interbank market

**Short Biography of the Lecturer:** Guido Caldarelli is Full Professor in Theoretical Physics at IMT School for Advanced Studies Lucca, and is Research associate at the European Centre for Living Technology, Venice. His main scientific activity is the study of networks, mostly analysis and modelling of financial networks. Author of more than 200 publication on the subject and three books, he is currently the president of the Complex Systems Society. He has been coordinator of the FET IP Project MULTIPLEX: Foundational Research on Multilevel Complex Networks and Systems (2012–2016), the FET OPEN Project FoC: Forecasting Financial Crises (2010–2014), and the FET OPEN Project COSIN: Coevolution and Self Organization in Complex

Networks (2002–2005). Guido Caldarelli received his Ph.D. from SISSA, after which he was a postdoc in the Department of Physics and School of Biology, University of Manchester. He then worked at the Theory of Condensed Matter Group, University of Cambridge. He returned to Italy as a lecturer at National Institute for Condensed Matter (INFM) and later as Primo Ricercatore in the Institute of Complex Systems of the National Research Council of Italy. In this period, he was also the coordinator of the Networks subproject, part of the Complexity Project, for the Fermi Centre. He also spent some terms at University of Fribourg (Switzerland) and in 2006 he has been visiting professor at École Normale Supérieure in Paris. More information and a complete CV are available at: http://www.guidocaldarelli.com.

# Data and Algorithmic Bias: Explaining the Network Effect in Opinion Dynamics and the Training Data Bias in Machine Learning

Dino Pedreschi

Università di Pisa, Istituto di Scienza e Tecnologie dell'Informazione,
Consiglio Nazionale delle Ricerche
http://kdd.isti.cnr.it

**Abstract.** Data science and network science are creating novel means to study the complexity of our societies and to measure, understand and predict social phenomena. My talk gives an overview of recent research at the Knowledge Discovery (KDD) Lab in Pisa within the SoBigData.eu research infrastructure, targeted at explaining the effects of data and algorithmic bias in different domains, using both data-driven and model-driven arguments. First, I introduce a model showing how algorithmic bias instilled in an opinion diffusion process artificially yields increased polarisation, fragmentation and instability in a population. Second, I focus on the urgent open challenge of how to construct meaningful explanations of opaque AI/ML black-box decision systems, introducing the local-to-global framework for the explanation of ML classifiers as a way towards explainable AI. The two cases show how the combination of data-driven and model-driven interdisciplinary research has a huge potential to shed new light on complex phenomena like discrimination and polarisation, as well as to explain how decision making black-boxes, both human and artificial, actually work. I conclude with an account of the open data science paradigm pursued in SoBigData.eu Research Infrastructure and its importance for inter-disciplinary data driven science that impacts societal challenges.

**Keywords:** Explainable AI · Data bias · Algorithmic bias

**Short Biography of the Lecturer:** Dino Pedreschi is a professor of computer science at the University of Pisa, and a pioneering scientist in data science. He co-leads the Pisa KDD Lab – Knowledge Discovery and Data Mining Laboratory http://kdd.isti.cnr.it, a joint research initiative of the University of Pisa and the Information Science and Technology Institute of the Italian National Research Council. His research focus is on big data analytics and mining and their impact on society. He is a founder of the Business Informatics MSc program at University of Pisa, a course targeted at the education of interdisciplinary data scientists, and of SoBigData.eu, the European H2020 Research Infrastructure "Big Data Analytics and Social Mining Ecosystem" www.sobigdata.eu. Dino has been a visiting scientist at Barabasi Lab (Center for

Complex Network Research) of Northeastern University, Boston, and earlier at the University of Texas at Austin, at CWI Amsterdam and at UCLA. In 2009, Dino received a Google Research Award for his research on privacy-preserving data mining. Dino is a member of the expert group in AI of the Italian Ministry of research and the director of the Data Science PhD program at Scuola Normale Superiore in Pisa. Dino is a co-PI of the 2019 ERC grant XAI – Science and technology for the explanation of AI decision making (PI: Fosca Giannotti).

# Representation Learning as a New Approach to Biomedical Research

Marinka Žitnik

Computer Science Department, School of Engineering, Stanford University

**Abstract.** Large datasets are being generated that can transform science and medicine. New machine learning methods are necessary to unlock these data and open doors for scientific discoveries. In this talk, I will argue that machine learning models should not be trained in the context of one particular dataset. Instead, we should be developing methods that combine data in their broadest sense into knowledge networks, enhance these networks to reduce biases and uncertainty, and then learn and reason over the networks. My talk will focus on two key aspects of this goal: representation learning and network science for knowledge networks. I will show how realizing this goal can set sights on new frontiers beyond classic applications of neural networks on biomedical image and sequence data. I will start by presenting a framework that learns deep models by embedding knowledge networks into compact embedding spaces whose geometry is optimized to reflect network topology, the essence of networks. I will then describe two applications of the framework to drug discovery and medicine. First, the framework allowed us to, for the first time, predict the safety of drug combinations at scale. We embedded a knowledge network of molecular, drug, and patient data at the scale of billions of interactions for all medications in the U.S. Using the embeddings, the approach can predict unwanted side effects for any combination of drugs that patients take, and we can validate predictions in the clinic using real patient data. Second, I will discuss how the framework enabled us to predict what diseases a new drug could treat. I will show how the new approach can make correct predictions for many recently repurposed drugs and can operate even on the hardest, yet critical, diseases for which no good treatments exist. I will conclude with future directions for learning over interaction data and translation of machine learning methods into solutions for biomedical problems.

**Keywords:** Biomedicine · Representation learning · Network science · Knowledge graphs

**Short Biography of the Lecturer:** Marinka Žitnik is a postdoctoral scholar in Computer Science at Stanford University. She will join Harvard University as a tenure-track assistant professor in December 2019. Her research investigates machine learning for sciences. Her methods have had a tangible impact in biology, genomics, and drug discovery, and are used by major biomedical institutions, including Baylor College of Medicine, Karolinska Institute, Stanford Medical School, and Massachusetts General Hospital. She received her Ph.D. in Computer Science from University of

Ljubljana while also researching at Imperial College London, University of Toronto, Baylor College of Medicine, and Stanford University. Her work received several best paper, poster, and research awards from the International Society for Computational Biology. She was named a Rising Star in EECS by MIT and also a Next Generation in Biomedicine by The Broad Institute of Harvard and MIT, being the only young scientist who received such recognition in both EECS and Biomedicine. She is also a member of the Chan Zuckerberg Biohub at Stanford.

# Contents

## Data and Knowledge Representation

## Feature Importance

## Interpretable Machine Learning

## Networks

## Pattern Discovery

## Time Series

# Advanced Machine Learning

# The CURE for Class Imbalance

Colin Bellinger[1]([✉]), Paula Branco[2], and Luis Torgo[2]

[1] National Research Council of Canada, Ottawa, Canada
colin.bellinger@nrc-cnrc.gc.ca
[2] Dalhousie University, Halifax, Canada
{pbranco,ltorgo}@dal.ca

**Abstract.** Addressing the class imbalance problem is critical for several real world applications. The application of pre-processing methods is a popular way of dealing with this problem. These solutions increase the rare class examples and/or decrease the normal class cases. However, these procedures typically only take into account the characteristics of each individual class. This segmented view of the data can have a negative impact. We propose a new method that uses an integrated view of the data classes to generate new examples and remove cases. ClUstered REsampling (CURE) is a method based on a holistic view of the data that uses hierarchical clustering and a new distance measure to guide the sampling procedure. Clusters generated in this way take into account the structure of the data. This enables CURE to avoid common mistakes made by other resampling methods. In particular, CURE prevents the generation of synthetic examples in dangerous regions and undersamples safe, non-borderline, regions of the majority class. We show the effectiveness of CURE in an extensive set of experiments with benchmark domains. We also show that CURE is a user-friendly method that does not require extensive fine-tuning of hyper-parameters.

**Keywords:** Imbalanced domains · Resampling · Clustering

## 1 Introduction

Class imbalance is a problem encountered in a wide variety of important classification tasks including oil spill, fraud detection, action recognition, text classification, radiation monitoring and wildfire prediction [4,17,21,22,24,27]. Prior research has shown that class imbalance has a negative impact on the performance of the learned binary classifiers. This problem becomes even more difficult when the underlying distribution is complex and when the minority class is rare [14,26]. Given the frequency of imbalanced learning problems and the possibility for negative impacts on learning, the study of class imbalance and methods for handling it have become important research topics. Indeed, it has been recognised as one of the ten challenging problems in data mining research [29].

The solutions proposed by the research community to solve the class imbalance problem include special-purpose learning methods, pre-processing and post-processing methods. Pre-processing (or resampling) methods transform the original training set making it more suitable for learning the important class(es).

© Springer Nature Switzerland AG 2019
P. Kralj Novak et al. (Eds.): DS 2019, LNAI 11828, pp. 3–17, 2019.
https://doi.org/10.1007/978-3-030-33778-0_1

This is accomplished by following a certain strategy for up- or down-sampling cases. Resampling methods are popular due to their versatility, effectiveness and ease of application. Moreover, they enable the use of any standard learning tool.

In addition to suffering from the class imbalance problem, many real-world domains are also complex by nature. They potentially include noisy instances and sub-concepts that exacerbate the imbalance problem. This complexity dictates that it is important to consider the inherent structure of the data. Failure to do so may negatively impact the effectiveness of the resampling strategy. The problem relates to: (i) the removal of majority class instances from sparse regions of the domain; (ii) the generation of synthetic minority cases between minority class sub-concepts (clusters); (iii) the reinforcement of noisy instances; and/or (iv) the obfuscation of overlapping regions.

To address these issues, we propose the ClUstered REsampling (CURE) method. CURE uses hierarchical clustering with a new class-sensitive distance measure prior to the resampling process. This allows the extraction of essential structural information that is used to guide the resampling. The advantages of this approach are: (i) meaningful clusters of the minority class are emphasised: (ii) the generation of minority class cases is avoided in error-prone regions between sub-concepts; (iii) only "safe" majority class samples are undersampled (*i.e.*, borderline cases are not removed.) In an extensive set of experiments, we show that the CURE algorithm is effective for tackling the class imbalance problem. We also show that CURE does not requires extensive fine-tuning of hyper-parameters to achieve good performance.

This paper is organised as follows. Section 2 provides an overview of the related work. In Sect. 3 the CURE algorithm is described. The results of an extensive experimental evaluation are presented and discussed in Sect. 4. Finally, Sect. 5 presents the main conclusions of the paper.

## 2   Related Work

Numerous resampling methods have been proposed and applied to address imbalanced classification problems [5]. Random oversampling and random undersampling (e.g. [16]) are the classic approaches to handling imbalance. They are well-known to suffer from the risk of overfitting the minority samples and discarding informative cases, respectively. The SMOTE algorithm [8] incorporates oversampling and undersampling, and was proposed to overcome the issues of over- and under-sampling. It attempts to do so by interpolating new synthetic instances between nearest neighbours rather than replicating instances of the minority class. Two key issues with SMOTE are: (*a*) it does not account for the structure of the training data when performing the under and oversampling, and (*b*) it uniformly applies oversampling and undersampling. On complex data domains, this generation process can reinforce noise and increase the class overlap.

Many variations of SMOTE have been proposed to either clean the data after synthetic oversampling or to preemptively avoid generating instances that would negatively impact classifier performance [3,7,11,12]. For instance, Tomek links (examples from different classes that are each other closest neighbours) can be removed from the training set after the application of SMOTE [3]. ADASYN [13] and Borderline-SMOTE [12] are examples of methods that apply SMOTE only in specific regions of the domain that are considered useful. ADASYN generates more synthetic examples from minority class cases that have more nearest neighbours from the majority class. Borderline-SMOTE generates more examples near the minority class border. However, Borderline-SMOTE applies uniform undersampling and may generate new cases between subconcepts of the minority class while ADASYN uses the local structure disregarding the global structure of the data. Resampling with a neighbourhood bias [6] is an alternative that introduces a bias in the selection of seed cases for both over and undersampling based on the class distribution in the local neighbourhood of each instance. Different biasing modes are proposed allowing to reinforce the most frontier or the most safe cases. Our proposal advances this idea by replacing the need for users to specify the $k$ value necessary for the $k$-nearest neighbours computation, which is applied homogeneously across all instances and may be difficult to determine *a-priori*. Alternatively, we utilise hierarchical clustering that automatically finds variable sized clusters in the underlying structure of the data for resampling.

Previous research has applied clustering plus random oversampling, clustering plus random undersampling, and clustering plus synthetic oversampling [2,15,18–20,28,30]. Of these methods, our proposal is most closely related to [30]. Whereas the other methods only cluster one class, our method and that of Yen *et al.* [30], clusters the complete training sets, and use the class distribution in each cluster to inform if, and how much, resampling should be applied. By clustering both classes instead of just one, we acquire a more complete view of the data structure. The work of Yen *et al.* [30], uses k-means clustering which has important limitations such as requiring the *a-priori* knowledge of the correct number of clusters. By using hierarchical clustering, we are able to dynamically discover the sub-clusters (clusters at different levels of the hierarchy) that best address our resampling objectives. In addition, our method differs in the fact that it applies both undersampling and synthetic oversampling which inflates the minority class space while smoothing over-represented concepts of the majority class.

## 3   The CURE Method

### 3.1   Overview

In this section, we present the ClUstered REsampling (CURE) method. The key feature of CURE is that it utilises the intrinsic structure of the training data from all of the problem classes to decide where and how to apply resampling. This way, CURE avoids resampling mistakes incurred by SMOTE-based methods. In particular, CURE reduces the risk of:

– Synthesising minority class samples deep inside the majority class space; and,
– Naively undersampling informative instances in the majority class.

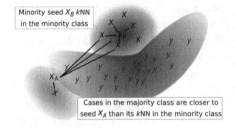

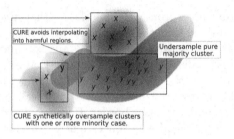

**Fig. 1.** Illustration of the intrusion into the majority class caused by SMOTE. (Color figure online)

**Fig. 2.** Illustration of the synthetic oversampling of natural minority class groupings discovery of CURE.

The oversampling issue of SMOTE-based methods is demonstrated in Fig. 1. Here, the nearest neighbours of $X_B$ are all minority class examples and thus interpolating between them is safe. However, between $X_A$ and some of its nearest minority class neighbours, there is an area populated with majority class examples. Interpolating between these neighbours risks generating new synthetic minority class case in the majority class space (the blue region).

The undersampling issue is highlighted in Fig. 3. Here, the resampler naïvely discards some user-specified percentage, $p$, of the majority class samples (the removed samples are shown as grey $y$) in order to balance the training set. The random removal process risks the loss of information from the edge of the majority class region, which could have a significant negative impact in the learned decision boundary.

CURE avoids the over/undersampling issues discussed above by ensuring that instances are generated in, and removed from, safe regions of the dataspace. This is achieved by applying hierarchical clustering and then resampling each cluster in a manner that is determined by the class makeup of the cluster.

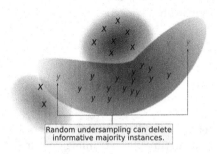

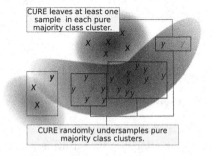

**Fig. 3.** Illustration of the removal of informative majority samples via random undersampling.

**Fig. 4.** Illustration of CURE keeping potentially information samples in mixed and small majority class clusters.

Figure 2 illustrates where and how resampling is applied to clusters involving minority class samples. Specifically, interpolation is only applied between minority class instances in the same cluster. This avoids the generation of samples deep inside the majority class (grey zone in figure). Figure 4 demonstrates how CURE randomly undersamples a percentage, $p$, of instances from each pure majority class cluster, rather than at random from the complete set of majority class instances. After undersampling, CURE will always leave at least one sample in each cluster to avoid wiping out information about edge cases and sub-concepts.

## 3.2 Hierarchical Clustering

A hierarchical clustering is formed by successively merging instances that are similar to each other. At the bottom of this hierarchy, we have the individual training cases, and at the top node we have a single cluster containing all of the cases. In between these extremes we have different groupings of the training data. Thus, the hierarchy specifies a set of possible clusterings of the data, where the clusters near the bottom of the hierarchy are smaller and more specific, and those nearer the top are larger and more general. It is up to the users to determine which clustering is best for their objectives.

The requirement to identify the "best" clusters from the hierarchy is a limitation in many pure clustering applications. For our purposes, however, it means we do not have to specify the number of clusters a-priori. Rather, we develop a method to automatically discover the clusters in the constructed hierarchy that are appropriate for resampling.

To produce the cluster hierarchy:

1. The pair-wise distance between each sample is calculated; and
2. A hierarchy is constructed by agglomeratively merging similar clusters.

The Ward variance minimisation algorithm [25] is used to construct the linkages in the hierarchy because it minimises the total within-cluster variance. This objective is appropriate for our goal of finding concise sub-concepts in the data to apply informed resampling on.

Given the set of clusters (also known as a forest) $C_i$ at level $i$ in the tree constructed thus far, the Ward variance minimisation algorithm search for clusters $s$ and $t$ in $C_i$ that have the minimum variance according to the Ward metric. The clusters $s$ and $t$ are then merged to form a new cluster $w = \{s \cup t\}$ at level $i - 1$. The linkage process halts when all samples are merged into a single cluster.

## 3.3 Supervised Distance Measure

Clustering is typically an unsupervised process. We postulate, however, that the discovery of natural groupings in the training data for the purposes of resampling should not be unsupervised. Our hypothesis is that the class labels should have some influence on the cluster formation, but this influence should not be absolute.

Given a seed instance $I_1 = \langle \mathbf{x}_1, A \rangle$ and two query instances $I_2 = \langle \mathbf{x}_2, A \rangle$ and $I_3 = \langle \mathbf{x}_3, B \rangle$, where Euclidean($\mathbf{x}_1, \mathbf{x}_2$) is equal to Euclidean($\mathbf{x}_1, \mathbf{x}_3$), then $I_2$ should be considered to be more similarly to $I_1$ because it is from the same class. Alternatively, if Euclidean($\mathbf{x}_1, \mathbf{x}_3$) is significantly less than Euclidean($\mathbf{x}_1, \mathbf{x}_2$), then $I_3$ should be considered to be more similarity regardless of its different class association.

To achieve this, we propose a new supervised measure named Distance with Class Label Integration ($DCLI_\alpha$). The $DCLI_\alpha$ measure is based on a standard user selected distance metric ($d$) and a parameter $\alpha$ that controls the importance of matching class labels. The $DCLI_\alpha$ measure is defined as,

$$DCLI_\alpha(\langle \mathbf{x}_i, y_i \rangle, \langle \mathbf{x}_j, y_j \rangle) = \begin{cases} m + \alpha(d(\mathbf{x}_i, \mathbf{x}_j) - m) & if \ y_i = y_j \\ d(\mathbf{x}_i, \mathbf{x}_j) & if \ y_i \neq y_j \end{cases} \qquad (1)$$

where $\langle \mathbf{x}_i, y_i \rangle$ represents an example with feature vector $\mathbf{x}_i$ and target class $y_i$, $d$ is a user selected distance metric, parameter $\alpha \in [0, 1]$ controls the influence of the class labels in the $DCLI_\alpha$ distance measure and $m$ is the minimum distance between instances in the training set measured using metric $d$.

The parameter $\alpha$ in Eq. 1 has the effect of weighting the significance of the class label agreement, i.e. instances with matching class labels are brought slightly closer together than their respective distances, $d$. Specifically, the $DCLI_\alpha$ distance between two instances $\mathbf{x}_i$, $\mathbf{x}_j$ with matching class labels is equal to some point, $p$, between $d(\mathbf{x}_i, \mathbf{x}_j)$ and the minimum distance is the data set $\arg\min_{\mathbf{x}_l, \mathbf{x}_k \in D} m = d(\mathbf{x}_l, \mathbf{x}_k)$. The proximity of $p$ to either extreme is controlled by the $\alpha$ parameter. In this paper, we have used the Euclidean distance for parameter $d$ in $DCLI_\alpha$. Figure 5 shows the effect of the $\alpha$ parameter on the $DCLI_\alpha$ distances for instances with matching class labels ($\mathbf{x}_1$, and $\mathbf{x}_2$), and instances with mismatched class labels ($\mathbf{x}_1$, and $\mathbf{x}_3$).

To summarise, the purpose of the measure is to promote the clustering of sparse groups of minority samples, even when a subset of those samples is slightly closer to the majority class. In Fig. 5, instances $\langle \mathbf{x}_1, A \rangle$ and $\langle \mathbf{x}_2, A \rangle$ will be linked in the hierarchy before $\langle \mathbf{x}_3, B \rangle$ for $\alpha < 0.8$. We discuss the sensitivity of $\alpha$ in Sect. 4.2.

## 3.4   CURE Algorithm

As previously stated, the CURE method consists of constructing a hierarchy using our proposed $DCLI_\alpha$ measure, and then automatically extracting clusters from the hierarchy for resampling. For clarity, we refer to the sets of instances at each level of the hierarchy as groups or groupings, and we refer to the subset of these groups automatically identified for resampling as clusters.

The groupings for each level of the hierarchy are stored in a data structure along with the corresponding intra-cluster distances, which we use for cluster formation. Each instance in the training set is assigned to a cluster defined by the

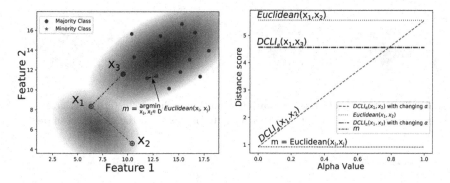

**Fig. 5.** Illustration of the impact of the $\alpha$ parameter on the $DCLI_\alpha$ score.

largest grouping to which it belongs that has an intra-cluster distance less than the threshold $\tau = \mu + s \times \sigma$, where $\mu$ and $\sigma$ are the mean and standard deviation of the intra-cluster distances, and $s$ is the number of standard deviations above the mean to set the threshold. Empirically, we found the intra-cluster distances to be approximately log-normal, thus it makes statistical sense to set the threshold in this manner.

We view the distribution of intra-cluster distances as a proxy for the overall spread of the data (variance is distances between training samples). As a result, $s$ should be set large enough to represent the variance in single sub-concepts (natural groupings) so as to form clusters around these sub-concepts, but not so large as to join multiple sub-concepts into one cluster. The setting of the $s$ parameter is simplified by the log-normal assumption. We postulate that approximately one standard deviation above the mean should achieve the required balance because it covers most of the variance in the data, whilst excluding exceptional levels of spread. The sensitivity of $s$ is discussed in more detail in Sect. 4.2. The details of the cluster method are shown in Algorithm 1, and Fig. 6 illustrates the process of generating these clusters.

We define three cluster composition for resampling: (i) the cluster includes only majority class cases; (ii) those with exactly one minority class case and zero or more majority class instances; and (iii) those with more than one minority class case and zero or more majority class. If a cluster contains more than one minority class case, we will interpolate between them generating new synthetic cases and will maintain the majority class examples. When the cluster contains exactly one minority class case, synthetic cases are generated by applying Gaussian jitter to it. Finally, if the cluster is formed exclusively by majority class examples, this means we randomly undersample the cluster. Algorithm 2 provides an high level overview of the proposed CURE method.

In summary, the main idea of CURE is to carry out case generation and undersampling inside regions of the input space that are safer. These regions are found by taking into account the intrinsic structure of the training data.

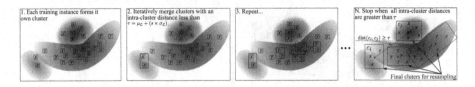

**Fig. 6.** Illustration of clusters generation using hierarchical clustering (Algorithm 1). The resampling strategy to apply in each cluster is based on the cluster examples.

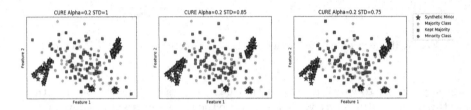

**Fig. 7.** Impact of changing CURE method hyper parameters in an artificial data set.

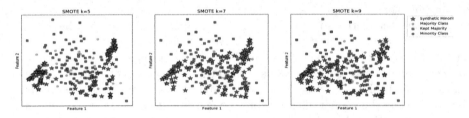

**Fig. 8.** Impact of changing the number of nearest neighbours considered in SMOTE algorithm in an artificial data set.

We achieve this using a new distance measure in the context of a hierarchical clustering process.

To better understand the way CURE avoids unsafe oversampling when compared to SMOTE, we prepared the 2-dimensional artificial data[1] in Figs. 7 and 8. These figures show the behaviour of each method with respect to their main hyper parameters. These figures illustrate that CURE is capable of detecting safe regions as opposed to SMOTE that generates new cases in regions that belong to the majority class.

---

[1] This is a hand curated 2-dimensional data set developed to demonstrate the strengths of CURE.

---

**Algorithm 1.** Generation of Clusters

---

**Input:** $\mathcal{D}$ - a classification data set
    $\alpha \in [0,1]$ - weights the influence of the class labels in DCLI distance
    $s$ - threshold on the standard deviation considered during clusters formation
**Output:** C - the clusters
1: **function** GENCLUSTERS($\mathcal{D}, \alpha, s$)
2:  $\mathcal{M}_\mathcal{D} \leftarrow$ pairwise distance matrix using $DCLI_\alpha$ measure
3:  $Z \leftarrow$ agglomerative linkage tree calculated over $\mathcal{M}_\mathcal{D}$
4:  $\mathcal{L} \leftarrow$ log transform of the inter-cluster distances obtained in $Z$
5:  $\mu_\mathcal{L} \leftarrow$ mean of the inter-cluster distances in $\mathcal{L}$
6:  $\sigma_\mathcal{L} \leftarrow$ standard deviation of the inter-cluster distances in $\mathcal{L}$
7:  $\tau \leftarrow \mu_\mathcal{L} + s \times \sigma_\mathcal{L}$  ▷ maximum inter-cluster distance for cluster formation
8:  $C \leftarrow$ Form clusters using $Z$ s.t. the inter-cluster distances of the new clusters is less or equal to $\tau$
9:  **return** $C$
10: **end function**

---

**Algorithm 2.** CURE Algorithm

---

**Input:** $\mathcal{D}$ - a classification data set
    $S_{min}, S_{maj}$ - number of minority and majority class instances to obtain in the new data set
    $\alpha$ - class labels weight parameter in $DCLI_\alpha$ distance
    $s$ - threshold on the standard deviation considered during clusters formation
**Output:** $\mathcal{D}'$ - new resampled data set
1: **function** CURE($\mathcal{D}, S_{min}, S_{maj}, \alpha, s$)
2:  $k \leftarrow S_{min}/|$minority class instances in $\mathcal{D}|$▷ minority class instances to generate for each instance
3:  $q \leftarrow [1 - (S_{maj}/|$majority class instances in $\mathcal{D}|)] \times 100$  ▷ % of majority class examples to remove
4:  $C \leftarrow$ GENCLUSTERS($\mathcal{D}, \alpha, s$)
5:  $\mathcal{D}' \leftarrow \mathcal{D}$
6:  **for** each cluster $c_i$ in $C$ **do**
7:    **if** $c_i$ contains only majority class instances **then**
8:      $\mathcal{D}' \leftarrow \mathcal{D}' \backslash \{$random selection of q% of the instances in $c_i\}$
9:    **else if** $c_i$ contains exactly one minority class instance **then**
10:      new $\leftarrow$ generate $k$ synthetic cases using Gaussian jitter
11:      $\mathcal{D}' \leftarrow \mathcal{D}' \bigcup new$
12:    **else**    ▷ several minority class instances in the cluster
13:      new $\leftarrow$ generate $k$ synthetic cases by interpolating minority cases in $c_i$
14:      $\mathcal{D}' \leftarrow \mathcal{D}' \bigcup new$
15:    **end if**
16:  **end for**
17:  **return** $\mathcal{D}'$
18: **end function**

---

# 4 Experimental Evaluation

## 4.1 Materials and Methods

We have selected a diverse set of 29 benchmark data sets from the KEEL repository [1]. In order to consider the effectiveness of CURE at different levels of absolute and relative imbalance, we process each original data set into three new versions for the purpose of our experiments. The new versions contain 10, 30 and 50 minority class cases, for which we use the notation of IR10, IR30 and IR50 to refer to these respectively. We conducted our experiments on 87 data sets ($29 \times 3$). The average imbalance ratios ($|min|/|maj|$) of the three versions range between 0.186 and 0.037. Therefore, they include a wide range of absolute and relative imbalance levels. Table 1 displays the main characteristics of the used data sets.

**Table 1.** Data sets name, dimensions (Dim), majority class cases ($|maj|$), and imbalance ratios when using 50, 30 and 10 minority class cases (IR50, IR30 and IR10).

| Data set | mammographic0vs1 | letterBFGMvsOPSTUW | dermatology1-3vs4-6 | contraceptive1vs2-3 | winequality-red1-5vs6-10 | movement-libras1-9vs10-15 | optdigits1-3vs5-8 | page-blocks1vs2-4 | texture2-6vs7-14 | marketing1-2vs7-9 | shuttle2-3vs5-7 | segment1-3vs5-7 | satimage1-2vs5-7 | letterBFGMvsOPSTUW | penbased3-4vs7-8 |
|---|---|---|---|---|---|---|---|---|---|---|---|---|---|---|---|
| Dim | 5 | 16 | 34 | 9 | 11 | 90 | 64 | 10 | 40 | 13 | 9 | 19 | 36 | 16 | 16 |
| \|maj\| | 214 | 1560 | 121 | 422 | 428 | 108 | 1134 | 2678 | 1750 | 1298 | 1645 | 660 | 1118 | 1560 | 1671 |
| IR50 | 0.234 | 0.032 | 0.413 | 0.118 | 0.117 | 0.463 | 0.044 | 0.019 | 0.029 | 0.039 | 0.030 | 0.076 | 0.045 | 0.032 | 0.030 |
| IR30 | 0.140 | 0.019 | 0.248 | 0.071 | 0.070 | 0.278 | 0.026 | 0.011 | 0.017 | 0.023 | 0.018 | 0.045 | 0.027 | 0.019 | 0.018 |
| IR10 | 0.047 | 0.006 | 0.083 | 0.024 | 0.023 | 0.093 | 0.009 | 0.004 | 0.006 | 0.008 | 0.006 | 0.015 | 0.009 | 0.006 | 0.006 |

| Data set | bands1vs0 | ionosphere0vs1 | vowell-5vs6-10 | wisconsin0vs1 | ring0vs1 | magic0vs1 | cleveland0vs1-4 | coil20000vs1 | heart0vs1 | spambase0vs1 | monk-20vs1 | wdbc0vs1 | vehicle1-2vs3-4 | thyroid2vs3 | Average |
|---|---|---|---|---|---|---|---|---|---|---|---|---|---|---|---|
| Dim | 19 | 33 | 13 | 9 | 20 | 10 | 13 | 85 | 13 | 57 | 6 | 30 | 18 | 21 | **25.3** |
| \|maj\| | 115 | 112 | 270 | 222 | 1019 | 2802 | 80 | 4618 | 75 | 1393 | 114 | 179 | 215 | 2967 | **1053.4** |
| IR50 | 0.435 | 0.446 | 0.185 | 0.225 | 0.049 | 0.018 | 0.625 | 0.011 | 0.667 | 0.036 | 0.439 | 0.279 | 0.233 | 0.017 | **0.186** |
| IR30 | 0.261 | 0.268 | 0.111 | 0.135 | 0.029 | 0.011 | 0.375 | 0.006 | 0.400 | 0.022 | 0.263 | 0.168 | 0.140 | 0.010 | **0.111** |
| IR10 | 0.087 | 0.089 | 0.037 | 0.045 | 0.010 | 0.004 | 0.125 | 0.002 | 0.133 | 0.007 | 0.088 | 0.056 | 0.047 | 0.003 | **0.037** |

We compare the performance of CURE to 7 state-of-the-art resampling methods, namely, random undersampling (RUS), random oversampling (ROS), the combined application of RUS and ROS simultaneously (ROS + RUS), adaptive synthetic oversampling (ADASYN), SMOTE algorithm, Borderline-SMOTE

(Borderline), and SMOTE with the removal of Tomek links (SMOTE + TL), and no resampling (None).

Support vector machines (SVM) with the radial basis function (RBF) kernel is selected for classification, because it is an effective non-parametric method that can be trained on small amounts of data relative to deep learning methods. Automatic parameter tuning is conducted after resampling via random search over the $\gamma \in [0.001, 20]$ and $C \in [0.001, 20]$. This promotes the discovery of the best SVM model for the resampled training set.

The evaluation is performed via $5 \times 2$-fold cross validation, because it has been observed that it has a lower probability of issuing a Type I error [9]. The performance is reported in terms of the geometric mean (g-mean) [16] and the $F_\beta$ measure [23]. Given the accuracy on the target class $a^+$ and the accuracy on the outlier class $a^-$, the g-mean for a classification model $f$ on test set $X$ is calculated as: $\text{g-mean}_{f(X)} = \sqrt{a^+ \times a^-}$. This metric enables us to evaluate whether the resampling methods are helping to improving the performance on the minority class, whilst having minimal impact on the majority class. The $F_\beta$ measure expresses the harmonic mean of precision and recall. We used $\beta = 1$ which assigns the same weight to precision and recall measures. The $F_\beta$ measure is popular in imbalanced domains as it provides a reliable assessment of the models effectiveness on the minority class (e.g. [10]).

Regarding the CURE algorithm, we have set parameters $S_{min}$ and $S_{maj}$ (c.f. Algorithm 2) as follows: $S_{min} = |min| + 0.5 \times |maj|$ and $S_{maj} = 0.5 \times |maj|$, where $|maj|$ and $|min|$ correspond respectively to the number of minority and majority class cases in the original data set. We apply this policy to the alternative resampling methods as well. To ensure an easy replication of our work all code and data sets used in the experiments are available at https://ltorgo.github.io/CURE/.

## 4.2    Results and Discussion

**Aggregated Results:** The first set of experiments focuses on the effectiveness of CURE for tackling the class imbalance problem. Figures 9 and 10 show the number of times each resampling method was the best (won) in terms of the average results during cross validation. The results are grouped according to the number of minority cases. Thus, Winner 10 FM in Fig. 9 specifies the number of times each resampling method won on the datasets with 10 minority class cases.

The figures illustrate that CURE has the highest number of wins in comparison to the 8 tested alternatives. Regarding the $F_1$ measure, CURE achieves 7, 12 and 10 wins for the IR10, IR30 and IR50 data sets respectively. The alternative that shows the most competitive results is Borderline with only 4, 3, and 2 wins for IR10, IR30 and IR50 data sets. Regarding the performance on G-Mean measure we observe that the advantage displayed by CURE method is overwhelming with 8, 14 and 12 wins on IR10, IR30 and IR50 data sets respectively. In this setting, the method showing the second most competitive performance is ADASYN displaying 6, 3 and 4 wins for the IR10, IR30 and IR50 data sets.

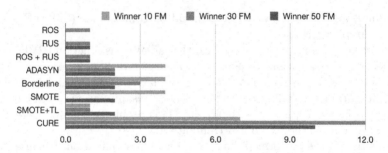

**Fig. 9.** Number of wins obtained by each tested resampling approach aggregated by data sets with 10, 30 and 50 minority class cases, for the $F_1$ measure.

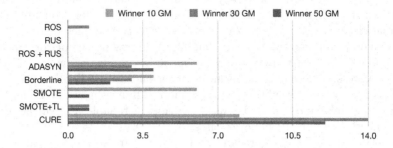

**Fig. 10.** Number of wins obtained by each tested resampling approach aggregated by data sets with 10, 30 and 50 minority class cases, for the G-Mean metric.

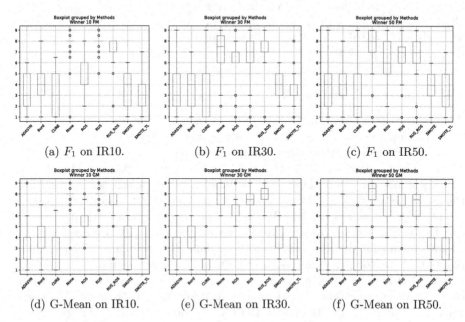

(a) $F_1$ on IR10.    (b) $F_1$ on IR30.    (c) $F_1$ on IR50.

(d) G-Mean on IR10.    (e) G-Mean on IR30.    (f) G-Mean on IR50.

**Fig. 11.** Ranks of each resampling approach on the three data set versions, IR10, IR30 and IR50, for both the $F_1$ (top row) and G-Mean (bottom row) metrics.

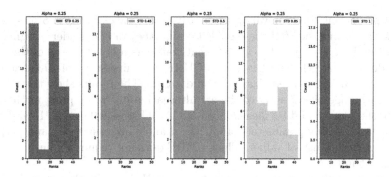

**Fig. 12.** CURE rankings for IR10 data set versions for: $\alpha = 0.25$ and $0.25 \leq s \leq 1$.

Figure 11 displays the boxplot of the rankings achieved by each resampling method on each performance assessment metric by data set version. The rankings shown were obtained using the average of the cross-validation results. These results clearly show the advantage of using CURE. Overall, the results obtained demonstrate the versatility of our proposed method over different class ratios, and demonstrates the benefit of utilising the inherent structure of data for resampling.

**Hyper-parameter Sensitivity:** CURE has two parameters: $\alpha$ and $s$. The $\alpha$ parameter determines the influence of matching class labels on the distance score ($DCLI_\alpha$). The second parameter, $s$, is number of standard deviations used in threshold for cluster formation.

Figure 12 shows the variation in the rankings of CURE method, on data sets from IR10 version, for parameter $\alpha$ fixed at 0.25 and parameter $s$ ranging between 0.25 and 1. Due to space constraints, we provide more figures that show the results for other parameter variations in: https://ltorgo.github.io/CURE/. The results obtained for $s \approx 1$ are concentrated around the lower (and thus better) rankings. As stated in Sect. 3, setting $s \approx 1$ makes good statistical sense, as well. The $\alpha$ parameter results suggest that values of $\alpha$ between 0.1 and 0.25 provides the best overall results. The good performance of CURE allied to this user-friendly perspective make CURE an excellent approach to tackle the problem of imbalanced domains.

## 5   Conclusion

We presented CURE, a novel method that uses the inherent structure of data to discover safer regions for resampling. These regions are found using a new class-sensitive distance measure and hierarchical clustering. A suitable resampling strategy is applied inside each cluster based on its characteristics. CURE aims at: (i) avoiding the generation of synthetic cases in unsafe regions of the data space, and (ii) preventing the removal of informative majority class cases. State-of-the-art resampling methods fail these goals because they only consider a segmented view of the data as opposed to CURE that considers a holistic view of the data.

We demonstrate the effectiveness of CURE on a diverse set of 29 benchmark domains and 87 imbalanced classification datasets. The results show that CURE has an advantage over 7 state-of-the-art alternatives for resampling methods in terms of the g-mean and $F_\beta$ measures on $5 \times 2$-fold cross-validation. In addition, we show that the key parameters of CURE, $\alpha$ and $s$ are easy to set and perform well over a large range of values. Thus, CURE does not require extensive hyper-parameter tuning.

As future work, we plan to demonstrate CURE in multi-class domains, and further improve the method for automatically detect the safe regions. Moreover, we also plan to explore the application of other resampling methods inside each safe region based on the regions characteristics.

# References

1. Alcalá-Fdez, J., et al.: Keel data-mining software tool: data set repository, integration of algorithms and experimental analysis framework. J. Multiple-Valued Logic Soft Comput. **17**, 255–287 (2011)
2. Barua, S., Islam, M.M., Murase, K.: A novel synthetic minority oversampling technique for imbalanced data set learning. In: Lu, B.-L., Zhang, L., Kwok, J. (eds.) ICONIP 2011. LNCS, vol. 7063, pp. 735–744. Springer, Heidelberg (2011). https://doi.org/10.1007/978-3-642-24958-7_85
3. Batista, G.E.A.P.A., Prati, R.C., Monard, M.C.: A study of the behavior of several methods for balancing machine learning training data. SIGKDD Explor. Newsl. **6**(1), 20–29 (2004)
4. Bellinger, C., Drummond, C., Japkowicz, N.: Manifold-based synthetic oversampling with manifold conformance estimation. Mach. Learn. **107**(3), 605–637 (2018)
5. Branco, P., Torgo, L., Ribeiro, R.P.: A survey of predictive modeling on imbalanced domains. ACM Comput. Surv. (CSUR) **49**(2), 31 (2016)
6. Branco, P., Torgo, L., Ribeiro, R.P.: Resampling with neighbourhood bias on imbalanced domains. Expert Syst. **35**(4), e12311 (2018). https://doi.org/10.1111/exsy.12311
7. Bunkhumpornpat, C., Sinapiromsaran, K., Lursinsap, C.: Safe-level-SMOTE: safe-level-synthetic minority over-sampling technique for handling the class imbalanced problem. In: Theeramunkong, T., Kijsirikul, B., Cercone, N., Ho, T.-B. (eds.) PAKDD 2009. LNCS (LNAI), vol. 5476, pp. 475–482. Springer, Heidelberg (2009). https://doi.org/10.1007/978-3-642-01307-2_43
8. Chawla, N.V., Bowyer, K.W., Hall, L.O., Kegelmeyer, W.P.: SMOTE: synthetic minority over-sampling technique. J. Artif. Intell. Res. **16**, 321–357 (2002)
9. Dietterich, T.G.: Approximate statistical tests for comparing supervised classification learning algorithms. Neural Comput. **10**(7), 1895–1923 (1998)
10. Estabrooks, A., Japkowicz, N.: A mixture-of-experts framework for learning from imbalanced data sets. In: Hoffmann, F., Hand, D.J., Adams, N., Fisher, D., Guimaraes, G. (eds.) IDA 2001. LNCS, vol. 2189, pp. 34–43. Springer, Heidelberg (2001). https://doi.org/10.1007/3-540-44816-0_4

11. Fernández, A., Garcia, S., Herrera, F., Chawla, N.V.: Smote for learning from imbalanced data: progress and challenges, marking the 15-year anniversary. J. Artif. Intell. Res. **61**, 863–905 (2018)

12. Han, H., Wang, W.-Y., Mao, B.-H.: Borderline-SMOTE: a new over-sampling method in imbalanced data sets learning. In: Huang, D.-S., Zhang, X.-P., Huang, G.-B. (eds.) ICIC 2005. LNCS, vol. 3644, pp. 878–887. Springer, Heidelberg (2005). https://doi.org/10.1007/11538059_91

13. He, H., Bai, Y., Garcia, E.A., Li, S.: ADASYN: adaptive synthetic sampling approach for imbalanced learning. In: IJCNN 2008, pp. 1322–1328. IEEE (2008)

14. He, H., Ma, Y.: Imbalanced Learning: Foundations, Algorithms, and Applications. Wiley, Hoboken (2013)

15. Jo, T., Japkowicz, N.: Class imbalances versus small disjuncts. ACM SIGKDD Explor. Newslett. **6**(1), 40–49 (2004). Special issue on learning from imbalanced datasets

16. Kubat, M., Matwin, S., et al.: Addressing the curse of imbalanced training sets: one-sided selection. In: ICML, Nashville, USA, vol. 97, pp. 179–186 (1997)

17. Lewis, D.D., Catlett, J., Hill, M.: Heterogeneous uncertainty sampling for supervised learning. In: International Conference on Machine Learning, pp. 148–156 (1994)

18. Lim, P., Goh, C.K., Tan, K.C.: Evolutionary cluster-based synthetic oversampling ensemble (eco-ensemble) for imbalance learning. IEEE Trans. Cybern. **47**(9), 2850–2861 (2017)

19. Lin, W.C., Tsai, C.F., Hu, Y.H., Jhang, J.S.: Clustering-based undersampling in class-imbalanced data. Inf. Sci. **409–410**, 17–26 (2017)

20. Nickerson, A.S., Japkowicz, N., Milios, E.: Using unsupervised learning to guide resampling in imbalanced data sets. In: Proceedings of the Eighth International Workshop on AI and Statistics, p. 5 (2001)

21. Oliveira, M., Torgo, L., Santos Costa, V.: Predicting wildfires. In: Calders, T., Ceci, M., Malerba, D. (eds.) DS 2016. LNCS (LNAI), vol. 9956, pp. 183–197. Springer, Cham (2016). https://doi.org/10.1007/978-3-319-46307-0_12

22. Provost, F., Fawcett, T.: Robust classification for imprecise environments. Mach. Learn. **42**(3), 203–231 (2001)

23. Rijsbergen, C.V.: Information retrieval, 2nd edition. Department of computer science, University of Glasgow (1979)

24. Slama, R., Wannous, H., Daoudi, M., Srivastava, A.: Accurate 3D action recognition using learning on the grassmann manifold. Pattern Recogn. **48**(2), 556–567 (2015)

25. Ward Jr., J.H.: Hierarchical grouping to optimize an objective function. J. Am. Stat. Assoc. **58**(301), 236–244 (1963)

26. Weiss, G.M.: Mining with rarity: a unifying framework. SIGKDD Explor. Newsl. **6**(1), 7–19 (2004). https://doi.org/10.1145/1007730.1007734

27. Williams, D., Myers, V., Silvious, M.: Mine classification with imbalanced data. IEEE Geosci. Remote Sens. Lett. **6**(3), 528–532 (2009)

28. Wu, J., Xiong, H., Chen, J.: COG: local decomposition for rare class analysis. Data Min. Knowl. Discov. **20**(2), 191–220 (2010)

29. Yang, Q., et al.: 10 challenging problems in data mining research. Int. J. Inf. Tech. Decis. **5**(4), 597–604 (2006)

30. Yen, S.J., Lee, Y.S.: Cluster-based under-sampling approaches for imbalanced data distributions. Expert Syst. Appl. **36**(3), 5718–5727 (2009)

# Mining a Maximum Weighted Set
# of Disjoint Submatrices

Vincent Branders[(✉)], Guillaume Derval, Pierre Schaus,
and Pierre Dupont

UCLouvain - ICTEAM/INGI, Louvain-la-Neuve, Belgium
{vincent.branders,guillaume.derval,pierre.schaus,
pierre.dupont}@uclouvain.be

**Abstract.** The objective of the maximum weighted set of disjoint submatrices problem is to discover $K$ disjoint submatrices that together cover the largest sum of entries of an input matrix. It has many practical data-mining applications, as the related biclustering problem, such as gene module discovery in bioinformatics. It differs from the maximum-weighted submatrix coverage problem introduced in [6] by the explicit formulation of disjunction constraints: submatrices must not overlap. In other words, all matrix entries must be covered by at most one submatrix. The particular case of $K = 1$, called the maximal-sum submatrix problem, was successfully tackled with constraint programming in [5]. Unfortunately, the case of $K > 1$ is more challenging to solve as the selection of rows cannot be decided in polynomial time solely from the selection of $K$ sets of columns. It can be proved to be $\mathcal{NP}$-hard. We introduce a hybrid column generation approach using constraint programming to generate columns. It is compared to a standard mixed integer linear programming (MILP) through experiments on synthetic datasets. Overall, fast and valuable solutions are found by column generation while the MILP approach cannot handle a large number of variables and constraints.

**Keywords:** Constraint programming · Maximum weighted
submatrix · Column generation · Maximum weighted set of disjoint
submatrices problem · Bi-cliques · Data-mining

## 1 Introduction

### 1.1 Problem Definition

We are interested in the mining of a numerical matrix to discover submatrices capturing a high total value. Precisely, we consider an input matrix $\mathcal{M}$ with $m$ rows and $n$ columns where element $\mathcal{M}_{i,j}$ is a given real value. The matrix is associated with a set of rows $R = \{r_1, \ldots, r_m\}$ and a set of columns $C = \{c_1, \ldots, n\}$. We use $(R; C)$ to denote the matrix $\mathcal{M}$.

If $I \subseteq R$ and $J \subseteq C$ are subsets of the rows and of the columns, respectively, the submatrix $(I; J)$ denotes all the elements $\mathcal{M}_{i,j}$ of $\mathcal{M}$ such that $i \in I \wedge j \in J$.

© Springer Nature Switzerland AG 2019
P. Kralj Novak et al. (Eds.): DS 2019, LNAI 11828, pp. 18–28, 2019.
https://doi.org/10.1007/978-3-030-33778-0_2

The max-sum submatrix problem (MSSP), introduced in [5], consists in identifying a subset of rows and of columns of an input matrix that maximizes the sum of the covered entries, which is the submatrix weight. The problem is formally stated below.

*The Max-Sum Submatrix Problem (MSSP)*: Given a matrix $\mathcal{M} \in \mathbb{R}^{m \times n}$, $R = \{1, \ldots, m\}$ and $C = \{1, \ldots, n\}$ the associated sets of rows and columns, respectively. The submatrix $(I^* \subseteq R, J^* \subseteq C)$ is of max-sum iff:

$$(I^*; J^*) = \underset{I,J}{\operatorname{argmax}} \sum_{i \in I, j \in J} \mathcal{M}_{i,j} \tag{1}$$

In this paper, we consider only the non-trivial problem matrices containing both positive and negative entries. Such a problem is both compelling and challenging to solve. A constraint programming (CP) implementation successfully tackled this difficult problem for matrices of thousands of rows and hundreds of columns, as is typical in several biological applications [5].

A natural extension of the MSSP is to identify $K$ submatrices. The maximum weighted submatrix coverage problem (MWSCP) proposed in [6] is an extension to the identification of $K$ possibly overlapping submatrices with maximal weight. It relies on a modification of the objective function such that covered entries contribute strictly once to the objective. However, it favors overlaps on negative entries: penalties are distributed among overlaps. Moreover, overlaps on positive entries will not improve the objective value.

In the present work, we consider an alternative extension to the identification of $K$ submatrices, relying on an objective function computed as the sum of submatrix weights, and the explicit addition of disjunction constraints. By allowing overlaps on the rows *or* the columns (but not both simultaneously due to the disjunction constraint) we avoid the unexpected behavior of the MWSCP. Moreover, the solution's interpretability by a domain expert is eased. Such a solution is usually called *nonoverlapping nonexclusive nonexhaustive* in the biclustering context [10].

**Definition 1.** *The Maximum Weighted Set of Disjoint Submatrices Problem (MWSDSP): Given a matrix $\mathcal{M} \in \mathbb{R}^{m \times n}$, $R = \{1, \ldots, m\}$ and $C = \{1, \ldots, n\}$ be the associated sets of rows and of columns, respectively, and $K$ be a target number of submatrices. The maximum weighted set of disjoint submatrices problem is to select a set of $K$ submatrices $(I^{k*}; J^{k*})$, with $I^{k*} \subseteq R$ and $J^{k*} \subseteq C$ for all $k \in \{1, \ldots, K\}$, such that each matrix entry is covered by at most one submatrix and the weight of the covered entries is maximal:*

$$(I^{1*}; J^{1*}), \cdots, (I^{K*}; J^{K*}) = \underset{(I^1;J^1),\cdots,(I^K;J^K)}{\operatorname{argmax}} \sum_{k=1}^{K} w_k \tag{2}$$

$$s.t. \quad (I^k \times J^k) \cap (I^{k'} \times J^{k'}) = \emptyset \quad \forall k, k' \in \{1, \ldots, K\}, k \neq k' \tag{3}$$

*where $w_k = \sum_{r \in I^k, c \in J^k} \mathcal{M}_{r,c}$ is the weight of submatrix $k$.*

Disjunction constraints (3) enforce that each matrix entry is selected by at most one submatrix. Restricting to $G$ ($> 1$) overlaps would result in $\lceil K/G \rceil$ groups of $G$ identical submatrices. While any submatrix pair may share rows *or* columns, the constraint prevents any pair from sharing rows *and* columns simultaneously. Note that the specific submatrix ordering is irrelevant.

## 1.2   Contributions

Our contributions are: (1) The introduction of the maximum weighted set of disjoint submatrices problem (MWSDSP) as a generalization of the max-sum submatrix (MSSP) problem; (2) A mathematical programming approach to solve the MWSDSP; (3) The formulation of the MWSDSP as an integer linear program (ILP) relying on constraint programming (CP) to produce relevant variables; (4) An evaluation of the performances of these two alternatives and the benefit of the ILP+CP over a greedy approach on synthetic datasets.

## 1.3   Motivation

The MWSDSP has many practical data-mining applications where one is interested in discovering $K$ specific relations between two groups of variables.

As an example, in gene expression analysis, $\mathcal{M}_{i,j}$ corresponds to the expression value of gene $i$ in sample $j$. One is typically interested in finding a subset of genes that present high expression value, i.e., an active biological pathway, in a subset of the samples. Finding multiple pathways specific to some samples is a common task in gene expression analysis. Submatrices overlaps would correspond to non-specific signal. In contrast, shared rows only would correspond to gene simultaneously active in multiple pathways, and shared columns only to subpopulations of samples exhibiting the same pathway activity.

## 1.4   Related Work

The max-sum submatrix problem (MSSP) and the maximum weighted submatrix coverage problem (MWSCP), presented in Sect. 1.1, are $\mathcal{NP}$-hard [6]. The present work and the MWSCP extend the MSSP to $K > 1$ by adding disjunction constraint and by adapting the objective function, respectively.

In the maximum subarray problem, introduced in [3], the aim is to find a subset of contiguous columns with maximal weight from an array. Polynomial-time complexity algorithms have been proposed for matrices [14]. This problem is simpler than the MWSDSP, however, as a single submatrix is required and it is constrained to be formed of contiguous subsets of rows and columns.

The biclustering problems are concerned with the discovery of homogeneous submatrices rather than maximizing the weight of covered entries. Madeira *et al.* provided a comprehensive review of biclustering problems [10].

The minimum sum-of-squares clustering problem involves the definition of non-overlapping sets of rows (or columns) covering all matrix entries. Although the problem differs, we use a similar approach as in [2]: the combination of an ILP and delayed column generation.

In the ranked tile mining problem, introduced in [9], entries are discrete ranks, corresponding to a permutation of column indices on each row. Moreover, the definition of a parametrized penalty for overlapping coverage discourages but allows identification of repetitive solutions.

## 2    Constraint Programming Approaches

### 2.1    Search Space

Let us define a set variable $T^k$ (resp. $U^k$) to represent the rows (resp. columns) included in submatrix $k$. The search space of the MSSP can be limited to searching on a single dimension, for instance the column set variable $U^1$. Indeed, optimal $T^1$ can be found in polynomial time: $\forall i \in R$: $\sum_{j \in U^1} \mathcal{M}_{i,j} > 0 \implies i \in T^1$.

Let us define the MWSDSP with fixed column selections formally.

**Definition 2. *The MWSDSP with fixed column selections*.** *The notations are the same as in Definition (1), but in this case the selections of columns for each submatrices (the $C^k$ sets) are given.*

$$R^{1*}, \cdots, R^{K*} = \underset{R^{1*}, \cdots, R^{K*}}{\mathrm{argmax}} \sum_{k=1}^{K} \sum_{r \in R^k, c \in C^k} \mathcal{M}_{r,c} \tag{4}$$

$$s.t. \quad (R^k \times C^k) \cap (R^{k'} \times C^{k'}) = \emptyset \quad \forall k, k' \in \{1, \ldots, K\}, k \neq k' \tag{5}$$

For $K > 1$, once all the column set variables $U^k$ are fixed, it remains to decide for each row $i$ and each submatrix $k$ whether $i$ is to be selected ($i \in T^k$) or not. These $K$ decisions per row cannot be optimally taken in polynomial time, as stated in Theorem (1). As a consequence, the search will have to assign both the row and column set variables, as opposed to the simpler $K = 1$ problem.

**Theorem 1.** *The MWSDSP with fixed column selections is $\mathcal{NP}$-Hard.*

*Proof.* We reduce the Maximum Weighted Independent Set (MWIS) problem to our problem. MWIS is $\mathcal{NP}$-Hard (by immediate reduction from the Independent Set problem [8]), and aims at finding, in a graph $G = <V, E>$ with weights $w_v$ on each vertex $v \in V$, the set of vertices with the maximum sum such that they do not share edges in $G$. For simplicity, we represent edges and vertices as numbers: $V = \{1, \ldots, |V|\}$ and $E = \{1, \ldots, |E|\}$. We reduce an instance of the MWIS to an instance of the MWSDSP with fixed column selections. We create a 1 by $(|V| + |E|)$ matrix $M$: $M_{1,i} = w_i$ if $i \in \{1, \ldots, |V|\}$, and $M_{1,i} = 0$ otherwise. The columns sets $C^1, \ldots, C^{|V|}$ are constructed as follows: $C^v = \{v\} \cup \{|V| + e \mid e \in E \wedge \text{edge } e \text{ has } v \text{ as origin or destination}\}$. Each vertex in the graph $G$ is transformed in a submatrix. If the single row of matrix $M$ is selected by a submatrix, then the vertex is included in the MWIS. The non-overlapping constraint of MWSDSP forbids two adjacent vertices (i.e., submatrices) to both be included in the solution (constructing an independent set), due to the way the column selections $C^1, \ldots, C^{|V|}$ are constructed. Resolving the MWSDSP then leads to the same optimal objective result as the original MWIS problem, and the selected rows $R^v$, $\forall v \in [1, \ldots, |V|]$, indicates, for each node $v$, if the node is inside the MWIS ($R^v = \{1\}$) or not ($R^v = \emptyset$). As computing the MWIS in general graphs is $\mathcal{NP}$-Hard, and as the MWSDSP with fixed column selections can encode the MWIS problem, we conclude that the MWSDSP with fixed column selections is $\mathcal{NP}$-Hard.                                     □

## 2.2   Greedy Approach

A simple approach to solving the MWSDSP is to solve the MSSP repeatedly. For each new max-sum submatrix found, the corresponding values are replaced by $-\infty$, forbidding subsequent iterations from selecting these entries again.

Each iteration is performed until optimality or absence of solution are proved; or at least one solution has been found.

## 2.3   Column Generation

We propose a column generation (CG) approach [7] to find solutions to the MWSDSP. It relies on CP[1] in an ILP setting. The CP part identifies candidate submatrices. The ILP efficiently combines submatrices and guides the CP part.

Let us represent the given matrix $\mathcal{M}$ of $m \times n$ entries as the vector $\mathcal{V}$ of $v = m \times n$ entries obtained by stacking the columns of the matrix $\mathcal{M}$ on top of one another. The MWSDSP is formulated using a $v \times 2^{m+n}$ binary matrix $\mathcal{B}$ representing all $2^{m+n}$ possible submatrices. Each column $l$ of $\mathcal{B}$ corresponds to a submatrix $l$ such that $\mathcal{B}_{i,l} = 1$ if and only if entry $\mathcal{V}_i$ is covered by the submatrix $l$. The weight $w_l$ of submatrix $l$ is the sum of its covered entries: $w_l = \sum_{i=1}^{v} \mathcal{V}_i \times \mathcal{B}_{i,l}$. Equations (2) and (3) can be formulated as an ILP:

$$\text{maximize} \quad \sum_{l \in L} w_l \times x_l \tag{6a}$$

$$\text{s.t.} \quad \sum_{l \in L} \mathcal{B}_{i,l} \times x_l \leq 1 \qquad \forall i \in \{1, \ldots, v\} \tag{6b}$$

$$\sum_{l \in L} x_l \leq K \tag{6c}$$

$$x_l \in \{0, 1\} \quad \forall l \in L \tag{6d}$$

where $L = \{1, \ldots, 2^{m+n}\}$ denotes all possible submatrices. The decision variable $x_l$ encodes the selection of submatrix $l$. Equation (6b) ensures submatrices disjunction and Eq. (6c) enforces the selection of at most $K$ submatrices.

Defining the matrix $\mathcal{B}$ before solving the ILP is computationally not feasible, even for small input matrices $\mathcal{M}$. In subproblem solving, the master problem

---

[1] See [11] for an introduction to CP.

(or ILP), in Eq. (6a)–(6d), is restricted to a subset $L' \subseteq L$ of submatrices effectively defining a restricted master problem (RMP). Iteratively, an RMP is solved, and one or multiple new submatrices (columns) are inserted in $L'$, defining a new RMP. Submatrices (columns) are candidates for insertion to an RMP if its insertion can improve the objective function of the RMP.

To find such candidate submatrices, we define a Linear Programming relaxation of the RMP (LP-RMP) which comes along the integrality constraints (6d) relaxation of the ILP (in an LP) and the subsetting of $L$. We use the dual of the LP-RMP to find submatrices with a positive reduced cost[2]. Such submatrix can improve the LP-RMP. If no such submatrix exists, the optimal solution to the LP-RMP is an optimal solution to the LP. The dual of the LP-RMP is:

$$\text{minimize} \quad \theta \times K + \sum_{i=1}^{v} \lambda_i \tag{7a}$$

$$\text{s.t.} \quad \theta + \sum_{i=1}^{v} \mathcal{B}_{i,l} \times \lambda_i \geq w_l \quad \forall l \in L' \tag{7b}$$

$$\lambda_i \geq 0 \quad \forall i \in \{1, \ldots, v\} \tag{7c}$$

$$\theta \geq 0 \tag{7d}$$

The dual values $\lambda_i$ and $\theta$ corresponding to the primal constraints defined in Eq. (6b) and (6c), respectively, are obtained by solving an LP-RMP. Each column $x_l$ of the RMP is associated with a constraint in the dual (Eq. 7b).

Finding a submatrix with a positive reduced cost is called pricing. Such a submatrix is defined as any submatrix $l \in L$ for which $-\theta - \sum_{i=1}^{v} \mathcal{B}_{i,l} \times \lambda_i + w_l < 0$. The LP-RMP is optimal if the pricing problem has no solution. Moreover, if the LP-RMP (being optimal) and the RMP have the same objective value, then the solution to the ILP is optimal.

The pricing problem can be reformulated as: $\sum_{i=1}^{v} [\mathcal{B}_{i,l} \times (\mathcal{V}_i - \lambda_i)] > \theta$.

Solving this pricing problem is not trivial: it amounts to identifying a submatrix in the input matrix modified by the $\lambda_i$ values such that its weight is larger than some $\theta$. While the pricing routine usually tries to identify a solution with maximum reduced cost, it can return any submatrix with positive reduced cost.

In practice, we use the greedy approach described earlier to find submatrices of weight larger than $\theta$ from an input matrix modified according to the $\lambda_i$ values. This provides solutions to the pricing problem.

*Implementation details* may have an important role in the effectiveness of the approach. Such details are present next.

To maximize the information given by the dual values, we avoid having redundant constraints, notably the constraints (6b). For example, if two submatrices overlap on more than one cell, we enforce only one constraint representing all the overlapping cells. Precisely, constraint (6b) is replaced by the following:

$$\sum_{l \in S} x_l \leq 1 \quad \forall S \in \left\{ \{l \mid \mathcal{B}_{i,l} = 1\} \mid i \in \{1, \ldots, v\} \right\}. \tag{8}$$

---

[2] Given that the problem is a maximization problem.

That is, we enforce one non-overlap constraint per group of entries sharing the same intersecting submatrices (an *overlapping group*)[3]. We then redistribute the dual value of the constraint equally (we divide it by the number of entries) over all the entries in this overlapping group. This allows the method to avoid a pitfall of most solvers: when facing multiple equivalent constraint, only one will be *tight*, i.e. having a non-zero dual value. Redistributing the duals on all the entries in an overlapping group allows the subproblem solver to find more interesting submatrices.

The LP-RMP does not necessarily provide a binary decision on the submatrix selection. To effectively identify a solution to the original MWSDSP, the RMP is solved for any solution to the LP-RMP. Observe that the objective value of the LP-RMP is an upper bound to the objective value of the RMP. All experiments present the results of the RMP solution.

The subset $L'$ defining the first RMP to solve is obtained using the greedy approach searching for $K$ submatrices. This serves as a **greedy hot-start** for the column generation approach.

Given the non-trivial pricing problem, there is no guarantee that the greedy subroutine identifies an optimal solution to the pricing problem. While it would be possible to use a branch-and-price algorithm [13], it would be non-trivial to solve the pricing problem to optimality. The running time needed to solve the LP-RMP to optimality (i.e. to the point where no new submatrix with positive reduced cost exists) is already quite high, as shown in the experiment section below. The authors consider that the use of a branch-and-price algorithm is outside of the paper's scope.

Guidance on the search for better submatrices requires many submatrices in the RMP with large weight. Moreover, the greedy subroutine may identify many solutions (i.e. submatrices) to the pricing problem. As the number of submatrices to find increases, the weight of these submatrices likely decreases. It is then more useful to seek multiple submatrices later in the column generation process. As a consequence, at iteration $p$ of the column generation, up to $p$ solutions, or submatrices, to the pricing problem are identified and are inserted in the RMP.

## 2.4   Mixed Integer Linear Programming

We propose a Mixed Integer Linear Programming model using the binary variables $T_i^k$ and $U_j^k$ to represent the selection of row $i$ and column $j$ for submatrix $k$. These decision variables are used to compute the contribution of the row $i$ for the submatrix $k$ ($r_i^{k+}$). The sum of the row contributions is the objective function to be maximized. The model presented below is based on a Big-M formulation of the MWSDSP where, $\forall i \in R$, constants $M_i^- = \sum_{j \in C} \min(0, \mathcal{M}_{i,j})$ and $M_i^+ = \sum_{j \in C} \max(0, \mathcal{M}_{i,j})$ are respectively the lower bound and upper bound on the sum of row $i$'s entries. The MILP model is formulated as follows:

---

[3] Equation (8) uses the set notation to implicitly remove duplicates.

$$\text{maximize} \qquad \sum_{i \in R, k \in \mathcal{K}} r_i^{k+} \qquad\qquad\qquad (9a)$$

$$\text{s.t.} \qquad r_i^{k+} \leq \sum_{j \in C} \left( \mathcal{M}_{i,j} \times U_j^k \right) + (T_i^k - 1) \times M_i^- \quad \forall i, k \qquad (9b)$$

$$r_i^{k+} \leq M_i^+ \times T_i^k \qquad\qquad\qquad\qquad \forall i, k \qquad (9c)$$

$$2 \times v_{i,j}^k \leq T_i^k + U_j^k \qquad\qquad\qquad\qquad \forall i, j, k \quad (9d)$$

$$T_i^k + U_j^k \leq 1 + v_{i,j}^k \qquad\qquad\qquad\qquad \forall i, j, k \quad (9e)$$

$$\sum_{k \in \mathcal{K}} v_{i,j}^k \leq 1 \qquad\qquad\qquad\qquad\qquad \forall i, j \qquad (9f)$$

Constraints (9b) and (9c) ensure that the row contribution $r_i^{k+}$ is computed correctly. If $T_i^k = 0$, constraint (9c) ensures the row contribution is zero, with the right hand side of constraint (9b) being always positive. Otherwise ($T_i^k = 1$), constraints (9b) and (9c) ensure $r_i^{k+} = \sum_{j \in C} \left( \mathcal{M}_{i,j} \times U_j^k \right)$, thus computing the effective value of the contribution.

Equations (9d) and (9e) linearize $v_{i,j}^k = T_i^k \times U_j^k$. The binary variable $v_{i,j}^k$ indicates if cell $(i,j)$ is selected by submatrix $k$ and ensures submatrices disjunction through constraint (9f).

This model is plagued by the number of variables and constraints which are both in $O(mnK)$, mainly due to the non-overlap constraints.

## 3    Experiments

This section describes experiments conducted to assess the performances of the proposed algorithms and to provide guidance on the selection of the appropriate solution. Given enough time and memory, both the column generation (CG) approach and the MILP approach converge to the optimal solution. Therefore comparing performances solely on the objective value of an approach is irrelevant. As a consequence, CG and MILP approaches are evaluated and compared given a budget of time, the time-out $TO$, on synthetic datasets with implanted submatrices using any-time profiles:

**Definition 3. *Any-Time Profile.*** *Let* f(a, i, t) *be the objective value of the best solution found so far by an algorithm* a *for an instance* i *at time* t. *Let* t$^{\max}$ *be the provided budget of time before breaking a run. The any-time profile of* a *is the solution quality* Q$_a$(t) *of* a *on all instances as a function of time:*

$$Q_a(t) = \frac{1}{|i|} \sum_i \frac{f(a, i, t)}{f(a_i^*, i, t^{\max})} \; with \; a_i^* = \underset{a}{\text{argmax}} \; f(i, i, t^{\max}) . \qquad (10)$$

All experiments are performed using Java 1.8.0 on an AMD Bulldozer clocked at 2.1 GHz; one core and 6 GB of RAM per instance and a time-out $TO$ of 2 h. MILP and CG approaches rely on Gurobi 8.1.0 [1]. The greedy hot-start of the CG process is given 5 min evenly split between each of its $K$ iterations of solving an MSSP. Solutions to the MSSP are carried out on OscaR [12] using a constraint programming approach relying on a global constraint (CPGC) provided in [5].

It is a depth-first search approach composed of major CP ingredients: (1) filtering rules, (2) bounding procedure, (3) dominance rules and (4) variable-value heuristic.

## 3.1   Datasets and Performances

Datasets are generated by implanting $K$ submatrices (called + entries) on a background noise (called − entries). In a first dataset, we consider alternative dispositions of + and − entries drawn from different distributions. Each combination defines a scenario presented in Fig. (1a–b). For each scenario, 14 different matrices are generated according to different input matrix size and number of implanted submatrices, as presented in Fig. (1c). These 70 instances are generated such that the hot-start is bound to find suboptimal solutions, giving very little information to the CG method. The benefit of CG is evaluated relative to the suboptimal hot-start solution through the objective value improvement.

Figure (2a) presents the any-time profile of each method for the first dataset. It clearly illustrates that CG can escape the suboptimal regions of the search space trapping the hot-start. Given roughly 25 times larger time-out than the suboptimal hot-start, MILP is outperformed by the greedy and the CG.

Local optimums (trapping the hot-start) are provided as starting solutions for CG. Such local optimum can be found before the given time-out. The shift between hot-start and CG curves in the first 300 s is explained by the fact that CG can refine solutions as soon as the hot-start subroutine is completed.

In the second dataset, 720 instances are generated according to the layout of scenarios 3 and 4 from Fig. (1a). It differs, however, by the size of the input matrix, the number, and size of implanted submatrices. More importantly, values are drawn from different distributions: − entries $\sim \mathcal{N}(-1, 1)$ and + entries $\sim \mathcal{N}(1, 0.5)$. Such matrices, generated following a similar protocol as in [6], are considered better representatives of gene expression matrices. Our script is available on Zenodo [4].

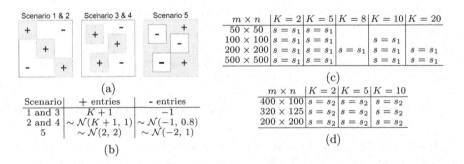

Fig. 1. Dataset construction. (a) Layout and (b) generative distribution of implanted + and − entries. (c) Parameters considered in the first dataset with $s_1 = \{1.0\}$. (d) Parameters considered in the second dataset with $s_2 = \{0.05, 0.01, 0.2, 0.5\}$. Implanted submatrices are of size $\left(\frac{m \times s}{K}; \frac{n \times s}{K}\right)$.

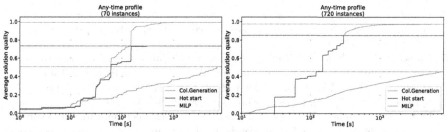

(a) Averaged results on the first dataset.   (b) Averaged results on the second dataset.

**Fig. 2.** Comparison of the different methods proposed to solve the MWSDSP. The graph presents the any-time profile described in Eq. (3). For each instance, the time-out is fixed at 2 h. The hot-start time-out equals 5 min. Col.Generation starts as soon as the hot-starts is completed.

Figure (2b) presents the any-time profile of CG and MILP on the second dataset. Whereas the average solution quality of CG and MILP should rise to 1, given enough time, it is clear that CG is significantly faster than MILP. The poor performances of MILP are explained by the number of variables and constraints required to model the problem: MILP obtains satisfactory results for the smaller problems, with $K = 2$, only (results not shown). In this experiment, the hot-start rarely ends before the allocated 5 min, explaining the near-perfect overlap between hot-start and CG curves.

## 4   Conclusions

We present a new optimization problem, called the Maximum Weighted Set of Disjoint Submatrix Problem (MWSDSP) along with two methods to solve it. One is based on mathematical programming, the other on constraint programming.

Our main contribution, the column generation (CG) method for the MWS-DSP, finds new candidate submatrices using dual variables of a linear relaxation of the submatrix selection problem. Experiments on synthetic datasets indicate that CG finds better solutions than the MILP approach.

The performances of the CG can be further improved by complementing the exploration with a branch-and-price algorithm [13]. Such improvement is non-trivial, however: the time taken to solve the underlying LP problem is already quite long but is nonetheless an attractive direction for future work.

## References

1. Gurobi Optimization, LLC (2018). http://www.gurobi.com
2. Aloise, D., Hansen, P., Liberti, L.: An improved column generation algorithm for minimum sum-of-squares clustering. Math. Program. **131**(1), 195–220 (2012)
3. Bentley, J.: Programming pearls: algorithm design techniques. Commun. ACM **27**(9), 865–873 (1984)

4. Branders, V., Derval, G., Schaus, P., Dupont, P.: Dataset generator for Mining a maximum weighted set of disjoint submatrices, August 2019. https://doi.org/10.5281/zenodo.3372282

5. Branders, V., Schaus, P., Dupont, P.: Combinatorial optimization algorithms to mine a sub-matrix of maximal sum. In: Appice, A., Loglisci, C., Manco, G., Masciari, E., Ras, Z.W. (eds.) NFMCP 2017. LNCS (LNAI), vol. 10785, pp. 65–79. Springer, Cham (2018). https://doi.org/10.1007/978-3-319-78680-3_5

6. Derval, G., Branders, V., Dupont, P., Schaus, P.: The maximum weighted submatrix coverage problem: a CP approach. In: Rousseau, L.-M., Stergiou, K. (eds.) CPAIOR 2019. LNCS, vol. 11494, pp. 258–274. Springer, Cham (2019). https://doi.org/10.1007/978-3-030-19212-9_17

7. Desaulniers, G., Desrosiers, J., Solomon, M.M.: Column Generation, vol. 5. Springer, Boston (2006). https://doi.org/10.1007/b135457

8. Garey, M.R., Johnson, D.S.: Computers and Intractability; A Guide to the Theory of NP-Completeness. W. H. Freeman & Co., New York (1990)

9. Le Van, T., van Leeuwen, M., Nijssen, S., Fierro, A.C., Marchal, K., De Raedt, L.: Ranked tiling. In: Calders, T., Esposito, F., Hüllermeier, E., Meo, R. (eds.) ECML PKDD 2014. LNCS (LNAI), vol. 8725, pp. 98–113. Springer, Heidelberg (2014). https://doi.org/10.1007/978-3-662-44851-9_7

10. Madeira, S.C., Oliveira, A.L.: Biclustering algorithms for biological data analysis: a survey. IEEE/ACM Trans. Comput. Biol. Bioinform. (TCBB) 1(1), 24–45 (2004)

11. Michel, L., Schaus, P., Van Hentenryck, P.: MiniCP: a lightweight solver for constraint programming (2018). https://minicp.bitbucket.io

12. OscaR Team: OscaR: Scala in OR (2012). https://bitbucket.org/oscarlib/oscar

13. Savelsbergh, M.: A branch-and-price algorithm for the generalized assignment problem. Oper. Res. 45(6), 831–841 (1997)

14. Takaoka, T.: Efficient algorithms for the maximum subarray problem by distance matrix multiplication. Electron. Notes Theoret. Comput. Sci. 61, 191–200 (2002)

# Dataset Morphing to Analyze the Performance of Collaborative Filtering

André Correia[1]([⊠]), Carlos Soares[1,2,3], and Alípio Jorge[3,4]

[1] Faculdade de Engenharia da Universidade do Porto, Porto, Portugal
{up200706629,csoares}@fe.up.pt
[2] LIACC, Porto, Portugal
[3] LIAAD-INESC TEC, Porto, Portugal
[4] Faculdade de Ciências da Universidade do Porto, Porto, Portugal
amjorge@fc.up.pt

**Abstract.** Machine Learning algorithms are often too complex to be studied from a purely analytical point of view. Alternatively, with a reasonably large number of datasets one can empirically observe the behavior of a given algorithm in different conditions and hypothesize some general characteristics. This knowledge about algorithms can be used to choose the most appropriate one given a new dataset. This very hard problem can be approached using metalearning. Unfortunately, the number of datasets available may not be sufficient to obtain reliable meta-knowledge. Additionally, datasets may change with time, by growing, shrinking and editing, due to natural actions like people buying in a e-commerce site. In this paper we propose dataset morphing as the basis of a novel methodology that can help overcome these drawbacks and can be used to better understand ML algorithms. It consists of manipulating real datasets through the iterative application of gradual transformations (morphing) and by observing the changes in the behavior of learning algorithms while relating these changes with changes in the meta features of the morphed datasets. Although dataset morphing can be envisaged in a much wider framework, we focus on one very specific instance: the study of collaborative filtering algorithms on binary data. Results show that the proposed approach is feasible and that it can be used to identify useful metafeatures to predict the best collaborative filtering algorithm for a given dataset.

**Keywords:** Recommender Systems · Metalearning

## 1 Introduction

In this paper, we propose an empirical methodology for improved understanding of the behavior of algorithms that combines a novel data manipulation approach, *dataset morphing*, with a Metalearning (MtL) approach. MtL consists of relating data characteristics (*metafeatures*) to the performance of learning algorithms. These metafeatures are expected to contain some useful information about the performance of the algorithms. To generalize the extracted (meta)

© Springer Nature Switzerland AG 2019
P. Kralj Novak et al. (Eds.): DS 2019, LNAI 11828, pp. 29–39, 2019.
https://doi.org/10.1007/978-3-030-33778-0_3

knowledge and, therefore, make it applicable to new problems, MtL approaches require a large collection of datasets, which is often not the case. Dataset morphing addresses this issue by iteratively transforming (morphing) one real dataset into another. If the two datasets display interesting contrasting behavior of algorithms (e.g. algorithm $A$ is better than $B$ on one dataset but not on the other) then interesting metaknowledge can be obtained (e.g. determine the turning point on the performance of the algorithms, and carefully analyzing what happens around the performance boundary in terms of data characteristics). The proposed methodology can be used to select the most appropriate algorithm for a new problem or to analyze algorithm behaviour with evolving data.

As a an example of application for Recommender Systems (RS), we instantiate the above proposed method to study two popular Collaborative Filtering (CF) algorithms: item-based and user-based neighborhood approaches. We focus on item recommendation (*top-N*), where the aim is to recommend ordered lists of items in a binary setting [12]. To automatically identify the most adequate CF algorithm for a given data set has proven challenging [2]. Empirical results about algorithm behavior are limited due to the absence of a large number of datasets [2] and purely artificially generated data does not entirely solve the problem because it is unlikely that it reflects real world distributions.

This work extends existing studies [2,3] by proposing dataset morphing as a process of generating multiple realistic datasets (viewed as meta-examples), that could be useful to enrich the metadata and, therefore, to improve the results of MtL processes that learn the relationship between the performance of RS algorithms and data characteristics. Despite the provided example with CF, dataset morphing is virtually applicable to any Machine Learning domain.

## 2    Metalearning

MtL studies how Machine Learning (ML) can be employed to understand the learning process and improve the use of ML in future applications [11]. A successful MtL approach can provide a solution to the problem of selecting an algorithm for a given dataset [2]. It allows the extraction of knowledge that explains why a suggested algorithm is a good choice. It uses ML techniques to obtain predictive models, which associate data characteristics to algorithm performance. The methodology involves extracting characteristics, named *metafeatures*, from multiple datasets and assessing the performance, which will be used as *metalabels*, of a group of algorithms. Afterwards, this data is used to induce a predictive model to represent the relationship between the metafeatures and metalabels. After obtaining an accurate MtL model we can predict the most promising algorithm without running a full-fledged empirical evaluation and also explain why an algorithms performs better or worse [3].

MtL has been used for algorithm selection in RS [2,3,6]. These authors manually define metafeatures, which aggregate information from datasets into single number statistics. For example, the number of instances in the dataset is a *simple* metafeature, the mean or kurtosis of a column is a *statistical* metafeature.

They then use supervised ML to learn the relationships between the metafeatures and the performance of recommendation algorithms on datasets, measured by standard metrics. Although the use of MtL for the selection of CF algorithms has already been investigated, the approaches proposed have limited scope: the set of datasets, recommendation algorithms and metafeatures studied was rather restricted An extensive overview of their positive and negative aspects can be seen in a recent survey [3].

## 3 Morphing Recommendation Datasets

Image morphing has proven to be a powerful tool for visual effects in film and television, enabling the fluid transformation of one digital image into another [14]. We believe that the principle behind image morphing can be applied to generate datasets that can be used to study the behavior of RS algorithms. Actually, we can generalize the morphing technique to any type of ML problem. Thus, dataset morphing can be defined, in general, as a process of gradually transforming a source dataset into a target dataset. That way, we can explore the space of datasets along trajectories and study the behavior of algorithms in regions of that space that are not currently available, particularly in regions where algorithms' performances change.

The approach proposed here consists in starting with two datasets—the source $(D_s)$ and the target $(D_t)$. The operational goal is to analyze the evolution in the behavior of one or more algorithms between two points of interest. In particular, we will pick up pairs of datasets where two RS algorithms $A$ and $B$ have contrasting relative performances. $A$ is better than $B$ in one dataset and viceversa. This set up will originate two regions of the space of datasets. We can study those two regions, their boundary and the trajectory that crosses that boundary. We get from one dataset to the other by sequentially applying transformations $\{T_1, T_2, ..., T_{n-1}, T_n\}$ (Fig. 1). The initial datasets have contrasting algorithm performance. The color gradient, illustrated in Space D, means that during the transformation process, (intermediate) datasets $\{D_1, D_2, ..., D_{n-2}, D_{n-1}\}$ will gradually become more similar to the target dataset $(D_t)$. As previously mentioned it is important to keep datasets as realistic as possible. To have that, intermediate datasets $\{D_1, D_2, ..., D_{n-2}, D_{n-1}\}$ are a mixture of real—source and target—datasets. In short, considering source $(D_s)$ and target $(D_t)$ datasets and a transformation function $(\tau)$, we define dataset morphing as a process of iteratively getting intermediate datasets $(D_j)$ such that:

$$\mathcal{D}_{morph} : \{D_j \mid D_0 = D_s, D_n = D_t, D_j = \tau(D_{j-1})\}, 1 \leq j < n \qquad (1)$$

where $\mathcal{D}_{morph}$ is the set of datasets and $n$ is the number of transformations needed to get from source $(D_s)$ to target $(D_t)$. The function $(\tau)$ is guaranteed to converge.

The upper layer of Fig. 1 illustrates the trajectory in the dataset space. The feature space $\mathcal{F}$, represented with vectors of metafeatures $\{MF_1, MF_2, ..., MF_f\}$, is the middle layer. These metafeatures are important to characterize the relative

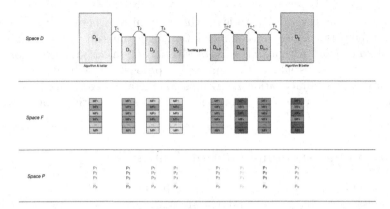

**Fig. 1.** Analysis of algorithm behavior between two points of interest using a metalearning approach based on dataset morphing. $D$ is the data space; $F$ is the metafeatures space; and $P$ is the algorithm performance space.

performance of the algorithms. As transformations are applied, some metafeatures may change a lot, others only slightly and others not at all. These different types of changes are illustrated in Fig. 1 with the magnitude of change in the colors between adjacent vectors. Finally, in the bottom layer, the performance space $\mathcal{P}$ with $\{P_1, P_2, ..., P_p\}$ is represented. The performance of the algorithms in the datasets of space $\mathcal{D}$ may be evaluated using several standard metrics. Once again, colors are intended to represent performance variation. To learn about algorithm behaviour, our focus will be on metafeatures that vary according to the changes in algorithm performance.

### 3.1 Dataset Transformations

Source and target datasets could be selected based on a specific property of interest. In this paper we focus on contrasting algorithm performance, where a given algorithm $A$ outperforms an algorithm $B$ in $(D_s)$ and the opposite happens in $(D_t)$. Other possibilities can be considered, such as different performances of the same algorithm, comparison with baselines or effects of parameters.

One of the key issues in the methodology is the definition of the transformation function. One important property it should have is *convergence*. After each application the resulting dataset is more similar to the target and less similar to the source. Another important property is *smoothness*. The difference between two consecutive datasets should be small both in terms of metafeatures and of performance. Other characteristics may depend on the task and the type of data available. In this paper, we will focus on CF with binary data. Transformations may be simple (e.g., random bit flipping, random rows/columns switching) or more complex (e.g., switch the most similar row/column first). They can also be applied in batches (e.g., flipping a portion of all bits, switching a portion of all rows/columns).

We also have natural morphing processes when users' preferences have a very volatile nature [9]. In real world RS, it is reasonable to approach ratings data as evolving datasets: ratings are continuously being generated, and we have no control over the data rate or the ordering of the arrival of new ratings. Actually, adding or removing a row means a new customer application or disassociation, respectively. Likewise, adding or removing a column denotes a new item arrival or removal, respectively.

## 4    Empirical Evaluation

The main aim of these experiments is to show that dataset morphing can be useful for identifying predictive metafeatures of the relative performance of CF algorithms using a limited number of original datasets. We intend to illustrate how dataset morphing enriches the metadata and improves the results of MtL processes.

### 4.1    Base-Level

In this study, we focus on the item recommendation CF task and evaluate $top-1$, 3, 5, 10, 15 and 20 recommendation lists. CF algorithms are evaluated using a 10-fold cross-validation scheme with the all-but-1 protocol to collect data about the behavior of two CF algorithms: user-based and item-based. As both are $k$ Nearest Neighbors (NN) algorithms, we considered $k = 20$ and $k = 50$ for user-based and item-based, respectively. The performance of these algorithms is estimated on each dataset, using $precision@k$. In terms of implementation we use on the *recommenderlab* package[1] since the comparison of recommender algorithms is readily available [7]. Other algorithms, parameters or platform could have been chosen, without loss of generality. Table 1 lists the 3 real-world datasets selected for this study. They were binarized by making items with a rating of 1, or higher, a positive rating. Due to the very large number of experiments needed for meta learning, to have feasible computational times we have used subsets of the original datasets, obtaining 60000 random samples from each dataset. Each sample has 250 rows (users) and 1000 columns (items).

**Table 1.** Datasets used in the base-level experiments.

| Dataset | #users | #items | #ratings | Ratings scale | Ref |
|---|---|---|---|---|---|
| amazon-movies | 73 k | 4 k | 111 k | [1;5] | [10] |
| movielens1m | 6 k | 4 k | 1 M | [0;5] | [5] |
| palcoprincipal-playlist | 4 k | 5 k | 37 k | [0;1] | [4] |

---

[1] https://cran.r-project.org/web/packages/recommenderlab/index.html.

Regarding the source $(D_s)$ and target $(D_t)$ datasets selection, we followed two main approaches: selecting two subsets from the same dataset or selecting source and target from different datasets. The first approach was applied both on *amazon-movies* and *palcoprincipal-playlist* datasets. The second approach was applied both on *movielens1m* and *palcoprincipal-playlist* datasets, selecting source $(D_s)$ datasets from *movielens1m*, and target $(D_t)$ datasets from *palcoprincipal-playlist*. As mentioned, the criterion considered in the experiments was the contrasting algorithm performance. This means that for the source dataset $(D_s)$ a given algorithm $A$ outperforms another given algorithm $B$ and, in target dataset $(D_t)$, $B$ outperforms $A$. In the experiments, the algorithms $A$ and $B$ are user-based CF and item-based CF, respectively. For each dataset, we evaluated both algorithms on each sample (out of the 60000 samples). For each algorithm, we selected the top-100 samples with the highest difference in precision (*delta*). Therefore, for each dataset selection approach described above, we formed 100 pairs, selecting, for each algorithm, the 100 samples with the highest *delta* values.

Regarding the dataset transformations, we decided to use random one row replacements. We iteratively replace rows in the source dataset $(D_s)$ with rows from the target $(D_t)$. This enforces smoothness and trivially guarantees convergence in a pre-defined number of steps. By way of illustration, to obtain dataset $D_1$ we start with source dataset $(D_s)$. We randomly sample, without replacement, one row index and copy that row from $(D_t)$ to $D_1$. Likewise, dataset $D_2$ has all but one row from dataset $D_1$. This procedure is repeated until intermediate and target datasets match. In the experiments, for each pair created we obtain 250 intermediate datasets. However, the wide variety of possibilities to get from source $(D_s)$ to target $(D_t)$ datasets, deserves future exploration. To minimize time and computational resources, in the experiments, for each pair created, we sampled 10 different trajectories between source $(D_s)$ to target $(D_t)$ datasets. This means that, for each pair, we applied the random row-wise transformations in 10 different ways.

## 4.2   Meta-level

One of the most important factors in the success of a MtL approach is the definition of a set of metafeatures that contain information about the performance of algorithms [1]. Part of the metafeatures used in this study are obtained procedurely [2] and are based on two different perspectives on their distribution: users and items. These distributions are aggregated, by row and by column, using simple, standard statistical functions (count and mean) and post-processing functions: maximum, minimum, mean, standard deviation, median, mode, entropy, Gini index, skewness and kurtosis. The notation used to represent metafeatures follows the format: *object.function.post function* (e.g., *column.mean.entropy*). Other metafeatures used in this study are based on [13]. That work organizes metafeatures in five groups i.e., subsets of data characterization measures that share similarities among them [1]. Since this study focuses only on binary rating-based CF datasets, we only considered metafeatures from the Simple and Information-theoretic groups [13]. From the Information-theoretic group, we used the attributes concentration (*attrConc*) and the attributes entropy (*attrEnt*) [13]. These metafeatures were implemented using the *mfe* package.[2] Lastly, we used two other metafeatures: number of zeros of the entire dataset and sparsity [8].

The techniques used in the meta-level are usually either classification or regression. This study focuses on classification tasks. Considering the 100 identified pairs of subset datasets, the 10 different trajectories between each pair and the 250 intermediate datasets for each trajectory as a result of dataset morphing process, we created 18 meta-datasets with 250.000 meta-examples. One for each of the 3 selected original dataset (Table 1), and for each of the 6 top-$N$. The algorithm selection problem is formulated as a classification task, where the class label is the best algorithm, according to the precision metric. Either $IB$ (item-based) or $UB$ (user-based). The predictive attributes are the metafeatures described above. We did some exploratory experiments with a set of classification algorithms: Adaboost, C5.0, Gradient Boosting Machine, Logistic Regression, Naive Bayes, Random Forest, rpart and XGBoost. Regarding the classification problem, we chose the following error measures: accuracy, recall for item-based class ($Recall_{IB}$), recall for user-based class ($Recall_{UB}$) and area under the curve (AUC). We performed tuning on algorithms, optimizing the AUC metric i.e., for each meta-level algorithm we considered different values for its hyperparameters. Despite considering many trajectories for each pair, intermediate datasets of same pair are very similar to each other. Therefore, the algorithms were evaluated in a leave 20 pairs out strategy. This means that we use 80 pairs of datasets for training and leave the remaining 20 pairs for testing. Meta-learning was done using the *caret* package,[3].

## 5    Experimental Results

**Base-Level:** as an example of the algorithm performance evaluation at the base-level, Fig. 2 illustrates the results for *palcoprincipal-playlist* dataset, for one pair and trajectory. The performance of user-based CF is represented in blue colour and the results of item-based CF are represented in red colour. We can observe that for top-3 and top-5 tasks item-based CF starts presenting higher values at approximately (intermediate) dataset $D_{100}$. This means that $D_{100}$ is on an interesting boundary and is worth looking into.

**Meta-level:** The exploratory data analysis allows to identify metafeatures that are good indicators of the relative performance of the algorithms. From the vast number of experiments we have performed, we show one figure, where the winning algorithm for each intermediate dataset is illustrated in different colours. Once again, user-based CF is in blue and item-based CF is in red. The presented results are only for one pair created and one trajectory of *palcoprincipal-playlist*

---

[2] https://cran.r-project.org/web/packages/mfe/index.html.
[3] https://cran.r-project.org/web/packages/caret/index.html.

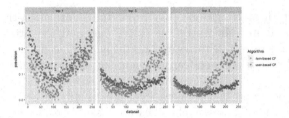

**Fig. 2.** Performance of $IB$ and $UB$ through morphing. (Color figure online)

dataset, and for the top-3 task. As an example, the metafeature illustrated in Fig. 3 seems to contain useful information about the relative performance of the algorithms i.e., it varies accordingly with algorithm performance. In fact, higher values of this metafeature indicate that user-based performs best. Item-based CF algorithm seems to be the winner when this metafeature has lower values. This is true for the following metafeatures: *attrConc.mean*, *column.count.mean*, *row.count.entropy*, *row.count.kurtosis* and *row.count.max*.

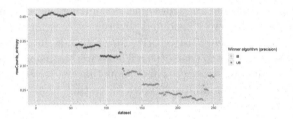

**Fig. 3.** Results of meta-level evaluation on *palcoprincipal-playlist* dataset—Entropy of row count. (Color figure online)

Regarding the meta-learning results, we firstly tested the meta-models against test sets obtained within the same dataset used for training. Then, in order to ensure that the extracted metaknowledge could be generalizable, we also tested each meta-model against the test set of remaining domains. Lastly, we present performance results of meta-models created with random samples. To serve as baseline, we did some experiments training meta-models with only the source and target datasets i.e., without the intermediate datasets.

Results on Table 2 show that the meta-models created help in intra-domain algorithm selection. The meta-models obtained from *amazon-movies* and *palco principal-playlist* datasets, seem to clearly identify the item-based CF instances. On the other hand, the *movielens1m/palcoprincipal-playlist* dataset yields a meta-model that seems to identify reasonably well both classes. The inter-domain test set results, partially support the conclusions of intra-domain results. In fact, meta-models obtained from *amazon-movies* and *palco principal-playlist* datasets, presented high values for $Recall_{IB}$ i.e., they seem to clearly identify the item-based CF instances. On the other hand, unlike the intra-domain results, the meta-model obtained from *movielens1m/palcoprincipal-playlist* dataset presented low values for $Recall_{IB}$ and high values for $Recall_{UB}$. This seems to mean that it clearly identifies the user-based CF instances, unlike the item-based CF ones. In short, the results of these experiments show that the extracted meta-knowledge could be generalized and transferred across the studied domains.

Concerning the performance results of meta-models trained without the intermediate datasets, and based on intra-domain results, it is possible to conclude that the meta-data obtained from morphing datasets leads to better meta-knowledge, when compared to meta-data obtained from random samples. Nevertheless, considering the inter-domain results, some performance metrics (e.g., $Recall_{IB}$) presented better overall results, on meta-models trained with trajectories, and some others don't (e.g., AUC).

**Table 2.** Summary of results. Green (red) dots indicate that meta-models from trajectories beat (loose against) random samples. Delta is the winning margin.

| Model/Test set | amazon-movies | | mlens/ palco | | palco | |
|---|---|---|---|---|---|---|
| | Metric | Delta | Metric | Delta | Metric | Delta |
| amazon-movies | AUC | 0.21 | AUC | 0.13 | AUC | 0.01 |
| | $Recall_{IB}$ | 0.36 | $Recall_{IB}$ | 0.07 | $Recall_{IB}$ | 0.33 |
| | $Recall_{UB}$ | 0.19 | $Recall_{UB}$ | 0.277 | $Recall_{UB}$ | 0.45 |
| movielens1m/ palcoprincipal- playlist | AUC | 0.07 | AUC | 0.07 | AUC | 0.03 |
| | $Recall_{IB}$ | 0.29 | $Recall_{IB}$ | 0.13 | $Recall_{IB}$ | 0.45 |
| | $Recall_{UB}$ | 0.18 | $Recall_{UB}$ | 0.17 | $Recall_{UB}$ | 0.37 |
| palcoprincipal- playlist | AUC | 0.01 | AUC | 0.01 | AUC | 0.11 |
| | $Recall_{IB}$ | 0.20 | $Recall_{IB}$ | 0.02 | $Recall_{IB}$ | 0.04 |
| | $Recall_{UB}$ | 0.22 | $Recall_{UB}$ | 0.25 | $Recall_{UB}$ | 0.22 |

To extract meta-knowledge we assess features frequency in the best models. The best metafeatures are: *row.count.entropy*, *row.count.max*, *row.count. kurtosis*, *attrConc.mean* and *attrEnt.mean*. We observe that both rows and columns hold important characteristics to solve the algorithm selection problem. This confirms previous studies [2,3]. Nevertheless, *row.count* is the most relevant distribution to be analyzed here. Actually, the choice between user-based and item-based depends on the ratio between the number of rows and the items.

## 6  Conclusions

In this study, we have proposed a methodology that generates new datasets by manipulating existing ones, for understanding algorithm behavior using MtL approaches. In the experiments, the proposed methodology was used to select CF algorithms. The algorithm selection problem was formulated as a classification task, where the target attribute is the best CF algorithm, according to precision metric. The results show that the proposed approach is feasible and that it can be used to identify useful metafeatures to predict the best collaborative filtering algorithm for a given dataset. Considering the majority of the scenarios studied, the results support the importance of dataset morphing to enrich the metadata and, therefore, to improve the results of MtL processes. As future work we intend to explore the multiple avenues offered to machine learning by dataset morphing.

**Acknowledgments.** This work is funded by ERDF through the Operational Programme of Competitiveness and Internationalization—COMPETE 2020—of Portugal 2020 within project PushNews—POCI-01-0247-FEDER-0024257.

# References

1. Brazdil, P., Carrier, C.G., Soares, C., Vilalta, R.: Metalearning: Applications to Data Mining. COGTECH. Springer, Heidelberg (2008). https://doi.org/10.1007/978-3-540-73263-1
2. Cunha, T., Soares, C., de Carvalho, A.C.P.L.F.: Selecting collaborative filtering algorithms using metalearning. In: Frasconi, P., Landwehr, N., Manco, G., Vreeken, J. (eds.) ECML PKDD 2016. LNCS (LNAI), vol. 9852, pp. 393–409. Springer, Cham (2016). https://doi.org/10.1007/978-3-319-46227-1_25
3. Cunha, T., Soares, C., de Carvalho, A.C.: Metalearning and recommender systems: a literature review and empirical study on the algorithm selection problem for collaborative filtering. Inf. Sci. **423**, 128–144 (2018)
4. Domingues, M.A., et al.: Combining usage and content in an online recommendation system for music in the long tail. Int. J. Multimed. Inf. Retrieval **2**, 3–13 (2013)
5. Maxwell, F., Konstan, J.A.: The MovieLens datasets: history and context. ACM Trans. Intell. Syst. Technol. (TIST) **5**, 1–19 (2015)
6. García-Saiz, D., Zorrilla, M.: A meta-learning based framework for building algorithm recommenders: an application for educational arena. J. Intell. Fuzzy Syst. **32**, 1–11 (2017)
7. Hahsler, M.: recommenderlab: a framework for developing and testing recommendation algorithms, November 2011
8. Massa, P., Avesani, P.: In: Proceedings of the 2007 ACM conference on Recommender systems - RecSys 2007 (2007)
9. Matuszyk, P., Vinagre, J., Spiliopoulou, M., Jorge, A.M., Gama, J.: Forgetting methods for incremental matrix factorization in recommender systems. In: Proceedings of the 30th Annual ACM Symposium on Applied Computing - SAC '15 (2015)
10. McAuley, J., Leskovec, J.: Hidden factors and hidden topics: understanding rating dimensions with review text. Proceedings of the 7th ACM conference on Recommender systems - RecSys 2013 (2015)
11. Prudêncio, R.B.C., Soares, C., Ludermir, T.B.: Combining meta-learning and active selection of datasetoids for algorithm selection. In: Corchado, E., Kurzyński, M., Woźniak, M. (eds.) HAIS 2011. LNCS (LNAI), vol. 6678, pp. 164–171. Springer, Heidelberg (2011). https://doi.org/10.1007/978-3-642-21219-2_22
12. Ricci, F., Rokach, L., Shapira, B.: Recommender systems: introduction and challenges. In: Ricci, F., Rokach, L., Shapira, B. (eds.) Recommender Systems Handbook, pp. 1–34. Springer, Boston, MA (2015). https://doi.org/10.1007/978-1-4899-7637-6_1

13. Rivolli, A., Garcia, L.P.F., Soares, C., Vanschoren, J., de Leon Ferreira de Carvalho, A.C.P.: Towards reproducible empirical research in meta-learning. CoRR abs/1808.1 (2018)
14. Wolberg, G.: Image morphing: a survey. Vis. Comput. **14**, 360–372 (1998)

# Construction of Histogram with Variable Bin-Width Based on Change Point Detection

Takayasu Fushimi[1]([⊠]), Kiyoto Iwasaki[2], Seiya Okubo[3], and Kazumi Saito[4,5]

[1] School of Computer Science, Tokyo University of Technology,
Hachioji 192-0982, Japan
takayasu.fushimi@gmail.com
[2] Industrial Research Institute of Shizuoka Prefecture, Shizuoka 421-1221, Japan
kiyoto1_iwasaki@pref.shizuoka.lg.jp
[3] School of Management and Information, University of Shizuoka,
Shizuoka 422-8526, Japan
s-okubo@u-shizuoka-ken.ac.jp
[4] Faculty of Science, Kanagawa University, Hiratsuka 259-1293, Japan
k-saito@kanagawa-u.ac.jp
[5] Center for Advanced Intelligence Project, RIKEN, Tokyo 103-0027, Japan
kazumi.saito@riken.jp

**Abstract.** For a given set of samples with a numeric variable and a set of nominal variables, we address a problem of constructing a histogram drawn by $K$ bins with variable widths, so as to have relatively large numbers of narrow bins for some ranges where numeric values distribute densely and change substantially, while small numbers of wide bins for the other ranges, together with the characteristic nominal values for describing these bins as annotation terms. For this purpose, we propose a new method, which incorporates a change point detection method to numeric values based on an $L1$ or $L2$ error criterion, and an annotation terms identification method for these bins based on the $z$-score with respect to the distribution of nominal values. In our experiments using four datasets of humidity deficit (HD) collected from vinyl greenhouses, we show that our proposed method can construct more natural histograms with appropriate variable bin widths than those with an equal bin width constructed by the standard method based on square-root choice or Sturges' formula, the histograms constructed with the $L1$ error criterion has more desirable property than those with the $L2$ error criterion, and our method can produce a series of naturally interpretable annotation terms for the constructed bins.

**Keywords:** Histogram · Change point detection · Variable bin-width · Visualization

© Springer Nature Switzerland AG 2019
P. Kralj Novak et al. (Eds.): DS 2019, LNAI 11828, pp. 40–50, 2019.
https://doi.org/10.1007/978-3-030-33778-0_4

# 1    Introduction

Histogram visualization has been widely used to analyze distribution of data. In general, the bin width of the histogram is a fixed size. It is important to set this bin width properly, and various methods and indicators have been proposed [3,6]. The agricultural environmental data dealt with in this study are often concentrated because they are controlled by farmers so as to take desirable values. In addition, concentrated parts might contain multiple agriculturally important sets. When such data are analyzed using a fixed-width histogram, important information is buried in one bin, and it is difficult to achieve further analysis.

To overcome this shortcoming, for a set of numeric data, we attempt to construct a histogram with variable bin-width where relatively large numbers of narrow bins for densely distributed and drastically changed values, while small numbers of wide bins for the other values. For this purpose, we propose a new method based on change point detection for the arranged values in ascending order. In our method, we produce a step function consisting of $K$ steps based on an $L1$ or $L2$ error criterion, and then by using these change points information, we construct a histogram drawn by $K$ bins with variable widths. In our experiments using real datasets collected from four vinyl greenhouses, we confirm that our proposed method can construct more natural histograms with appropriate variable bin widths than those with an equal bin width.

The paper is organized as follows. Section 2 describes related work. Section 3 gives our problem setting and proposed method. In Sects. 4 and 5, we report and discuss experimental results using real world data. Finally, Sect. 6 concludes this paper and address the future work.

# 2    Related Work

In this study, for the purpose of revealing the underlying mechanism of agricultural environment data, we employ the change points detection method [5] to individually determine adequate bin-widths for our histogram construction, where this detection method is formulated as a regime-switching problem (e.g., [4,5]). This problem setting is different from anomaly detections for statistically significant short-term outliers compared to stationary models and the statistical machine learning frameworks that set up stationary models as mixed models of probability distributions [2].

There are also some works about histogram with variable bin-width. For example, a method of constructing the equal-area histogram (also called the percentile mesh) of Scott et al., and a Denby & Mallows method to construct an intermediate histogram of equal-width bins and equal-areas were proposed [3,6]. In particular, these techniques are said to be effective in identifying sharp peaks, etc. These methods are based solely on the data distribution attributed to the bins. On the other hand, in the proposed method, since the histogram is constructed based on the minimization of the empirical error by the $L2$ and $L1$ distance scales, the proposed method and the conventional method are essentially different.

There are some works about evaluating environmental control technology by collecting various agricultural data and analyzing those data [1,7]. For example, there are works that draw environmental changes as a graph and give reasons to each change for a specific day. On the other hand, there is no discussion about the cause by automatically dividing and visualizing biased data.

## 3  Proposed Method

For a given set of samples described by both a numeric variable and a set of nominal variables, we propose a new method for producing a histogram with variable bin-width, some of whose bins are annotated by characteristic nominal variables. More specifically, we first construct a histogram with variable bin-width from the numeric values, and then provide annotation terms with some of the obtained bins by using the nominal variables. In what follows, we describe the details of our proposed algorithm.

### 3.1  Histogram Construction

For a given set of samples described by a numeric variable, $\mathcal{X} = \{x_t \mid t = 1, \cdots, T\}$, we first construct a histogram with variable bin-width. Similarly to the case of a standard histogram with a fixed bin-width, its horizontal and vertical axes correspond to the range of numeric values and the number of samples in the range, respectively. More specifically, we divide the entire range of numeric values into a series of adjacent intervals, and then compute the frequency of samples that fall into each interval. In this paper, we also express such a histogram as a step function where a variable $s$ and a function $h(s)$ are used for representing a numeric value and frequency, respectively. For instance, when the numbers of samples in intervals, $0 \le s \le 10$ and $11 \le s \le 20$, are 20 and 30, respectively, this step function returns the frequency $h(s) = 20$ if $0 \le s \le 10$, and $h(s) = 30$ if $11 \le s \le 20$.

For a predetermined number of bins denoted by $K$, we describe an algorithm for constructing a standard histogram with a fixed bin-width using the above notations. For a given set $\mathcal{X}$, we first arrange the numeric values in ascending

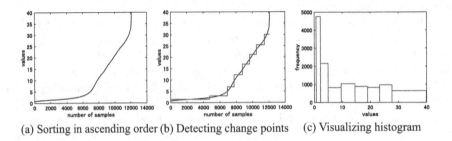

(a) Sorting in ascending order  (b) Detecting change points    (c) Visualizing histogram

**Fig. 1.** Procedure of our proposed method

---

**Algorithm 1.** Variable Bin-width Histogram

---

1: **Input:** A numeric data set $\mathcal{X} = \{x_1, \ldots, x_T\}$, the number of bins $K$
2: **Output:** Variable bin-width histogram $h(s)$
3: **Initialize:** Sort the elements of $\mathcal{X}$ in ascending order so as to satisfy $x_t \leq x_{t+1}$
4: Find change-points $\mathcal{G} = \{0 \leq G(k) \leq T; \ \ 0 \leq k \leq K\}$ by minimizing the objective function $\ell_K^2(\mathcal{G})$ or $\ell_K^1(\mathcal{G})$
5: Set end-points $\mathcal{F} = \{F(k) \leftarrow x_{G(k)}; \ \ 0 \leq k \leq K\}$
6: Enumerate the number of elements $h(s) \leftarrow |\mathcal{T}_k|, \ \ \mathcal{T}_k = \{s \in [x_0, x_T]; \ \ F(k-1) < s \leq F(k)\}$

---

order so as to satisfy $x_t \leq x_{t+1}$ for each $t(< T)$. Then, in order to assign each sample to one of $K$ adjacent intervals, after setting the bin-width $\delta$ to $\delta = (x_T - x_1)/K$, we produce the end-point for the $k$-th interval as $F(k) = x_1 + k\delta$ where $k \in \{1, \cdots, K-1\}$. Here, by using two additional values, $F(0) = x_0$ and $F(K) = x_T$, we consider a set of end-points defined by $\mathcal{F} = \{F(0), \cdots, F(K)\}$, where $x_0$ means some value smaller than $x_1$. Finally, since the samples belonging to the $k$-th bin are obtained as $\mathcal{T}_k = \{t \mid F(k-1) < x_t \leq F(k)\}$, we can construct a histogram as a step function defined by $h(s) = |\mathcal{T}_k|$, where $k = \lceil (s - x_1)/\delta \rceil$ and $s \in [x_0, x_T]$. Hereafter, this method is referred to as NM (Naive histograM). Evidently, the NM method might have a severe limitation when the distribution of values $\mathcal{X}$ contains both coarse and dense parts. Namely, we want to have relatively large numbers of narrow bins for some ranges when values distribute densely and change substantially, and to have small numbers of wide bins for the other ranges.

The idea of the proposed method is shown in Fig. 1. First, we sort the numeric data set $\mathcal{X}$ in ascending order like Fig. 1(a), where the horizontal and the vertical axes respectively show the order and the value. Next, we detect change points from the sorted values by minimizing the errors between the value of data points and approximated step functions like Fig. 1(b), where the vertical red line is the change point. When analyzing data, it is important how the number of elements changes before and after the change point. Now, data between change points can be interpreted as a gradual change in value. Therefore, by separating elements at change points, elements can be divided into major $K$ sets. Finally, we count the number of elements between the change points, and draw a histogram as shown in Fig. 1(c). The overview of our algorithm is as follows.

In Algorithm 1, $\mathcal{G}$ is a set of change points, $\ell_K^2(\mathcal{G})$ and $\ell_K^1(\mathcal{G})$ are objective functions used to find change-points, and $F(k) \leftarrow x_{G(k)}$ is a value of each change-point $G(k)$. The details of the algorithm are as follows.

As mentioned above, we produce a step function so as to minimize the sum of errors with respect to $\mathcal{X}$. For this purpose, by employing either L2-norm or L1-norm as a standard error criterion, we can derive two different change-point detection methods, which are simply referred to as the L2 and L1 methods, respectively. First, we consider the case that there is no change-point, which means that the sum of errors is minimized by only one value. Then, in case of the L2 method, the L2-norm error $\ell_0^2 = \sum_{t=1}^{T}(x_t - \mu(1, T))^2$ is minimized by using the mean value $\mu(1, T)$, where the mean function $\mu(a, b)$ is defined by $\mu(a, b) = (b - a + 1)^{-1} \sum_{t=a}^{b} x_t$. On the other hand, in case of the L1 method, the L1-norm error $\ell_0^1 = \sum_{t=1}^{T} |x_t - \nu(1, T)|$ is minimized by using the median value $\nu(1, T)$, where the median function $\nu(a, b)$ is defined by $\nu(a, b) = x_{(a+b)/2}$ if $a + b$ is even; otherwise $(x_{(a+b)/2} + x_{(a+b)/2+1})/2$, Here note that we can efficiently obtain $\nu(a, b)$ in case that the values in $\mathcal{X}$ are sorted in advance. Next, we consider the case that there exists only one change-point expressed by a sample index $\tau$. Then, the following formulae respectively minimize the L2-norm error $\ell_1^2(\tau)$ and L1-norm error $\ell_1^1(\tau)$:

$$\ell_1^2(\tau) = \sum_{t=1}^{\tau} (x_t - \mu(1, \tau))^2 + \sum_{t=\tau+1}^{T} (x_t - \mu(\tau + 1, T))^2$$

$$\ell_1^1(\tau) = \sum_{t=1}^{\tau} |x_t - \nu(1, \tau)| + \sum_{t=\tau+1}^{T} |x_t - \nu(\tau + 1, T)|.$$

Evidently, we also need to minimize $\ell_1^2(\tau)$ and $\ell_1^1(\tau)$ with respect to $\tau$.

Now, we generalize these error functions, $\ell_1^2(\tau)$ and $\ell_1^1(\tau)$. Namely, in case that the number of change-points is $K - 1$, let $G(k)$ be a sample index which corresponds to the $k$-th change point. Again, by using two additional indices, $G(0) = 0$ and $G(K) = T$, we can consider a set of sample indices defined by $\mathcal{G} = \{G(0), \cdots, G(K)\}$. Then, we can express the generalized error functions for $\ell_K^2(\mathcal{G})$ and $\ell_K^1(\mathcal{G})$ as follows:

$$\ell_{K-1}^2(\mathcal{G}) = \sum_{k=1}^{K} \sum_{t=G(k-1)+1}^{G(k)} (x_t - \mu(G(k-1) + 1, G(k)))^2$$

$$\ell_{K-1}^1(\mathcal{G}) = \sum_{k=1}^{K} \sum_{t=G(k-1)+1}^{G(k)} |x_t - \nu(G(k-1) + 1, G(k))|.$$

Therefore, we can formalize our change-point detection problem as the minimization problem of $\ell_K(\mathcal{G})$ with respect to $\mathcal{G}$. In order to obtain $\mathcal{G}$, we employ an efficient local improvement algorithm described in [5]. After obtaining the set of sample indices, $\mathcal{G}$, we can produce a set of end-points $\mathcal{F}$ by setting $F(k) = x_{G(k)}$ for $k \in \{0, \cdots, K\}$, where recall that $x_0$ means a value smaller than $x_1$. Thus, we can construct the following histogram with variable bin-widths:

$$h(s) = |\mathcal{T}_k|, \quad \text{where} \quad F(k-1) < s \leq F(k), \quad s \in [x_0, x_T].$$

Evidently, we can analyze rough structure of histogrmas by setting the number $K$ of bins to a small one, while some details of them by relatively large ones.

## 3.2   Annotation Generation

After obtaining the histogram with variable bin-width from a set of samples with a numeric variable, we generate annotation terms for these obtained bins from a set of the same samples with nominal variables. Let $\mathcal{Y}^{(i)} = \{y_t^{(i)} \mid t = 1, \cdots, T\}$ be a set of nominal values for the $i$-th variable, where $i \in \{1, \cdots, I\}$, $I$ denotes the number of variables, and we assume that each nominal variable has only one of $J^{(i)}$ categories identified by positive integers from 1 to $J^{(i)}$, i.e., $y_t^{(i)} \in \{1, \cdots, J^{(i)}\}$. Moreover, we assume that each sample with numeric value $x_t$ has the corresponding nominal variables $\{y_t^{(1)}, \cdots, y_t^{(I)}\}$.

For a pair of the $i$-th variable and its category $j \in \{1, \cdots, J^{(i)}\}$, we can define the set of samples such that $y_t^{(i)} = j$ by $\mathcal{T}^{(i,j)} = \{t \mid y_t^{(i)} = j\}$, and compute the empirical probability $p^{(i,j)}$ by $p^{(i,j)} = |\mathcal{T}^{(i,j)}|/T$. For the $k$-th bin of the obtained histogram, we can compute the expected number of samples such that $y_t^{(i)} = j$ and its standard deviation by $p^{(i,j)}|\mathcal{T}_k|$ and $\sqrt{p^{(i,j)}(1 - p^{(i,j)})|\mathcal{T}_k|}$. Thus, we can compute the following $z$-score $z_k^{(i,j)}$ of appearing the samples to be $y_t^{(i)} = j$ in the $k$-th bin.

$$z_k^{(i,j)} = \frac{|\mathcal{T}^{(i,j)} \cap \mathcal{T}_k| - p^{(i,j)}|\mathcal{T}_k|}{\sqrt{p^{(i,j)}(1 - p^{(i,j)})|\mathcal{T}_k|}}. \tag{1}$$

In case that the $z$-score $z_k^{(i,j)}$ is substantially large, we can consider that the $i$-th variable with the $j$-th category appears characteristically in the $k$-th bin. In our proposed method, for a predetermined number $H$, we output the top-$H$ pairs of the $i$-th variable and $j$-th category as an annotation term to the $k$-th bin, according to the $z$-score $z_k^{(i,j)}$.

## 4    Experimental Evaluations

### 4.1    Datasets and Settings

In this section, we confirm the validity of the proposed method by experimental evaluation using real datasets. Specifically, the effectiveness is evaluated by analyzing environmental data obtained from vinyl greenhouses of four rose farmers, which we call House A, B, C and D, in Shizuoka prefecture. In this paper, we employed the humidity deficit (HD), which is an indicator of how much water vapor can be contained at a particular temperature and humidity. Controlling HD within a specific range is considered important for the growth of agricultural products. The IoT device we used does not have a sensor that measures HD directly, so it was calculated using the following formula:

$$HD = (100 - Humi) * \frac{217 * \frac{6.1078 * 10^{\frac{7.5 * Temp}{Temp + 237.3}}}{Temp + 273.15}}{100}. \tag{2}$$

where $Humi$ and $Temp$ represents humidity and Celsius temperature, respectively, and HD values of 3 to $6\,\mathrm{g/m^3}$ are considered to be optimal. The data consists of HD observed from 00:00 on March 27, 2018, to 24:00 on May 7, 2018, i.e. observed for 42 days. 288 data points are contained per day so the data of each house is represented as 12,096 values.

## 4.2    Results of Fixed Bin-Width Histogram

First, we look over the fixed bin-width histograms shown in Fig. 2, where histograms of HD observed in each rose-farm-vinyl-house are depicted for each number of bins, $K = 16, 32, 128$. According to Sturges' formula and Square-root choice, for the data with the number of data points, $T = 12,096$, the appropriate number of bins is $\lceil \log_2 T + 1 \rceil \approx 15$ and $\sqrt{T} \approx 100$, in an existing fixed-width histogram. Figure 2(a) shows the histograms with the number of bins as $K = 16$, which is nearly corresponding to the results of Starges' formula, and Fig. 2(b) is those with the number of bins as $K = 32$. Figure 2(c) includes histograms with $K = 128$, which is roughly corresponding to the results of Square-root choice criteria. From these figures, we can observe that many data points are distributed relatively densely in the range HD < 10 than in the range 10 < HD. That is, at $K = 128$, in the range of 10 < HD where the data points are sparse, the division is too fine and the redundancy is high. On the other hand, at $K = 16$ and $K = 32$, the division is coarse and the resolution is low in the range of HD < 10 where the data points are dense. From these results, it is necessary to increase the resolution of densely distributed data and to reduce the resolution of sparsely distributed data.

## 4.3    Comparing Histograms by Changing the Number of Bins

Next, we compared histograms by changing the number of bins. Figures 3 and 4 show the variable bin-width histograms obtained by the L1 and L2 methods, respectively. When the number of bins $K$ is set to 8, as shown in Figs. 3(a) and 4(a), we can see that almost half of the bins are constructed in the rang of HD < 10, and these numbers of bins increase when $K$ becomes large as shown in Figs. 3 and 4. These characteristics are inherently dofferent those obtained in Fig. 2. In addition, as to House D, although the NM method cannot detect outlier points like HD $\simeq$ 0 at $K = 16$ as shown in Fig. 2, the $L2$ and the $L1$ method can detect some outlier points as separated bin, at all the number of bins.

To summarize the findings obtained from these figures: (1) histograms have similar tendency regardless of the number of bins, $K$, (2) data points densely distributed, like 2 < HD < 7, are finely divided in the variable bin-width histogram, and (3) in the proposed histogram, outlier data points can be detected as a separated bin.

Even when the data points of the data set $\mathcal{X}$ are distributed in a biased manner, the proposed method can automatically adjust the bin width appropriately so that the resolution is low in the sparse part and the resolution is high in the dense part. In order to quantitatively confirm that the data points are assigned to each bin without concentrating on some bins, we calculate the entropy defined by the following equation for the constructed histogram $h(\cdot)$:

$$E(h(\cdot)) = -\sum_{k=1}^{K} \frac{|\mathcal{X}_k|}{T} \log \frac{|\mathcal{X}_k|}{T} = -\sum_{k=1}^{K} \frac{h(F(k))}{T} \log \frac{h(F(k))}{T}. \tag{3}$$

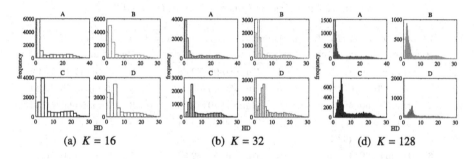

**Fig. 2.** Results of fixed bin-width histogram

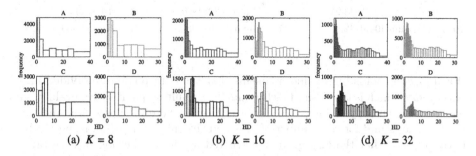

**Fig. 3.** Results of variable bin-width histogram based on $L1$ norm

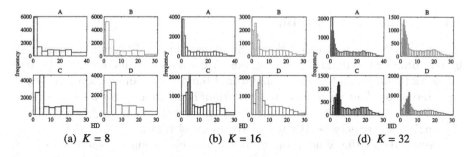

**Fig. 4.** Results of variable bin-width histogram based on $L2$ norm

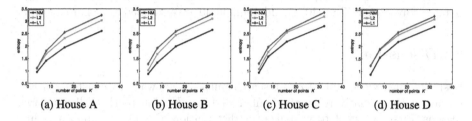

**Fig. 5.** Entropy: Concentration degree of data points

**Table 1.** Annotation terms of each bin (House C)

| Rank | Proposed method ($L1$, $K = 8$) | | | Proposed method ($L2$, $K = 8$) | | | NM method ($K = 8$) | | |
|------|-----|--------|---------|-----|--------|---------|-----|--------|---------|
|      | Bin | term | $z$-score | Bin | term | $z$-score | Bin | term | $z$-score |
| 1 | 1 | Humi80-90 | 81.56 | 1 | Humi80-90 | 72.68 | 8 | Temp35-40 | 70.99 |
| 2 | 2 | Humi70-80 | 67.06 | 5 | Humi50-60 | 66.80 | 6 | Temp30-35 | 53.44 |
| 3 | 4 | Humi50-60 | 58.22 | 8 | Humi20-30 | 59.68 | 3 | Humi50-60 | 50.34 |
| 4 | 7 | Temp30-35 | 53.66 | 8 | Temp30-35 | 59.14 | 5 | Humi40-50 | 49.65 |
| 5 | 3 | Humi60-70 | 49.61 | 4 | Humi60-70 | 52.80 | 1 | Temp20-25 | 47.53 |
| 6 | 6 | Humi40-50 | 48.14 | 6 | Humi40-50 | 52.23 | 1 | Humi80-90 | 46.74 |
| 7 | 8 | Humi20-30 | 46.71 | 3 | Humi70-80 | 50.33 | 6 | Humi20-30 | 46.08 |
| 8 | 7 | Humi30-40 | 46.55 | 8 | Humi30-40 | 45.95 | 6 | Humi30-40 | 46.00 |
| 9 | 2 | Temp20-25 | 43.21 | 5 | Temp25-30 | 42.78 | 3 | Temp25-30 | 43.90 |
| 10 | 5 | Humi40-50 | 42.46 | 7 | Humi40-50 | 40.68 | 4 | Humi40-50 | 42.02 |

When $E(h(\cdot))$ holds high value, it means that the data points are divided into bins evenly, which is the desired histogram. The result of quantitative evaluation based on the entropy of Eq. (3) is shown in Fig. 5, where the horizontal and the vertical axes stand for the number of bins and the entropy $E(h(\cdot))$, respectively. It can be seen that the histogram constructed by the $L1$ method has desirable properties because the value of $E(h(\cdot))$ is larger than the $L2$ method, also larger than the NM method.

### 4.4   Annotating to Each Bin

Finally, we generated annotation terms for each bin of histograms constructed by our proposed and the naive methods. We employed time-window (Hour), humidity (Humi), and Celsius temperature (Temp) as an annotation term, e.g., eight terms for Hour form "Hour0-3" to "Hour21-24", ten terms for Humi from "Humi0-10" to "Humi90-100", and ten terms for Temp from "Temp0-5" to "Temp45-". where eacg postfix means the range of values.

Table 1 shows the annotation result for the histogram of House C, where 10 terms with the highest $z$-score values for each bin are shown. The value of the bin column in these tables means what number bin from the left (lowest). For example, when the value of HD is large as in the 7th and 8th bins, higher temperature such as Temp30-35 and Temp35-40 are extracted as a characteristic.

## 5   Discussion

Basically, in a viewpoint of our research purpose, we want to construct a histogram with a relatively small number of bins, $K$, due to the following two reasons. First, we want to visualize the distribution of numeric values as a histogram consisting of high and coarse resolution parts by virtue of individual

variable bin-widths. Second, we want to generate statistically significant annotation terms for the obtained bins according to the $z$-score. Below we discuss that
our constructed histograms with annotation terms meet our research purpose.

From the results shown in Fig. 2 by the NM method, we can observe that
the numeric values are concentrated in a narrow range between 0 and 7. More
specifically, in the case of $K = 16$, we can observe that the concentrated peaks
locate around left parts in this range for House A and B, while around right
parts for House C and D. On the other hand, in the case of $K = 128$, we can
clearly observe that the concentrated peaks for House A, B, C, and D have
uniquely different characteristics. This means that the histograms by the NM
method with $K = 16$ are not enough for analyzing these situations. Besides, it
is not easy to visually identify each bin in the case of $K = 128$, together with
bringing about the difficulty of identifying annotation terms for many bins.

From the results shown in Figs. 3 and 4 by the proposed method, we can also
observe that the numeric values are concentrated in a narrow range between
0 and 7. More speciffically, even in the case of $K = 8$, from the results by
our proposed methods, we can roughly observe that the concentrated peaks for
House A, B, C, and D have uniquely different characteristics. Moreover, in the
case of $K = 16$, we can observe quite similar characteristics for this range, which
are observed from the results by the NM method with $K = 128$ shown in Fig. 2.
These experimental results indicate that in a viewpoint of our research purpose,
we can successfully visualize the distribution of numeric values as a histogram
consisting of high and coarse resolution parts by virtue of individual variable
bin-widths.

Finally, we discuss the validity of generated annotation terms by using the
results from House C shown in Table 1. From this table, for the constructed
bins by each method, we can consistently observe that the annotation terms
indicating higher $Humi$ values are generated for the bins with lower numbers
indicating lower HD values, and the annotation terms using the $Hour$ variable
are not generated. Here, we can easily confirm that the validity of these observations from the definition of HD shown in the Eq. (2), i.e., the HD values have
a negative correlation to the $Humi$ values, and are independent of the $Hour$
values. On the other hand, from Figs. 3(a) and 4(a), we can see that the second
bin from the left constructed by the $L1$ method can be regarded as unique one
not constructed by neither the NM method nor the $L2$ method. It might be
notable that the annotation term "Humi70-80" is generated for this unique bin,
which naturally interpolates the degradation of the $Humi$ values. We believe
that these experimental results support the vitality of the $L1$ method equipped
with annotation generation.

## 6   Conclusion

In this study, for the purpose of constructing a histogram drawn by $K$ bins with
variable widths and extracting the characteristic nominal values for describing
these bins as annotation terms, we proposed a new method, which applies a

change points detection method based on an $L1$ or $L2$ error criterion, and identifies some annotation terms for the bins in our constructed histogram based on the $z$-score with respect to the distribution of nominal values. From experimental evaluations using the dataset of humidity deficit (HD) collected from vinyl greenhouses, we confirmed the following results: (1) our proposed method can construct more natural histograms with appropriate variable bin widths compared to histograms with an equal bin width; (2) the histograms constructed with the $L1$ error criterion has more desirable property than those with the $L2$ error criterion; and (3) our method can produce a series of naturally interpretable annotation terms for the constructed bins. As a future task, further experiments are needed where we utilize various types of datasets obtained from other domain such as educational field, also including multivariate data, and compare to existing methods proposed in previous studies. In addition, assuming application to agricultural scenarios, we plan to extend our method so that it can be applied to continuously obtained streaming data and add discussions with practical utility.

**Acknowledgments.** This material is based upon work supported by JSPS Grant-in-Aid for Scientific Research (C) (No. 18K11441), (B) (No. 17H01826) and Early-Career Scientists (No. 19K20417).

# References

1. Saishin nogyo gijutsu yasai. 8. Rural Culture Association Japan (2015). http://amazon.co.jp/o/ASIN/454015057X/
2. Chandola, V., Banerjee, A., Kumar, V.: Anomaly detection: a survey. ACM Comput. Surv. **41**(3), 15:1–15:58 (2009). https://doi.org/10.1145/1541880.1541882
3. Denby, L., Mallows, C.: Variations on the histogram. J. Comput. Graph. Stat. **18**, 21–31 (2009)
4. Kim, C.J., Piger, J., Startz, R.: Estimation of Markov regime-switching regression models with endogenous switching. J. Econom. **143**(2), 263–273 (2008)
5. Saito, K., Ohara, K., Kimura, M., Motoda, H.: Change point detection for burst analysis from an observed information diffusion sequence of tweets. J. Intell. Inf. Syst. **44**(2), 243–269 (2015). https://doi.org/10.1007/s10844-013-0283-2
6. Scott, D.W.: Multivariate Density Estimation: Theory, Practice, and Visualization, 2nd edn. Wiley, New York (1992)
7. Yamada, H., Watanabe, C.: Approach feature extraction of nature image with observation report and transition of histogram. Technical Report 16 (2010)

# A Unified Approach to Biclustering Based on Formal Concept Analysis and Interval Pattern Structure

Nyoman Juniarta$^{(\boxtimes)}$ ⓘ, Miguel Couceiro, and Amedeo Napoli

Université de Lorraine, CNRS, Inria, LORIA, 54000 Nancy, France
{nyoman.juniarta,miguel.couceiro,amedeo.napoli}@loria.fr

**Abstract.** In a matrix representing a numerical dataset, a bicluster is a submatrix whose cells exhibit similar behavior. Biclustering is naturally related to Formal Concept Analysis (FCA) where concepts correspond to maximal and closed biclusters in a binary dataset. In this paper, a unified characterization of biclustering algorithms is proposed using FCA and pattern structures, an extension of FCA for dealing with numbers and other complex data. Several types of biclusters – constant-column, constant-row, additive, and multiplicative – and their relation to interval pattern structures is presented.

**Keywords:** Biclustering · FCA · Gene expression · Pattern structures

## 1 Introduction

Given a numerical dataset represented as a table or a matrix with objects in rows and attributes in columns, the objective of clustering is to group a set of objects according to all attributes using a similarity or distance measure. By contrast, biclustering simultaneously operates on the set of objects and the set of attributes, where a subset of objects can be grouped w.r.t. a subset of attributes, based on user-defined constraints such as having constant values, constant values within columns or rows. Then, if a cluster represents object relations at a global scale, a bicluster represents it at a local scale w.r.t. the set of attributes. More generally, biclustering searches in a data matrix for sub-matrices or biclusters composed of a subset of objects (rows) and a subset of attributes (columns) which exhibit a specific behavior w.r.t. some criteria.

Biclustering is an important tool in many domains, e.g. bioinformatics and gene expression data, recommendation and collaborative filtering, text mining, social networks, dimensionality reduction, etc. As surveyed in [17], biclustering received a lot of attention in biology, and especially, for analyzing gene expression data, where biologists are searching for a set of genes whose behavior is consistent across certain experiments/conditions [3,4,20]. Biclustering is still actively studied in biology [9,18,19]. Biclustering is also actively studied in recommendation systems [12,13], where the objective is to retrieve a set of users sharing similar interest across a subset of items instead of the set of all possible items.

© Springer Nature Switzerland AG 2019
P. Kralj Novak et al. (Eds.): DS 2019, LNAI 11828, pp. 51–60, 2019.
https://doi.org/10.1007/978-3-030-33778-0_5

**Table 1.** Examples of some bicluster types.

| 4 | 2 | 5 | 3 | 4.0 | 2.5 | 5.7 | 3.1 | 2 | 3 | 5 | 1 | 2 | 6 | 12 | 4 |
|---|---|---|---|-----|-----|-----|-----|---|---|---|---|---|---|----|---|
| 4 | 2 | 5 | 3 | 3.9 | 2.4 | 5.6 | 3.0 | 5 | 6 | 8 | 4 | 1 | 3 | 6 | 2 |
| 4 | 2 | 5 | 3 | 4.1 | 2.4 | 5.9 | 2.9 | 1 | 2 | 4 | 0 | 4 | 12 | 24 | 8 |
| 4 | 2 | 5 | 3 | 4.1 | 2.6 | 5.8 | 2.9 | 4 | 5 | 7 | 3 | 3 | 9 | 18 | 6 |

Following the lines of [8–10], in this paper we are interested in biclustering algorithms based on "pattern-mining" techniques [1]. These techniques allow an exhaustive and flexible search with efficient algorithms. Moreover, authors in [9] discuss the benefits of using pattern-based biclustering w.r.t. scalability requirements, and mostly w.r.t. generality and diversity of the types of biclusters which are mined. In addition, they point out the fact that pattern-based biclustering algorithms can naturally take into account overlapping biclusters, and as well, additive, multiplicative and symmetric assumptions concerning biclusters.

In this paper, we revisit all these aspects and propose an alternative framework for pattern-based biclustering based on Formal Concept Analysis (FCA [7]). In [21], authors directly reuse the FCA framework and adapt the algorithms for biclustering. By contrast, in this paper, we go further and we consider the so-called "pattern-structures", an extension of FCA for dealing with complex values such as numbers, sequences, or graphs [6]. We especially reuse "interval pattern structures" – which are detailed in the following – for defining a unique framework for pattern-based biclustering. In this way, we introduce an alternative approach than [9], as we do not need to apply any scaling, discretization, or transformation procedures over the data to discover biclusters.

This paper is organized as follows. First we describe some types of biclustering in Sect. 2 and basic definitions about FCA in Sect. 3. We then propose our approach of biclustering based on interval pattern structures in Sect. 4 and present the empirical experiments in Sect. 5. Finally, we conclude our work and give some future works in Sect. 6.

## 2    Biclustering

In this section, we recall the basic background and discuss illustrative examples of the different types of biclusters [17]. We consider that a dataset is a matrix $(G, M)$ where $G$ is a set of objects and $M$ is a set of attributes. The value of $m \in M$ for object $g \in G$ is written as $m(g)$. In this paper, we work with numerical datasets. In such a dataset, it may be interesting to find which subset of objects have the same values w.r.t. a subset of attributes. Regarding the matrix representation, this is equivalent to the problem of finding a submatrix where all elements have the same value. This task is called *biclustering with constant values*, which is a simultaneous clustering of the rows and columns of a matrix.

**Table 2.** A numerical context and an SC bicluster in gray.

| $G$ | $m_1$ | $m_2$ | $m_3$ | $m_4$ | $m_5$ |
|-----|-------|-------|-------|-------|-------|
| $g_1$ | 1 | 2 | 2 | 1 | 6 |
| $g_2$ | 2 | 1 | 1 | 0 | 6 |
| $g_3$ | 2 | 2 | 1 | 7 | 6 |
| $g_4$ | 8 | 9 | 2 | 6 | 7 |

Moreover, given a dataset $(G, M)$, a pair $(A, B)$ (where $A \subseteq G$, $B \subseteq M$) is a *constant-column* (CC) bicluster iff $\forall m \in B, \forall g, h \in A, m(g) = m(h)$. An example of CC bicluster is illustrated in Table 1a. CC biclustering have more relaxed variations, namely similar-column (SC) biclustering. With these relaxations, instead of finding biclusters with exactly constant columns, we can obtain biclusters whose columns have similar values as shown in Table 1b. These types of biclusters are widely used in recommendation systems to detect a set of users sharing similar preference over a set of items.

An additive bicluster is illustrated in Table 1c. Here we see that there is a constant difference between any two columns. For example, each value in the second column is two more than the corresponding value in the fourth row. Therefore, given a dataset $(G, M)$, a pair $(A, B)$ (where $A \subseteq G$, $B \subseteq M$) is an *additive* bicluster iff $\forall g, h \in A, \forall m, n \in B, m(g) - n(g) = m(h) - n(h)$; or a *multiplicative* bicluster iff $\forall g, h \in A, \forall m, n \in B, m(g)/n(g) = m(h)/n(h)$. Both additive and multiplicative biclusters were studied in the domain of gene expression dataset [4,5,16]. They represent a set of genes having similar expression patterns across a set of experiments.

Bicluster discovery is naturally related to FCA. In this paper, we show that an extension of FCA called partition pattern structures can be used for discovering biclusters. In the following section, we explain some basic theories about FCA and pattern structures.

## 3    FCA and Pattern Structure

In a binary matrix, FCA tries to find maximal submatrices with a constant value across all of its cells. Therefore, a formal concept is a bicluster with constant value. More precisely, FCA is a mathematical framework based on lattice theory and used for classification, data analysis, and knowledge discovery [7]. From a formal context, FCA detects all formal concepts, and arranges them in a concept lattice. FCA is restricted to specific datasets where each attribute is binary (e.g. has only yes/no value). This limitation prohibits FCA to work in more complex datasets, e.g. a user-rating matrix or a gene expression dataset, which are not binary. Therefore, FCA is then generalized into pattern structures [6].

A *pattern structure* is a triple $(G, (D, \sqcap), \delta)$, where $G$ is a set of objects, $(D, \sqcap)$ is a complete meet-semilattice (of descriptions), and $\delta : G \rightarrow D$ maps an object to a description. The operator $\sqcap$ is a similarity operation that returns the common

**Table 3.** Example of additive column alignments. (a) Original table and the additive bicluster in gray, (b) alignment on $m_1$, (c) alignment on $m_2$.

| | $m_1$ | $m_2$ | $m_3$ | $m_4$ |
|---|---|---|---|---|
| $g_1$ | 4 | 1 | 3 | 0 |
| $g_2$ | 6 | 4 | 6 | 3 |
| $g_3$ | 2 | 3 | 5 | 2 |
| $g_4$ | 1 | 6 | 1 | 7 |

(a)

| | $m_1$ | $m_2$ | $m_3$ | $m_4$ |
|---|---|---|---|---|
| $g_1$ | 4 | 1 | 3 | 0 |
| $g_2$ | 4 | 2 | 4 | 1 |
| $g_3$ | 4 | 5 | 7 | 4 |
| $g_4$ | 4 | 9 | 4 | 10 |

(b)

| | $m_1$ | $m_2$ | $m_3$ | $m_4$ |
|---|---|---|---|---|
| $g_1$ | 4 | 1 | 3 | 0 |
| $g_2$ | 3 | 1 | 3 | 0 |
| $g_3$ | 0 | 1 | 3 | 0 |
| $g_4$ | −4 | 1 | −4 | 2 |

(c)

elements between any two descriptions. It is verified that $c \sqcap d = c \Leftrightarrow c \sqsubseteq d$. A description can be a number, a set, a sequence, a tree, a graph, or another complex structure. The Galois connection for a pattern structure $(G, (D, \sqcap), \delta)$ is defined as:

$$A^\diamond = \bigsqcap_{g \in A} \delta(g), \qquad\qquad A \subseteq G, \qquad (1)$$

$$d^\diamond = \{g \in G | d \sqsubseteq \delta(g)\}, \qquad\qquad d \in D. \qquad (2)$$

A pattern concept is a pair $(A, d)$, $A \subseteq G$ and $d \in D$, where $A^\diamond = d$ and $d^\diamond = A$.

FCA can be understood as a particular pattern structure. The description of an object is a set of attributes, and the $\sqcap$ operator between two description is the intersection of two sets of attributes.

## 4  Biclustering Using Interval Pattern Structure

In gene expression data, we often have a numerical matrix. Biclustering in such matrix should find submatrices whose cells present regularities, e.g. each column has similar value in the case of similar-column (SC) biclustering. SC biclustering task is similar to FCA in the sense that FCA also searches consistent submatrix. But since SC biclustering works on a numerical matrix, we need to generalize FCA to a pattern structure. One such generalization is where the description of each object is a set of numerical values and the similarity between any two descriptions is the intervals that encompass those values. This kind of pattern structure is called an *interval pattern structure*.

Interval pattern structures (IPS) was introduced by Kaytoue et al. [14] to analyze gene expression data (GED). A GED is typically represented as a 2-D numerical matrix with genes as rows and conditions as columns, as shown in Table 2. In this matrix, the submatrix $(\{g_1, g_2, g_3\}, \{m_1, m_2, m_3, m_5\})$ is an SC bicluster, defined by the parameter $\theta = 1$. It means that the range of values of each column in the submatrix has the length of at most 1.

### 4.1  Interval Pattern Structure

In IPS, a description is several intervals describing the values of every column. For example, the description of $g_1$ – denoted by $\delta(g_1)$ – in Table 2 is

**Table 4.** Some interval pattern concepts with $\theta = 1$ from Table 2.

| Extent | Intent |
|--------|--------|
| $\{g_1\}$ | $\langle[1,1][2,2][2,2][1,1][6,6]\rangle$ |
| $\{g_1, g_3\}$ | $\langle[1,2][2,2][1,2] * [6,6]\rangle$ |
| $\{g_1, g_4\}$ | $\langle * * [2,2] * [6,7]\rangle$ |
| $\{g_1, g_2, g_3\}$ | $\langle[1,2][1,2][1,2] * [6,6]\rangle$ |
| $\{g_1, g_2, g_3, g_4\}$ | $\langle * * [1,2] * [6,7]\rangle$ |

$\langle[1,1][2,2][2,2][1,1][6,6]\rangle$. The similarity operator ($\sqcap$) for IPS is defined as the convex hull of two intervals. Therefore, the similarity of $\delta(g_1)$ and $\delta(g_4)$ – denoted by $\delta(g_1) \sqcap \delta(g_4)$ – is $\langle[1,8][2,9][2,2][1,6][6,7]\rangle$.

Given a subset of objects $A \subseteq G$, Eq. 1 says that $A^\diamond$ is the similarity of the description of all objects in $A$. Therefore, in IPS the corresponding $A^\diamond$ is the convex hull of the descriptions of all objects in $A$. For example, with $A = \{g_1, g_2, g_4\}$, $A^\diamond = \langle[1,8][1,9][1,2][0,6][6,7]\rangle$.

Furthermore, given a description $d \in D$, Eq. 2 indicates that $d^\diamond$ is the set of all objects whose description subsumes $d$. In IPS, a description $d_1$ is subsumed by another description $d_2$ – denoted by $d_1 \sqsubseteq d_2$ – if every interval in $d_2$ is a sub interval in the corresponding interval in $d_1$. Notice that in IPS, a sub interval subsumes a larger interval. Therefore, if $d_x = \langle[1,8][1,9][1,2][0,6][6,7]\rangle$, then $d_x^\diamond = \{g_1, g_2, g_4\}$. Since $\delta(g_3) = \langle[2,2][2,2][1,1][7,7][6,6]\rangle$, $g_3$ is not included in $d_x^\diamond$ because the fourth interval ($[7,7]$) is not sub interval of the fourth interval of $d_x$ ($[0,6]$).

Following the definition of a concept of any pattern structure (in Sect. 3), an interval pattern concept is a pair $(A, d)$, for $A \subseteq G$ and $d \in D$, where $A^\diamond = d$ and $d^\diamond = A$. Furthermore, the set of interval pattern concepts are partially ordered, and can be depicted as a lattice. An interval pattern concept $(A_1, d_1)$ is a subconcept of $(A_2, d_2)$ if $A_1 \subseteq A_2$ (equivalently $d_2 \sqsubseteq d_1$).

## 4.2 Similar-Column Biclustering

A similar-column (SC) bicluster can be found in an interval pattern concept by introducing a parameter $\theta$. This parameter acts as the maximum difference between any two values to be considered as similar. For example, with $\theta = 1$, the value 1 is similar to 2, but not similar to 3.

In calculating the similarity between any two descriptions, if the length of an interval is larger than $\theta$, then the star sign ($*$) is put as the interval. From Table 2, $\delta(g_2) \sqcap \delta(g_4)$ without $\theta$ is $\langle[2,8][1,9][1,2][0,6][6,7]\rangle$, and with $\theta = 1$ is $\langle * * [1,2] * [6,7]\rangle$.

The similarity $\sqcap$ between $*$ and any other interval is $*$. For example, suppose that we have two descriptions $d_x = \langle[1,1][2,3]\rangle$ and $d_y = \langle[2,2]*\rangle$. Then, $d_x \sqcap d_y = \langle[1,2]*\rangle$. This also means that $*$ is subsumed by any other interval. Therefore, the description of each object in Table 2 subsumes $\langle * * [1,2] * [6,7]\rangle$. With $\theta = 1$, $(\{g_1, g_2, g_3, g_4\}, \langle * * [1,2] * [6,7]\rangle)$ is an interval pattern concept. Some interval pattern concepts from Table 2 are listed in Table 4.

**Table 5.** Example of multiplicative column alignments. (a) Original table and the multiplicative bicluster in gray, (b) alignment on $m_2$.

|       | $m_1$ | $m_2$ | $m_3$ | $m_4$ |       | $m_1$ | $m_2$ | $m_3$ | $m_4$ |
|-------|-------|-------|-------|-------|-------|-------|-------|-------|-------|
| $g_1$ | 3     | 1     | 2     | 3     | $g_1$ | 3     | 1     | 2     | 3     |
| $g_2$ | 1     | 3     | 6     | 9     | $g_2$ | 0.3   | 1     | 2     | 3     |
| $g_3$ | 2     | 2     | 4     | 6     | $g_3$ | 1     | 1     | 2     | 3     |
| $g_4$ | 1     | 2     | 6     | 8     | $g_4$ | 0.5   | 1     | 3     | 4     |
|       |       | (a)   |       |       |       |       | (b)   |       |       |

From an interval pattern concept, an SC bicluster can be formed by the concept's extent and the set of columns where the interval is not $*$ in the concept's intent. For example, from the concept $(\{g_1, g_2, g_3\}, \langle [1,2][1,2][1,2]*[6,6]\rangle)$, $(\{g_1, g_2, g_3\}, \{m_2, m_2, m_3, m_5\})$ is an SC bicluster with $\theta = 1$.

By using IPS with parameter $\theta$, constant-column biclustering is a specific case of SC biclustering. It can be noticed that with $\theta = 0$, we obtain intervals with length 0, and that corresponds to constant-column biclusters.

### 4.3 Additive and Multiplicative Biclustering

An additive bicluster is a submatrix where there is a constant (or similar) difference between any two columns across all of its rows (see Sect. 2). Constant (or similar) column biclustering is a specific case of additive biclustering. Using this fact, we can obtain additive biclusters by aligning (similar to [9]) each column, and then find interval pattern concepts on the alignments.

Table 3 provides an example of column alignment for additive biclustering. The original matrix is shown in Table 3a, having 4 rows and 4 columns. The submatrix $(\{g_1, g_2, g_3\}, \{m_2, m_3, m_4\})$ is an additive bicluster in the original matrix. This bicluster can be found by applying constant-column or similar-column biclustering to the column alignments. Table 3b shows the first column alignment, can be seen by the consistency of the first column $(m_1)$. In this example, each object value is converted such that its $m_1$ value is equal to the value of $m_1$ in $g_1$. This means that the values 0, $-2$, 2, and 3 are added to $g_1$, $g_2$, $g_3$, and $g_4$ respectively. This alignment is repeated for every column. Table 3c is the alignment of $m_2$, by adding 0, $-3$, $-2$, and $-5$ to $g_1$, $g_2$, $g_3$, and $g_4$ respectively.

Constant-column (or similar-column) biclustering is applied to every column alignment to find additive biclusters. In the second column alignment (Table 3c), we obtain $(\{g_1, g_2, g_3\}, \{m_2, m_3, m_4\})$ as a constant-column bicluster. This corresponds to the additive bicluster $(\{g_1, g_2, g_3\}, \{m_2, m_3, m_4\})$ in the original matrix (Table 3a).

Multiplicative biclusters can also be obtained using similar column alignment. In multiplicative column alignment, instead of adding values to each row, we multiply each row such that a column has a constant value. Table 5b shows the second column alignment of the original matrix in Table 5a. Here, a constant value is achieved for $m_2$ by multiplying $g_1$, $g_2$, $g_3$, and $g_4$ by 1, $\frac{1}{3}$, $\frac{1}{2}$, and $\frac{1}{2}$ respectively. Then, by applying IPS to each alignment, we can obtain the multiplicative biclusters. For example, constant-column biclustering using

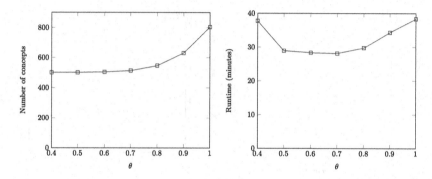

**Fig. 1.** Effect of $\theta$ on a $500 \times 60$ dataset with $min\_col = 20$ and $min\_row = 1$.

IPS in Table 5b returns $(\{g_1, g_2, g_3\}, \{m_2, m_3, m_4\})$, which is the corresponding multiplicative bicluster in Table 5a.

### 4.4 Concept Mining

Being a generalization of FCA, the mining of interval pattern concepts can be performed using some existing algorithms that generate a complete list of formal concepts. In this paper, we use CloseByOne (CbO) [15] since it requires us to only define the similarity ($\sqcap$) and subsumption relation ($\sqsubseteq$) of any two descriptions.

In a given numerical matrix, we may obtain an exponential number of interval pattern concepts. To reduce the number of concepts, we should introduce some parameters that can filter out some uninteresting concepts.

The first parameter, $\theta$, is previously mentioned in Sect. 4.2. It limits the length of intervals, and later in Sect. 5 we demonstrate the effect of $\theta$ on the runtime and number of concepts.

The second parameter $min\_col$ is the minimum number of columns in the retrieved biclusters. The number of columns in a bicluster corresponds to the number of non-star intervals in the concept's intent. For example, the concept with intent $\langle * * [2, 2] * [6, 7] \rangle$ gives us a bicluster with two columns (the third and the fifth). To take into account the $min\_col$ parameter, it is necessary to modify the definition of similarity between any two descriptions. In addition to the definition of $\sqcap$ in Sect. 4.1, we verify if the number of non-star intervals in the description. The number of non-star intervals should be more than $min_{col}$. If not, we "skip" the concept, by converting each interval to $*$. In Table 2 with $\theta = 1$, $g_1 \sqcap g_4$ is $\langle * * [2, 2] * [6, 7] \rangle$. Using $min\_col = 3$ for example, $g_1 \sqcap g_4$ becomes $\langle * * * * * \rangle$.

Related to $min\_col$ is $min\_row$, a parameter that put a constraint on the number of rows in a bicluster. It corresponds to the number of objects in a concept's extent. With the inclusion of $min\_row$, the calculation of $Y^\diamond$ (all objects whose description subsumes $Y$) is performed only if the number of objects in $Z$ (extent of the candidate concept) is at least $min\_row$.

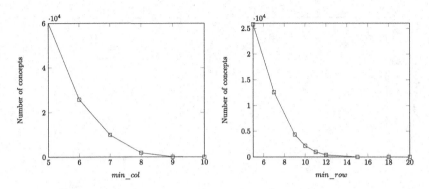

**Fig. 2.** Effect of $min\_col$ (with $\theta = 1$ and $min\_row = 5$) and $min\_row$ (with $\theta = 1$ and $min\_col = 6$) on a $500 \times 60$ dataset.

## 5   Experiments

In this section, we report some experimental results to show the scalability of IPS in the task of biclustering. By using CbO as concept miner, the space/time complexity of IPS follows CbO (see [15]). We use the synthetic datasets provided by Henriques and Madeira [9]: $500 \times 60$ and $1000 \times 100$, with hidden SC biclusters.

First, we investigate the effect of $\theta$ on the runtime and the number of concepts. The results are illustrated in Fig. 1. The left figure confirms that the larger $\theta$ generates more interval pattern concepts, and generally longer runtime as it can be seen in the right figure. The $\theta = 0.4$ requires longer runtime than $\theta = 0.5$ to 0.9. This is normal since for similar number of concepts, the probability of smaller $\theta$ obtaining a concept is smaller than the larger $\theta$. Using CbO with smaller $\theta$, a candidate concept will have shorter intervals in its intent, hence smaller number of objects whose description subsumes this interval.

The effect of $min\_col$ is shown in Fig. 2 left. Lesser $min\_col$ produces more concepts, and therefore longer runtime. Similarly, Fig. 2 right shows that larger $min\_row$ generates more concepts.

In the previous experiments, the CbO was terminated until all interval pattern concepts were retrieved. In the following experiment, CbO is terminated until 500 concepts are found. We compare them to BicPAM [9] that uses a discretization parameter (as a number of alphabet/items), while IPS uses the length of intervals as $\theta$. After the mapping step (normalization, discretization, and missing values and noise handling), BicPAM applies a pattern mining method (F2G [11] as default), and the closing step (extension, merging, and filtering) is performed. Results in Table 6 show a similar performance of both methods. It should be noted that the number of biclusters from BicPAM is lower due to the merging and/or filtering.

Furthermore, still from Table 6, the runtime of IPS is not exactly correlated with $\theta$ (especially with $\theta = 2$), similar to our previous experiment shown in Fig. 1. Overall, with similar runtime, biclustering with IPS can return similar number of biclusters without discretization.

**Table 6.** Comparison with BicPAM on $1000 \times 100$ dataset. For the IPS, the parameters $min\_row = 10$ and $min\_col = 5$ are used, with varying $\theta$.

| Method | Parameter | Runtime (s) | Number of biclusters |
|--------|-----------|-------------|----------------------|
| BicPAM | alphabet = 20 | <15 | ~100 |
|        | alphabet = 10 | <15 | <200 |
|        | alphabet = 7 | <15 | <200 |
|        | alphabet = 5 | <30 | ~200 |
| IPS | $\theta = 1$ | 37 | 500 |
|     | $\theta = 2$ | >500 | 500 |
|     | $\theta = 4$ | 47 | 500 |
|     | $\theta = 8$ | 39 | 500 |

# 6    Conclusion

In this paper, we propose an alternative method of biclustering in numerical datasets. Discretization is a general preprocessing step while working with numerical values. Here we explore the possibility of working directly on numerical datasets without discretization. This can be achieved using interval pattern structures, where a bicluster can be found from any interval pattern concept. To filter the number of concepts (which can be very large) it is necessary to provide some parameters, like the length of intervals, minimum number of rows and columns, or even minimum number of biclusters. Our experiments show that these parameters can reduce the computation to a reasonable runtime. Another way to reduce the number of biclusters is to develop post-processing techniques similar to BicPAM, which include merging, filtering, and extension.

We use the CbO algorithm, a formal concept generator that can be generalized to interval pattern structures. In-Close 2 [2] in particular is faster than CbO in formal concept mining, but its efficiency in interval pattern concept mining should be studied. Another future research is to extend our FCA-based approach to other types of biclusters, e.g. coherent-evolution, coherent-sign-changes, etc. Furthermore, the existence of missing values and/or outliers should be considered in improving the proposed biclustering method.

# References

1. Aggarwal, C.C., Han, J. (eds.): Frequent Pattern Mining. Springer, Cham (2014). https://doi.org/10.1007/978-3-319-07821-2
2. Andrews, S.: In-Close2, a high performance formal concept miner. In: Andrews, S., Polovina, S., Hill, R., Akhgar, B. (eds.) ICCS 2011. LNCS (LNAI), vol. 6828, pp. 50–62. Springer, Heidelberg (2011). https://doi.org/10.1007/978-3-642-22688-5_4
3. Ben-Dor, A., Chor, B., Karp, R., Yakhini, Z.: Discovering local structure in gene expression data: the order-preserving submatrix problem. J. Comput. Biol. **10**(3–4), 373–384 (2003)

4. Cheng, Y., Church, G.M.: Biclustering of expression data. In: ISMB, vol. 8, pp. 93–103 (2000)
5. Duarte, R.P., Simões, Á., Henriques, R., Neto, H.C.: FPGA-based OpenCL accelerator for discovering temporal patterns in gene expression data using biclustering. In: Proceedings of the 6th International Workshop on Parallelism in Bioinformatics, pp. 53–62. ACM (2018)
6. Ganter, B., Kuznetsov, S.O.: Pattern structures and their projections. In: Delugach, H.S., Stumme, G. (eds.) ICCS-ConceptStruct 2001. LNCS (LNAI), vol. 2120, pp. 129–142. Springer, Heidelberg (2001). https://doi.org/10.1007/3-540-44583-8_10
7. Ganter, B., Wille, R.: Formal Concept Analysis: Mathematical Foundations, 2nd edn. Springer, Heidelberg (1999)
8. Henriques, R., Ferreira, F.L., Madeira, S.C.: BicPAMS: software for biological data analysis with pattern-based biclustering. BMC Bioinform. **18**(1), 82 (2017)
9. Henriques, R., Madeira, S.C.: BicPAM: pattern-based biclustering for biomedical data analysis. Algorithms Mol. Biol. **9**(1), 27 (2014)
10. Henriques, R., Madeira, S.C.: BicSPAM: flexible biclustering using sequential patterns. BMC Bioinform. **15**(1), 130 (2014)
11. Henriques, R., Madeira, S.C., Antunes, C.: F2G: efficient discovery of full-patterns. In: ECML/PKDD nfMCP, pp. 1–9 (2013)
12. Ignatov, D.I., Kuznetsov, S.O., Poelmans, J.: Concept-based biclustering for internet advertisement. In: 2012 IEEE 12th International Conference on Data Mining Workshops (ICDMW), pp. 123–130. IEEE (2012)
13. Ignatov, D.I., Poelmans, J., Zaharchuk, V.: Recommender system based on algorithm of bicluster analysis RecBi. arXiv preprint arXiv:1202.2892 (2012)
14. Kaytoue, M., Kuznetsov, S.O., Napoli, A., Duplessis, S.: Mining gene expression data with pattern structures in formal concept analysis. Inf. Sci. **181**(10), 1989–2001 (2011)
15. Kuznetsov, S.O., Obiedkov, S.A.: Comparing performance of algorithms for generating concept lattices. J. Exp. Theor. Artif. Intell. **14**(2–3), 189–216 (2002)
16. Li, G., Ma, Q., Tang, H., Paterson, A.H., Xu, Y.: QUBIC: a qualitative biclustering algorithm for analyses of gene expression data. Nucleic Acids Res. **37**(15), e101–e101 (2009)
17. Madeira, S.C., Oliveira, A.L.: Biclustering algorithms for biological data analysis: a survey. IEEE/ACM Trans. Comput. Biol. Bioinform. (TCBB) **1**(1), 24–45 (2004)
18. Pio, G., Ceci, M., D'Elia, D., Loglisci, C., Malerba, D.: A novel biclustering algorithm for the discovery of meaningful biological correlations between microRNAs and their target genes. BMC Bioinform. **14**(7), S8 (2013)
19. Pontes, B., Giráldez, R., Aguilar-Ruiz, J.S.: Biclustering on expression data: a review. J. Biomed. Inform. **57**, 163–180 (2015)
20. Tanay, A., Sharan, R., Shamir, R.: Discovering statistically significant biclusters in gene expression data. Bioinformatics **18**(suppl-1), S136–S144 (2002)
21. Veroneze, R., Banerjee, A., Von Zuben, F.J.: Enumerating all maximal biclusters in numerical datasets. Inf. Sci. **379**, 288–309 (2017)

# A Sampling-Based Approach
# for Discovering Subspace Clusters

Sandy Moens[1], Boris Cule[1(✉)], and Bart Goethals[1,2]

[1] University of Antwerp, Antwerp, Belgium
{sandy.moens,boris.cule,bart.goethals}@uantwerpen.be
[2] Monash University, Melbourne, Australia

**Abstract.** Subspace clustering aims to discover clusters in projections of highly dimensional numerical data. In this paper, we focus on discovering small collections of interesting subspace clusters that do not try to cluster all data points, leaving noisy data points unclustered. To this end, we propose a randomised method that first converts the highly dimensional database to a binarised one using projected samples of the original database. This database is then mined for frequent itemsets, which we show can be translated back to subspace clusters. In our extensive experimental analysis, we show on synthetic as well as real world data that our method is capable of discovering highly interesting subspace clusters.

## 1 Introduction

The main task of clustering is to group similar objects together, while keeping sufficiently different objects apart. However, due to the *curse of dimensionality*, traditional clustering methods struggle with high-dimensional data. In short, with high-dimensional data, the distances between pairs of objects, measured over all dimensions, become increasingly similar. As a result, no proper clusters can be formed, as all objects end up almost equally distant from each other.

Subspace clustering attempts to solve this problem by discovering clusters of objects that are similar in a limited number of dimensions. However, given the exponential complexity of the search space, identifying the relevant set of dimensions is computationally demanding, which is why existing subspace clustering methods suffer from long run-times [1]. Furthermore, some existing approaches produce full clusterings, thereby ensuring that each object is assigned to exactly one cluster. This is not always desirable: (1) the data may contain a lot of noise, that should ideally not be assigned to any cluster and (2) there is no reason why a particular object should not be assigned to multiple clusters, especially if the sets of dimensions that define these clusters are entirely different.

In this paper, we take a similar approach as CARTICLUS [2]: we first convert a numeric database to a transactional one and then use frequent pattern mining to extract subspace clusters. Our method can efficiently produce highly interesting subspace clusters, along with the dimensions that define them. We avoid the

© Springer Nature Switzerland AG 2019
P. Kralj Novak et al. (Eds.): DS 2019, LNAI 11828, pp. 61–71, 2019.
https://doi.org/10.1007/978-3-030-33778-0_6

computational complexity of existing methods by deploying a randomised algorithm. We first take a large number of samples from the original data, such that each sample consists of a number of objects in a fixed (random) set of dimensions. In each sample, we then cluster the objects, and subsequently assign all objects in the original data to the nearest cluster centroid. This produces a set of objects per centroid, which we interpret as a transaction. By merging the transactions produced for all different samples, we obtain a transaction database. We then sample maximal frequent itemsets from this database to obtain potential clusters. Finally, we identify the relevant dimensions for each discovered cluster.

The main contributions of this paper can be summarised as follows: we propose a randomised sampling algorithm that efficiently identifies localised clusters and their relevant dimensions, we allow data objects to be part of multiple clusters, and we leave noise objects unclustered, and we perform a theoretical evaluation to show the efficiency of our method and an extensive experimental evaluation to demonstrate the quality of our output.

## 2   Background

**Subspace Clustering.** Let $\mathcal{D} = \{D_1, \ldots, D_m\}$ be a set of $m$ dimensions. Each dimension $D_i$ comes with a domain $dom(D_i)$. An $m$-dimensional data point $p = (d_1, \ldots, d_m)$ is a tuple of values over $\mathcal{D}$, such that $d_i \in dom(D_i)$ for each $i = \{1, \ldots, m\}$. The input database $\mathcal{P} = (p_1, \ldots, p_q)$ contains a collection of $q$ such $m$-dimensional data points. Furthermore, each dimension $D_i$ comes with a distance function $\delta_{D_i} : dom(D_i) \times dom(D_i) \to \mathbb{R}$. Additionally, we assume that for any subset of dimensions $D = \{D_1, \ldots, D_l\}$, with $1 \leq l \leq m$ and $D \subseteq \mathcal{D}$ there exists a distance function $\delta_D : (dom(D_1) \times \ldots \times dom(D_l)) \times (dom(D_1) \times \ldots \times dom(D_l)) \to \mathbb{R}$. All used distance functions must satisfy the usual conditions (non-negativity, identity, symmetry and the triangle inequality). Given a subset of dimensions $D \subseteq \mathcal{D}$, we denote by $p^D$ a data point, and by $P^D$ a set of data points, projected onto the given dimensions. A *subspace cluster* $S$ is a tuple containing a subset of datapoints and dimensions, i.e., $S = (P, D)$, with $P \subseteq \mathcal{P}$ and $D \subseteq \mathcal{D}$.

**Frequent Itemset Mining.** Let $\mathcal{I} = (i_1, \ldots i_n)$ be a finite set of $n$ items. A *transaction* $t$ is a subset of items. We denote by $\mathcal{T} = (t_1, \ldots, t_o)$ a database of $o$ transactions. An *itemset* I is also a subset of items. A transaction $t$ is said to support an itemset I if $I \subseteq t$. The set of all transactions that support an itemset is called the *cover* of that itemset, i.e., $cov(I) = \{t \mid t \in \mathcal{T} \wedge I \subseteq t\}$. The *support* of an itemset is the size of its cover, i.e., $sup(I) = |cov(I)|$. Given a minimal support threshold $\sigma \geq 0$, an itemset I is considered *frequent* if its support is larger than or equal to $\sigma$, i.e., $sup(I) \geq \sigma$. An itemset I is called *maximal* if there exists no superset of I that is also frequent with respect to $\sigma$. The anti-monotonic property of the support of itemsets guarantees that all subsets of a frequent itemset are also frequent.

# 3    Randomised Subspace Clusters

Existing methods for discovering subspace clusters from numeric data often focus on the complete raw dataset to compute subspace clusters using a bottom-up [1,3] or a top-down approach [4]. In this paper we introduce RASCL, which uses randomised subsets of the data (both in the data points and in the dimensions) as a starting point for detecting subspace clusters. The discovered clusters are then checked for occurrence in multiple subsamples of the data. If a cluster occurs frequently enough in the set of samples we output it as a subspace cluster. Our algorithm relies on two simple premises: (1) higher dimensional subspace clusters also form subspace clusters in lower dimensions; (2) if we take enough samples and use them to detect clusters, a lot of similar subclusters of the same true cluster will be found in different projections. Moreover, by repeating such a randomised procedure many times we end up with a stable solution.

## 3.1    Randomised Data Transformation

**Data Binarisation.** To binarise a numeric database $\mathcal{P}$ into a transaction database $\mathcal{T}$ we use the indices of data points as the items for $\mathcal{T}$, resulting in $|\mathcal{P}|$ items. In addition, we obtain a mapping between data points and items. Ideally, a transaction contains data points that are close together in some set of dimensions. Then an itemset (essentially a set of data points) that occurs in a large fraction of transactions can be seen as a subspace cluster over some set of dimensions.

We define a randomised process for constructing a single transaction database. We repeat this process $n$ times and concatenate all transactions into a single database $\mathcal{T}^*$. We first sample a small subset of data points P and a small subset of dimensions D (the sampling strategy is explained below). The data points are projected onto the subset of dimensions and used as input for the $K$-means clustering algorithm. The resulting cluster centroids are used to partition the original data points, assigning each data point to the closest centroid. As such, each centroid represents one transaction and its items are the data points assigned to it. Formally, for a set of centroids $C^D$ the closest centroid for a projected data point $p^D$ is given by $c^{p^D} = \mathrm{argmin}_{c^D \in C^D}(\delta_D(p^D, c^D))$.

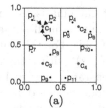

| centroid id | items |
|---|---|
| $c_1$ | 1, 2, 3 |
| $c_2$ | 4, 5, 6 |
| $c_3$ | 7, 8, 9 |
| $c_4$ | 10, 11 |

(a)                                         (b)

**Fig. 1.** (a) A fictitious example dataset with 2 dimensions, 11 data points (black dots) and 4 centroids (red circles). (b) Binarised dataset in short format for the toy dataset. (Color figure online)

*Example 1: Figure* 1(a) *shows a toy database of 11 data points in a 2D space. A red circle represents a synthetic cluster centroid and the surrounding square visually shows data points closest to that centroid. When constructing the binarised database, the index of a data point is added to the transaction of the nearest cluster centroid. The resulting transaction database is shown in Fig.* 1(b).

**Generating Data Samples.** As mentioned previously, our binarisation strategy requires a sample of data points and a sample of dimensions. The main question now is how we can bias the sampling procedure to obtain samples that will have a higher potential to contain cluster structures.

For the data points, we can sample $k$ data points uniformly at random. By repeating this a large number of times, we expect each cluster to be represented by a sufficient number of data points in a high enough number of samples.

For the dimensions, a naive solution would be to sample uniformly at random a subset of dimensions of size $x$, with $1 \leq x \leq |\mathcal{D}|$. However, since the number of combinations larger than 2 can blow up, a random sample of dimensions will likely be too large to contain a meaningful cluster. Sampling just one dimension may result in discovering cluster structures that do not span multiple dimensions. Our empirical results (omitted due to space constraints) have shown that sampling 2 dimensions results in higher quality clusters. We apply weighted sampling to boost the probability of sampling dimensions that contain cluster structures. Similar to Moise et al. [1], we assume that uniformly distributed dimensions do not contain any cluster structure. As such, to detect non-uniformity of a dimension we create a histogram using the Freedman-Diaconis' rule [5] to compute an appropriate number of bins for the data. This rule is robust to outliers and does not assume data to be normally distributed. Let us denote by $B^D$ the bins for a given dimension using the Freedman-Diaconis' rule and let $|b|$ denote the number of data points falling in bin $b$. We compute how many bins contain less than the number of expected data points under uniform data distribution. The unnormalised sampling potential $\mathcal{W}$ of a dimension is given by

$$\mathcal{W}(D) = \sqrt{\frac{|\{b \mid b \in B^D \wedge |b| \leq \frac{|\mathcal{P}|}{|B^D|}\}|}{|B^D|}}. \tag{1}$$

The resulting distribution favours dimensions with more cluster potential.

**Time Complexity.** The worst case complexity of our binarisation method is mostly dependent on $K$-means. However, we use only a small subset of data points, typically $|\text{P}| \ll |\mathcal{P}|$, to compute cluster centroids. For this small subset the complexity for clustering is $\mathcal{O}(n \times (|\text{P}| \times |\text{D}| \times K \times i))$ with $n$ the number of database samples and $i$ the number of iterations. Generation of samples for both data points can be done in $\mathcal{O}(|\mathcal{P}|)$ and for dimensions can be done in $\mathcal{O}(|\mathcal{D}|)$. Assignment of data points to cluster centroids is done in a single sweep, i.e., $\mathcal{O}(K \times |\mathcal{P}|)$. The total time complexity for generating samples and binarising the database is $\mathcal{O}(K \times |\mathcal{P}| + |\mathcal{D}| + n \times (|\text{P}| \times |\text{D}| \times K \times i))$.

## 3.2 Extracting Subspace Clusters

We previously constructed a binarised database $\mathcal{T}^*$ by concatenating $n$ binarised ones built using random samples of data points and dimensions. The premise is that transactions represent cluster centroids and their items are indices of data points in their close proximity for the set of dimensions. Since we generated $n$ samples, we know that each index occurs $n$ times within $\mathcal{T}^*$. If then a set of items occurs often together in the database, i.e., it is a frequent itemset with high support, then we know that in many sets of dimensions the same set of data points occur in close proximity, which is exactly the objective for a subspace cluster. However, typically the number of frequent itemsets is huge (largely because all subsets of frequent itemsets are frequent). To alleviate this problem we use maximal itemsets and, more particularly, our algorithm samples $\mu$ maximal frequent itemsets from the binarised database. The resulting itemsets are the data points for subspace clusters. An effective method for sampling maximal frequent itemsets was introduced by Moens and Goethals [6]. It iteratively extends an itemset with new items, until the set is found to be maximal given a threshold $\tau$ and a monotonic quality measure (e.g., support).

After extracting a collection of data points, we have to discover the dimensions in which the data points form a cluster. In contrast to some existing methods [1,4], we do not require to go back to the data itself to check each dimension individually, since our binarisation process preserved some essential information that can guide us here. That is, our algorithm previously sampled collections of dimensions which can be reused to determine a valid subset of dimensions. We denote by $dims(t)$ a map that for a transaction returns its *linked dimensions*, i.e., the dimensions that were used for its construction in the binarisation process. For a maximal itemset I we can use the transactions in its cover to determine its *relevant dimensions*, i.e., the set containing all linked dimensions for transactions in $cov(\mathtt{I})$. Formally, $dims(\mathtt{I}) = \{d | d \in \mathcal{D} \wedge d \in dims(t) \wedge t \in cov(\mathtt{I})\}$. An itemset I, mapped to the data points P, forms together with its relevant dimensions the subspace cluster $S = (\mathtt{P}, dims(\mathtt{I}))$.

## 3.3 Selecting the Best Subspace Clusters

After discovering a large number of subspace clusters (depending on parameter $\mu$), we finally select a small collection of $r$ clusters that can be deemed the most interesting subspace clusters. The number of data points that is present in the subspace cluster is an indication that the same set of data points are often related even in different subsets of dimensions (experiments omitted due to space constraints). In our method we will employ this heuristic (i.e., the larger the cluster, the better) for sorting discovered subspace clusters. Finally, to reduce redundancy in the cluster results, we sequentially evaluate each cluster and select those clusters that have less than 25% cluster overlap with previously selected ones. Note that when sorting clusters using the number of objects, this results in smaller clusters as $r$ increases. Finally, we exclude very small clusters with less than 10 data points.

# 4    Experiments

In our experiments, we use synthetic data provided by Günnemann et al. [7]. The dataset characteristics are shown in Table 1. To measure the performance we use PRECISION and RECALL scores on object level, as well as their harmonic mean F1. Additionally we use their dimensionality aware counterparts which are indicated by the subscripts D (for scores about the dimensions) and SC (for scores about the combination of objects and dimensions) [7]. Finally, we also use $M^{E4SC}$ [7], a measure to assess the quality of clusterings. Note that unless stated otherwise, we assign just one discovered cluster to each ground truth cluster.

We compare two variants[1]: RASCL sets $k > K$ and RASCL$_R$ sets $k = K$ which essentially skips the clustering step. For each dataset we use the ground truth and select the ground truth cluster with the largest overlap with the cluster being evaluated to compute its quality. We run each experiment 10 times and report the average results for the first $r$ subspace clusters. Less than $r$ clusters may be reported. Unless stated otherwise, we fix the following parameters: $n = 1000$, $k = 100$, $K = 20$, $\sigma = 200$, $\mu = 100$ and $r = 10$. We provide the following guidance: $n$ should be set high enough to obtain a representative sample, $k$ should be sufficiently larger than $K$ for the clustering to make sense, $r$ should be set to the desired number of clusters, $\mu$ should be high enough so no information is lost due to randomisation. $K$ and $\sigma$ are more difficult to set, but we show that the performance of RASCL is not overly sensitive to changes in their values.

**Table 1.** Main characteristics of the synthetic datasets.

| | #rows | #dimensions | #clusters | #objects/cluster | #dimensions/cluster |
|---|---|---|---|---|---|
| dbsizescale$_{s1500}$ | 1,595 | 20 | 10 | 166.3 | 14.0 |
| dbsizescale$_{s2500}$ | 2,658 | 20 | 10 | 276.5 | 14.0 |
| dbsizescale$_{s3500}$ | 3,722 | 20 | 10 | 385.8 | 14.0 |
| dbsizescale$_{s4500}$ | 4,785 | 20 | 10 | 496.2 | 14.0 |
| dbsizescale$_{s5500}$ | 5,848 | 20 | 10 | 608.5 | 14.0 |
| dimscale$_{d05}$ | 1,595 | 5 | 10 | 182.6 | 3.5 |
| dimscale$_{d10}$ | 1,595 | 10 | 10 | 181.5 | 6.7 |
| dimscale$_{d25}$ | 1,595 | 25 | 10 | 180.9 | 16.9 |
| dimscale$_{d50}$ | 1,595 | 50 | 10 | 181.6 | 33.5 |
| dimscale$_{d75}$ | 1,595 | 75 | 10 | 181.9 | 50.4 |
| noisescale$_{n10}$ | 1,611 | 20 | 10 | 166.5 | 14.6 |
| noisescale$_{n30}$ | 2,071 | 20 | 10 | 166.1 | 14.6 |
| noisescale$_{n50}$ | 2,900 | 20 | 10 | 166.3 | 14.6 |
| noisescale$_{n70}$ | 4,833 | 20 | 10 | 166.8 | 14.6 |

---

[1] Source code and experiments are available via https://gitlab.com/adrem/rascl.

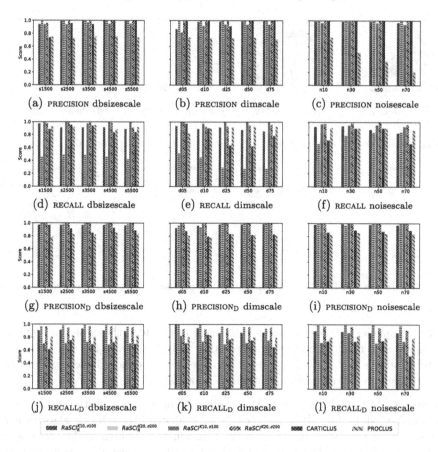

**Fig. 2.** Object quality (a–f) and dimension quality (g–l) scores for different datasets.

**Cluster Quality.** We compare our methods to CARTICLUS [2] and PROCLUS [4]. We used different instantiations of RASCL and RASCL$_R$ by varying $K$ and $\sigma$. For CARTICLUS we use the parameter settings as selected by the authors [2] as basis for this experiment. For PROCLUS we set parameters following the ground truth.

Object quality results are shown in Fig. 2(a–f). All algorithms perform very well with respect to PRECISION except PROCLUS, and setting $K = 20$ and $\sigma = 200$ we slightly outperform CARTICLUS. RASCL$_R^{K10\sigma100}$, RASCL$^{K10\sigma100}$ and RASCL$^{K20\sigma200}$ outperform the competitors on RECALL, while RASCL$_R^{K20\sigma200}$ often fails to deliver good results. This is due to the introduced randomness: using random centroids leads to more partially similar transactions. Combined with a high support this results in small subclusters of the true ground truth clusters.

Results for the dimension quality are shown in Fig. 2(g–l). We see that our algorithms generally outperform the competitors by quite a margin and we see that our simple solution of using linked dimensions (Sect. 3.2) works really

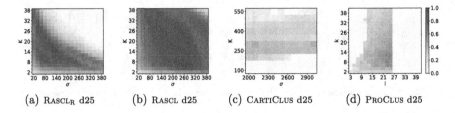

(a) RASCL$_R$ d25     (b) RASCL d25     (c) CARTICLUS d25     (d) PROCLUS d25

**Fig. 3.** Grid search quality results for various methods on the dimscale$_{d25}$ dataset.

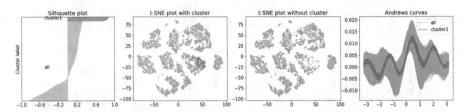

**Fig. 4.** Subspace cluster 1 for the pendigits datasets using RASCL. (Color figure online)

well. For smaller subspace clusters with lower PRECISION, the dimension quality decreases as a result. Comparing ($K = 10$, $\sigma = 100$) to ($K = 20$, $\sigma = 200$), the latter produces better results, mostly because there are more linked dimensions tied to the cover of the maximal itemset, boosting RECALL$_D$.

**Parameter Sensitivity.** We test the influence of $K$ and $\sigma$ on the dimscale$_{d25}$ dataset and show the $M^{E4SC}$ scores. Note that even though our algorithm is not meant for producing full clusterings, it produces very high quality results on this metric. Figure 3 shows scores for the first 10 subspace clusters using various parameter settings for the algorithms. For our algorithm we use a window around the default parameters. For CARTICLUS we use a grid around the optimal parameters and for PROCLUS we define sensible grids. We see that RASCL is not overly susceptible to parameter changes and that, in general, the default parameters produce good and stable results. In contrast, RASCL$_R$ can still produce very good results, but the quality diminishes quickly when the parameters are not too far from the optimal parameters. Increasing $K$ or $\sigma$ results in subclusters of the true clusters, thus decreasing the overall score. We see that finding good settings for CARTICLUS and PROCLUS is much harder. For PROCLUS $l$ cannot exceed the number of dimensions in the data, resulting in lots of 0 scores in the figures. The experiments on other datasets produced similar results.

**Real World Datasets.** We tested our method on the pendigits dataset, a classification dataset found in the UCI machine learning repository[2]. Using RASCL with $n = 1000$, $k = 100$, $K = 10$, $\sigma = 100$, $\mu = 100$ and $r = 10$ we discover multiple subspace clusters for each class. A general trend we found was that the discovered clusters have a very high PRECISION of approx. 91%, but they

---

[2] https://archive.ics.uci.edu/ml/datasets.html.

have rather low RECALL averaging around 20%. We evaluate the largest subspace cluster in Fig. 4. The silhouette plot shows high similarity for points in the cluster (red) and a much lower score for points outside the cluster (blue). A similar trend is found in the scatter plots, which are obtained using *t-SNE* transformation [8] based on the relevant dimensions. The left scatter plot shows all data (blue) together with the subspace cluster (red), while the right scatter plot shows only data points not in the cluster. This shows that using our method we do not miss many data points that are within the region of the subspace cluster according to this transformation. The plot on the right is the Andrews plot [9], which is a smoothed parallel coordinates plot, showing cluster structures more clearly. Similar plots are found for the remaining clusters.

## 5   Related Work

Subspace clustering attempts to find clusters in subsets of dimensions. However, some traditional clustering models, unsuited to this setting, have been adapted for this purpose. PROCLUS [4], one of the first methods for subspace clustering, adapts $K$-means [10] to this setting. The analogy to our work is the initialisation: a two-step randomised procedure is used to obtain an approximation to a *piercing set*, i.e., a set of points each from a different cluster, which are refined to clusters.

DOC [11] is an algorithm that finds subspace clusters using a Monte Carlo method to sample a random point from a cluster as well as a *discriminating* set of points. It then extends the random point to a full subspace cluster using a bounding box around that point. Its extension MINECLUS [12] uses the same medoid points for expanding the cluster, but it drops the randomised procedure. Similar to our approach, it also converts the data to a binarised dataset. Other clustering algorithms, such as DBSCAN [13], have also been adapted for the subspace clustering task [14]. Recently, more general techniques have been proposed for searching the subspace [15], where the discovery of clusters is left to specialised algorithms. However, all of the above methods are computationally very expensive as they search in an exponential set of subspaces.

FIRES [16] is a generic framework for finding subspace clusters, employing existing clustering techniques to compute a set of base clusters in single dimensions. These base clusters are then merged based on their similarity, and the resulting clusters are then pruned and refined to optimise accuracy. The CARTICLUS algorithm [2], like our method, creates a binarised dataset. However, in CARTICLUS, the dimensions are defined during the construction of transactions (or *carts*), such that all carts rely on the same dimension sets. Finally, the carts are mined for frequent itemsets which are then translated back to subspace clusters. Bi-clustering [17] also simultaneously clusters rows and columns of numeric matrices. However, bi-clusters allow for more general clusters as they, for instance, group rows with constant values for a set of columns or group columns that decrease similarly over a set of rows. Typically, such methods are used for analysis of biological data such as gene expression data.

# 6    Conclusion

In this paper, we present a novel method for discovering interesting clusters in high-dimensional data. We started by converting the original data into a transaction database by selecting a small number of random data objects, projecting them to a small number of random dimensions, then clustering them, and, finally, building transactions by assigning all data objects to their closest cluster centroids. We repeat this procedure many times and merge the results. We then sample maximal itemsets randomly from the resulting transaction database, and consider each such itemset to be a potentially interesting cluster of objects. Finally, for each discovered cluster, we identify a relevant set of dimensions.

A major advantage of our method is that, by using the two randomised procedures, we avoid both the combinatorial explosion of possible dimension sets, and the computational cost of frequent itemset mining. In addition, we do not attempt to produce full clusterings, and we allow data objects to be part of multiple clusters, while noise objects will not be part of any cluster at all. Experimentally, we demonstrate that our method produces quality clusters and is not overly sensitive to changes in the parameter settings, which is crucial for an unsupervised learning task.

# References

1. Moise, G., Sander, J., Ester, M.: P3C: a robust projected clustering algorithm. In: Sixth International Conference on Data Mining (ICDM 2006), pp. 414–425. IEEE (2006)
2. Aksehirli, E., Goethals, B., Muller, E., Vreeken, J.: Cartification: a neighborhood preserving transformation for mining high dimensional data. In: 2013 IEEE 13th International Conference on Data Mining (ICDM), pp. 937–942. IEEE (2013)
3. Agrawal, R., Gehrke, J., Gunopulos, D., Raghavan, P.: Automatic subspace clustering of high dimensional data for data mining applications, vol. 27, no. 2. ACM (1998)
4. Aggarwal, C.C., Wolf, J. L., Yu, P.S., Procopiuc, C., Park, J.S.: Fast algorithms for projected clustering. In: ACM SIGMoD Record, vol. 28, no. 2, pp. 61–72. ACM (1999)
5. Freedman, D., Diaconis, P.: On the histogram as a density estimator: L 2 theory. Probab. Theory Relat. Fields **57**(4), 453–476 (1981)
6. Moens, S., Goethals, B.: Randomly sampling maximal itemsets. In: KDD Workshop on Interactive Data Exploration and Analytics, pp. 79–86. ACM (2013)
7. Günnemann, S., Färber, I., Müller, E., Assent, I., Seidl, T.: External evaluation measures for subspace clustering. In: Proceedings of the 20th ACM International Conference on Information and Knowledge Management, pp. 1363–1372. ACM (2011)
8. Maaten, L.v.d., Hinton, G.: Visualizing data using t-SNE. J. Mach. Learn. Res. **9**, 2579–2605 (2008)
9. Andrews, D.F.: Plots of high-dimensional data. Biometrics **29**, 125–136 (1972)
10. MacQueen, J., et al.: Some methods for classification and analysis of multivariate observations. In: Proceedings of the Fifth Berkeley Symposium on Mathematical Statistics and Probability, Oakland, CA, USA, vol. 1, no. 14, pp. 281–297 (1967)

11. Procopiuc, C.M., Jones, M., Agarwal, P.K., Murali, T.: A Monte Carlo algorithm for fast projective clustering. In: Proceedings of the 2002 ACM SIGMOD International Conference on Management of Data, pp. 418–427. ACM (2002)

12. Yiu, M.L., Mamoulis, N.: Frequent-pattern based iterative projected clustering. In: Third IEEE International Conference on Data Mining, ICDM 2003, pp. 689–692. IEEE (2003)

13. Ester, M., Kriegel, H.-P., Sander, J., Xu, X., et al.: A density-based algorithm for discovering clusters in large spatial databases with noise. In: KDD, vol. 96, no. 34, pp. 226–231 (1996)

14. Kailing, K., Kriegel, H.-P., Kröger, P.: Density-connected subspace clustering for high-dimensional data. In: Proceedings of the 2004 SIAM International Conference on Data Mining, pp. 246–256. SIAM (2004)

15. Nguyen, H.V., Müller, E., Vreeken, J., Keller, F., Böhm, K.: CMI: an information-theoretic contrast measure for enhancing subspace cluster and outlier detection. In: SIAM International Conference on Data Mining, pp. 198–206. SIAM (2013)

16. Kriegel, H.-P., Kroger, P., Renz, M., Wurst, S.: A generic framework for efficient subspace clustering of high-dimensional data. In: Fifth IEEE International Conference on Data Mining (ICDM 2005), 8-pp. IEEE (2005)

17. Madeira, S.C., Oliveira, A.L.: Biclustering algorithms for biological data analysis: a survey. IEEE/ACM Trans. Comput. Biol. Bioinform. (TCBB) **1**(1), 24–45 (2004)

# Epistemic Uncertainty Sampling

Vu-Linh Nguyen[1(✉)], Sébastien Destercke[2], and Eyke Hüllermeier[1]

[1] Heinz Nixdorf Institute, Department of Computer Science, Paderborn University,
Paderborn, Germany
`vu.linh.nguyen@uni-paderborn.de, eyke@upb.de`
[2] UMR CNRS 7253 Heudiasyc, Sorbonne Universités, Université de Technologie
de Compiègne, Compiègne, France
`sebastien.destercke@hds.utc.fr`

**Abstract.** Various strategies for active learning have been proposed in
the machine learning literature. In uncertainty sampling, which is among
the most popular approaches, the active learner sequentially queries the
label of those instances for which its current prediction is maximally
uncertain. The predictions as well as the measures used to quantify the
degree of uncertainty, such as entropy, are almost exclusively of a probabilistic nature. In this paper, we advocate a distinction between two
different types of uncertainty, referred to as *epistemic* and *aleatoric*, in
the context of active learning. Roughly speaking, these notions capture
the reducible and the irreducible part of the total uncertainty in a prediction, respectively. We conjecture that, in uncertainty sampling, the
usefulness of an instance is better reflected by its epistemic than by its
aleatoric uncertainty. This leads us to suggest the principle of "epistemic
uncertainty sampling", which we instantiate by means of a concrete approach for measuring epistemic and aleatoric uncertainty. In experimental
studies, epistemic uncertainty sampling does indeed show promising performance.

**Keywords:** Active learning · Uncertainty sampling · Epistemic
uncertainty · Aleatoric uncertainty

## 1 Introduction

The goal in standard supervised learning, such as binary or multi-class classification, is to learn models with high predictive accuracy from labelled training
data [7,22]. However, labelled data does normally not come for free. On the contrary, labelling can be expensive, time-consuming, and costly. The ambition of
*active learning*, therefore, is to exploit labelled data in the most effective way.
More specifically, the idea is to let the learning algorithm itself decide which
examples it considers to be most informative. Compared to random sampling,
the hope is to achieve better performance with the same amount of training
data, or to reach the same performance with less data [6,20].

© Springer Nature Switzerland AG 2019
P. Kralj Novak et al. (Eds.): DS 2019, LNAI 11828, pp. 72–86, 2019.
https://doi.org/10.1007/978-3-030-33778-0_7

The selection of training examples is often done in an iterative manner, i.e., the active learner alternates between re-training and selecting new examples. In each iteration, the usefulness of a candidate example is estimated in terms of a *utility score*, and the one with the highest score is queried. In this regard, the notion of utility typically refers to uncertainty reduction: To what extent will the knowledge about the label of a specific instance help to reduce the learner's uncertainty about the sought model? In *uncertainty sampling* [20], which is among the most popular approaches, utility is quantified in terms of predictive uncertainty, i.e., the active learner selects those instances for which its current prediction is maximally uncertain. The predictions as well as the measures used to quantify the degree of uncertainty, such as entropy, are almost exclusively of a probabilistic nature. Such approaches indeed proved to be successful in many applications.

Yet, as pointed out by [21], existing approaches can be criticized for not informing about the *reasons* for why an instance is considered uncertain, although this might be relevant for judging the usefulness of an example. In this paper, we advocate a distinction between two different types of uncertainty, referred to as *epistemic* and *aleatoric*—roughly speaking, these capture the reducible and the irreducible part of the total uncertainty in a prediction, respectively. The conjecture that, in uncertainty sampling, the usefulness of an instance is better reflected by its epistemic than by its aleatoric uncertainty leads us to the idea of "epistemic uncertainty sampling". Our approach, which builds on a formalization of epistemic and aleatoric uncertainty as proposed by [19], is generic in the sense that is can be instantiated for any learning algorithm; concretely, we present instantiations for a Parzen window classifier, decision tree learning, and logistic regression.

The rest of this paper is organized as follows. In the next section, we recall the general framework of uncertainty sampling and provide a brief survey of related work on active learning. In Sect. 3, we recall the approach of [19] for modeling epistemic and aleatoric uncertainty, and then present our idea of generalizing uncertainty sampling on the basis of this approach. Instantiations of our approach for local learning (Parzen window classifier), decision tree learning and logistic regression are presented in Sect. 4. Experimental evaluations are given in the Sect. 5. The paper concludes with a short summary and an outlook on future work in Sect. 6.

## 2    Uncertainty Sampling

As usual in active learning, we assume to be given a labelled set of training data $\mathbf{D}$ and a pool of unlabeled instances $\mathbf{U}$ that can be queried by the learner:

$$\mathbf{D} = \big\{(\boldsymbol{x}_1, y_1), \ldots, (\boldsymbol{x}_N, y_N)\big\}, \quad \mathbf{U} = \big\{\boldsymbol{x}_1, \ldots, \boldsymbol{x}_J\big\}$$

Instances are represented as features vectors $\boldsymbol{x}_i = \big(x_i^1, \ldots, x_i^d\big) \in \mathcal{X} = \mathbb{R}^d$. In this paper, we only consider the case of binary classification, where labels $y_i$ are taken from $\mathcal{Y} = \{0, 1\}$, leaving the more general case of multi-class classification for future work. We denote by $\mathcal{H} \subset \mathcal{Y}^{\mathcal{X}}$ the underlying hypothesis space, i.e., the class of candidate models $h : \mathcal{X} \longrightarrow \mathcal{Y}$ the learner can choose from. Often, hypotheses are parametrized by a parameter vector $\theta \in \Theta$; in this case, we equate a hypothesis $h = h_\theta \in \mathcal{H}$ with the parameter $\theta$, and the model space $\mathcal{H}$ with the parameter space $\Theta$.

In uncertainty sampling, instances are queried in a greedy fashion. Given the current model $\theta$ that has been trained on $\mathbf{D}$, each instance $\boldsymbol{x}_j$ in the current pool $\mathbf{U}$ is assigned a *utility* score $s(\theta, \boldsymbol{x}_j)$, and the next instance to be queried is the one with the highest score [11,20,21]. The chosen instance is labelled (by an oracle or expert) and added to the training data $\mathbf{D}$, on which the model is then re-trained. The active learning process for a given budget $B$ (i.e., the number of unlabelled instances to be queried) is summarized in Algorithm 1.

---

**Algorithm 1:** Uncertainty sampling

**Input: U, D,** $\theta$- initial pool, training data, classifier, and $B$-budget
**Output: U, D,** $\theta$ - updated pool, training data, classifier
1  initialize $b = 0$;
2  **while** $b < B$ **do**
3  |  **foreach** $\boldsymbol{x} \in \mathbf{U}$ **do**
4  |  |  compute $s(\theta, \boldsymbol{x})$
5  |  query the label of the optimal instance $\boldsymbol{x}^*$ with respect to $s(\theta, \boldsymbol{x})$
   |  $\mathbf{D} = \mathbf{D} \cup \{\boldsymbol{x}^*, y^*\}$ ;
6  |  $\mathbf{U} = \mathbf{U} \setminus \{\boldsymbol{x}^*, y^*\}$ ;
7  |  train $\theta$ from $\mathbf{D}$;
8  |  $b = b + 1$;
9  **Return U, D,** $\theta$;

---

Assuming a probabilistic model producing predictions in the form of probability distributions $p_\theta(\cdot \,|\, \boldsymbol{x})$ on $\mathcal{Y}$, the utility score is typically defined in terms of a measure of uncertainty. Thus, instances on which the current model is highly uncertain are supposed to be maximally informative [20,21]. Popular examples of such measures include

– the entropy:

$$s(\theta, \boldsymbol{x}) = - \sum_{\lambda \in \mathcal{Y}} p_\theta(\lambda \,|\, \boldsymbol{x}) \log p_\theta(\lambda \,|\, \boldsymbol{x}), \tag{1}$$

– the least confidence:

$$s(\theta, \boldsymbol{x}) = 1 - \max_{\lambda \in \mathcal{Y}} p_\theta(\lambda \,|\, \boldsymbol{x}), \tag{2}$$

– the smallest margin:

$$s(\theta, \boldsymbol{x}) = p_\theta(\lambda_n \,|\, \boldsymbol{x}) - p_\theta(\lambda_m \,|\, \boldsymbol{x}), \tag{3}$$

where $\lambda_m = \arg\max_{\lambda \in \mathcal{Y}} p_\theta(\lambda \,|\, \boldsymbol{x})$ and $\lambda_n = \arg\max_{\lambda \in \mathcal{Y} \setminus \lambda_m} p_\theta(\lambda \,|\, \boldsymbol{x})$.

All the three measures ought to be maximized. In the case of binary classification, i.e., $\mathcal{Y} = \{0, 1\}$, all these measures rank unlabelled instances in the same order and look for instances with small difference between $p_\theta(0 \mid \boldsymbol{x})$ and $p_\theta(1 \mid \boldsymbol{x})$.

## 3  Epistemic and Aleatoric Uncertainty

A main building block of our approach to active learning is the distinction between the *epistemic* and *aleatoric* uncertainty involved in the prediction for an instance $\boldsymbol{x}$. Although this distinction is well accepted in the literature on uncertainty [8], it has been considered in machine learning only very recently [9,13,19]. Here, we adopt the formal model proposed by [19], which is based on the use of relative likelihoods, historically proposed by [2] and then justified in other settings such as possibility theory [23]. For the sake of completeness and self-containedness, we briefly recall the essence of this approach.

As before, we proceed from an instance space $\mathcal{X}$, an output space $\mathcal{Y} = \{0, 1\}$ encoding the two classes, and a hypothesis space $\mathcal{H}$ consisting of probabilistic classifiers $h : \mathcal{X} \longrightarrow [0, 1]$. We denote by $p_h(1 \mid \boldsymbol{x}) = h(\boldsymbol{x})$ and $p_h(0 \mid \boldsymbol{x}) = 1 - h(\boldsymbol{x})$ the (predicted) probability that instance $\boldsymbol{x} \in \mathcal{X}$ belongs to the positive and negative class, respectively. Given a set of training data $\mathbf{D} = \{(\boldsymbol{x}_i, y_i)\}_{i=1}^N \subset \mathcal{X} \times \mathcal{Y}$, the normalized likelihood of a model $h$ is defined as

$$\pi_{\mathcal{H}}(h) = \frac{L(h)}{L(h^{ml})} = \frac{L(h)}{\max_{h' \in \mathcal{H}} L(h')}, \tag{4}$$

where $L(h) = \prod_{i=1}^N p_h(y_i \mid \boldsymbol{x}_i)$ is the likelihood of $h$, and $h^{ml} \in \mathcal{H}$ the maximum likelihood estimation on the training data. For a given instance $\boldsymbol{x}$, the degrees of support (plausibility) of the two classes are defined as follows:

$$\pi(1 \mid \boldsymbol{x}) = \sup_{h \in \mathcal{H}} \min \left[ \pi_{\mathcal{H}}(h), p_h(1 \mid \boldsymbol{x}) - p_h(0 \mid \boldsymbol{x}) \right], \tag{5}$$

$$\pi(0 \mid \boldsymbol{x}) = \sup_{h \in \mathcal{H}} \min \left[ \pi_{\mathcal{H}}(h), p_h(0 \mid \boldsymbol{x}) - p_h(1 \mid \boldsymbol{x}) \right]. \tag{6}$$

So, $\pi(1 \mid \boldsymbol{x})$ is high if and only if a highly plausible model supports the positive class much stronger (in terms of the assigned probability mass) than the negative class (and $\pi(0 \mid \boldsymbol{x})$ can be interpreted analogously)[1]. Note that, with $f(a) = 2a - 1$, we can also rewrite (5)–(6) as follows:

$$\pi(1 \mid \boldsymbol{x}) = \sup_{h \in \mathcal{H}} \min \left[ \pi_{\mathcal{H}}(h), f(h(\boldsymbol{x})) \right], \tag{7}$$

$$\pi(0 \mid \boldsymbol{x}) = \sup_{h \in \mathcal{H}} \min \left[ \pi_{\mathcal{H}}(h), f(1 - h(\boldsymbol{x})) \right]. \tag{8}$$

Given the above degrees of support, the degrees of epistemic uncertainty $u_e$ and aleatoric uncertainty $u_a$ are defined as follows:

$$u_e(\boldsymbol{x}) = \min \left[ \pi(1 \mid \boldsymbol{x}), \pi(0 \mid \boldsymbol{x}) \right], \tag{9}$$

$$u_a(\boldsymbol{x}) = 1 - \max \left[ \pi(1 \mid \boldsymbol{x}), \pi(0 \mid \boldsymbol{x}) \right]. \tag{10}$$

---

[1] Technically, we assume that, for each $\boldsymbol{x} \in \mathcal{X}$, there are hypotheses $h, h' \in \mathcal{H}$ such that $h(\boldsymbol{x}) \geq 0.5$ and $h'(\boldsymbol{x}) \leq 0.5$, which implies $\pi(1 \mid \boldsymbol{x}) \geq 0$ and $\pi(0 \mid \boldsymbol{x}) \geq 0$.

Thus, epistemic uncertainty refers to the case where both the positive and the negative class appear to be plausible, while the degree of aleatoric uncertainty (10) is the degree to which none of the classes is supported. These uncertainty degrees are completed with degrees $s_1(\boldsymbol{x})$ and $s_0(\boldsymbol{x})$ of (strict) preference in favor of the positive and negative class, respectively:

$$s_1(\boldsymbol{x}) = \begin{cases} 1 - (u_a(\boldsymbol{x}) + u_e(\boldsymbol{x})) & \text{if } \pi(1\,|\,\boldsymbol{x}) > \pi(0\,|\,\boldsymbol{x}), \\ \frac{1-(u_a(\boldsymbol{x})+u_e(\boldsymbol{x}))}{2} & \text{if } \pi(1\,|\,\boldsymbol{x}) = \pi(0\,|\,\boldsymbol{x}), \\ 0 & \text{if } \pi(1\,|\,\boldsymbol{x}) < \pi(0\,|\,\boldsymbol{x}). \end{cases}$$

With an analogous definition for $s_0(\boldsymbol{x})$, we have $s_0(\boldsymbol{x})+s_1(\boldsymbol{x})+u_a(\boldsymbol{x})+u_e(\boldsymbol{x}) \equiv 1$. Besides, it has the following properties:

- $s_1(\boldsymbol{x})$ $(s_0(\boldsymbol{x}))$ will be high if and only if, for all plausible models, the probability of the positive (negative) class is significantly higher than the one of the negative (positive) class;
- $u_e(\boldsymbol{x})$ will be high if class probabilities strongly vary within the set of plausible models, i.e., if we are unsure how to compare these probabilities. In particular, it will be 1 if and only if we have $h(\boldsymbol{x}) = 1$ and $h'(\boldsymbol{x}) = 0$ for two totally plausible models $h$ and $h'$;
- $u_a(\boldsymbol{x})$ will be high if class probabilities are similar for all plausible models, i.e., if there is strong evidence that $h(\boldsymbol{x}) \approx 0.5$. In particular, it will be close to 1 if all plausible models allocate their probability mass around $h(\boldsymbol{x}) = 0.5$.

Roughly speaking, aleatoric uncertainty is due to influences on the data-generating process that are inherently random, whereas epistemic uncertainty is caused by a lack of knowledge. Or, stated differently, $u_e$ and $u_a$ measure the *reducible* and the *irreducible* part of the total uncertainty, respectively. It thus appears reasonable to assume that epistemic uncertainty is more relevant for active learning: While it makes sense to query additional class labels in regions where uncertainty can be reduced, doing so in regions of high aleatoric uncertainty appears to be less reasonable. This leads us to the principle of *epistemic uncertainty sampling*, which prescribes the selection

$$\boldsymbol{x}^* = \arg\max_{\boldsymbol{x}\in\mathbf{U}} u_e(\boldsymbol{x}). \tag{11}$$

For comparison, we will also consider an analogous selection rule based on the aleatoric uncertainty, i.e.,

$$\boldsymbol{x}^* = \arg\max_{\boldsymbol{x}\in\mathbf{U}} u_a(\boldsymbol{x}). \tag{12}$$

Let us note that the above approach is completely generic and can in principle be instantiated with any hypothesis space $\mathcal{H}$. The uncertainty measures (11–12) can be derived very easily from the support degrees (7–8). The computation of the latter may become difficult, however, as it requires the solution of an optimization problem, the properties of which depend on the choice of $\mathcal{H}$.

# 4   Instantiations of the General Approach

We are going to present practical methods to determine (7–8) for the cases of local learning and logistic regression in Sects. 4.1 and 4.2, respectively.

## 4.1   Local Learning

This section presents an instantiation of our approach for the case of local learning using a Parzen window classifier [4]. The method is then adapted to the case where the decision tree classifier [16,18] is employed as the based learner.

As already said, instantiating the approach essentially means to address the question of how to compute the degrees of support (7–8), from which everything else can easily be derived.

By local learning, we refer to a class of non-parametric models that derive predictions from the training information in a local region of the instance space, for example the local neighborhood of a query instance [3,5]. As a simple example, we consider the Parzen window classifier [4], to which our approach can be applied in a quite straightforward way. To this end, for a given instance $x$, define the set of its neighbours as follows:

$$R(x, \epsilon) = \{(x_i, y_i) \in \mathbf{D} \mid \|x_i - x\| \leq \epsilon\}, \tag{13}$$

where $\epsilon$ is the width of the Parzen window (a practical method to determine such a width will be given latter).

In binary classification, a local region $R$ can be associated with a constant hypothesis $h_\theta$, $\theta \in \Theta = [0, 1]$, where $h_\theta(x) \equiv \theta$ is the probability of the positive class in the region; thus, $h_\theta$ predicts the same probabilities $p_h(1 \mid x) = \theta$ and $p_h(0 \mid x) = 1 - \theta$ for all $x \in R$. The underlying hypothesis space is given by $\mathcal{H} = \{h_\theta \mid 0 \leq \theta \leq 1\}$. With $n$ and $p$ the number of positive and negative instances, respectively, within a Parzen window $R(x, \epsilon)$, the likelihood and the maximum likelihood estimate of $\theta$ are respectively given by

$$L(\theta) = \binom{n + p}{n} \theta^n (1 - \theta)^p \text{ and } \hat{\theta} = \frac{n}{n + p}. \tag{14}$$

Therefore, the degrees of support for the positive and negative classes are

$$\pi(1 \mid x) = \sup_{\theta \in [0,1]} \min \left( \frac{\theta^p (1 - \theta)^n}{\left(\frac{p}{n+p}\right)^p \left(\frac{n}{n+p}\right)^n}, 2\theta - 1 \right), \tag{15}$$

$$\pi(0 \mid x) = \sup_{\theta \in [0,1]} \min \left( \frac{\theta^p (1 - \theta)^n}{\left(\frac{p}{n+p}\right)^p \left(\frac{n}{n+p}\right)^n}, 1 - 2\theta \right). \tag{16}$$

Solving (15) and (16) comes down to maximizing a scalar function over a bounded domain, for which standard solvers can be used. We applied Brent's method[2] (which is a variant of the golden section method) to find a local minimum in the interval $\theta \in [0, 1]$. From (15–16), the epistemic and aleatoric uncertainty associated with the region $R$ can be derived according to (11) and (12), respectively. For different combinations of $n$ and $p$, these uncertainty degrees can be pre-computed (cf. Fig. 1).

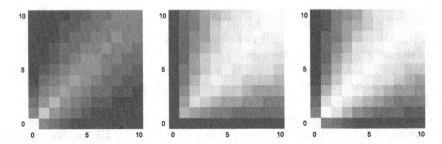

**Fig. 1.** From left to right: Epistemic, aleatoric, and total uncertainty (epistemic + aleatoric) as a function of the numbers $p, n \in \{0, 1, \ldots, 10\}$ of positive and negative examples in a region (Parzen window) of the instance space (lighter colors indicate higher values).

How to determine the width $\epsilon$ of the Parzen window? This value is difficult to assess, and an appropriate choice strongly depends properties of the data and the dimensionality of the instance space. Intuitively, it is even difficult to say in which range this value should lie. Therefore, instead of fixing $\epsilon$, we fixed an absolute number $K$ of neighbors in the training data, which is intuitively more meaningful and easier to interpret. A corresponding value of $\epsilon$ is then determined in such a way that the average number of nearest neighbours of instances $x_i$ in the training data $\mathbf{D}$ is just $K$ (see Algorithm 2). In other words, $\epsilon$ is determined indirectly via $K$.

Since $K$ is an average, individual instances may have more or less neighbors in their Parzen windows. In particular, a Parzen window may also be empty. In this case, we set $u_e(x) = 1$ by definition, i.e., we consider this as a case of full epistemic uncertainty. Likewise, the uncertainty is considered to be maximal for all other sampling techniques. If the accuracy of the Parzen classifier needs to be determined, we assume that it yields a wrong prediction.

In a similar way, the approach can be applied to decision tree learning [16,18]. In fact recall that a decision tree partitions the instance space $\mathcal{X}$ into (rectangular) regions $R_1, \ldots, R_L$ (i.e., $\bigcup_{i=1}^{L} R_i = \mathcal{X}$ and $R_i \cap R_j = \emptyset$ for $i \neq j$) associated with corresponding leafs of the tree (each leaf node defines a region $R$). Again, in the case of binary classification, we can assume each region $R$ to be associated with a constant hypothesis $h_\theta$, $\theta \in \Theta = [0, 1]$, where $h_\theta(x) \equiv \theta$ is the probability

---

[2] For an implementation in Python, see https://docs.scipy.org/doc/scipy-0.19.1/reference/generated/scipy.optimize.minimize_scalar.html.

---

**Algorithm 2:** Determining the width $\epsilon$.

---

**Input:** D-normalized data, $K$-number
**Output:** the local width $\epsilon_K$

1  **foreach** $x_n \in \mathbf{D}$ **do**
2  $\quad$ **foreach** $x_m \neq x_n$ **do**
3  $\quad\quad$ $\lfloor$ compute $d(x_n, x_m)$;
4  $\quad$ form $1 \times (n-1)$ vector $\mathbf{d}_n = \big(d(x_n, x_m) \,|\, n \neq m\big)$;
5  $\quad$ $\lfloor$ sort $\mathbf{d}_n$ by increasing order and determine the $K$-th element $\mathbf{d}_n^K$;
6  **return** $\epsilon_K = \frac{\sum_{n=1}^{|\mathbf{D}|} \mathbf{d}_n^K}{|\mathbf{D}|}$;

---

of the positive class. Therefore, degrees of epistemic and aleatoric uncertainty degrees can be derived in the same way as described above.

## 4.2 Logistic Regression

In this section, we present another instantiation of our approach for a commonly used learning algorithm, namely logistic regression. In contrast to nonparametric, local learning methods such as the Parzen window classifier, logistic regression is a parametric class of linear models, and hence coming with comparatively restrictive assumptions.

Recall that logistic regression assumes posterior probabilities to depend on feature vectors $x = (x^1, \ldots, x^d) \in \mathbb{R}^d$ in the following way:

$$h(x) = p(1 \,|\, x) = \frac{\exp\left(\theta_0 + \sum_{i=1}^{d} \theta_i\, x^i\right)}{1 + \exp\left(\theta_0 + \sum_{i=1}^{d} \theta_i\, x^i\right)} \tag{17}$$

This means that learning the model comes down to estimating a parameter vector $\theta = (\theta_0, \ldots, \theta_d)$, which is commonly done through likelihood maximization [12]. To avoid numerical issues (e.g, having to deal with the exponential function for large $\theta$) when maximizing the target function, we employ $L_2$-regularization. The corresponding version of the log-likelihood function (18) is strictly concave [17]:

$$l(\theta) = \log L(\theta) = \sum_{n=1}^{N} y_n \left(\theta_0 + \sum_{i=1}^{d} \theta_i x_n^i\right) \tag{18}$$

$$- \sum_{n=1}^{N} \ln\left(1 + \exp\left(\theta_0 + \sum_{i=1}^{d} \theta_i x_n^i\right)\right) - \frac{\gamma}{2} \sum_{i=0}^{d} \theta_i^2,$$

where the regularization term $\gamma$ will be fixed to 1.

We now focus on determining the degree of support (7) for the positive class, and then summarize the results for the negative class (which can be determined in a similar manner). Associating each hypothesis $h \in \mathcal{H}$ with a vector $\theta \in \mathbb{R}^{d+1}$, the degree of support (7) can be rewritten as follows:

$$\pi(1 \,|\, \boldsymbol{x}) = \sup_{\theta \in \mathbb{R}^{d+1}} \min \left[ \pi(\theta), 2h(\boldsymbol{x}) - 1 \right] \tag{19}$$

It is easy to see that the target function to be maximized in (19) is not necessarily concave. Therefore, we propose the following approach.

Let us first note that whenever $h(\boldsymbol{x}) < 0.5$, we have $2h(\boldsymbol{x}) - 1 \le 0$ and $\min \left[ \pi_{\mathcal{H}}(h), 2h(\boldsymbol{x}) - 1 \right] \le 0$. Thus the optimal value of the target function (7) can only be achieved for some hypotheses $h$ such that $h(\boldsymbol{x}) \in [0.5, 1]$. For a given value $\alpha \in [0.5, 1]$, the set of hypotheses $h$ such that $h(\boldsymbol{x}) = \alpha$ corresponds to the convex set

$$\theta^{\alpha} = \left\{ \theta \,\Big|\, \theta_0 + \sum_{i=1}^{d} \theta_i x^i = \ln \left( \frac{\alpha}{1 - \alpha} \right) \right\}. \tag{20}$$

The optimal value $\pi_{\alpha}^{*}(1 \,|\, \boldsymbol{x})$ that can be achieved within the region (20) can be determined as follows:

$$\pi_{\alpha}^{*}(1 \,|\, \boldsymbol{x}) = \sup_{\theta \in \theta^{\alpha}} \min \left[ \pi(\theta), 2\alpha - 1 \right] = \min \left[ \sup_{\theta \in \theta^{\alpha}} \pi(\theta), 2\alpha - 1 \right]. \tag{21}$$

Thus, to find this value, we maximize the concave log-likelihood over a convex set:

$$\theta_{\alpha}^{*} = \arg \sup_{\theta \in \theta^{\alpha}} l(\theta) \tag{22}$$

As the log-likelihood function (18) is concave and has second-order derivatives, we tackle the problem with a Newton-CG algorithm [14]. Furthermore, the optimization problem (22) can be solved using sequential least squares programming[3] [15]. Since regions defined in (20) are parallel hyperplanes, the solution of the optimization problem (7) can then be obtained by solving the following problem:

$$\sup_{\alpha \in [0.5,1)} \pi_{\alpha}^{*}(1|\boldsymbol{x}) = \sup_{\alpha \in [0.5,1)} \min \left[ \pi(\theta_{\alpha}^{*}), 2\alpha - 1 \right]. \tag{23}$$

Following a similar procedure, we can estimate the degree of support for the negative class (8) as follows:

$$\sup_{\alpha \in (0,0.5]} \pi_{\alpha}^{*}(0|\boldsymbol{x}) = \sup_{\alpha \in (0,0.5]} \min \left[ \pi(\theta_{\alpha}^{*}), 1 - 2\alpha \right] \tag{24}$$

Note that limit cases $\alpha = 1$ and $\alpha = 0$ cannot be solved, since the region (20) is then not well-defined (as $\ln(\infty)$ and $\ln(0)$ do not exist). For the purpose of practical implementation, we handle (23) by discretizing the interval over $\alpha$. That is, we optimize the target function for a given number of values $\alpha \in [0.5, 1)$

---

[3] For an implementation in Python, see https://docs.scipy.org/doc/scipy/reference/generated/scipy.optimize.minimize.html.

and consider the solution corresponding to the $\alpha$ with the highest optimal value of the target function $\pi_\alpha^*(1\,|\,x)$ as the maximum estimator. Similarly, (24) can be handled over the domain $(0, 0.5]$.

In practice, we evaluate (23) and (24) on uniform discretizations of cardinality 50 of $[0.5, 1)$ and $(0, 0.5]$, respectively. We can further increase efficiency by avoiding computations for values of $\alpha$ for which we know that $2\alpha - 1$ and $1 - 2\alpha$ are lower than the current highest support value given to class 1 and 0, respectively. See Algorithm 3 for a pseudo-code description of the whole procedure.

---

**Algorithm 3:** Degrees of support for logistic regression

---

    **Input:** $Q$, $\mathbf{D}$, $\theta^{ml}$, $x$- initial pool, training data, classifier, unlabelled instance
    **Output:** $\pi(1\,|\,x)$, $\pi(0\,|\,x)$ - degrees of support
1  initialize subsets $Q_p$, $Q_n$ of cardinality $Q$;
2  $\pi(1\,|\,x) = \max(2h^{ml}(x) - 1, 0)$ , $\pi(0\,|\,x) = \max(1 - 2h^{ml}(x), 0)$ ;
3  **for** $q = 1, \ldots, Q$ **do**
4     $\alpha_p = \max(Q_p)$; $\alpha_n = \min(Q_n)$ ;
5     **if** $2\alpha_p - 1 > \pi(1\,|\,x)$ **then**
6         solve (22) for $x$, $\alpha_p$ and return $\theta$;
7         $\pi(1\,|\,x) = \max(\pi(1\,|\,x), \min(\pi_{\mathcal{H}}(\theta), 2\alpha_p - 1))$ ;
8     **if** $1 - 2\alpha_n > \pi(0\,|\,x)$ **then**
9         solve (22) for $x$, $\alpha_n$ and return $\theta$;
10        $\pi(0\,|\,x) = \max(\pi(0\,|\,x), \min(\pi_{\mathcal{H}}(\theta), 1 - 2\alpha_p))$ ;
11     $Q_p = Q_p \setminus \{\alpha_p\}, Q_n = Q_n \setminus \{\alpha_n\}$ ;
12 **Return** $\pi(1\,|\,x)$, $\pi(0\,|\,x)$ ;

---

## 5 Experimental Results

To illustrate the performance of our uncertainty measures in active learning, we conducted experiments on data sets from the UCI repository[4], the main properties of which are summarized in Table 1.

### 5.1 Local Learning

We follow a 10-fold cross-validation procedure, considering each fold as the test set, while the other folds are used for learning. The latter is randomly split into a training data set and a pool set. The proportions of training/pool/test sets are 10/80/10% and accuracies are averaged. The budget of the active learner is fixed to be 30% of the original data.

---

[4] http://archive.ics.uci.edu/ml/index.php.

**Table 1.** Data sets used in the experiments

| # | Name | # instances | # features | Attributes |
|---|------|-------------|-----------|------------|
| 1 | Parkinsons | 197 | 22 | Real |
| 2 | Vertebral-column | 310 | 6 | Real |
| 3 | Ionosphere | 351 | 34 | Real |
| 4 | Climate-model | 540 | 18 | Real |
| 5 | Breast-cancer | 569 | 30 | Real |
| 6 | Blood-transfusion | 748 | 5 | Real |
| 7 | QSAR | 1055 | 41 | Integer, real |
| 8 | Banknote-authentication | 1372 | 4 | Real |

After each query, we update the data sets and, correspondingly, the classifiers. The improvements of the classifiers are compared for four different uncertainty measures, i.e., uncertainty sampling (following the strategy presented in Algorithm 1) based on four measures for selecting unlabelled instances: random sampling, standard uncertainty (2), epistemic uncertainty (9), aleatoric uncertainty (10).

To reduce the computational efforts, in each iteration, the learner is allowed to evaluate and query instances from a randomly selected subset consisting of 10% of the data in the pool. Since we are not, in the first place, interested in maximizing performance, but in analyzing the effectiveness of active learning approaches, we simply fix the neighborhood size $K$ as the square root of the size of the data set (number of instances in the initial training set and pool) [10].

As can be seen in Fig. 2, the results are nicely in agreement with our expectations: Epistemic uncertainty sampling performs the best and aleatoric uncertainty sampling the worst. Moreover, standard uncertainty sampling and random sampling are in-between the two. This supports our conjecture that, from an active learning point of view, epistemic uncertainty is the more useful information. Even if the improvements compared to standard uncertainty sampling are not huge, they are still visible and quite consistent.

The results for decision tree learning (cf. Fig. 3) are quite similar and again in agreement with our expectations.

## 5.2 Logistic Regression

For logistic regression, we start with a relatively small amount of initial training data, thereby making improvements in the beginning more visible. More specifically, the proportions of training/pool/test set are 1/89/10%, and the accuracies are averaged. The budget is fixed to be 20% of the original data, and in each iteration, the learner is allowed to evaluate and query instances from a (randomly) subset consisting of 10% data of the pool.

In the case of logistic regression, the improvements through epistemic uncertainty sampling are less pronounced—on the contrary, the performance of epis-

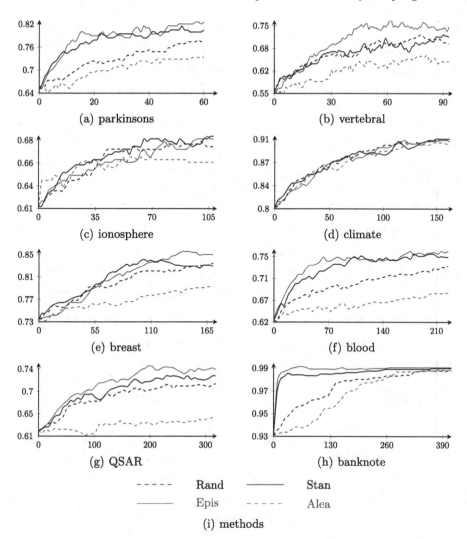

**Fig. 2.** Average accuracies (y-axis) for the Parzen window classifier as a function of the number of examples queried from the pool (x-axis).

temic and standard uncertainty sampling is quite comparable. Two examples, which are quite representative, are shown in Fig. 4. As a plausible explanation, note that logistic regression comes with a very strong learning bias in the form of a linearity assumption. Therefore, the epistemic (or model) uncertainty disappears quite quickly.

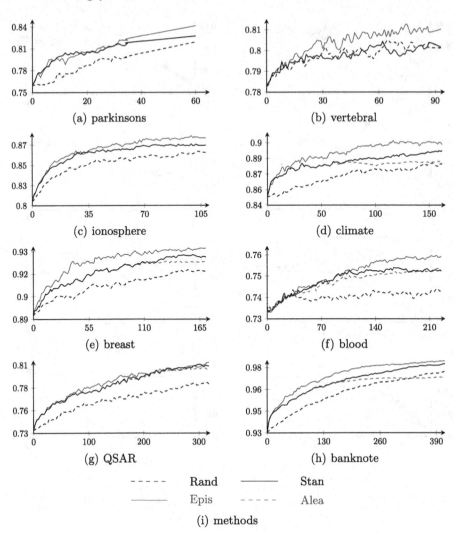

**Fig. 3.** Average accuracies (y-axis) for the decision tree classifier as a function of the number of examples queried from the pool (x-axis).

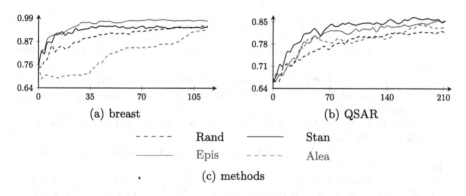

**Fig. 4.** Average accuracies (y-axis) for logistic regression as a function of the number of examples queried from the pool (x-axis).

## 6  Conclusion

This paper reconsiders the principle of uncertainty sampling in active learning from the perspective of uncertainty modeling. More specifically, it starts from the supposition that, when it comes to the question of which instances to select from a pool of candidates, a learner's predictive uncertainty due to "not knowing" should be more relevant than its uncertainty due to inherent randomness.

To corroborate this conjecture, we proposed *epistemic uncertainty sampling*, in which standard uncertainty measures such as entropy are replaced by a novel measure of epistemic uncertainty. The latter is borrowed from a recent framework for uncertainty modeling, in which epistemic uncertainty is distinguished from aleatoric uncertainty [19]. We interpret our experimental results, especially those for local learning (Parzen window classifier and decision trees) as evidence in favor of our conjecture. They clearly show that a separation of the total uncertainty (into epistemic and aleatoric) is effective, and that the epistemic part is the better criterion for selecting instances to be queried. This was the main purpose of the paper.

Given this affirmation, we are now encouraged to elaborate on epistemic uncertainty sampling in more depth, and to develop it in more sophistication. This includes an extension to other learning algorithms and more general learning problems (such as multi-class classification), as well as a comparison to other variants of uncertainty sampling, such as [1] and [21].

**Acknowledgements.** This work was supported by the German Research Foundation (DFG) and the French National Agency for Research (Labex MS2T).

## References

1. Antonucci, A., Corani, G., Gabaglio, S.: Active learning by the naive credal classifier. In: Proceedings of the Sixth European Workshop on Probabilistic Graphical Models (PGM), pp. 3–10 (2012)

2. Birnbaum, A.: On the foundations of statistical inference. J. Am. Stat. Assoc. **57**(298), 269–306 (1962)
3. Bottou, L., Vapnik, V.: Local learning algorithms. Neural Comput. **4**(6), 888–900 (1992)
4. Chapelle, O.: Active learning for Parzen window classifier. In: Proceedings of the Tenth International Workshop on Artificial Intelligence and Statistics (AISTATS), vol. 5, pp. 49–56 (2005)
5. Cover, T., Hart, P.: Nearest neighbor pattern classification. IEEE Trans. Inf. Theory **13**(1), 21–27 (1967)
6. Fu, Y., Zhu, X., Li, B.: A survey on instance selection for active learning. Knowl. Inf. Syst. **35**(2), 249–283 (2013)
7. Hastie, T., Tibshirani, R., Friedman, J., Franklin, J.: The elements of statistical learning: data mining, inference and prediction. Math. Intelligencer **27**(2), 83–85 (2005)
8. Hora, S.C.: Aleatory and epistemic uncertainty in probability elicitation with an example from hazardous waste management. Reliab. Eng. Syst. Saf. **54**(2–3), 217–223 (1996)
9. Kendall, A., Gal, Y.: What uncertainties do we need in Bayesian deep learning for computer vision? In: Proceedings of the 31st Conference on Neural Information Processing Systems (NIPS) (2017)
10. Lall, U., Sharma, A.: A nearest neighbor bootstrap for resampling hydrologic time series. Water Resour. Res. **32**(3), 679–693 (1996)
11. Lewis, D.D., Gale, W.A.: A sequential algorithm for training text classifiers. In: Croft, B.W., van Rijsbergen, C.J. (eds.) SIGIR 1994, pp. 3–12. Springer, London (1994). https://doi.org/10.1007/978-1-4471-2099-5_1
12. Menard, S.: Applied Logistic Regression Analysis, vol. 106. Sage, Thousand Oaks (2002)
13. Nguyen, V.L., Destercke, S., Masson, M.H., Hüllermeier, E.: Reliable multi-class classification based on pairwise epistemic and aleatoric uncertainty. In: Proceedings of the 27th International Joint Conference on Artificial Intelligence (IJCAI), pp. 5089–5095. AAAI Press (2018)
14. Nocedal, J., Wright, S.: Numerical Optimization. Springer Series in Operations Research and Financial Engineering. Springer, New York (2006). https://doi.org/10.1007/978-0-387-40065-5
15. Philip, E., Elizabeth, W.: Sequential quadratic programming methods. UCSD Department of Mathematics, Technical report NA-10-03 (2010)
16. Quinlan, J.R.: Induction of decision trees. Mach. Learn. **1**(1), 81–106 (1986)
17. Rennie, J.D.: Regularized logistic regression is strictly convex. Technical report, MIT (2005)
18. Safavian, S.R., Landgrebe, D.: A survey of decision tree classifier methodology. IEEE Trans. Syst. Man. Cybern. **21**(3), 660–674 (1991)
19. Senge, R., et al.: Reliable classification: Learning classifiers that distinguish aleatoric and epistemic uncertainty. Inf. Sci. **255**, 16–29 (2014)
20. Settles, B.: Active learning literature survey. Technical report, University of Wisconsin, Madison, vol. 52, no. 55–66, p. 11 (2010)
21. Sharma, M., Bilgic, M.: Evidence-based uncertainty sampling for active learning. Data Min. Knowl. Disc. **31**(1), 164–202 (2017)
22. Vapnik, V.N.: An overview of statistical learning theory. IEEE Trans. Neural Networks **10**(5), 988–999 (1999)
23. Walley, P., Moral, S.: Upper probabilities based only on the likelihood function. J. Roy. Stat. Soc. Ser. B (Stat. Methodol.) **61**(4), 831–847 (1999)

# Utilizing Hierarchies in Tree-Based Online Structured Output Prediction

Aljaž Osojnik[1(✉)] ⓘ, Panče Panov[1,2], and Sašo Džeroski[1,2]

[1] Department of Knowledge Technologies, Jožef Stefan Institute, Ljubljana, Slovenia
aljaz.osojnik@ijs.si
[2] Jožef Stefan International Postgraduate School, Ljubljana, Slovenia

**Abstract.** Methods for online prediction of structured values are becoming more and more popular. However, hierarchical prediction, which has recently been shown to produce good results in terms of predictive performance in the batch learning setting, has not yet been applied in the online learning setting. We address the recently introduced task of hierarchical multi-target regression. To this end, we propose a hierarchical extension of iSOUP-Tree, which can address online multi-target regression. The extension weighs the split evaluation heuristic according to the location of the targets in the hierarchy. We design the experimental setup to ascertain whether the additional information contained in the hierarchy can be utilized to improve the predictive performance in the leaf targets. The proposed method shows promising results, producing potential improvements that should be investigated further.

**Keywords:** Online hierarchical prediction · Hierarchical multi-target regression

## 1 Introduction

The recent popularity of online predictive modeling has also extended toward tasks where the predicted values are composed of multiple components. Commonly, this encompasses the prediction of multiple nominal or continuous values. Interestingly, hierarchical prediction, a fairly popular approach used in the batch learning setting, has not yet been applied to online learning. In hierarchical prediction, the nominal or continuous values to be predicted are arranged in a hierarchy, which can be utilized to provide superior predictions over the unstructured case [12,17].

In this paper, we seek to explore how hierarchical prediction can be applied to online predictive modeling. In online predictive modeling, data examples arrive one by one and the predictive model must be periodically updated. In particular, we seek to answer the question of whether the use of a hierarchy on the target space can improve the predictive performance over using a flat set of targets in the online setting. To this end, we propose a hierarchical extension [12,17] of

© Springer Nature Switzerland AG 2019
P. Kralj Novak et al. (Eds.): DS 2019, LNAI 11828, pp. 87–95, 2019.
https://doi.org/10.1007/978-3-030-33778-0_8

the iSOUP-Tree method [13,14] and explore how well its various configurations utilize the information provided in the hierarchy.

In particular, we focus on the online predictive task of multi-target regression and its hierarchical extension called hierarchical multi-target regression. *Multi-target regression* (MTR) is concerned with predicting multiple continuous target variables. Furthermore, we define the hierarchical prediction task pf *hierarchical multi-target regression* (HMTR), where the individual targets are arranged into a hierarchy. As the base method we extend to address hierarchical prediction addresses online MTR, we focus on HMTR, though we briefly discuss the related task of hierarchical multi-label classification and how to approach it.

The hierarchy of targets is generally represented by a graph. We distinguish between tree hierarchies and directed acyclic graph (DAG) hierarchies. In a *tree hierarchy*, each target only has one parent, while in a *DAG hierarchy* a target can have multiple parents. In this paper, we evaluate the proposed approach on datasets with tree hierarchies, but, the proposed method also works with DAG hierarchies.

Before properly defining HMTR, we start with a definition of *hierarchical multi-label classification* (HMLC), as some concepts are introduced more easily in this context. In multi-label classification (MLC), each example is annotated with a set of labels and in HMLC, the binary targets/labels are arranged in a hierarchy, e.g., as in Fig. 1. A key property of the HMLC task is the *hierarchy constraint*, which is satisfied when, for each label that is present in an example, all its ancestors are also present. In other words, labels lower in the hierarchy are refinements of their ancestor labels.

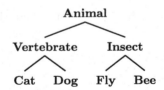

**Fig. 1.** A sample (tree) hierarchy for use HMLC.

*Hierarchical multi-target regression* (HMTR) is a hierarchical variant of multi-target regression. In place of the binary values, now continuous targets are arranged in a hierarchy. In hierarchical multi-target regression, the hierarchy constraint is not as straightforward as in hierarchical multi-label classification. In HMLC, the value of a non-leaf label is based on the values of its children. If any of its children are present, so must be the observed label according to the hierarchy constraint. In other words, the label is present if we take the disjunction of its children's labels' presence, i.e.,

$$\text{label } \lambda \text{ is present} \iff \bigvee_{\lambda' \text{ is a child of } \lambda} (\text{label } \lambda' \text{ is present}).$$

The observed label's presence is an aggregate of the presences of its children. This prompts us to define the hierarchy constraint in hierarchical multi-target regression in a similar way. In HMTR, a non-leaf target is assumed to have a value that is an aggregate of its children's values. The aggregate can be a sum, minimum, maximum, etc. However, due to the targets being continuous, the enforcement of the hierarchy constraint is not as simple as in HMLC. Instead of expecting the aggregate values to be matched exactly, we instead expect the predictions to be as close to the aggregate values as possible according to the chosen evaluation measure.

Notably, when we observe a single example $(x, y)$, where $x$ and $y$ are the descriptive and target vectors, respectively, we do not encode the hierarchy explicitly in $y$. Instead, we use the standard vector representation as flat MTR and encode the hierarchy as a relation of the components of the target vector. We then define evaluation measures that take the hierarchy into account, as shown later in the paper.

This paper continues with an overview of related work in Sect. 2. The proposed hierarchical iSOUP-Tree method is presented in Sect. 3. In Sect. 4, we describe the experimental setup, including the evaluation measures and datasets. We present and discuss the experimental results in Sect. 5 and conclude the paper with plans for further work in Sect. 6.

## 2   Related Work

Hierarchical prediction problems are found in many application domains, most notably in text classification [15], functional genomics [1] and object recognition [16]. Historically, the only hierarchical prediction task that was addressed for was for a long time hierarchical multi-label classification. The task of hierarchical multi-target regression was introduced recently [12]. To the best of our knowledge, there are no methods that address online hierarchical prediction. Hence, we present related work for batch HMTR and online MTR.

**Batch HMTR.** The hierarchical multi-target regression task has only been introduced recently by Mileski et al. [12]. They address hierarchical multi-target regression by using predictive clustering trees, which had previously been used for batch MTR and hierarchical multi-label classification [17].

**Online MTR.** In the online setting, some attention has been given to multi-target regression, exclusively based on the Hoeffding bound. Namely, Ikonomovska et al. [10] proposed the FIMT-DD method for online regression, which was extended to the multi-target regression setting in the FIMT-MT method [9].

iSOUP-Tree [14] extends FIMT-MT with support for nominal input attributes and multi-target leaf models. iSOUP-Tree has also been applied to online MLC [13]. Recently, Duarte and Gama [5] implemented a rule-based learning method for multi-target regression.

## 3    iSOUP-Tree for Online Hierarchical Prediction

iSOUP-Tree [13,14] is a tree-based method for online multi-target regression and online multi-label classification. It uses a similar learning mechanism to Hoeffding trees [4] for classification and FIMT-DD [10] for regression; that is, it periodically checks if enough data instances have accumulated in any leaves and, based on the appropriate heuristic and the Hoeffding bound [8], determines weather the best split is statistically supported. To evaluate a split candidate $S$, iSOUP-Tree uses as a heuristic the *intra-cluster variance reduction* (ICVR), defined as

$$ICVR(S) = \frac{1}{M} \sum_{j=1}^{M} \left( 1 - \frac{|S_\top|}{|S|} \frac{\text{Var}^j(S_\top)}{\text{Var}^j(S)} - \frac{|S_\perp|}{|S|} \frac{\text{Var}^j(S_\perp)}{\text{Var}^j(S)} \right),$$

where $j$ indexes the target variables, $M$ is the number of targets, $S$ is the set of examples accumulated in the given leaf and $S_\top$ and $S_\perp$ are the post-split subsets of $S$ and for which the considered split test is evaluated either as true and false, respectively. $\text{Var}^j$ is the variance of the $j$-th target:

$$\text{Var}^j(S) = \frac{1}{|S|} \sum_{i=1}^{|S|} \left( y_i^j - \overline{y}^j \right)^2,$$

where $i$ indexes the examples from (sub)set $S$, $y_i^j$ is the value of the $j$-th target of the $i$-th example, and $\overline{y}^j$ is the average value of the $j$-th target in the set $S$. For further details on the iSOUP-Tree algorithm, see Osojnik et al. [14].

In order to address online hierarchical prediction, we utilize a weighted splitting heuristic, commonly used in the batch setting for hierarchical multi-label classification [2,17], as well as in hierarchical multi-target regression [12]. The weighted heuristic assigns a weight to each target based on its location in the hierarchy.

The weighted ICVR heuristic is calculated as above, however instead of using regular variance, we use the weighted variance

$$\text{wVar}^j(S) = w_j \frac{\sum_{i=1}^{|S|} (y_i^j - \overline{y}^j)}{|S|},$$

where $w_j$ is the weight of the $j$-th target, which is calculated as $w_j = w_0^{\text{depth}(j)}$ where $w_0 \in \mathbb{R}^+$ is the weight of the root node and depth($j$) is the average depth of the $j$-th target over all paths from the root to it in the hierarchy. In the case of a tree hierarchy, this depth($j$) coincides with the standard definition of depth. When the weight of the root node is less than one, i.e., $w_0 < 1$, a larger emphasis is placed on the variances of the targets higher in the hierarchy, i.e., nodes which are closer to the root of the hierarchy. This aims to address the fact that a wrong prediction higher in the hierarchy is more detrimental than a mistake lower in the hierarchy, i.e., wrongly predicting a high-level concept results in wrong predictions of the lower-level concepts, due to the hierarchy constraint.

On the other hand, when $w_0 > 1$, variances of the targets deeper in the hierarchy, in particular of leaf targets, are emphasized. This directly prioritizes splits which reduce the variances of the leaf targets. The variances of non-leaf targets are then used as a discrimination factor for splits with similar variances in the leaf targets.

# 4    Experimental Setup

In the experiments for hierarchical tasks, we consider the task of online multi-target regression, presented here. To this aim, we use HMTR datasets. More specifically, we are exploring if and how the hierarchically adjusted splitting heuristic [17] affects the predictive performance of the iSOUP-Tree method.

**Experimental Scenarios.** In hierarchical prediction tasks, we are often interested primarily in the predictions for the leaf targets. In these experiments, we examine how the hierarchy and the adapted method affect the predictive performance in the leaf targets. To this end we define three scenarios. The first scenario serves as a "control group". We remove all non-leaf targets in the observed datasets and keep only the leaf targets. Essentially, this yields an online MTR task that we can address with tje regular iSOUP-Tree approach. This scenario is named **leaves-only**. In the second scenario, we use a **bottom-weighted** hierarchical iSOUP-Tree, as described in Sect. 3, by selecting a root node weight of $w_0 = 2$. This method places greater emphasis on the homogeneity of the leaf targets when selecting splits. In the third scenario, we use a **top-weighted** hierarchical iSOUP-Tree, which places emphasis on the targets that are closer to the root node of the hierarchy by selecting a root node weight of $w_0 = 0.5$.

**Performance Evaluation.** There are several approaches for evaluating the predictive performance of hierarchical prediction models. They differ in what they are trying to measure, and are different for different hierarchical tasks.

In this paper, we wish to observe whether the addition of the hierarchy can improve the prediction in the leaves. We can think of the hierarchy as a tool to improve the predictive performance and we are evaluating whether it can improve the performance and what its effectiveness is in this regard.

As we are most interested in the predictive performance in the leaves, we calculate the evaluation measures only on the leaf targets. To obtain the predictions of all the models, we use the *prequential* evaluation approach [7]. The evaluation measure we use is the *average relative mean absolute error* ($\overline{\text{RMAE}}$) measure, defined on an evaluation sample $S$ as

$$\overline{\text{RMAE}}(S) = \frac{1}{M} \sum_{j=1}^{M} \frac{\sum_{i=1}^{n} \left| y_j^i - \hat{y}_i^j \right|}{\sum_{i=1}^{n} \left| y_i^j - \overline{y}^j(i) \right|},$$

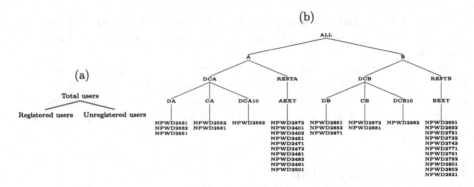

**Fig. 2.** The hierarchy of targets for the (a) Bicycles and (b) Mars Express datasets.

where $y_i^j$ is a true value of the target $j$ for example $i$, $\hat{y}_i^j$ are the predictions of the evaluated model, and $\overline{y}^j(i)$ is the value predicted by the $j$-th mean regressor for the $i$-th example. Lower values of $\overline{\text{RMAE}}$ mean better performance.

**Datasets.** For our experimental evaluation, we use two datasets. The first is a modified *Bicycles* [6] dataset, in which we arrange the targets into a very simple hierarchy, where the registered users and unregistered users targets are children of the total users target. The second dataset is the *Mars Express* dataset, where the task is to predict the power consumption at 33 locations in the Mars Express satellite [3]. The targets have been arranged into a hierarchy according to their physical location within the satellite [11]. The hierarchies of the Bicycles and the Mars Express dataset are shown in Fig. 2.

## 5   Experimental Results and Discussion

The results of applying hierarchical and leaves-only variants of iSOUP-Tree to the two datasets described in Sect. 4 are shown in Fig. 3. In both datasets, the bottom-weighted hierarchical models outperform the top-weighted models. The bottom-weighted models start out slightly worse than the leaves-only tree; however, at some point, the bottom-weighted models reach and even slightly outperform the leaves-only tree. On the other hand, top-weighted models perform worse than the leaves-only model, with the exception of a few intervals in both datasets, where their predictive performances are comparable. Both the difference between the leaves-only model and the bottom-weighted model and the difference between the leaves-only model and the top-weighted models might be a consequence of slower growth of the model trees.

Above, we clearly see a difference between the top- and bottom-weighted hierarchical methods, where the bottom-weighted method appears superior in terms of predictive performance. However, we must consider that we have chosen an evaluation procedure that is solely focused on the predictive performance in the leaf targets. Intuitively, by putting a larger weight on the variance of the leaves

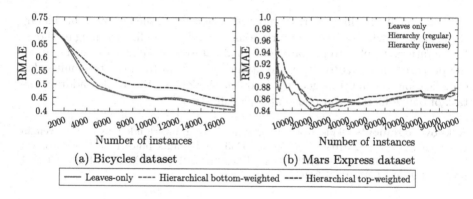

**Fig. 3.** The evolution of the performance ($\overline{\text{RMAE}}$) of iSOUP-Tree using the different heuristics on two HMTR datasets.

of the target hierarchy, as in the bottom-weighted iSOUP-Tree, we are selecting splits which first and foremost reduce the variance of the leaf targets. If we were to use a different evaluation methodology which would consider hierarchical evaluation measures where the "cost" of an error higher up in the hierarchy is higher than for targets lower in the hierarchy, the top-weighted models might outperform the bottom-weighted ones.

With regard to the question of whether the use of target hierarchy improves predictive performance, these experiments do not conclusively confirm this is the case. In our results, bottom-weighted models do eventually outperform the leaves-only ones. However, we must also consider whether adding additional targets to a multi-target regression problem inhibits the growth of the models. As we are averaging more and more individual variances in the calculation of the ICVR heuristic, we might encounter the effects of the central limit theorem, which states that the normalized sum of independent random variables tends toward a normal distribution. In particular, as we average more values, the resulting heuristics of the split candidates will be distributed closer and closer to the normal distribution, with a prescribed mean. To address this problem, we might explore the use of option trees, as they try to address some of the shortcomings of Hoeffding inequality-based approaches.

## 6   Further Work

The most needed extension of this paper concerns the evaluation methodology. Due to our specific inquiry into whether a hierarchy can improve the predictive performance in the leaves, we have focused exclusively on the predictions in the leaves. However, in many applications, the remaining non-leaf targets are at least as important as the leaf targets. As such, it would be prudent to extend the evaluation measures and methodology to explore how the proposed method performs in these scenarios.

Additionally, in this paper, we rely rather heavily on the few available hierarchical datasets which are of appropriate size for use with online learning methods. In the future, we plan to prepare additional datasets of sufficient size and apply the proposed methods to them. Additionally, we plan to apply the multi-label classification via multi-target regression methodology [13] to also address the task of online hierarchical multi-label classification.

**Acknowledgements.** This work is supported by grants funded by the Slovenian Research Agency (P2-0103, N2-0056, J2-9230) and grants funded by the European Commission (H2020 785907 HBP SGA2, H2020 635201 LANDMARK, H2020 769661 SAAM, H2020 833671 RESILOC).

# References

1. Barutcuoglu, Z., Schapire, R.E., Troyanskaya, O.G.: Hierarchical multi-label prediction of gene function. Bioinformatics **22**(7), 830–836 (2006)
2. Blockeel, H., Schietgat, L., Struyf, J., Džeroski, S., Clare, A.: Decision trees for hierarchical multilabel classification: a case study in functional genomics. In: Fürnkranz, J., Scheffer, T., Spiliopoulou, M. (eds.) PKDD 2006. LNCS (LNAI), vol. 4213, pp. 18–29. Springer, Heidelberg (2006). https://doi.org/10.1007/11871637_7
3. Breskvar, M., et al.: Predicting thermal power consumption of the Mars Express satellite with machine learning. In: 6th International Conference on Space Mission Challenges for Information Technology (SMC-IT 2017), pp. 88–93. IEEE (2017)
4. Domingos, P., Hulten, G.: Mining high-speed data streams. In: International Conference on Knowledge Discovery and Data Mining (KDD 2000), pp. 71–80. ACM (2000)
5. Duarte, J., Gama, J.: Multi-target regression from high-speed data streams with adaptive model rules. In: International Conference on Data Science and Advanced Analytics (DSAA 2015), pp. 1–10. IEEE (10 2015)
6. Fanaee-T, H., Gama, J.: Event labeling combining ensemble detectors and background knowledge. Prog. Artif. Intell. **2**(2–3), 113–127 (2013)
7. Gama, J.: Knowledge Discovery from Data Streams. CRC Press, Boca Raton (2010)
8. Hoeffding, W.: Probability inequalities for sums of bounded random variables. J. Am. Stat. Assoc. **58**(301), 13–30 (1963)
9. Ikonomovska, E., Gama, J., Džeroski, S.: Incremental multi-target model trees for data streams. In: Symposium on Applied Computing (SAC 2011), pp. 988–993. ACM (2011)
10. Ikonomovska, E., Gama, J., Džeroski, S.: Learning model trees from evolving data streams. Data Min. Knowl. Disc. **23**(1), 128–168 (2011)
11. Mileski, V.: Tree methods for hierarchical multi-target regression. Master's thesis, Jožef Stefan International Postgraduate School, Ljubljana, Slovenia (2017)
12. Mileski, V., Džeroski, S., Kocev, D.: Predictive clustering trees for hierarchical multi-target regression. In: Adams, N., Tucker, A., Weston, D. (eds.) IDA 2017. LNCS, vol. 10584, pp. 223–234. Springer, Cham (2017). https://doi.org/10.1007/978-3-319-68765-0_19
13. Osojnik, A., Panov, P., Džeroski, S.: Multi-label classification via multi-target regression on data streams. Mach. Learn. **106**(6), 745–770 (2017)

14. Osojnik, A., Panov, P., Džeroski, S.: Tree-based methods for online multi-target regression. J. Intell. Inf. Syst. **50**(2), 315–339 (2017)
15. Rousu, J., Saunders, C., Szedmak, S., Shawe-Taylor, J.: Kernel-based learning of hierarchical multilabel classification models. J. Mach. Learn. Res. **7**, 1601–1626 (2006)
16. Stenger, B., Thayananthan, A., Torr, P.H., Cipolla, R.: Estimating 3D hand pose using hierarchical multi-label classification. Image Vis. Comput. **25**(12), 1885–1894 (2007)
17. Vens, C., Struyf, J., Schietgat, L., Džeroski, S., Blockeel, H.: Decision trees for hierarchical multi-label classification. Mach. Learn. **73**(2), 185–214 (2008)

# On the Trade-Off Between Consistency and Coverage in Multi-label Rule Learning Heuristics

Michael Rapp$^{(\boxtimes)}$, Eneldo Loza Mencía, and Johannes Fürnkranz

Knowledge Engineering Group, TU Darmstadt, Darmstadt, Germany
{mrapp,eneldo,juffi}@ke.tu-darmstadt.de

**Abstract.** Recently, several authors have advocated the use of rule learning algorithms to model multi-label data, as rules are interpretable and can be comprehended, analyzed, or qualitatively evaluated by domain experts. Many rule learning algorithms employ a heuristic-guided search for rules that model regularities contained in the training data and it is commonly accepted that the choice of the heuristic has a significant impact on the predictive performance of the learner. Whereas the properties of rule learning heuristics have been studied in the realm of single-label classification, there is no such work taking into account the particularities of multi-label classification. This is surprising, as the quality of multi-label predictions is usually assessed in terms of a variety of different, potentially competing, performance measures that cannot all be optimized by a single learner at the same time. In this work, we show empirically that it is crucial to trade off the consistency and coverage of rules differently, depending on which multi-label measure should be optimized by a model. Based on these findings, we emphasize the need for configurable learners that can flexibly use different heuristics. As our experiments reveal, the choice of the heuristic is not straight-forward, because a search for rules that optimize a measure locally does usually not result in a model that maximizes that measure globally.

**Keywords:** Multi-label classification · Rule learning · Heuristics

## 1 Introduction

As many real-world classification problems require to assign more than one label to an instance, multi-label classification (MLC) has become a well-established topic in the machine learning community. There are various applications of MLC such as text categorization [16], the annotation of images [4,18] and music [24, 26], as well as use cases in bioinformatics [8] and medicine [20].

Rule learning algorithms are a well-researched approach to solve classification problems [13]. In comparison to complex statistical methods, like for example support vector machines or artificial neural networks, their main advantage is the interpretability of the resulting models. Rule-based models can easily be

© Springer Nature Switzerland AG 2019
P. Kralj Novak et al. (Eds.): DS 2019, LNAI 11828, pp. 96–111, 2019.
https://doi.org/10.1007/978-3-030-33778-0_9

understood by humans and form a structured hypothesis space that can be analyzed and modified by domain experts. Ideally, rule-based approaches are able to yield insight into the application domain by revealing patterns and regularities hidden in the data and allow to reason why individual predictions have been made by a system. This is especially relevant in safety-critical domains, such as medicine, power systems, or financial markets, where malfunctions and unexpected behavior may entail the risk of health damage or financial harm.

**Motivation and Goals.** To assess the quality of multi-label predictions in terms of a single score, several commonly used performance measures exist. Even though some of them originate from measures used in binary or multi-class classification, different ways to aggregate and average the predictions for individual labels and instances—most prominently *micro-* and *macro-averaging*—exist in MLC. Some measures like *subset accuracy* are even unique to the multi-label setting. No studies that investigate the effects of using different rule learning heuristics in MLC and discuss how they affect different multi-label performance measures have been published so far.

In accordance with previous publications in single-label classification, we argue that all common rule learning heuristics basically trade off between two aspects, *consistency* and *coverage* [12]. Our long-term goal is to better understand how these two aspects should be weighed to assess the quality of candidate rules during training if one is interested in a model that optimizes a certain multi-label performance measure. As a first step towards this goal, we present a method for flexibly creating rule-based models that are built with respect to certain heuristics. Using this method, we empirically analyze how different heuristics affect the models in terms of predictive performance and model characteristics. We demonstrate how models that aim to optimize a given multi-label performance measure can deliberately be trained by choosing a suitable heuristic. By comparing our results to a state-of-the-art rule learner, we emphasize the need for configurable approaches that can flexibly be tailored to different multi-label measures. Due to space limitations, we restrict ourselves to micro-averaged measures, as well as to Hamming and subset accuracy.

**Structure of This Work.** We start in Sect. 2 by giving a formal definition of multi-label classification tasks as well as an overview of inductive rule learning and the rule evaluation measures that are relevant to this work. Based on these foundations, in Sect. 3, we discuss our approach for flexibly creating rule-based classifiers that are built with respect to said measures. In Sect. 4, we present the results of the empirical study we have conducted, before we provide an overview of related work in Sect. 5. Finally, we conclude in Sect. 6 by recapitulating our results and giving an outlook on planned future work.

## 2    Preliminaries

MLC is a supervised learning problem in which the task is to associate an instance with one or several labels $\lambda_i$ out of a finite label space $\mathbb{L} = \{\lambda_1, \ldots, \lambda_n\}$, with $n = |\mathbb{L}|$ being the total number of predefined labels. An individual instance $\boldsymbol{x}_j$ is represented in attribute-value form, i.e., it consists of a vector $\boldsymbol{x}_j = (v_1, \ldots, v_l) \in \mathbb{D} = A_1 \times \cdots \times A_l$, where $A_i$ is a numeric or nominal attribute. Additionally, each instance $\boldsymbol{x}_j$ is associated with a binary label vector $\boldsymbol{y}_j = (y_1, \ldots, y_n) = \{0, 1\}^n$, where $y_i$ indicates the presence (1) or absence (0) of label $\lambda_i$. Consequently, the training data set of a MLC problem can be defined as a set of tuples $T = \{(\boldsymbol{x}_1, \boldsymbol{y}_1), \ldots, (\boldsymbol{x}_m, \boldsymbol{y}_m)\}$, with $m = |T|$ being the number of available training instances. The classifier function $g(.)$, that is deduced from a given training data set, maps an instance $\boldsymbol{x}$ to a predicted label vector $\hat{\boldsymbol{y}} = (\hat{y}_1, \ldots, \hat{y}_n) = \{0, 1\}^n$.

### 2.1    Classification Rules

In this work, we are concerned with the induction of conjunctive, propositional rules $\boldsymbol{r} : H \leftarrow B$. The body $B$ of such a rule consists of one or several conditions that compare an attribute-value $v_i$ of an instance to a constant by using a relational operator such as $=$ (in case of nominal attributes), or $<$ and $\geq$ (in case of numerical attributes). On the one hand, the body of a conjunctive rule can be viewed as a predicate $B : \boldsymbol{x} \to \{\text{true}, \text{false}\}$ that states whether an instance $\boldsymbol{x}$ satisfies all of the given conditions, i.e., whether the instance is *covered* by the rule or not. On the other hand, the head $H$ of a (single-label head) rule consists of a single label assignment ($\hat{y}_i = 0$ or $\hat{y}_i = 1$) that specifies whether the label $\lambda_i$ should be predicted as present (1) or absent (0).

### 2.2    Binary Relevance Method

In the present work, we use the *binary relevance* transformation method (cf. [4]), which reduces MLC to binary classification by treating each label $\lambda_i \in \mathbb{L}$ of a MLC problem independently. For each label $\lambda_i$, we aim at learning rules that predict the minority class $t_i \in \{0, 1\}$, i.e., rules that contain the label assignment $\hat{y}_i = t_i$ in their head. We define $t_i = 1$, if the corresponding label $\lambda_i$ is associated with less than 50% of the training instances, or $t_i = 0$ otherwise.

A rule-based classifier—also referred to as a *theory*—combines several rules into a single model. In this work, we use (unordered) rule sets containing all rules that have been induced for the individual labels. Such a rule set can be considered as a disjunction of conjunctive rules (DNF). At prediction time, all rules that cover a given instance are taken into account to determine the predicted label vector $\hat{\boldsymbol{y}}$. An individual element $\hat{y}_i \in \hat{\boldsymbol{y}}$, that corresponds to the label $\lambda_i$, is set to the minority class $t_i$ if at least one of the covering rules contains the label assignment $\hat{y}_i = t_i$ in its head. Otherwise, the element is set to the majority class $1 - t_i$. As all rules that have been induced for a label $\lambda_i$ have the same head, no conflicts may arise in the process.

## 2.3   Bipartition Evaluation Functions

To assess the quality of individual rules, usually bipartition evaluation functions $\delta : \mathbb{N}^{2\times 2} \to \mathbb{R}$ are used [25]. Such functions—also called *heuristics*—map a two-dimensional confusion matrix to a heuristic value $h \in [0, 1]$. A confusion matrix consists of the number of *true positive* (*TP*), *false positive* (*FP*), *true negative* (*TN*), and *false negative* (*FN*) labels that are predicted by a rule. We calculate the example-wise aggregated confusion matrix $C_r$ for a rule $r : \hat{y}_i \leftarrow B$ as

$$C_r := \begin{pmatrix} TP \; FP \\ FN \; TN \end{pmatrix} = C_i^1 \oplus \cdots \oplus C_i^j \oplus \cdots \oplus C_i^m \tag{1}$$

where $\oplus$ denotes the cell-wise addition of atomic confusion matrices $C_i^j$ that correspond to label $\lambda_i$ and instance $x_j$.

Further, let $y_i^j$ and $\hat{y}_i^j$ denote the absence (0) or presence (1) of label $\lambda_i$ for an instance $y_j$ according to the ground truth and a rule's prediction, respectively. Based on these variables, we calculate the elements of $C_i^j$ as

$$\begin{array}{ll} TP_i^j = [\![y_i^j = t_i \wedge \hat{y}_i^j = t_i]\!] & FP_i^j = [\![y_i^j \neq t_i \wedge \hat{y}_i^j = t_i]\!] \\ FN_i^j = [\![y_i^j = t_i \wedge \hat{y}_i^j \neq t_i]\!] & TN_i^j = [\![y_i^j \neq t_i \wedge \hat{y}_i^j \neq t_i]\!] \end{array} \tag{2}$$

where $[\![x]\!] = 1$, if $x$ is true, 0 otherwise.

## 2.4   Rule Learning Heuristics

A good rule learning heuristic should (among other aspects) take both, the *consistency* and *coverage* of a rule, into account [13,15]. On the one hand, rules should be consistent, i.e., their prediction should be correct for as many of the covered instances as possible. On the other hand, rules with great coverage, i.e., rules that cover a large number of instances, tend to be more reliable, even though they may be less consistent.

The *precision* metric exclusively focuses on the consistency of a rule. It calculates as the fraction of correct predictions among all covered instances:

$$\delta_{prec}(C) := \frac{TP}{TP + FP} \tag{3}$$

In contrast, *recall* focuses on the coverage of a rule. It measures the fraction of covered instances among all—covered and uncovered—instances for which the label assignment in the rule's head is correct:

$$\delta_{rec}(C) := \frac{TP}{TP + FN} \tag{4}$$

The *F-measure* calculates as the (weighted) harmonic mean of precision and recall. It allows to trade off the consistency and coverage of a rule depending on the user-configurable parameter $\beta$:

$$\delta_F(C) := \frac{\beta^2 + 1}{\frac{\beta^2}{\delta_{rec}(C)} + \frac{1}{\delta_{prec}(C)}}, \text{ with } \beta \in [0, +\infty] \tag{5}$$

As an alternative to the F-measure, we use different parameterizations of the *m-estimate* in this work. It is defined as

$$\delta_m\left(C\right) := \frac{TP + m \cdot \frac{P}{P+N}}{TP + FP + m}, \text{ with } m \geq 0 \tag{6}$$

where $P = TP + FN$ and $N = FP + TN$. Depending on the parameter $m$, this measure trades off precision and *weighted relative accuracy* (WRA). If $m = 0$, it is equivalent to precision and therefore focuses on consistency. As $m$ approaches $+\infty$, it converges to WRA and puts more emphasis on coverage, respectively [13].

## 3   Induction of Rule-Based Theories

For our experimental study, we implemented a method that allows to generate a large number of rules for a given training data set in a short amount of time (cf. Sect. 3.1).[1] The rules should ideally be unbiased, i.e., they should not be biased in favor of a certain heuristic, and they should be diverse, i.e., general rules should be included as well as specific rules. Given that these requirements are met, we consider the generated rules to be representative samples for the space of all possible rules, which is way too large to be explored exhaustively. We use the generated candidate rules as a starting point for building different theories. They consist of a subset of rules that are selected with respect to a specific heuristic (cf. Sect. 3.2) and filtered according to a threshold (cf. Sect. 3.3). Whereas the first step yields a theory with great coverage, the threshold selection aims at improving its consistency.

### 3.1   Generation of Candidate Rules

As noted in Sect. 2.2, we consider each label $\lambda_i \in \mathbb{L}$ of a MLC problem independently. For each of the labels we train multiple random forests [5], using varying configuration parameters, and extract rules from their decision trees.[2] As illustrated in Algorithm 1, we repeat the process until a predefined number of rules $\gamma$ has been generated.

Each random forest consists of a predefined number of decision trees (we specify $I = 10$). To ensure that we are able to generate diverse rules later on, we vary the configuration parameter $depth \in [0, 8]$ that specifies the maximum depth of trees (unrestricted, if $depth = 0$) (cf. Algorithm 1, trainForest). For building individual trees, we only take a subset of the available training instances and attributes into account, which guarantees a diverse set of trees. Bagging is used for sampling the training instances, i.e., if $m$ instances are available in total, $m \cdot P$ instances ($P = 100\%$, by default) are drawn randomly with replacement. Additionally, each time a new node is added to a decision tree, only a random selection of $K$ out of $l$ attributes ($K = \log_2\left(l - 1\right) + 1$, by default) is considered.

---

[1] Source code available at https://github.com/mrapp-ke/RuleGeneration.

[2] We use the random forest implementation provided by Weka 3.9.3, which is available at https://www.cs.waikato.ac.nz/ml/weka.

---

**Algorithm 1:** Iterative generation of rules from random forests

---

**input** : min. number of rules to be generated $\gamma$
**output**: rule set $R$
$R = \emptyset$
**while** $|R| < \gamma$ **do**
    **foreach** $\lambda_i \in \mathbb{L}$ **and** $depth \in [0, 8]$ **do**
        $rf = $ trainForest$(\lambda_i, depth)$
        $R = R \cup$ extractRules$(rf)$
**return** $R$

---

To extract rules from a random forest (cf. Algorithm 1, extractRules), we traverse all paths from the root node to a leaf in each of its decision trees. We only consider paths that lead to a leaf where the minority class $t_i$ is predicted. As a consequence, all rules that are generated with respect to a certain label $\lambda_i$ have the same head $\hat{y}_i = t_i$. The body of a rule consists of a conjunction of all conditions encountered on the path from the root to the corresponding leaf.

### 3.2  Candidate Subset Selection

Like many traditional rule learning algorithms, we use a *separate-and-conquer* (SeCo) strategy for selecting candidate rules, i.e., new rules are added to the theory until all training instances are covered (or until it describes the training data sufficiently according to some stopping criterion). Whenever a new rule is added to the theory the training instances it covers are removed ("separate" step), and the next rule is chosen according to its performance on the remaining instances ("conquer" step).

To create different theories, we select subsets of the rules that have been generated earlier (cf. Sect. 3.1). We therefore apply the SeCo strategy for each label independently, i.e., for each label $\lambda_i$ we take all rules with head $\hat{y}_i = t_i$ into account. Among these candidates we successively select the best rule according to a heuristic $\delta$ (cf. Sect. 2.4) until all *positive* training instances $P_i = \{(\boldsymbol{x}, \boldsymbol{y}) \in T \mid y_i = t_i\}$, with respect to label $\lambda_i$, are covered. To measure the quality of a candidate $r$ according to $\delta$, we only take yet uncovered instances into account for computing the confusion matrix $C_r$. If two candidates evaluate to the same heuristic value, we prefer the one that (a) covers more true positives, or (b) contains fewer conditions in its body. Whenever a new rule is added, the overall coverage of the theory increases, as more positive training instances are covered. However, the rule may also cover some of the *negative* instances $N_i = T \setminus P_i$. As the rule's prediction is incorrect in such cases, the consistency of the theory may decrease.

### 3.3  Threshold Selection

As described in Sect. 3.2, we use a SeCo strategy to select more rules until all positive training instances are covered for each label. In this way, the coverage of the resulting theory is maximized at the expense of consistency, because each rule contributes to the overall coverage, but might introduce wrong predictions for some instances. To trade off between these aspects, we allow to (optionally) specify a threshold $\phi$ that aims at diminishing the effects of inconsistent rules. It is compared to a heuristic value that is calculated for each rule according to the heuristic $\delta$. For calculating the heuristic value, the rule's predictions on the entire training data set are taken into account. This is different from the candidate selection discussed in Sect. 3.2, where instances that are already covered by previously selected rules are not considered. Because the candidate selection aims at selecting non-redundant rules, that cover the positive training instances as uniformly as possible, it considers rules in the context of their predecessors. In contrast, the threshold $\phi$ is applied at prediction time when no order is imposed on the rules, i.e., all rules whose heuristic value exceeds the threshold equally contribute to the prediction.

## 4   Evaluation

In this section, we present an empirical study that emphasises the need to use varying heuristics for candidate selection and filtering to learn theories that are tailored to specific multi-label measures. We further compare our method to different baselines to demonstrate the benefits of being able to flexibly adjust a learner to different measures, rather than employing a general-purpose learner.

### 4.1   Experimental Setup

We applied our method to eight different data sets taken from the Mulan project.[3] We set the minimum number of rules to be generated to 300.000 (cf. Algorithm 1, parameter $\gamma$). For candidate selection according to Sect. 3.2, we used the m-estimate (cf. Eq. 6) with $m = 0, 2^1, 2^2, \ldots, 2^{19}$. For each of these variants, we applied varying thresholds $\phi$ according to Sect. 3.3. The thresholds have been chosen such that they are satisfied by at least $100\%, 95\%, \ldots, 5\%$ of the selected rules. All results have been obtained using 10-fold cross validation.

In addition to the m-estimate, we also used the F-measure (cf. Eq. 5) with varying $\beta$-parameters. As the conclusions drawn from these experiments are very similar to those for the m-estimate, we focus on the latter at this point.

Among the performance measures that we report are micro-averaged precision and recall. Given a global confusion matrix $C := C_1^1 \oplus \cdots \oplus C_i^j \oplus \cdots \oplus C_n^m$ that consists of the $TP$, $FP$, $TN$, and $FN$ aggregated over all test instances $x_j$ and labels $\lambda_i$, these two measures are calculated as defined in Eqs. 3 and 4. Moreover, we report the micro-averaged F1 score (cf. Eq. 5 with $\beta = 1$) as well as Hamming and subset accuracy. Hamming accuracy calculates as

$$\delta_{Hamm}(C) := \frac{TP + TN}{TP + FP + TN + FN} \tag{7}$$

---

[3] Data sets and detailed statistics available at http://mulan.sourceforge.net/datasets-mlc.html.

whereas subset accuracy differs from the other measures, because it is computed instance-wise. Given true label vectors $Y = (\boldsymbol{y}_1, \ldots, \boldsymbol{y}_m)$ and predicted label vectors $\hat{Y} = (\hat{\boldsymbol{y}}_1, \ldots, \hat{\boldsymbol{y}}_m)$, it measures the fraction of perfectly labeled instances:

$$\delta_{acc}\left(Y, \hat{Y}\right) := \frac{1}{m}\sum_j [\![\boldsymbol{y}_j = \hat{\boldsymbol{y}}_j]\!] \tag{8}$$

## 4.2   Analysis of Different Parameter Settings

For a broad analysis, we trained $20^2 = 400$ theories per data set using the same candidate rules, but selecting and filtering them differently by using varying combinations of the parameters $m$ and $\phi$ as discussed in Sect. 4.1. We visualize the performance and characteristics of the resulting models as two-dimensional matrices of scores (cf. e.g. Fig. 1). One dimension corresponds to the used $m$-parameter, the other refers to the threshold $\phi$, respectively.

Some of the used data sets (CAL500, FLAGS, and YEAST) contain very frequent labels for which the minority class $t_i = 0$. This is rather atypical in MLC and causes the unintuitive effect that the removal of individual rules results in a theory with greater recall and/or lower precision. To be able to compare different parameter settings across multiple data sets, we worked around this effect by altering affected data sets., i.e., inverting all labels for which $t_i = 0$.

**Predictive Performance.** In Figs. 1 and 2 the average ranks of the tested configurations according to different performance measures are depicted. The rank of each of the 400 parameter settings was determined for each data set separately and then averaged over all data sets. The depicted standard deviations show that the optimal parameter settings for a respective measure may vary depending on the data set. However, for each measure there is an area in the parameter space where a good setting can be found with high certainty.

As it can clearly be seen, precision and recall are competing measures. The first is maximized by choosing small values for $m$ and filtering extensively, the latter benefits from large values for $m$ and no filtering. Interestingly, setting $m = 0$, i.e., selecting candidates according to the precision metric, does not result in models with the highest overall precision. This is in accordance with Fig. 3, where the models with the highest F1 score do not result from using the F1-measure for candidate selection. Instead, optimizing the F1 score requires to choose small values for $m$ to trade off between consistency and coverage. The same applies to Hamming and subset accuracy, albeit both of these measure demand to put even more weight on consistency and filtering more extensively compared to F1.

**Model Characteristics.** Besides the predictive performance, we are also interested in the characteristics of the theories. Figure 4 shows how the number of rules in a theory as well as the average number of conditions are affected by varying parameter settings. The number of rules independently declines when

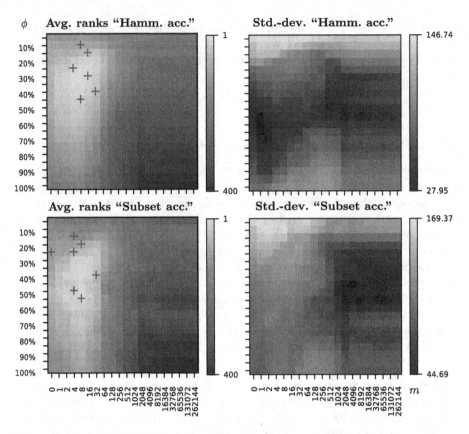

**Fig. 1.** Ranks and standard deviation of average ranks over all data sets according to Hamming and subset accuracy using different parameters $m$ (horizontal axis) and $\phi$ (vertical axis). Best parameters for different data sets specified by red + signs. (Color figure online)

using greater values for the parameter $m$ and/or smaller values for $\phi$. resulting in less complex theories that can be comprehended by humans more easily. The average number of conditions is mostly affected by the parameter $m$.

Figure 5 provides an example of how different parameters affect the model characteristics. It shows the rules for predicting the same label as induced by two fundamentally different approaches. The first approach ($m = 16, \phi = 0.3$) reaches high scores according to the F1-measure, Hamming accuracy, and subset accuracy, whereas the second one ($m = 262144, \phi = 1.0$) results in high recall.

### 4.3   Baseline Comparison

Although the goal of this work is not to develop a method that generally outperforms existing rule learners, we want to ensure that we achieve competitive results. For this reason, we compared our method to JRip, Weka's re-

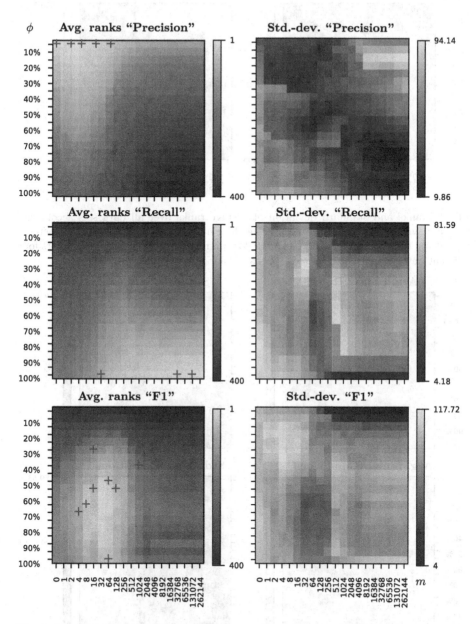

**Fig. 2.** Ranks and standard deviation of average ranks over all data sets according to micro-averaged precision, recall, and F1-measure. Best parameters for different data sets specified by red + signs. (Color figure online)

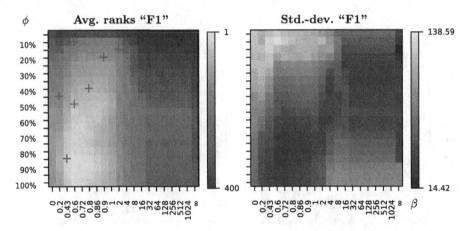

**Fig. 3.** Ranks and standard deviation of average ranks over all data sets according to micro-averaged F1-measure, when using the F-measure with varying $\beta$-parameters (horizontal axis) instead of the m-estimate for candidate selection. Best parameters for different data sets specified by red + signs. (Color figure online)

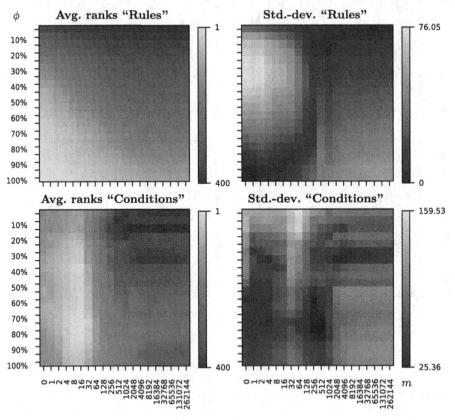

**Fig. 4.** Ranks and standard deviation of average ranks over all data sets regarding the number of rules and conditions. A smaller rank means more rules or conditions.

| $m = 16, \phi = 0.3$ | Mi. Precision = 74.07%, Mi. Recall = 78.26% |
|---|---|

$Cough \leftarrow$ "$cough$" $\wedge$ "$aldrich$" $\wedge$ "$opacity$" $\wedge$ "$tachypnea$" $\wedge$ "$streaky$" $\wedge$ "$side$" $\wedge$
     "$distal$" $\wedge$ "$diaphragm$"

$Cough \leftarrow$ "$cough$" $\wedge$ "$x - rays$" $\wedge$ "$vomiting$" $\wedge$ "$proximity$" $\wedge$ "$hematuria$" $\wedge$
     "$focal$"

$Cough \leftarrow$ "$cough$" $\wedge$ $\overline{\text{"}group\text{"}}$ $\wedge$ $\overline{\text{"}edema\text{"}}$ $\wedge$ $\overline{\text{"}fever\text{"}}$

$Cough \leftarrow$ "$cough$" $\wedge$ $\overline{\text{"}lobe\text{"}}$ $\wedge$ $\overline{\text{"}breathing\text{"}}$

$Cough \leftarrow$ "$coughing$"

| $m = 262144, \phi = 1.0$ | Mi. Precision = 65.61%, Mi. Recall = 89.57% |
|---|---|

$Cough \leftarrow$ "$cough$" $\wedge$ $\overline{\text{"}ureteral\text{"}}$ $\wedge$ "$stones$" $\wedge$ "$contrast$"

$Cough \leftarrow$ "$coughing$"

$Cough \leftarrow$ "$code$"

$Cough \leftarrow$ "$substance$"

**Fig. 5.** Exemplary rule sets predicting the label *786.2:Cough* of the data set MEDICAL, which contains textual radiology reports that were categorized into diseases.

**Table 1.** Predictive performance of Ripper using IREP and post-processing ($R_3$), without using post-processing ($R_2$), and using neither IREP nor post-processing ($R_1$) compared to approaches trying to optimize micro-averaged F1 ($M_F$), Hamming accuracy ($M_H$), and subset accuracy ($M_S$).

| | F1 | | | | Hamming acc. | | | | Subset acc. | | | |
|---|---|---|---|---|---|---|---|---|---|---|---|---|
| | $R_1$ | $R_2$ | $R_3$ | $M_F$ | $R_1$ | $R_2$ | $R_3$ | $M_H$ | $R_1$ | $R_2$ | $R_3$ | $M_S$ |
| BIRDS | 43.65 | 41.12 | **46.01** | 45.33 | 94.39 | 94.48 | **95.17** | 95.10 | 44.20 | 45.57 | **51.48** | 48.85 |
| CAL500 | 33.63 | 33.18 | 33.76 | **40.10** | 82.14 | 83.66 | 85.39 | **86.02** | 0.00 | 0.00 | 0.00 | 0.00 |
| EMOTIONS | 56.96 | 58.68 | 60.97 | **65.20** | 75.12 | 75.38 | 77.21 | **77.65** | 18.04 | 20.40 | **23.60** | 22.42 |
| ENRON | 50.57 | 53.05 | **55.33** | 51.07 | 94.35 | 94.70 | **94.93** | 94.54 | 6.17 | 7.99 | **9.16** | 7.81 |
| FLAGS | 71.81 | 72.96 | **74.85** | 72.83 | 73.02 | 74.08 | **75.20** | 73.39 | 15.47 | 17.05 | **21.00** | 9.82 |
| GENBASE | 98.83 | 98.68 | 98.68 | **99.14** | 99.89 | 99.88 | 99.88 | **99.92** | 97.28 | 96.83 | 96.83 | **97.89** |
| MEDICAL | 81.40 | 83.67 | **84.81** | 81.67 | 99.01 | 99.10 | **99.15** | 98.98 | 66.74 | 69.91 | **72.16** | 66.43 |
| SCENE | 63.97 | 63.25 | 64.55 | **67.44** | 87.87 | 87.25 | 88.03 | **88.93** | 46.61 | 44.54 | 46.24 | **49.73** |
| YEAST | 58.65 | 60.41 | 61.19 | **64.25** | 78.50 | 78.29 | 78.77 | **79.24** | 8.73 | 7.86 | 9.18 | **11.75** |
| **Avg. rank** | 3.44 | 3.00 | 1.67 | 1.78 | 3.44 | 2.89 | 1.67 | 1.89 | 2.89 | 2.67 | 1.56 | 2.11 |

implementation of Ripper [7], using the binary relevance method. By default, Ripper uses *incremental reduced error pruning* (IREP) and post-processes the induced rule set. Although our approach could make use of such optimizations, this is out of the scope of this work. For a fair comparison, we also report the results of JRip without using IREP ($P = false$) and/or with post-processing turned off ($O = 0$).

Note that we do not consider the random forests from which we generate rules (cf. Sect. 3.1) to be relevant baselines. This is, because random forests use voting for making a prediction, which is fundamentally different than rule learners that model a DNF. Also, we train random forests consisting of a very large number of trees with varying depths to generate diverse rules. In our experience, these random forests perform badly compared to commonly used configurations.

We tested three different configurations of our approach. The parameters $m$ and $\phi$ used by these approaches have been determined on a validation set by using nested 5-fold cross validation on the training data. For the approach $M_F$, the parameters have been chosen such that the F1-measure is maximized. $M_H$ and $M_S$ were tuned with respect to Hamming and subset accuracy, respectively.

According to Table 1, our method is able to achieve reasonable predictive performances. With respect to the measure they try to optimize, our approaches generally rank before JRip with optimizations turned off ($R_1$), which is the competitor that is conceptually closest to our method. Although IREP definitely has a positive effect on the predictive performance, our approaches also tend to outperform JRip with IREP enabled, but without using post-processing ($R_2$). Despite the absence of advanced pruning and post-processing techniques, our approaches are even able to surpass the fully fledged variant of JRip ($R_1$) on some data sets. We consider these results as a clear indication that it is indispensable to be able to flexibly adapt the heuristic used by a rule learner if one aims at deliberately optimizing a specific multi-label performance measure.

## 5 Related Work

Several rule-based approaches to multi-label classification have been proposed in the literature. On the one hand, there are methods based on descriptive rule learning, such as association rule discovery [17,18,22,23], genetic algorithms [1,6], or evolutionary classification systems [2,3]. On the other hand, there are algorithms that adopt the separate-and-conquer strategy used by many traditional rule learners for binary or multi-class classification, e.g. by Ripper [7], and transfer it to MLC [19,21]. Whereas in descriptive rule learning one does usually not aim at discovering rules that minimize a certain (multi-label) loss, the latter approaches employ a heuristic-guided search for rules that optimize a given rule learning heuristic and hence could benefit from the results of this work.

Similar to our experiments, empirical studies aimed at discovering optimal rule learning heuristics have been published in the realm of single-label classification [14,15]. Moreover, to investigate the properties of bipartition evaluation functions, ROC space isometrics have been proven to be a helpful tool [9,10]. They have successfully been used to study the effects of using different heuristics in separate-and-conquer algorithms [12], or for ranking and filtering rules [11].

## 6 Conclusions

In this work, we presented a first empirically study that thoroughly investigates the effects of using different rule learning heuristics for candidate selection and filtering in the context of multi-label classification. As commonly used multi-label measures, such as micro-averaged F1, Hamming accuracy, or subset accuracy, require to put more weight on the consistency of rules rather than on their coverage, models that perform well with respect to these measures are usually small and tend to contain specific rules. This is beneficial in terms of interpretability as less complex models are assumed to be easier to understand by humans.

As our main contribution, we emphasise the need to flexibly trade off the consistency and coverage of rules, e.g., by using parameterized heuristics like the m-estimate, depending on the multi-label measure that should be optimized by the model. Our study revealed that the choice of the heuristic is not straightforward, because selecting rules that minimize a certain loss functions locally does not necessarily result in that loss being optimized globally. E.g., selecting rules according to the F1-measure does not result in the overall F1 score to be maximized. For optimal results, the trade-off between consistency and coverage should be fine-tuned depending on the data set at hand. However, our results indicate that, even across different domains, the optimal settings for maximizing a measure can often be found in the same region of the parameter space.

In this work, we restricted our study to DNFs, i.e., models that consist of non-conflicting rules predicting the same outcome for a label. This restriction simplifies the implementation and comprehensibility of the learner, as no conflicts may arise at prediction time. However, we expect that including both, rules that model the presence as well as the absence of labels, could be beneficial in terms of robustness and could have similar, positive effects on the consistency of the models as the threshold selection used in this work. Furthermore, we leave the empirical analysis of macro-averaged performance measures for future work.

**Acknowledgments.** This research was supported by the German Research Foundation (DFG) (grant number FU 580/11).

# References

1. Allamanis, M., Tzima, F.A., Mitkas, P.A.: Effective rule-based multi-label classification with learning classifier systems. In: Tomassini, M., Antonioni, A., Daolio, F., Buesser, P. (eds.) ICANNGA 2013. LNCS, vol. 7824, pp. 466–476. Springer, Heidelberg (2013). https://doi.org/10.1007/978-3-642-37213-1_48
2. Arunadevi, J., Rajamani, V.: An evolutionary multi label classification using associative rule mining for spatial preferences. In: IJCA Special Issue on Artificial Intelligence Techniques-Novel Approaches and Practical Applications (2011)
3. Ávila-Jiménez, J.L., Gibaja, E., Ventura, S.: Evolving multi-label classification rules with gene expression programming: a preliminary study. In: Corchado, E., Graña Romay, M., Manhaes Savio, A. (eds.) HAIS 2010. LNCS (LNAI), vol. 6077, pp. 9–16. Springer, Heidelberg (2010). https://doi.org/10.1007/978-3-642-13803-4_2
4. Boutell, M.R., Luo, J., Shen, X., Brown, C.M.: Learning multi-label scene classification. Pattern Recogn. **37**(9), 1757–1771 (2004)
5. Breiman, L.: Random forests. Mach. Learn. **45**(1), 5–32 (2001)

6. Cano, A., Zafra, A., Gibaja, E.L., Ventura, S.: A grammar-guided genetic programming algorithm for multi-label classification. In: Krawiec, K., Moraglio, A., Hu, T., Etaner-Uyar, A.Ş., Hu, B. (eds.) EuroGP 2013. LNCS, vol. 7831, pp. 217–228. Springer, Heidelberg (2013). https://doi.org/10.1007/978-3-642-37207-0_19

7. Cohen, W.W.: Fast effective rule induction. In: Proceedings of the 12th International Conference on International Conference on Machine Learning (1995)

8. Diplaris, S., Tsoumakas, G., Mitkas, P.A., Vlahavas, I.: Protein classification with multiple algorithms. In: Bozanis, P., Houstis, E.N. (eds.) PCI 2005. LNCS, vol. 3746, pp. 448–456. Springer, Heidelberg (2005). https://doi.org/10.1007/11573036_42

9. Flach, P.A.: The geometry of ROC space: understanding machine learning metrics through ROC isometrics. In: Proceedings of the 20th International Conference on Machine Learning (2003)

10. Fürnkranz, J., Flach, P.A.: An analysis of rule evaluation metrics. In: Proceedings of the 20th International Conference on Machine Learning (2003)

11. Fürnkranz, J., Flach, P.: An analysis of stopping and filtering criteria for rule learning. In: Boulicaut, J.-F., Esposito, F., Giannotti, F., Pedreschi, D. (eds.) ECML 2004. LNCS (LNAI), vol. 3201, pp. 123–133. Springer, Heidelberg (2004). https://doi.org/10.1007/978-3-540-30115-8_14

12. Fürnkranz, J., Flach, P.A.: ROC 'n' rule learning-towards a better understanding of covering algorithms. Mach. Learn. **58**(1), 39–77 (2005)

13. Fürnkranz, J., Gamberger, D., Lavrač, N.: Foundations of Rule Learning. Springer, Heidelberg (2012). https://doi.org/10.1007/978-3-540-75197-7

14. Janssen, F., Fürnkranz, J.: An empirical investigation of the trade-off between consistency and coverage in rule learning heuristics. In: Jean-Fran, J.-F., Berthold, M.R., Horváth, T. (eds.) DS 2008. LNCS (LNAI), vol. 5255, pp. 40–51. Springer, Heidelberg (2008). https://doi.org/10.1007/978-3-540-88411-8_7

15. Janssen, F., Fürnkranz, J.: On the quest for optimal rule learning heuristics. Mach. Learn. **78**(3), 343–379 (2010)

16. Klimt, B., Yang, Y.: The enron corpus: a new dataset for email classification research. In: Boulicaut, J.-F., Esposito, F., Giannotti, F., Pedreschi, D. (eds.) ECML 2004. LNCS (LNAI), vol. 3201, pp. 217–226. Springer, Heidelberg (2004). https://doi.org/10.1007/978-3-540-30115-8_22

17. Lakkaraju, H., Bach, S.H., Leskovec, J.: Interpretable decision sets: a joint framework for description and prediction. In: Proceedings of the 22nd ACM SIGKDD International Conference on Knowledge Discovery and Data Mining (2016)

18. Li, B., Li, H., Wu, M., Li, P.: Multi-label classification based on association rules with application to scene classification. In: The 9th International Conference for Young Computer Scientists (2008)

19. Mencía, E.L., Janssen, F.: Learning rules for multi-label classification: a stacking and a separate-and-conquer approach. Mach. Learn. **105**(1), 77–216 (2016)

20. Pestian, J.P., et al.: A shared task involving multi-label classification of clinical free text. In: Proceedings of the Workshop on BioNLP 2007: Biological, Translational, and Clinical Language Processing (2007)

21. Rapp, M., Loza Mencía, E., Fürnkranz, J.: Exploiting anti-monotonicity of multi-label evaluation measures for inducing multi-label rules. In: Phung, D., Tseng, V.S., Webb, G.I., Ho, B., Ganji, M., Rashidi, L. (eds.) PAKDD 2018. LNCS (LNAI), vol. 10937, pp. 29–42. Springer, Cham (2018). https://doi.org/10.1007/978-3-319-93034-3_3

22. Thabtah, F.A., Cowling, P., Peng, Y.: MMAC: a new multi-class, multi-label associative classification approach. In: 4th IEEE International Conference on Data Mining (2004)

23. Thabtah, F.A., Cowling, P., Peng, Y.: Multiple labels associative classification. Knowl. Inf. Syst. **9**(1), 109–129 (2006)

24. Trohidis, K., Tsoumakas, G., Kalliris, G., Vlahavas, I.P.: Multi-label classification of music into emotions. In: International Society for Music Information Retrieval (2008)

25. Tsoumakas, G., Katakis, I., Vlahavas, I.: Mining multi-label data. In: Maimon, O., Rokach, L. (eds.) Data Mining and Knowledge Discovery Handbook. Springer, Boston (2009). https://doi.org/10.1007/978-0-387-09823-4_34

26. Turnbull, D., Barrington, L., Torres, D., Lanckriet, G.: Semantic annotation and retrieval of music and sound effects. IEEE Trans. Audio Speech Lang. Process. **16**(2), 467–476 (2008)

# Hyperparameter Importance
# for Image Classification
# by Residual Neural Networks

Abhinav Sharma[1]([⊠]), Jan N. van Rijn[1,2]([⊠]), Frank Hutter[3],
and Andreas Müller[1]

[1] Columbia University, New York City, USA
{as5414,acm2248}@columbia.edu
[2] Leiden University, Leiden, The Netherlands
j.n.van.rijn@liacs.leidenuniv.nl
[3] Albert-Ludwigs-Universität Freiburg, Freiburg, Germany
fh@cs.uni-freiburg.de

**Abstract.** Residual neural networks (ResNets) are among the state-of-the-art for image classification tasks. With the advent of automated machine learning (AutoML), automated hyperparameter optimization methods are by now routinely used for tuning various network types. However, in the thriving field of deep neural networks, this progress is not yet matched by equal progress on rigorous techniques that yield information beyond performance-optimizing hyperparameter settings. In this work, we aim to answer the following question: Given a residual neural network architecture, what are generally (across datasets) its most important hyperparameters? In order to answer this question, we assembled a benchmark suite containing 10 image classification datasets. For each of these datasets, we analyze which of the hyperparameters were most influential using the functional ANOVA framework. This experiment both confirmed expected patterns, and revealed new insights. With these experimental results, we aim to form a more rigorous basis for experimentation that leads to better insight towards what hyperparameters are important to make neural networks perform well.

**Keywords:** Hyperparameter importance · Residual neural networks

## 1 Introduction

Residual neural networks [10] are among the state-of-the-art for image classification tasks. Given sufficient data and proper hyperparameter settings, residual neural networks can achieve remarkable results, but their performance (and that of other neural networks) highly depends on their hyperparameter settings. As a consequence, there has been a lot of recent work and progress on hyperparameter optimization, with methods including Bayesian optimization [1,29], meta-learning [4] and bandit-based methods [18]; see [8] for a review.

Despite impressive results both on common benchmarks and various application domains, the experiments in many academic machine learning papers are

© Springer Nature Switzerland AG 2019
P. Kralj Novak et al. (Eds.): DS 2019, LNAI 11828, pp. 112–126, 2019.
https://doi.org/10.1007/978-3-030-33778-0_10

designed to answer *which* particular method works better, typically by introducing a new algorithm and demonstrating success over a limited set of baselines or benchmarks [31]. In a recent paper, Sculley et al. (2018) identify this as a problem: 'Empirical studies have become challenges to be won, rather than a process for developing insight and understanding' [27]. Additionally, many advances in deep learning have been evaluated on a small number of datasets. It has long been recognized that small-scale studies can create a false sense of progress [9]. Recht et al. (2018) speculate that by overly using the same test set, reported results tend to overfit and demonstrate that performance results of many introduced models does not generalize to other (newly assembled) test sets [24].

In this work, we aim to provide a more rigorous approach to the following question: Given a residual neural network architecture, what are generally (across datasets) its most important hyperparameters? In order to answer this question, we assembled an image classification benchmark suite consisting of 10 popular datasets from the literature. On each of these datasets we obtained performance results with varying hyperparameter settings. Although the aim of this paper is not to improve predictive performance, we compare the results with state-of-the-art results reported by other researchers, to ensure that the results are credible and applicable. We see this as a first step towards creating more rigorous insights about the conditions under which residual neural networks perform well and which hyperparameters influence this.

Our contributions are the following: (i) We assembled a benchmark suite of 10 well-known image classification datasets, allowing researchers to draw conclusions across datasets. We made all code, data and results publicly available;[1] (ii) we apply functional ANOVA [30] on performance results of residual neural networks, to identify the importance of the various hyperparameters to predictive accuracy; and (iii) we verified expected behaviour regarding hyperparameter interactions, and gained new insights regarding typical marginal curves and hyperparameter interactions. Most notable is the observation that for the concerning datasets the marginals of important hyperparameters exhibit very similar landscapes. Overall, this work is the first to provide large-scale quantitative evidence for which hyperparameters of residual neural networks are important, providing a better scientific basis for the field than previous knowledge based on small-scale studies and intuition.

## 2   Related Work

In this section we review related work on residual neural networks, hyperparameter importance and landscape analysis.

**Residual Neural Networks.** Deep residual neural networks were introduced in [10] and have set the benchmark for image recognition tasks in recent years. They provide good predictive accuracy while maintaining an affordable model size. Their defining characteristic is the use of *residual learning*, in which deeper layers of the network are linked to shallower layers directly using 'shortcut connections' skipping several layers in between. These shortcut connections perform

---

[1] https://www.github.com/janvanrijn/openml-pimp

an identity mapping which ensures the convergence of the deep network is at least as good as its shallower counterpart and hence limit divergence during training. In this way, the residual learning framework eases the training of networks that are substantially deeper. Furthermore, empirical evidence suggests that these residual neural networks are easier to optimize, and can gain accuracy from considerably increased depth. On the ImageNet dataset the residual nets were evaluated with a depth of up to 152 layers – 8 times deeper than VGG nets [28] but still having lower complexity. Furthermore, residual learning can be used on networks of varying depth to fit the task at hand. Smaller residual neural networks, like 'ResNet18' (as the name suggests, consisting of 18 layers), provide great performance while being very efficient in terms of size and speed [10].

**Hyperparameter Importance.** When using a new algorithm on a given dataset, it is typically a priori unknown which hyperparameters should be tuned, what are the good ranges for these hyperparameters to sample from, and which values in these ranges are most likely to yield high performance. Various techniques exist that allow for the assessment of hyperparameter importance. These techniques generally consider either local importance (dependent on a specific setting for other hyperparameters) or global importance (independent of specific hyperparameter settings).

*Forward selection* [12] is based on the assumption that important attributes in a dataset have high impact on the performance of classifiers trained on it. It trains a model which predicts the performance of a configuration based on a subset of hyperparameters. This set is initialized empty and greedily filled with the next most important hyperparameter. *Ablation analysis* [2] requires a default setting and an optimized setting and calculates a so-called ablation trace, which embodies how much the hyperparameters contributed towards the difference in performance between the two settings. *Local Parameter Importance* [3] studies the performance changes of a configuration along each parameter using an *empirical performance model* (sometimes also called a 'surrogate' model). *Functional ANOVA* [30] is a global hyperparameter importance framework that can detect the importance of both individual hyperparameters and interaction effects between arbitrary subsets of hyperparameters. It is the key technique upon which this research is built.

Functional ANOVA depends on the concept of the *marginal* of a hyperparameter, i.e., how a given value for a hyperparameter performs, averaged over all possible combinations of the other hyperparameters' values. While there are an exponential number of combinations, the authors of [13] showed how this can be calculated efficiently using tree-based surrogate models.

All the aforementioned techniques are post-hoc techniques, i.e., when confronted with a new dataset, these do not reveal what hyperparameters are important prior to experimenting on that particular dataset. Contrary, various researchers argued that it is more useful to generalize the notion of hyperparameter importance across datasets [22,25,26]. In particular, it has been shown how to apply functional ANOVA across datasets for a given algorithm [25,26]. These works build upon the assumption that if this hyperparameter importance quantification method is applied on a large enough set of datasets, we can draw conclusions regarding which hyperparameters are generally important. However, neither of these studies applied this methodology to convolutional neural networks. To the best of our knowledge, this is the first work that addresses hyperparameter importance for residual neural networks.

**Landscape Analysis.** The interaction between configurations and the respective results can be seen as an high-dimensional landscape, which in turn can be analyzed for mathematical properties [23]. Although this particular study is executed on 'satisfiability', 'mixed integer programming' and 'traveling salesman problems' benchmarks, it shows evidence that configuration landscapes are often uni-model and even convex.

# 3    Background and Methods

We follow the notation that was introduced in [13]. We assume that a given residual neural network model has $n$ hyperparameters with domains $\Theta_1, \dots, \Theta_n$ and *configuration space* $\boldsymbol{\Theta} = \Theta_1 \times \dots \times \Theta_n$. Let $N = \{1, \dots, n\}$ be the set of all hyperparameters of the classifier. An instantiation (or configuration) of the classifier is a vector $\boldsymbol{\theta} = \langle \theta_1, \dots, \theta_n \rangle$ with $\theta_i \in \Theta_i$. A partial instantiation is a vector $\boldsymbol{\theta}_U = \langle \theta_i, \dots, \theta_j \rangle$ with a subset $U \subseteq N$ of the hyperparameters fixed, and the values for other hyperparameters unspecified. The *marginal* $\hat{a}_U^p(\boldsymbol{\theta}_U)$ is defined as the average performance on measure $p$ of all complete instantiations $\boldsymbol{\theta}$ that agree with $\boldsymbol{\theta}_U$ in the instantiations of hyperparameters $U$. The variance of $\hat{a}_U^p(\boldsymbol{\theta}_U)$ is denoted as $\mathbb{V}_U^p$. Intuitively, if the marginal $\hat{a}_U^p(\boldsymbol{\theta}_U)$ has a high variance, this means that hyperparameter was of high importance to performance measure $p$, and vice versa. For a more complete description, the reader is referred to [13].

In this research, we address the following problem. Given (i) a residual neural network architecture with configuration space $\boldsymbol{\Theta}$, (ii) a set of datasets $\mathcal{D}^{(1)}, \dots, \mathcal{D}^{(M)}$, with $M$ being the number of datasets, and (iii) for each of the datasets, a set of empirical performance measurements $\langle \boldsymbol{\theta}_i, Y_i \rangle_{i=1}^K$ for different hyperparameter settings $\boldsymbol{\theta}_i \in \boldsymbol{\Theta}$, where $Y_i$ is a tuple of all relevant performance measures (in this case, predictive accuracy), we aim to determine which hyperparameters affect the algorithm's empirical performance most, and which values are likely to yield good performance.

For a given dataset, we use the performance data $\langle \boldsymbol{\theta}_i, Y_i \rangle_{i=1}^K$ collected on this dataset to fit an internal tree-based surrogate model, in this case, a random forest with 16 trees. Functional ANOVA then uses this surrogate model to calculate the variance contribution $\mathbb{V}_j^p / \mathbb{V}^p$ of every hyperparameter $j \in N$, with high values indicating high importance. We then study the distribution of these variance contributions across datasets to obtain empirical data regarding which hyperparameters tend to be most important.

It is possible that a hyperparameter is responsible for a high variance on many datasets, but that its best value is the same across all of them. We note that functional ANOVA will flag such hyperparameters as important, although it could be argued that they have appropriate defaults and do not need to be tuned [22, 26]. For example, it is reasonable to expect that for any type of neural network the marginal of the number of epochs has a high variance, where obviously better performances are achieved for higher values (at the cost of additional run-time). For this reason, it is always important to consider the individual marginals, as well as the generalizations across datasets.

## 4 Experimental Setup

Section 4.1 describes the training procedure of the residual neural network, Sect. 4.2 the configuration space from which we sampled the various configurations, and Sect. 4.3 the datasets that we included in this study.

### 4.1 Models

In this work, we focus on the fixed architecture of 'ResNet18'. This model gives good predictive accuracy for datasets while being small in size which allows for faster training [10]. As the datasets in this research all contain images, with relatively similar dimensions, we could use the same architecture for all of them (see Sect. 4.3). The optimizer is fixed to Stochastic Gradient Descent (SGD), parameterized by momentum and weight decay. The training starts with an initial learning rate. Thereafter an adaptive learning rate scheduler is used which decays the learning rate by a factor (hyperparameter: learning rate decay) when the test accuracy plateaus for a given number (hyperparameter: patience) of epochs. The details of the hyperparameter space are described in Sect. 4.2.

We record the time taken (in seconds) and the accuracy on the test set after every epoch. The goal is not to identify an optimum parameter setting, as using a test-set simply computing the maximum would result in overly optimistic evaluation. If one would be interested in using the results for hyperparameter optimization a proper nested cross-validation procedure should be applied [5]. We performed all runs on single NVIDIA P100 GPU.

### 4.2 Configuration Space

We selected twelve hyperparameters. This selection was made based on visual inspection of the modules in the 'Torch' package, as well as personal experience. Even though it feels natural to fix the values for some hyperparameters to seemingly good values, in this work we aim to verify the applicability of these values.

Our hyperparameter space contains six hyperparameters for the SGD optimizer (number of epochs, initial learning rate, learning rate decay, momentum, batch size, and whether to shuffle the data), two early stopping hyperparameters (tolerance and patience), and four regularization hyperparameters (weight decay, and data augmentation by resize crop, horizontal flips, and vertical flips). Note that since we keep a fixed network structure, we do not modify any architectural hyperparameters (this would be very interesting, but is out of scope for the current study).

**Table 1.** Overview of the hyperparameters used in this research.

| Hyperparameter | Range | Description |
|---|---|---|
| Batch size | $2^{\{3,4,5,6,7,8,9\}}$ | Number of samples in one batch of gradient descent used during training |
| Epochs | [1–200] | The number of times each training observation is passed to the network |
| Horizontal flip | Boolean | Whether to apply data augmentation by flipping the image horizontally |
| Vertical flip | Boolean | Whether to apply data augmentation by flipping the image vertically |
| Learning rate | $[10^{-6}$–1] (log) | The learning rate with which the network starts training |
| Learning rate decay | [2–1000] (log) | Factor to reduce the learning rate with, if no improvement is obtained after several epochs |
| Momentum | [0–1] | Value of momentum multiplier used during gradient descent |
| Patience | [2–200] | Number of epochs without improvements that are being tolerated before learning rate is reduced |
| Shuffle | Boolean | Whether to shuffle the train set before an epoch |
| Resize crop | Boolean | Whether to apply data augmentation by resizing and then cropping the image |
| Tolerance | $[10^{-5}$–$10^{-2}]$ (log) | Tolerance for early stopping criterion |
| Weight decay | $[10^{-6}$–$10^{-2}]$ (log) | L2 loss on the weights |

We note that some of the hyperparameters we tune are sometimes rather chosen manually on a per-dataset basis based on domain knowledge (e.g., certain data augmentations don't make sense for some types of images, and the batch size is often set to the maximum feasible given the GPU's memory). We still included these in our study to study how large their impact is on performance.

Table 1 lists all the hyperparameters and their maximal ranges we considered. In order to obtain reasonable performance for datasets of different input sizes, we had to use slightly different hyperparameter spaces across datasets; in particular, the datasets with large input size (Fruits 360, Flower, STL-10 and Dog vs. Cat) would have led to memory issues with batch sizes of 256 or 512, and we therefore only considered batch size values of $2^{\{3,4,5,6,7\}}$ in those cases and of $2^{\{5,6,7,8,9\}}$ in all other cases. Also, owing to limited computational resources, some other manual modifications were made to speed up experiments and focus on a region of good hyperparameters: for datasets MNIST, Fashion MNIST and Fruits, the maximum number of epochs was set to 50, for datasets Dog vs. Cat the maximum number of epochs was set to 100, and for datasets Fruits 360, Flower, Dog vs. Cat, MNIST and Fashion MNIST, the maximum learning rate was set to 0.1.

For each dataset, we sampled $K = 200$ configurations uniformly from this configuration space, with the maximum number of epochs. For each run we stored performance results after every epoch, allowing functional ANOVA to model the marginal of this hyperparameter more accurately.

**Table 2.** Overview of the datasets used in this research.

| Name | Description | Dimensions | Class | Train | Test | Ref |
|------|-------------|------------|-------|-------|------|-----|
| MNIST | Handwritten digits | $28 \times 28$ | 10 | 60,000 | 10,000 | [17] |
| Fashion MNIST | Gray-scale objects | $28 \times 28$ | 10 | 60,000 | 10,000 | [34] |
| CIFAR-10 | Colored objects | $32 \times 32 \times 3$ | 10 | 50,000 | 10,000 | [16] |
| CIFAR-100 | Colored objects | $32 \times 32 \times 3$ | 100 | 50,000 | 10,000 | [16] |
| STL-10 | Colored objects | $96 \times 96 \times 3$ | 10 | 5,000 | 8,000 | [6] |
| HAM10000 | Skin cancer images | $28 \times 28 \times 3$ | 7 | 9,013 | 1,002 | [32] |
| SVHN | House number images | $32 \times 32 \times 3$ | 10 | 73,257 | 26,032 | [21] |
| Flower | Flower images | $96 \times 96 \times 3$ | 5 | 3,888 | 435 | [19] |
| Fruits 360 | Fruit images | $96 \times 96 \times 3$ | 82 | 41,814 | 14,041 | [20] |
| Dog vs. Cats | Dog and cat images | $96 \times 96 \times 3$ | 2 | 22,500 | 2,500 | [15] |

When computing hyperparameter importances across different datasets, the question arises how to treat differing hyperparameter spaces. For the important learning rate hyperparameter we felt it to be important to use identical ranges everywhere and therefore used a reduced range of $[10^{-6}, 0.1]$. However, for the batch size hyperparameter, no single range makes sense for all datasets, and we therefore simply computed hyperparameter importance separately based on the range used for the dataset at hand. Likewise, for the maximum number of epochs, we used 200 throughout; this is justified by the assumption that the internal random forest model would correctly model the plateaued performance.

## 4.3  Datasets

This section reviews the datasets that were used in this research. We assembled a diverse set of image classification datasets, including often used datasets (e.g., MNIST and CIFAR-100). We excluded the common benchmark ImageNet to keep the computational costs reasonable.

All our datasets, listed in Table 2, are classification tasks (the respective task is briefly described in column 'Description'). For example, in the MNIST dataset the task is to classify hand-written digits, whereas for the CIFAR-10 dataset the task is to identify images into 10 classes (airplane, automobile, bird, cat, deer, dog, frog, horse, ship or truck). Column 'Dimensions' represents the size and number of channels of the training images. Black and white or gray-scale datasets have two dimensions (width and height), whereas colored datasets have three dimensions (width, height and number of color channels; in this study the number of color channels is always 3). Column 'Class.' represents the number of classes, column 'Train' the number of train observations and column 'Test' the number of test observations. Finally, column 'Ref.' contains a reference to the publication where the dataset was introduced.

As the dimensions of these datasets are all approximately the same, we could use the same architecture for all of them. There are minor modifications to the

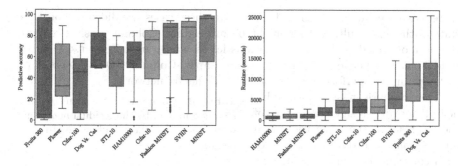

**Fig. 1.** Performance results of the various configurations per dataset, sorted by median performance. Complementary, Table 3 shows the best obtained result per dataset.

input and output layers due to different input dimensions and output classes of each dataset. For datasets with gray-scale images (i.e., MNIST and Fashion MNIST) the pixel values are duplicated over three dimensions during pre-processing. Whether data augmentation techniques like random crops and random flips were performed is controlled by the respective hyperparameter.

## 5   Results

In this section we analyze the results of the experiments. In Sect. 5.1 we discuss some basic performance characteristics compared to state-of-the-art algorithms. In Sects. 5.2 and 5.3 we discuss the main contribution of this work, the importance of hyperparameters according to functional ANOVA. Finally, Sect. 5.4 discusses limitations that could inspire future work.

### 5.1   Performance Results

We explore some basic characteristics about the performance results obtained on the datasets. As mentioned before, obtaining state-of-the-art performance is neither the aim nor the contribution of this paper, but in order for the results to be credible and applicable, it is important to verify that the results are in the same ballpark as good results reported in literature. Figure 1 shows the predictive accuracy (left) and run-time (right) of all hyperparameter configurations $\theta$ grouped per dataset in a box-plot. Both plots include only the measured performance (run-time or accuracy) after the final epoch, and thus not the recorded intermediate results.

We made a best effort to find established state-of-the-art results for existing datasets. Table 3 compares the best results of the conducted experiments with the best found result in literature. We obtained the results of state-of-the-art methods from public sources on the internet[2,3]. For some lesser known datasets,

---

[2]   https://benchmarks.ai/

[3]   https://github.com/RedditSota/state-of-the-art-result-for-machine-learning-problems

there was no established state-of-the-art. In these cases, we did not report any state-of-the-art result, as that might be misleading. Column 'ResNet' denotes the best obtained performance (optimistic, as was argued in Sect. 4.1) of the residual neural network through random search. Column 'SOTA' denotes the (also optimistic) performance of the state-of-the-art network.

The comparison between residual neural network results and state-of-the-art results contains various conditions that need to be accounted for. As a consequence, this comparison is somewhat biased. However, it serves the purpose of providing context to the obtained results. In some cases the best results obtained by the trained residual neural networks are close to the best reported results in literature (e.g., for MNIST and Fashion MNIST), while for others the differences are bigger (e.g., CIFAR-100 and STL-10). Overall, the performance results are good enough to expect that some conclusions drawn may carry over to state-of-the-art models.

**Table 3.** Comparison between the best results obtained by the residual neural networks in this study and state-of-the-art results.

| Dataset | ResNet | SOTA | Source |
|---|---|---|---|
| MNIST | 99.62 | 99.79 | [33] |
| Fashion MNIST | 94.18 | 96.35 | [35] |
| CIFAR-10 | 93.29 | 99.00 | [11] |
| CIFAR-100 | 72.66 | 91.30 | [11] |
| STL-10 | 79.91 | 88.80 | [14] |
| HAM10000 | 82.83 | - | |
| SVHN | 96.66 | 98.98 | [7] |
| Flower | 89.20 | - | |
| Fruits 360 | 99.38 | - | |
| Dog vs. Cats | 96.52 | - | |

### 5.2   Marginals per Dataset

This section details the results of the hyperparameter importance experiment. Figure 2 shows the predictive accuracy marginals of important (combinations) of hyperparameters, per dataset.

For each of the 10 datasets, we plotted the marginal of several important hyperparameters and one pair of hyperparameters. From left to right this image displays the marginals of the 'number of epochs', the 'initial learning rate', 'weight decay', 'momentum' and the combination of 'number of epochs and initial learning rate'. Note that we calculated the marginals for all 12 hyperparameters and all 65 hyperparameter pairs, but only show this subset due to space reasons. The $x$-axis shows the value of the hyperparameter and the $y$-axis shows the marginal performance (predictive accuracy). The epochs marginals are most detailed, presumably because of the recorded performance after every epoch.

The marginals reveal several patterns about the interaction between hyperparameter values and final performance of the network. Firstly, as the number of epochs increases, so does the performance of the network. Although this is quite obvious, it is important to verify that the proposed methodology can discover known and expected behaviors. Secondly, the marginals for the initial learning rate reveal that there is no perfect default across datasets, but that values between $10^{-3}$ and $10^{-2}$ generally perform quite well, whereas setting it much

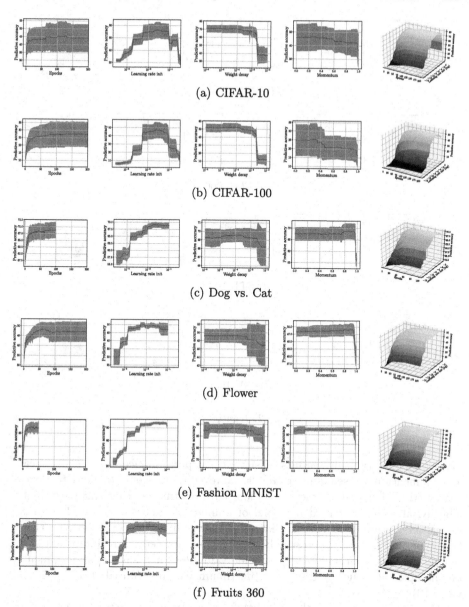

(a) CIFAR-10

(b) CIFAR-100

(c) Dog vs. Cat

(d) Flower

(e) Fashion MNIST

(f) Fruits 360

**Fig. 2.** Marginal plots for predictive accuracy on the test set per dataset. The first four columns show the marginal of a single hyperparameter. The blue line represents the marginal, the red area represents the standard deviation. The fifth column shows the combined marginal of two hyperparameters (note that the ranges per dataset differ). (Color figure online)

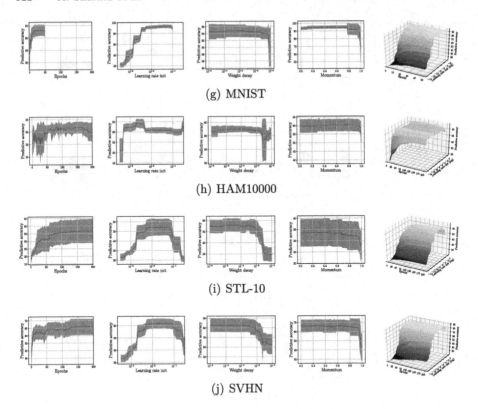

(g) MNIST

(h) HAM10000

(i) STL-10

(j) SVHN

**Fig. 2.** *(continued)*

lower or higher typically results in suboptimal performance. For weight decay and momentum we see similar trends, however there seems to be a tendency that setting their values too low is less harmful than setting them too high. Thirdly, also the combined marginal of number of epochs and initial learning rate is interesting, as it reveals that there is very little interplay between these two hyperparameters. Interestingly, both hyperparameters are important, but setting one hyperparameter to a specific value does not have a large influence on the optimal value for the other (maximum variance contribution: 0.026 on Cifar-10).

Most interestingly, we observe that for each hyperparameter the marginals follow similar trends across the datasets. Although the marginals exhibit a rough and edgy pattern, based on visual inspection we conclude that after some smoothing the landscapes would be uni-modal and convex. Even though the methodology and application domain are slightly different, these results seem to be in line with earlier reported findings [23].

## 5.3   Importance Across Datasets

Figure 3(a) shows box-plots for the variance per hyperparameter, presented similarly to [26]. For each partial configuration $\boldsymbol{\theta}_U$ with $U = 1$ and the three most important partial configurations with $U = 2$, we record variance of the marginal $V_U^p$ per dataset, and present these across datasets in box-plots. We observe various expected results. Hyperparameters related to the optimizer seem generally most important, i.e., 'weight decay', 'momentum', and 'learning rate init'. The data augmentation hyperparameters are among the least important hyperparameters. We note that functional ANOVA is meant as a tool for assessing global hyperparameter importance; data augmentation techniques are generally used to be applied on already good performing models, in order to further improve the performance. As such, the utility of data augmentation techniques might not be detected by functional ANOVA but can most likely be measured with local hyperparameter importance tools, such as Ablation Analysis [2].

Furthermore, we observe that the number of epochs, a hyperparameter which we expected to be important, ranks only 5th when analyzing the marginals. We note that the variance of the marginal (upon which functional ANOVA is built) is highly dependent on the selected ranges. To alleviate this problem, Fig. 3(b) shows the results in an alternative way. For each partial configuration $\boldsymbol{\theta}_U$ with $U = 1$, we record the maximum of marginal (i.e., $max(\hat{a}_U^p(\boldsymbol{\theta}_U))$) and the minimum of the marginal (i.e., $min(\hat{a}_U^p(\boldsymbol{\theta}_U))$) per dataset, and present the difference between these across datasets in box-plots. We observe that this plot confirms the importance of the epochs hyperparameter, making it the second most important hyperparameter after the learning rate. Also note that the other hyperparameters with the highest median variances according to Fig. 3(a) are still ranked as important in Fig. 3(b), albeit in a slightly changed order.

Finally, based on Fig. 3(a) we note that most of the variance can be explained by the effect of single hyperparameters. Apart from the variance contribution of single hyperparameters ($U = 1$) it shows the 3 most important combinations of hyperparameters ($U = 2$). The variance contribution of combined

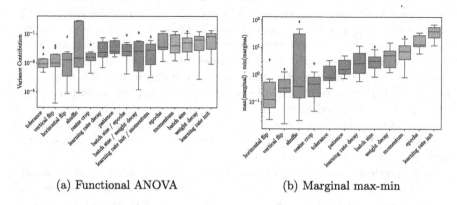

(a) Functional ANOVA                  (b) Marginal max-min

**Fig. 3.** Importance per (combination of) hyperparameters across datasets

hyperparameters seems rather low, as none are ranked highly compared to the variance contribution of single hyperparameters. However, like the data augmentation techniques, we speculate that even though the combined effect is relatively small, it will still be important to consider when optimizing for performance.

### 5.4 Limitations

Looking at the results in Fig. 3, the following result stands out. The shuffle hyperparameter value has a rather low median but a very high tail. This indicates that for most datasets the marginal is not particularly affected by this hyperparameter, however for some datasets (i.e., Flower, Fruits 360, Dog vs. Cat and HAM10000) it seems extremely important.

Furthermore, Fig. 1 reveals that the median performance is quite low, despite the decent maximal performance. This is confirmed by the marginals in Fig. 2. For example, none of the marginals for CIFAR-10 exceed the 80% accuracy threshold, whereas the best found configuration obtained an accuracy of 93.29% (according to Table 3). This gives rise to the question whether a hyperparameter tool like functional ANOVA can still reveal hyperparameters that are important for fine-tuning models (such as data augmentation), or whether only global trends are detected.

Finally, functional ANOVA highly relies on a proper configuration space. A seemingly important hyperparameter like 'number of epochs' will account for a relative low variance if the range is selected in such a way that it exceeds the values for which the performance reaches the plateau. It is currently an open question how to construct the configuration space to avoid this problem.

## 6   Conclusions

This work was motivated by the call for more rigor in hyperparameter optimization and neural network research [24,27]. We assembled a benchmark suite with corresponding performance results of residual neural networks, and made it publicly available. Our hyperparameter importance experiment confirmed existing beliefs about which hyperparameters are most influential across datasets, i.e., the initial learning rate and the number of epochs. Other important hyperparameters are the weight decay and momentum. Most of the other hyperparameters did not have a large variance of the marginal, however we note that in many image classification benchmarks the devil is in the detail. In order to go from a reasonable performance to state-of-the-art performance, also hyperparameters with a small effect should be set to adequate values.

We confirmed some well expected patterns, for example the form of the marginals for the number of epochs and the volatility across datasets of the marginals for the initial learning rate.

We acknowledge that this is only a first step towards more rigorous results in neural network research. While this research focused specifically on residual neural networks with a fixed architecture, future work should focus on other network types and also important parameters in architecture search.

# References

1. Bergstra, J., Bardenet, R., Bengio, Y., Kégl, B.: Algorithms for hyper-parameter optimization. In: Advances in Neural Information Processing Systems, vol. 24, pp. 2546–2554. Curran Associates, Inc. (2011)
2. Biedenkapp, A., Lindauer, M., Eggensperger, K., Fawcett, C., Hoos, H.H., Hutter, F.: Efficient parameter importance analysis via ablation with surrogates. In: Proceedings of the Thirty-First AAAI Conference on Artificial Intelligence, pp. 773–779. AAAI Press (2017)
3. Biedenkapp, A., Marben, J., Lindauer, M., Hutter, F.: CAVE: configuration assessment, visualization and evaluation. In: Battiti, R., Brunato, M., Kotsireas, I., Pardalos, P.M. (eds.) LION 12 2018. LNCS, vol. 11353, pp. 115–130. Springer, Cham (2019). https://doi.org/10.1007/978-3-030-05348-2_10
4. Brazdil, P., Giraud-Carrier, C., Soares, C., Vilalta, R.: Metalearning. Applications to Data Mining, 1st edn. Springer, Heidelberg (2008). https://doi.org/10.1007/978-3-540-73263-1
5. Cawley, G.C., Talbot, N.L.: On over-fitting in model selection and subsequent selection bias in performance evaluation. J. Mach. Learn. Res. **11**, 2079–2107 (2010)
6. Coates, A., Ng, A., Lee, H.: An analysis of single-layer networks in unsupervised feature learning. In: Proceedings of the Fourteenth International Conference on Artificial Intelligence and Statistics, vol. 15, pp. 215–223. PMLR (2011)
7. Cubuk, E.D., Zoph, B., Mane, D., Vasudevan, V., Le, Q.V.: AutoAugment: learning augmentation strategies from data. In: The IEEE Conference on Computer Vision and Pattern Recognition (CVPR) (2019)
8. Feurer, M., Hutter, F.: Hyperparameter optimization. In: Hutter, F., Kotthoff, L., Vanschoren, J. (eds.) Automated Machine Learning: Methods, Systems, Challenges. TSSCML, pp. 3–33. Springer, Cham (2019). https://doi.org/10.1007/978-3-030-05318-5_1
9. Hand, D.J.: Classifier technology and the illusion of progress. Stat. Sci. **21**(1), 1–14 (2006)
10. He, K., Zhang, X., Ren, S., Sun, J.: Deep residual learning for image recognition. In: 2016 IEEE Conference on Computer Vision and Pattern Recognition (CVPR), pp. 770–778 (2016)
11. Huang, Y., et al.: GPipe: efficient training of giant neural networks using pipeline parallelism. arXiv preprint arXiv:1811.06965 (2018)
12. Hutter, F., Hoos, H.H., Leyton-Brown, K.: Identifying key algorithm parameters and instance features using forward selection. In: Nicosia, G., Pardalos, P. (eds.) LION 2013. LNCS, vol. 7997, pp. 364–381. Springer, Heidelberg (2013). https://doi.org/10.1007/978-3-642-44973-4_40
13. Hutter, F., Hoos, H., Leyton-Brown, K.: An efficient approach for assessing hyperparameter importance. In: Proceedings of the 31st International Conference on Machine Learning, vol. 32, pp. 754–762. PMLR (2014)
14. Ji, X., Henriques, J.F., Vedaldi, A.: Invariant information clustering for unsupervised image classification and segmentation. arXiv preprint arXiv:1807.06653 (2018)
15. Kaggle: Dogs vs. Cats Redux: Kernels Edition (2016). https://www.kaggle.com/c/dogs-vs-cats-redux-kernels-edition. Accessed December 2018
16. Krizhevsky, A.: Learning multiple layers of features from tiny images. Technical report, University of Toronto (2009)

17. LeCun, Y.: The MNIST database of handwritten digits (1998). http://yann.lecún. com/exdb/mnist/. Accessed December 2018
18. Li, L., Jamieson, K.G., DeSalvo, G., Rostamizadeh, A., Talwalkar, A.: Hyperband: bandit-based configuration evaluation for hyperparameter optimization. In: 5th International Conference on Learning Representations, ICLR 2017. OpenReview.net (2017)
19. Mamaev, A.: Flowers Recognition (version 2). https://www.kaggle.com/alxma maev/flowers-recognition. Accessed December 2018
20. Mureşan, H., Oltean, M.: Fruit recognition from images using deep learning. Acta Universitatis Sapientiae, Informatica 10(1), 26–42 (2018)
21. Netzer, Y., Wang, T., Coates, A., Bissacco, A., Wu, B., Ng, A.Y.: Reading digits in natural images with unsupervised feature learning. In: NIPS Workshop on Deep Learning and Unsupervised Feature Learning (2011)
22. Probst, P., Boulesteix, A.L., Bischl, B.: Tunability: importance of hyperparameters of machine learning algorithms. J. Mach. Learn. Res. 20(53), 1–32 (2019)
23. Pushak, Y., Hoos, H.: Algorithm configuration landscapes: more benign than expected? In: Auger, A., Fonseca, C.M., Lourenço, N., Machado, P., Paquete, L., Whitley, D. (eds.) PPSN 2018. LNCS, vol. 11102, pp. 271–283. Springer, Cham (2018). https://doi.org/10.1007/978-3-319-99259-4_22
24. Recht, B., Roelofs, R., Schmidt, L., Shankar, V.: Do CIFAR-10 Classifiers Generalize to CIFAR-10? arXiv preprint arXiv:1806.00451 (2018)
25. van Rijn, J.N., Hutter, F.: An empirical study of hyperparameter importance across datasets. In: AutoML@ PKDD/ECML, pp. 91–98 (2017)
26. van Rijn, J.N., Hutter, F.: Hyperparameter importance across datasets. In: Proceedings of the 24th ACM SIGKDD International Conference on Knowledge Discovery and Data Mining, pp. 2367–2376. ACM (2018)
27. Sculley, D., Snoek, J., Wiltschko, A., Rahimi, A.: Winner's curse? on pace, progress, and empirical rigor. In: Proceedings of ICLR 2018 (2018)
28. Simonyan, K., Zisserman, A.: Very deep convolutional networks for large-scale image recognition. arXiv preprint arXiv:1409.1556 (2014)
29. Snoek, J., Larochelle, H., Adams, R.P.: Practical Bayesian optimization of machine learning algorithms. In: Advances in Neural Information Processing Systems, vol. 25, pp. 2951–2959. ACM (2012)
30. Sobol, I.M.: Sensitivity estimates for nonlinear mathematical models. Math. Model. Comput. Exp. 1(4), 407–414 (1993)
31. Strang, B., Putten, P., Rijn, J.N., Hutter, F.: Don't rule out simple models prematurely: a large scale benchmark comparing linear and non-linear classifiers in OpenML. In: Duivesteijn, W., Siebes, A., Ukkonen, A. (eds.) IDA 2018. LNCS, vol. 11191, pp. 303–315. Springer, Cham (2018). https://doi.org/10.1007/978-3-030-01768-2_25
32. Tschandl, P., Rosendahl, C., Kittler, H.: The HAM10000 dataset, a large collection of multi-source dermatoscopic images of common pigmented skin lesions. Sci. Data (2018)
33. Wan, L., Zeiler, M., Zhang, S., Le Cun, Y., Fergus, R.: Regularization of neural networks using DropConnect. In: International Conference on Machine Learning, pp. 1058–1066 (2013)
34. Xiao, H., Rasul, K., Vollgraf, R.: Fashion-MNIST: a novel image dataset for benchmarking machine learning algorithms. arXiv preprint arXiv:1708.07747 (2017)
35. Zhong, Z., Zheng, L., Kang, G., Li, S., Yang, Y.: Random erasing data augmentation. arXiv preprint arXiv:1708.04896 (2017)

# Applications

# Cellular Traffic Prediction and Classification: A Comparative Evaluation of LSTM and ARIMA

Amin Azari[1(✉)], Panagiotis Papapetrou[1], Stojan Denic[2], and Gunnar Peters[2]

[1] Department of Computer and Systems Sciences, Stockholm University,
Stockholm, Sweden
{amin.azari,panagiotis}@dsv.su.se
[2] Huawei, Stockholm, Sweden
{stojan.denic,gunnar.peters}@huawei.com

**Abstract.** Prediction of user traffic in cellular networks has attracted profound attention for improving the reliability and efficiency of network resource utilization. In this paper, we study the problem of cellular network traffic prediction and classification by employing standard machine learning and statistical learning time series prediction methods, including long short-term memory (LSTM) and autoregressive integrated moving average (ARIMA), respectively. We present an extensive experimental evaluation of the designed tools over a real network traffic dataset. Within this analysis, we explore the impact of different parameters on the effectiveness of the predictions. We further extend our analysis to the problem of network traffic classification and prediction of traffic bursts. The results, on the one hand, demonstrate the superior performance of LSTM over ARIMA in general, especially when the length of the training dataset is large enough and its granularity is fine enough. On the other hand, the results shed light onto the circumstances in which, ARIMA performs close to the optimal with lower complexity.

**Keywords:** Statistical learning · Machine learning · LSTM · ARIMA · Cellular traffic · Predictive network management

## 1 Introduction

A major driver for the beyond fifth-generation (5G) wireless networks consists in offering the wide set of cellular services in an energy and cost-efficient way [22]. Toward this end, the legacy design approach, in which resource provisioning and operation control are performed based on the peak traffic scenarios, are substituted with predictive analysis of mobile network traffic and proactive network resource management [5,9,22]. Indeed, in cellular networks with limited and highly expensive time-frequency radio resources, precise prediction of user traffic arrival can contribute significantly in improving the resource utilization [5]. As a result, in recent years, there has been an increasing interest in leveraging

© Springer Nature Switzerland AG 2019
P. Kralj Novak et al. (Eds.): DS 2019, LNAI 11828, pp. 129–144, 2019.
https://doi.org/10.1007/978-3-030-33778-0_11

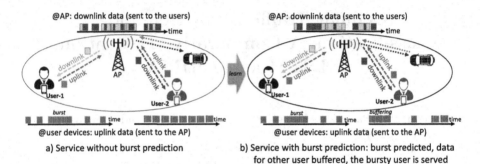

**Fig. 1.** A communication network including access point, users, and uplink and downlink data communications. (a) service is offered without prediction of bursts, (b) service is adapted to the probability of occurrence of bursts.

machine learning tools in analyzing the aggregated traffic served in a service area for optimizing the operation of the network [1, 28, 30, 32]. Scaling of fronthaul and backhaul resources for 5G networks has been investigated in [1] by leveraging methods from recurrent neural networks (RNNs) for traffic estimation. Analysis of cellular traffic for finding anomaly in the performance and provisioning of on-demand resources for compensating such anomalies have been investigated in [32]. Furthermore, prediction of light-traffic periods, and saving energy for access points (APs) through sleeping them in the respective periods has been investigated in [28, 30]. Moreover, Light-weight reinforcement learning for figuring out statistics of interfering packet arrival over different wireless channels has been recently explored [4]. While one observes that analysis of the aggregated traffic at the network side is an established field, there is lack of research on the analysis and understanding at the user level, i.e., of the specific users' traffic arrival. In 5G-and-beyond networks, the (i) explosively growing demand for radio access, (ii) intention for serving battery- and radio-limited devices requiring low-cost energy-efficient service [4], and (iii) intention for supporting ultra-reliable low-latency communications [5], mandate studying not only the aggregated traffic arrival from users, but also studying the features of traffic arrival in all users, or at least for critical users. A critical user could be defined as a user whose quality-of-service (QoS) is at risk due to the traffic behavior of other devices, or its behavior affects the QoS of other users. Let us exemplify this challenge in the sequel in the context of cellular networks.

**Example.** Figure 1(a) represents a communication network in which, an AP is serving users in the uplink (towards AP) and downlink (towards users). One further observes that traffic from user-2 represents a semi-stable shape, which is usually the case in video streaming, while the traffic from user-1 represents a bursty shape, which could be the case in surfing and on-demand file download. One observes that once a burst in traffic of user-1 occurs, the server (i.e. AP) will have difficulty in serving both users in a timely manner, and hence, QoS degradation occurs. Figure 1(a) represents a similar network in which, AP predicts the arrival of burst to user-1, immediately fills the buffer of user-2. Thus, at the time of arrival of burst for user-1, user-2 will require minimal data transfer from the AP, and hence, QoS degradation for user-2 will be prevented. Backed to this motivation, the remainder of this paper is dedicated to investigating the feasibility of exploiting the traffic history at the user level and employing it for future traffic prediction via machine learning and statistical learning approaches.

**Research Problem.** Let us assume time in our problem is quantized into intervals of length $\tau$ seconds. The research problem tackled in this work could be stated as follows: *Given the history of traffic arrival for a certain number of time intervals, how accurately can we estimate (a) the intensity of traffic in the next time intervals, (b) the occurrence of burst in future time intervals (c) the application which is generating the traffic?*

This problem can be approached as a time series forecasting problem, where for example, the number of packet arrivals in each unit of time constitutes the value of the time series at that point. While the literature on time series forecasting using statistical and machine learning approaches is mature, e.g., refer to [24,31] and references herein, finding patterns in the cellular traffic and making the decision based on such prediction is never an easy task due to the following reasons [33]. First, the traffic per device originates from different applications, e.g. surfing, video and audio calling, video streaming, gaming, and etc. Each of these applications could be mixed with another, and could have different modes, making the time series seasonal and mode switching. Second, each application can generate data at least in two modes, in active use and in the background, e.g. for update and synchronization purposes. Third, each user could be in different modes in different hours, days, and months, e.g. the traffic behavior in working days differs significantly from the one in the weekends. Fourth, and finally, the features in the traffic, e.g., the inter-arrival time of packets, vary significantly in traffic -generating applications and activity modes.

**Contributions.** Our contributions in this paper are summarized as follows:

- We present a comprehensive comparative evaluation for prediction and classification of network traffic; autoregressive integrated moving average (ARIMA) against the long short-term memory (LSTM);
- We investigate how a deep learning model compares with a linear statistical predictor model in terms of short-term and long-term predictive performance, and how additional engineered features, such as the ratio of uplink to downlink packets and protocol used in packet transfer, can improve the predictive performance of LSTM;
- Within these analyses, the impact of different design parameters, including the length of training data, length of future prediction, the feature set used in machine learning, and traffic intensity, on the performance are investigated;
- We further extend our analysis to the classification of the application generating the traffic and prediction of packet and burst arrivals. The results presented in this work pave the way for the design of traffic-aware network planning, resource management, and network security.

The remainder of this paper is organized as follows: In Sect. 2, we outline the related work in the area and introduce the knowledge gaps of state-of-the-art. In Sect. 3, we formulate the problem studied in this paper, while Sect. 4 presents the two methods used for solving it. Section 5 presents the experimental evaluation results for different methods and feature sets, as well as provides a conclusive discussion on the results. Finally, concluding remarks and future direction of research are provided in Sect. 6.

## 2    Related Work and Research Gap

We summarize state-of-the-art research on cellular traffic prediction and classification, and introduce the research gaps which motivate our work.

**Cellular Traffic Prediction.** Understanding dynamics of cellular traffic and prediction of future demands are, on the one hand, crucial requirements for improving resource efficiency [5], and on the other hand, are complex problems due to the diverse set of applications that are behind the traffic. Dealing with network traffic prediction as a time series prediction, one may categorize the state-of-the-art proposed schemes into three categories: statistical learning [8,19], machine learning [26,27], and hybrid schemes [12]. ARIMA and LSTM, as two popular methods of statistical learning and machine learning time series forecasting, have been compared in a variety of problems, from economics [10,19,23] to network engineering [6]. A comprehensive survey on cellular traffic prediction schemes, including convolutional and recurrent neural networks, could be found in [13,15]. A deep learning-powered approach for prediction of overall network demand in each region of cities has been proposed in [2]. In [18,27], the spatial and temporal correlations of the cellular traffic in different time periods and neighboring cells, respectively, have been explored using neural networks in order to improve the accuracy of traffic prediction. In [14], convolutional and recurrent neural networks have been combined in order to further capture dynamics of time series, and enhance the prediction performance. In [6,26], preliminary results on network traffic prediction using LSTM have been presented, where the set of features used in the experiment and other technical details are missing. Reviewing the state-of-the-art, one observes there is a lack of research of leveraging advanced learning tools for cellular traffic prediction, selection of adequate features, especially when it comes to each user with a specific set of applications and behaviors.

**Cellular Traffic Classification.** Traffic classification has been a hot topic in computer/communication networks for more than two decades due to its vastly diverse applications in resource provisioning, billing and service prioritization, and security and anomaly detection [20,29]. While different statistical and machine learning tools have been used till now for traffic classification, e.g. refer to [16] and references herein, most of these works are dependent upon features which are either not available in encrypted traffic, or cannot be extracted in real-time, e.g. port number and payload data [16,20]. In [25], classification of traffic using convolutional neural network using 1400 packet-based features as well as network flow features has been investigated for classification of encrypted traffic, which is too complex for a cellular network to be used for each user. Reviewing the state-of-the-art reveals that there is a need for investigation of low-complex scalable cellular traffic classification schemes (i) without looking into the packets, due to encryption and latency, (ii) without analyzing the inter-packet arrival for all packets, due to latency and complexity, and (iii) with a few numbers of features as possible. This research gap is addressed in this work (Fig. 3).

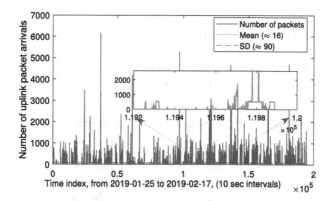

**Fig. 2.** The number of uplink packet arrivals for 24 days in 10-s intervals

## 3    Problem Description and Traffic Prediction Framework

In this section, we first provide our problem setup and formulate the research problem addressed in the paper. Then, we present the overall structure of the traffic prediction framework, which is introduced in this work.

We consider a cellular device, on which a set of applications, denoted by $\mathbf{A}$, are running, e.g., User-1 in Fig. 1. At a given time interval $[t, t + \tau]$ of length $\tau$, each application could be in an *active* or *background* mode, based on the user behaviour. We further consider a set of features describing the aggregated cellular traffic in $[t, t+\tau]$ for a specific user, such as the overall number of uplink/downlink packets and the overall size of uplink/downlink packets, which don't require decoding the packets. Let vector $\mathbf{x}_i(t)$ denote the set of features describing the traffic in interval $[t - i\tau, t - (i - 1)\tau]$ for $i \geq 1$, and in interval $[t - (i+1)\tau, t - i\tau]$ for $i < 0$ respectively. Furthermore, $\mathbf{X}_m(t)$ is a matrix containing $m$ feature vectors of the traffic, including $\mathbf{x}_1(t){:}\mathbf{x}_m(t)$ for $m > 0$, and $\mathbf{x}_{-1}(t){:}\mathbf{x}_{-m}(t)$ for $m < 0$. Further, denote by $\mathbf{s}$ an indicator vector, with elements either 0 or 1. Then, given a matrix $\mathbf{X}_m(t)$ and a binary indicator vector $\mathbf{s}$, we define $\mathbf{X}_m^s(t)$ the submatrix of $\mathbf{X}_m(t)$, such that all respective rows, for which $\mathbf{s}$ indicates a zero value, are removed. For example, let $\mathbf{X}_m(t) = [1, 2; 3, 4]$ and $s = [1, 0]$. Then, $\mathbf{X}_m^s(t) = [1, 2]$.

Now, the research question in Sect. 1 could be rewritten as:

$$\text{Given} \quad \mathbf{X}_m(t), m \geq 1;$$
$$\text{minimize} \quad L\big(\mathbf{X}^{\mathrm{s}}_{-n}(t), \mathbf{Y}(t)\big) \tag{1}$$

where $n > 0$ is the length of the future predictions, e.g., $n = 1$ for one $\tau$ future prediction, $\mathbf{Y}(t)$ is of the same size as $\mathbf{X}^{\mathrm{s}}_{-n}(t)$ and represents the predicted matrix at time $t$, while $L(\cdot)$ is the desired error function, e.g., it may compute the mean squared error between $\mathbf{X}^{\mathrm{s}}_{-n}(t)$ and $\mathbf{Y}(t)$.

## 4 Time Series Prediction

In this section, we give a short description of the two methods benchmarked in this paper to be used within the proposed prediction framework in Sect. 4.1.

### 4.1 The Proposed Traffic Prediction Framework

Recall the challenges described in the previous section on the prediction of cellular traffic, where the major challenge consists of dependency of traffic arrival to user behavior and type of the application(s) generating the traffic. Then, as part of the solution to this problem, one may first predict the application(s) in use and behavior of the user, and then use them as extra features in the solution. This approach for solving (1) has been illustrated in Fig. 3. In order to realize such a framework, it is of crucial importance to first evaluate the traffic predictability and classification using only statistics of traffic with granularity $\tau$, and then, to investigate hybrid models for augmenting predictors by online classifications, and finally to investigate traffic-aware network management design.

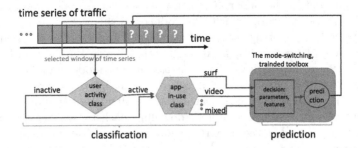

**Fig. 3.** The proposed framework for cellular traffic prediction

## 4.2   Statistical Learning: ARIMA

The first method we consider in our work is Autoregressive integrated moving average (ARIMA), which is essentially a statistical regression model. The predictions performed by ARIMA are based on considering the lagged values of a given time series, while at the same time accommodating non-stationarity. ARIMA is one of the most popular linear models in statistical learning for time series forecasting, originating from three models: the autoregressive (AR) model, the moving average (MA) model, and their combination, ARMA [7].

More concretely, let $\mathcal{X} = X_1, \ldots, X_n$ define a uni-variate time series, with $X_i \in \mathbb{R}$, for each $i \in [1, n]$. A $p$-$order$ AR model, $AR(p)$, is defined as follows:

$$X_t = c + \alpha_1 X_{t-1} + \alpha_2 X_{t-2} + \ldots + \alpha_p X_{t-p} + \epsilon_t, \qquad (2)$$

where $X_t$ is the predicted value at time $t$, $c$ is a constant, $\alpha_1, \ldots, \alpha_p$ are the parameters of the model and $\epsilon_t$ corresponds to a white noise variable.

In a similar, a $q$-order moving average process, $MA(q)$, expresses the time series as a linear combination of its current and $q$ previous values:

$$X_t = \mu + \epsilon_t + \beta_1 \epsilon_{t-1} + \beta_2 \epsilon_{t-2} + \ldots + \beta_q \epsilon_{t-q}, \qquad (3)$$

where $\mu$ is the mean of $X$, $\beta_1, \ldots, \beta_q$ are the model parameters and $\epsilon_i$ corresponds to a white noise random variable. The combination of an $AR$ and an $MA$ process coupled with their corresponding $p$ and $q$ order parameters, respectively, defines an ARMA process, denoted as $ARMA(p, q)$, and defined as $X_t = AR(p) + MA(q)$. The original limitation of ARMA is that, by definition, it can only be applied to stationary time series. Nonetheless, non-stationary time series can be stationarized using the $d^{th}$ differentiation process, where the main objective is to eliminate any trends and seasonality, hence stabilizing the mean of the time series. This process is simply executed by computing pairwise differences between consecutive observations. For example, a first-order differentiation is defined as $X_t^{(1)} = X_t - X_{t-1}$, and a second order differentiation is defined as $X_t^{(2)} = X_t^{(1)} - X_{t-1}^{(1)}$. Finally, an ARIMA model, $ARIMA(p, d, q)$, is defined by three parameters $p, d, q$ [17], where $p$ and $q$ correspond to the AR and MA processes, respectively, while $d$ is the number of differentiations performed to the original time series values, that is $X_t$ is converted to $X_t^{(d)} = \nabla^d X_t$, with $X_t^{(d)}$ being the time series value at time $t$, with differentiation applied $d$ times. The full $ARIMA(p, d, q)$ model is computed as $X_t^{(d)} = \sum_{i=1}^{p} \alpha_i X_{t-i}^{(d)} + \epsilon_t + c + \sum_{i=1}^{q} \beta_q \epsilon_{t-q} + \mu$.

**Finding Optimized Parameters.** In this study, the ARIMA parameters, including $p$, $d$, and $q$, are optimized by carrying out a grid search over potential values in order to locate the best set of parameters. In experimental results, Fig. 5, we represent the root mean square error (RMSE) results for different ARIMA $(p, d, q)$ configurations, when they are applied to the dataset for prediction of the number of future packet arrivals. From these results and Bayesian information criterion (BIC), the best performance is achieved by ARIMA(6,1,0).

### 4.3  Machine Learning: LSTM

Next, we consider is a long short-term memory (LSTM) architecture based on a Recurrent Neural Network (RNN), a generalization of the feed forward network model for dealing with sequential data, with the addition of an ongoing internal state serving as a memory buffer for processing sequences. Let $\{X_1, \ldots, X_n\}$ define the input (features) of the RNN, $\{Y_1, \ldots, Y_n\}$ be the set of outputs, and let $\{Y_1', \ldots, Y_n'\}$ denote the actual time series observations that we aim to predict. For this study the internal state of the network is processed by Gated Recurrent Units (GRU) [11] defined by iterating the following three equations:

$$r_j = sigm([W_r X]_j + [U_r h_{t-1}]_j), \tag{4}$$

$$z_j = sigm([W_z X]_j + [U_z h_{t-1}]_j)), \tag{5}$$

$$h_j^t = z_j h_j^{t-1} + (1 - z_j) h_{new}, \tag{6}$$

$$h_{new}^t = tanh([W X]_j + [U(r \circ h_{t-1})]_j), \text{where} \tag{7}$$

- $r_j$: a reset gate showing if a previous state is ignored for the $j^{th}$ hidden unit,
- $h_{t-1}$; the previous hidden internal state $h_{t-1}$,
- $W$ and $U$: parameter matrices containing weights to be learned,
- $z_j$: an update gate that determines if a hidden state should be updated,
- $h_j^t$: the activation function of hidden unit $h_j$,
- $sigm(\cdot)$: the sigmod function, and
- $\circ$: the Hadamard product.

Finally, the loss function we optimize is the squared error, defined for all inputs as $\mathcal{L} = \sum_{t=1}^{n} (Y_t - Y_t')^2$. The RNN tools leveraged in this work for traffic prediction consist of 3 layers, including the LSTM layer, with 100 hidden elements, the fully connected (FC) layer, and the regression layer. The regression layer is substituted with the softmax layer in the classification experiments.

## 5  Experimental Evaluation

In this section, we investigate the performance of the proposed prediction and classification tools over a real cellular dataset.

### 5.1  Dataset

We generated our own cellular traffic dataset and made part of it available online [3]. The data generation was done by leveraging a packet capture tool, e.g. Wire-Shark, at the user side. Using these tools, packets are captured at the Internet protocol (IP) level. One must note that the cellular traffic is encrypted in layer 2, and hence, the payload of captured traffic is neither accessible nor intended for analysis. The latter is due to the fact that for the realization of a low-complexity low-latency traffic prediction/classification tool, we are interested in achieving the objectives just by looking at the traffic statistics. For generating labels for part of the dataset, to be used for classification, a controlled environment at the user-side is prepared in which, we filter internet connectivity for all applications unless a subset of applications, e.g., Skype. Then, the traffic labels will be generated based on the different filters used at different time intervals. In our study, we focus on **seven packet features**: (i) time of packet arrival/departure, (ii) packet length, (iii) whether the packet is uplink or downlink, (iv) the source IP address, (v) the destination IP address, (vi) the communication protocol, e.g., UDP, and (vii) the encrypted payload, where only the first three features are derived without looking into the header of packets. We experimented with different values for the interval length parameter $\tau$, and for most of our experiments $\tau$ was set to 10 s. Table 1 provides the set of features for each time interval in rows, and the subsets of features used in different feature sets (FSs). It is straightforward to infer that $\tau$ tunes a tradeoff between complexity and reliability of the prediction. If $\tau$ tends to zero, i.e., $\tau = 1$ ms, one can predict traffic arrival for the next $\tau$ interval with high reliability at the cost of extra effort for keeping track of data with such a fine granularity. On the other hand, when $\tau$ tends to seconds or minutes, the complexity and memory needed for prediction decrease, which also results in lower predictive performance during the next intervals.

## 5.2 Setup

The experimental results in the following sections are presented within 3 categories, i.e., (i) prediction of the number of packet arrivals in future time intervals, (ii) prediction of burst occurrence in future intervals, and (iii) classification of applications generating the traffic. In the first two categories, we performed a comprehensive set of Monte Carlo MATLAB simulations [21], over the data set, varying different data parameters, such as length of the training set, length of future prediction, feature sets used in learning and prediction. Each RMSE result in Fig. 5 for each scheme has been derived by averaging over 37 experiments. In each experiment, each scheme is trained using a training dataset, and then tested over 2000 future time intervals (non-overlapping with the training dataset). For the classification performance evaluation, we have leveraged 16 labeled datasets, each containing traffic from 4 mobile applications. Then, we constructed 16 tests, where in each test, one dataset is used for performance evaluation. The notation of the schemes used in the experiments, extracted from the basic ARIMA and LSTM methods described in Sect. 4, is as follows: (i) AR(1), representing the traffic prediction based on the last observation; (ii) optimized ARIMA, in which the number of lags and coefficients of ARIMA are optimized using a grid search for RMSE minimization; and (iii) LSTM(FS-$x$), in which FS-$x$ for $x \in \{1, \cdots, 6\}$ represents the feature set used in the LSTM prediction/classification tool. The overall configuration of experiments can be found in Table 2.

**Reproducibility.** All experiments can be reproduced using the anonymized GDPR-compliance traffic dataset available at the supporting repository [3].

## 5.3  Empirical Results

In this section, we present the prediction and classification performance results.

**Prediction of Traffic Intensity.** Figure 5 depicts the RMSE results for differ-ent ARIMA and LSTM configurations versus AR(1), when the number of uplink packets in intervals of 10 s is to be estimated. Towards this end, the right $y$-axis represents the absolute RMSE of AR(1), the left $y$-axis represents the relative performance of other schemes versus AR(1), and the $x$-axis represents the stan-dard deviation (SD) of the test dataset. The results are insightful and shed light to the regions in which ARIMA and LSTM perform favorably, as follows. When the SD of traffic from its average value is more than 30% of the long-term SD of the dataset[1], which is almost the case in the active mode of phone usage by human users, LSTM outperforms the benchmark schemes. On the other hand, when there is only infrequent light background traffic, which is the case on the right-end side of Fig. 5, ARIMA outperforms the benchmark schemes. When we average the performance over a 24-days dataset, we observe that LSTM(FS-6), LSTM(FS-5), LSTM(FS-3), and optimized ARIMA outperform the AR(1) by 16%, 14.5%, 14%, and 12%, respectively, for $\tau = 10$ s. Recall that LSTM(FS-6) keeps track of the number of uplink and downlink packets, as well as statis-tics of the communication protocol used by packets in each time interval, while LSTM(FS-5) does not care about the protocol used by packets. The superior performance of LSTM(FS-6) with regards to LSTM(FS-5), as depicted in Fig. 5, represents that how adding features to the LSTM predictor can further improve the prediction performance in comparison with the linear predictors.

**Table 1.** Feature sets.

| Feature sets (FSs) | 1 | 2 | 3 | 4 | 5 | 6 |
|---|---|---|---|---|---|---|
| Num. of UL packets | 1 | 1 | 1 | 1 | 1 | 1 |
| Num. of DL packets | 1 | 0 | 0 | 1 | 1 | 1 |
| Size of UL packets | 1 | 0 | 0 | 0 | 0 | 0 |
| Size of DL packets | 1 | 0 | 0 | 0 | 0 | 0 |
| UL/DL packets | 1 | 1 | 0 | 1 | 0 | 0 |
| Comm. protocol | 0 | 0 | 0 | 0 | 0 | 1 |

**Table 2.** Parameter configuration.

| Parameters | Description |
|---|---|
| Traffic type | Cellular traffic |
| Capture point | IP layer, device side |
| Length of dataset | 48 days traffic |
| RNN for prediction (eq. classification) | [LSTM, FC, regression(eq. softmax)] |
| Time granularity, $\tau$ | Default: 10 s |

We investigate if LSTM can further outperform the benchmark schemes by increasing time-granularity of the dataset, decreasing length of future obser-vation, and increasing length of the training set. First, let us investigate the performance impact of $\tau$, i.e. the time granularity of dataset. Figure 5 (left) represents the absolute (left $y$-axis) and rational (right $y$-axis) RMSE results for the proposed and benchmark schemes as a function of time granularity of

---

[1] The long-term SD of the dataset is 90.

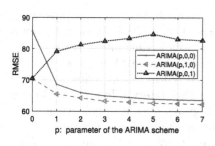 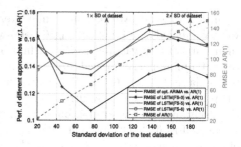

**Fig. 4.** Prediction of the future number of uplink packets ($\tau = 10$ s). (left) Finding optimized ARIMA($p, d, q$) configuration; (right) The RMSE performance comparison of LSTM and ARIMA.

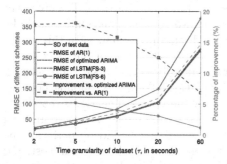 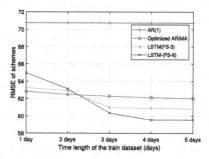

**Fig. 5.** (left) RMSE of prediction as a function of $\tau$ (time granularity of dataset); (right) RMSE of prediction of number of uplink packets as a function of length of the training dataset (as well as length of future prediction).

dataset ($\tau$, the $x$-axis). One must further consider the fact that $\tau$ not only represents how fine we have access to the history of the traffic, but also represents the length of future prediction. It is clear that the best results for the lowest $\tau$, e.g. when $\tau = 1$, the LSTM (FS-6) outperforms the optimized ARIMA by 5% and the AR(1) by 18%. One further observes that by increasing the $\tau$, not only the RMSE increases but also the merits of leveraging predictors decrease, e.g. for $\tau = 60$, LSTM(FS-6) outperforms AR(1) by 7%. Now, we investigate the performance impact of the length of the training set on the prediction in Fig. 5 (right). One observes that the LSTM(FS-6) with poor training (1 day) even performs worse than optimized ARIMA. However, as the length of training data set increases, the RMSE performance for the LSTM predictors, especially for LSTM(FS-3) with further features, decreases significantly.

**Prediction of Event Bursts.** We investigate the usefulness of the proposed schemes for burst prediction in future time intervals. For the following experiments, we label a subset of time intervals as *bursts*, based on the underlying traffic intensity, i.e., the number and length of packets. Then, based on this training dataset, we aim at predicting whether a burst will occur in the next time interval. As a benchmark to the LSTM predictors, we compare the performance against AR(1), i.e., we estimate a time interval as burst if the previous time interval was labeled as a burst. In Fig. 6 (left) we see the recall of bursty and normal (non-bursty) intervals for a burst definition in which, time intervals with more than 90 uplink packet arrivals are treated as burst when the SD of packet arrivals in the dataset is 90. The LSTM predictor developed in this experiment returns the probability of burst occurrence in the next time interval. In order to declare the decision as burst or non-burst, we set a probability threshold value. The $x$-axis of Fig. 6 (left) represents the decision threshold, which tunes the weight of recall and accuracy of decisions. In this figure, we observe that the probability of missing a burst is very low on the left side, while the accuracy of decisions is low (it can be inferred from the recall of normal intervals). Furthermore, on the right side of the figure, the probability of missing a burst has decreased, however, the accuracy has increased (high recall of normal intervals). The crossover point, where the recall values of bursty and non-bursty intervals match, could be an interesting point for investigating the prediction performance. In this figure, one observes that when the decision threshold is 0.02, 91% of bursts could be predicted, while only 9% of normal intervals are labeled as bursty (false alarm).

In Fig. 6 (right) we observe some insightful results on the coupling between recall of predictions and degree of rareness of the bursts. The $x$-axis represents the definition of bursts, e.g. for $x = 90$, we label time intervals with more than 90 packets as a burst. From this figure, it is clear that LSTM outperforms the benchmarks in recalling the burst with a reasonable non-burst recall cost. For example, for $x = 1(\approx 0.01SD)$, we aim at predicting if the next time interval will contain a packet or not, i.e., time intervals with a packet transmission are defined as bursts. One observes that 78% of bursts could be predicted using LSTM(FS-5), while only 28% of non-bursts are declared as bursts. Having the information that 20% of time internals contain bursts, we infer that the accuracy of prediction is 78%. As the frequency of burst occurrence decreases, i.e., we move to the right side of the figure, the recall performance of LSTM increases slightly up to some point beyond which, it starts decreasing. On the other hand, the accuracy of prediction by moving from left to right decreases substantially due to the rareness of the bursts. Clearly, LSTM outperforms AR(1), especially when bursts are occurring infrequently.

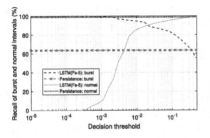 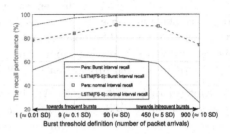

**Fig. 6.** (left) Prediction of bursts as a function of decision threshold; (right) Prediction of bursts as a function of frequency of occurrence of bursts. ($\tau = 10$ s)

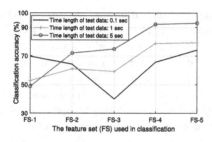 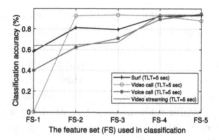

**Fig. 7.** (left) The overall accuracy of classification as a function of the feature set used in the experiment; (right) Per application accuracy of classification.

**Traffic Classification.** We investigate leveraging machine learning schemes for classification of the application generating the cellular traffic in this subsection. For the classification purpose, a controlled experiment at the user-side has been carried out in which, 4 popular applications including surfing, video calling, voice calling, and video streaming have been used by the user. Figure 7 (left) represents the overall accuracy of classification for different feature sets used in the machine learning tool. One observes that the LSTM(FS-5) and LSTM(FS) outperform the others significantly in the accuracy of classification. Furthermore, in this Fig. 3 curves for different lengths of the test data, to be classified, have been depicted. For example, when the length of the test data is 0.1 s, the time granularity of dataset ($\tau$) is 0.1 s, and we also predict labels of intervals of length 0.1 s. It is clear here that as the length of $\tau$ increases, the classification performance increase because we will have more evidence from the data in the test set to be matched to each class. To further observe the recall of classification for different applications, Fig. 7 (right) represents the accuracy results per each application. One observes that the LSTM(FS-4) and LSTM(FS-5) outperform the others. It is also insightful that adding the ratio of uplink to downlink packets to FS-5, and hence constructing FS-4 (based on Table 1), can make the prediction performance more fair for different applications. It is further insightful to observe that the choice of feature set to be used is sensitive to the application used in the traffic dataset. In other words, FS-3, which benefits from one feature, outperforms the others in the accuracy of classification for video calling, while it results in classification error for other traffic types.

## 5.4   Discussion

The experimental results represent that the accuracy of prediction strongly depends on the length of the training dataset, time granularity of dataset, length of future prediction, mode of activity of the user (standard deviation of test dataset), and the feature set used in the learning scheme. The results, for example, indicate that the proposed LSTM(FS-3) is performing approximately 5% better than optimized ARIMA, and 18% better than AR(1) for $\tau = 10$ s. The results further indicated that the performance of LSTM could be further improved by designing more features related to the traffic, e.g. the protocol in use for packets, and the ratio of uplink to downlink packets. Moreover, our experiments indicated that the design of a proper loss function, and equivalently the decision threshold, can significantly impact the recall and accuracy performance. Furthermore, we observed that the frequency of occurrence of bursts (definition of burst), the time granularity of dataset, and length of future prediction, can also significantly impact the prediction performance. The results, for example, indicated that a busy interval, i.e. an interval with at least one packet, could be predicted by 78% accuracy as well as recall. The experimental results represented the facts that, first, accuracy and recall performance of classification is highly dependent on the feature set used in the classification. For example, a feature set that can achieve an accuracy of 90% for classification of one application may result in a recall of 10% for another application. Then, the choice of feature set should be in accordance with the set of applications used by the user. Second, if we can tolerate delay in the decision, e.g. 5 s, the classification performance will be much more accurate when we gather more information and decide on longer time intervals. The overall accuracy performance for different applications using the developed classification tool is approximately 90%.

## 6   Conclusions

In this work, the feasibility of per-user traffic prediction for cellular networks has been investigated. Towards this end, a framework for cellular traffic prediction has been introduced, which leverages statistical/machine learning units for traffic classification and prediction. A comprehensive comparative analysis of prediction tools based on statistical learning, ARIMA, and the one based on machine learning, LSTM, has been carried out, under different traffic circumstances and design parameter selections. The LSTM model, in particular, when the length of training data is long enough and the model is augmented by additional features like the ratio of uplink to downlink packets and the communication protocol used in prior packet transfers, exhibited demonstrable improvement over the benchmark schemes for future traffic predictions. Furthermore, the usefulness of the developed LSTM model for classification of cellular traffic has been investigated, where the results represent high sensitivity of accuracy and recall of classification to the feature set in use. Additional investigations could be performed in the future works regarding making the prediction tool mode-switching, in order to reconfigure the feature set and prediction parameters based on the changes in the behavior of user/applications in an hourly/daily basis.

## References

1. Alawe, I., et al.: Smart scaling of the 5G core network: an RNN-based approach. In: Globecom 2018-IEEE Global Communications Conference, pp. 1–6 (2018)

2. Assem, H., Caglayan, B., Buda, T.S., O'Sullivan, D.: ST-DenNetFus: a new deep learning approach for network demand prediction. In: Brefeld, U., et al. (eds.) ECML PKDD 2018. LNCS (LNAI), vol. 11053, pp. 222–237. Springer, Cham (2019). https://doi.org/10.1007/978-3-030-10997-4_14
3. Azari, A.: Cellular traffic analysis (2019). https://github.com/AminAzari/cellular-traffic-analysis
4. Azari, A., Cavdar, C.: Self-organized low-power IoT networks: a distributed learning approach. In: IEEE Globecom 2018 (2018)
5. Azari, A., et al.: Risk-aware resource allocation for URLLC: challenges and strategies with machine learning. IEEE Commun. Mag. **57**, 42–48 (2019)
6. Azzouni, A., Pujolle, G.: NeuTM: a neural network-based framework for traffic matrix prediction in SDN. In: NOMS 2018–2018 IEEE/IFIP Network Operations and Management Symposium, pp. 1–5 (2018)
7. Box, G.E., Jenkins, G.M.: Time Series Analysis: Forecasting and Control, Revised edn. Holden-Day, San Francisco (1976)
8. Brockwell, P.J., Davis, R.A., Calder, M.V.: Introduction to Time Series and Forecasting, vol. 2. Springer, New York (2002). https://doi.org/10.1007/b97391
9. Chen, M., et al.: Machine learning for wireless networks with artificial intelligence: a tutorial on neural networks. arXiv preprint arXiv:1710.02913 (2017)
10. Chen, T., et al.: Multivariate arrival times with recurrent neural networks for personalized demand forecasting. arXiv preprint arXiv:1812.11444 (2018)
11. Cho, K., van Merrienboer, B., Gülçehre, Ç., Bougares, F., Schwenk, H., Bengio, Y.: Learning phrase representations using RNN encoder-decoder for statistical machine translation. CoRR abs/1406.1078 (2014). http://arxiv.org/abs/1406.1078
12. Choi, H.K.: Stock price correlation coefficient prediction with ARIMA-LSTM hybrid model. arXiv preprint arXiv:1808.01560 (2018)
13. Huang, C.W., Chiang, C.T., Li, Q.: A study of deep learning networks on mobile traffic forecasting. In: 2017 IEEE 28th Annual International Symposium on Personal, Indoor, and Mobile Radio Communications (PIMRC), pp. 1–6 (2017)
14. Lai, G., Chang, W.C., Yang, Y., Liu, H.: Modeling long-and short-term temporal patterns with deep neural networks. In: The 41st International ACM SIGIR Conference on Research & Development in Information Retrieval, pp. 95–104 (2018)
15. Li, R., et al.: The prediction analysis of cellular radio access network traffic. IEEE Commun. Mag. **52**(6), 234–240 (2014)
16. Lopez-Martin, M., Carro, B., Sanchez-Esguevillas, A., Lloret, J.: Network traffic classifier with convolutional and recurrent neural networks for internet of things. IEEE Access **5**, 18042–18050 (2017)
17. Mills, T.C.: Time Series Techniques for Economists. Cambridge University Press, New York (1991)
18. Qiu, C., Zhang, Y., Feng, Z., Zhang, P., Cui, S.: Spatio-temporal wireless traffic prediction with recurrent neural network. IEEE Wirel. Commun. Lett. **7**(4), 554–557 (2018)
19. Rebane, J., Karlsson, I., Denic, S., Papapetrou, P.: Seq2seq RNNs and ARIMA Models for cryptocurrency Prediction: A Comparative Study (2018)
20. Rezaei, S., Liu, X.: Deep learning for encrypted traffic classification: an overview. arXiv preprint arXiv:1810.07906 (2018)
21. Rubinstein, R.Y., Kroese, D.P.: Simulation and the Monte Carlo Method, vol. 10. Wiley, New York (2016)
22. Saad, W., et al.: A vision of 6G wireless systems: applications, trends, technologies, and open research problems. arXiv preprint arXiv:1902.10265 (2019)

23. Skehin, T., et al.: Day ahead forecasting of FAANG stocks using ARIMA, LSTM networks and wavelets. In: CEUR Workshop Proceedings (2018)
24. Tealab, A.: Time series forecasting using artificial neural networks methodologies. Future Comput. Inf. J. **3**(2), 334–340 (2018)
25. Tong, V., Tran, H.A., Souihi, S., Mellouk, A.: A novel QUIC traffic classifier based on convolutional neural networks. In: IEEE International Conference on Global Communications (GlobeCom), pp. 1–6 (2018)
26. Trinh, H.D., Giupponi, L., Dini, P.: Mobile traffic prediction from raw data using LSTM networks. In: 2018 IEEE 29th Annual International Symposium on Personal, Indoor and Mobile Radio Communications (PIMRC), pp. 1827–1832 (2018)
27. Wang, J., et al.: Spatio-temporal modeling and prediction in cellular networks: a big data enabled deep learning approach. In: IEEE Conference on Computer Communications, pp. 1–9 (2017)
28. Wang, K., et al.: Modeling and optimizing the LTE discontinuous reception mechanism under self-similar traffic. IEEE Trans. Veh. Tech. **65**(7), 5595–5610 (2016)
29. Williams, N., Zander, S., Armitage, G.: A preliminary performance comparison of five machine learning algorithms for practical IP traffic flow classification. ACM SIGCOMM Comput. Commun. Rev. **36**(5), 5–16 (2006)
30. Ye, J., Zhang, Y.J.A.: DRAG: Deep reinforcement learning based base station activation in heterogeneous networks. arXiv preprint arXiv:1809.02159 (2018)
31. Zhang, G.P.: Time series forecasting using a hybrid ARIMA and neural network model. Neurocomputing **50**, 159–175 (2003)
32. Zhang, Q., et al.: Machine learning for predictive on-demand deployment of UAVs for wireless communications. arXiv preprint arXiv:1805.00061 (2018)
33. Zhang, Y., Årvidsson, A.: Understanding the characteristics of cellular data traffic. In: Proceedings of the 2012 ACM SIGCOMM Workshop on Cellular Networks: Operations, Challenges, and Future Design, pp. 13–18 (2012)

# Main Factors Driving the Open Rate of Email Marketing Campaigns

Andreia Conceição[1]($\boxtimes$) and João Gama[1,2]($\boxtimes$)

[1] Faculty of Economics, University of Porto, Porto, Portugal
andreia.g.conceicao@hotmail.com, jgama@fep.up.pt
[2] LIAAD-INESC TEC, University of Porto, Porto, Portugal

**Abstract.** Email Marketing is one of the most important traffic sources in Digital Marketing. It yields a high return on investment for the company and offers a cheap and fast way to reach existent or potential clients. Getting the recipients to open the email is the first step for a successful campaign. Thus, it is important to understand how marketers can improve the open rate of a marketing campaign. In this work, we analyze what are the main factors driving the open rate of financial email marketing campaigns. For that purpose, we develop a classification algorithm that can accurately predict if a campaign will be labeled as *Successful* or *Failure*. A campaign is classified as *Successful* if it has an open rate higher than the average, otherwise it is labeled as *Failure*. To achieve this, we have employed and evaluated three different classifiers. Our results showed that it is possible to predict the performance of a campaign with approximately 82% accuracy, by using the Random Forest algorithm and the redundant filter selection technique. With this model, marketers will have the chance to sooner correct potential problems in a campaign that could highly impact its revenue. Additionally, a text analysis of the subject line and preheader was performed to discover which keywords and keyword combinations trigger a higher open rate. The results obtained were then validated in a real setting through A/B testing.

**Keywords:** Digital Marketing · Email Marketing · Marketing campaigns · Open rate

## 1  Introduction

The introduction of the Internet allowed a new form of communication, known as Digital Marketing. One of the most important traffic sources in Digital Marketing is Email Marketing. A recent study showed that 59% of the marketers[1] inquired stated that Email Marketing is the source that brings the highest return on investment (ROI) for the firm [8]. Thus, it´s is crucial for email marketers to know how to improve the performance of their marketing campaigns.

---

[1] A person or company that advertises or promotes something.

© Springer Nature Switzerland AG 2019
P. Kralj Novak et al. (Eds.): DS 2019, LNAI 11828, pp. 145–154, 2019.
https://doi.org/10.1007/978-3-030-33778-0_12

In this work, we analyze what are the main factors driving the open rate of financial email marketing campaigns. Getting the recipients to open the email is the first step for a successful campaign, since it determines the reach of the campaign [6, 7]. Therefore, it´s important for marketers to first understand how they can improve the email open rate. With that purpose, we developed a classification algorithm that can accurately predict if a campaign will be classified as *Successful* or *Failure*. A campaign is labeled as *Successful* if it has an open rate higher than the average, otherwise it is classified as *Failure* Additionally, we did a text analysis of the subject line and preheader to discover which keywords and keyword combinations are associated with a higher email open rate. To validate the results obtained in a real setting, we performed A/B testing in the deployment stage. This framework was applied in a Portuguese Digital Marketing company, as a case study.

By using data-driven models, advertisers can predict the performance of a campaign before even sending it. In fact, if a marketer knows in advance if a campaign is going to be successful or not, it provides the opportunity to sooner correct problems that could strongly impact its revenue. To our knowledge, this is the first publication that does an extensive qualitative analysis of the main factors driving the opening behavior of financial marketing campaigns. Nowadays, financial institutions are using Email Marketing as an important source to reach their clients. Thus, this work will guide marketers on how to implement successful campaigns in this field.

This paper follows the CRISP-DM methodology [15], which consists of the following stages: Business Understanding, Data Understanding, Data Preparation, Modelling, Evaluation and Deployment.

## 2 Related Work

The existent research studies on email open rate prediction assume an approach at the recipient level [13] or at the campaign level [2, 11]. In this work, we studied the email open rate at the campaign level because we didn´t have access to data at the recipient level, due to the General Data Protection Regulation (GDPR) requirements. We treated this work as a classification problem because our objective was to analyze qualitatively the main factors contributing to marketing campaigns with an email open rate above the average.

In 2014, Balakrishnan and Parekh [2] proposed a method for predicting the open rate of an email subject line, by learning from past subject lines. They used syntactical, historical and derived features of each keyword in the subject line and of the entire subject line. The model developed for the prediction was the Random Forest regression model, which predictions improved over the baseline. For the baseline, the open rate prediction was equal to the mean open rate of past emails that used the same subject line. For new subject lines, the open rate was predicted as the average open rate of all the subject lines.

In 2015, Luo *et al.* [13] developed a classification algorithm to predict if a targeted email will be open or not. For each email recipient, the model classified the email in "open" or "unopen". The model used features extracted from the emails and from the email recipients' profiles. For the prediction phase they used two different classifiers, Support Vector Machine and Decision Tree, on two different datasets using different feature selection methods (include or not include the recipient's domains). The Decision Tree outputted the other classifier, achieving a F1-measure rate of approximately 80% on the "opens", in the case of considering all features. In the case where the recipient's domains were not considered, the performance of both classifiers dropped, which indicates this component is important to predict the email open behavior.

In 2018, Jaidka *et al.* [11] also studied the problem of predicting email opens, based on the subject line. They explored the differences in the recipient's preferences for subject lines sent by different business areas (Finance, Cosmetics and Television). The methodology used was a Data Mining model to predict the open rate of different email subject lines, a regression analysis to study the effect of different subject line language styles in the open rate and a domain adaptation method. The learning model used was a five–fold cross–validated weighted linear regression, which predictions improved over the baselines - state-of-the-art model [2] and the mean open rate of the entire dataset. The use of the domain adaptation method improved the prediction of the model for unseen domains and business. They concluded that using certain styling strategies in the subject line, according to the business area of the campaign, can strongly impact the email open rate.

The contributions of our work to these papers are the inclusion of the preheader and the email sender as features in the prediction task. Before opening an email, the recipient has also information of these components; therefore, we decided to test if these features are important to predict the open behavior of a campaign.

## 3  Business Understanding

The objective of the Portuguese company involved in this study was to understand how they could improve the open rate of their email marketing campaigns. The company has an online publishing business that sends Spanish marketing campaigns, mainly from financial institutions. To satisfy that business objective, we developed a classification algorithm that can accurately predict if a campaign will be classified as *Successful* or *Failure*. In addition, we applied a text analysis of the subject line and preheader to find which keywords and keyword combinations trigger a higher open rate. To validate the results obtained in the company business, we conducted A/B testing.

The tools that were used in this study were the software *Knime* [3] and *SPSS* [10].

## 4  Data Understanding

### 4.1  Data Extraction

For this work, we collected data of 217 Spanish email campaigns from the company in study, sent since February 2018 until February 2019. The features extracted were the following: campaign name; email sender[2]; subject line[3]; preheader[4]; number of emails sent; number of emails delivered; number of emails opened.

**Keyword Extraction Process:** To calculate the number of keywords in each email subject line and preheader, we first had to extract the set of keywords. That process is shown in Fig. 1 and Table 1.

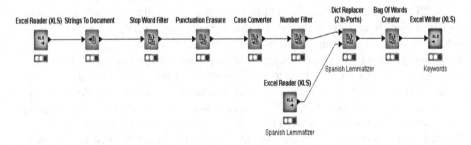

**Fig. 1.** Keyword extraction process.

– Excel Reader: Dataset composed by the campaign name, the subject line and pre-header of each campaign, the respective email open rate and classification.
– Stop Word Filter: This node removes the terms of the input documents which are in the Spanish Stop Word list.
– Dict Replacer: This node replaces the terms of the input documents that match with the specified dictionary terms by the corresponding specified value. The dictionary file used was an external source Spanish Lemmatizer (*GitHub* source), as the *Knime* software only allows Lemmatization for English terms.

**Table 1.** Keywords.

| Campaign id | Subject line & preheader | Keywords |
|---|---|---|
| 4 | "Ahorra un 30% en el seguro de tu hogar. Sin renunciar a ninguna cobertura." | *ahorrar* |
| | | *seguro* |
| | | *hogar* |
| | | *renunciar* |
| | | *cobertura* |

---

[2] Person or entity that sends the email.

[3] Short description of the email content.

[4] Description that complements the email subject line.

## 4.2  Data Description

The variables used for this study are described in Table 2. These variables were extracted and derived from email campaigns of the studied company.

**Table 2.** Data description.

| Variables | | Description |
|---|---|---|
| Campaign statistics | Sent *(Sent)* | Number of emails sent |
| | Sent days *(Sent_days)* | Number of days the campaign was sent |
| Syntactical variables | Length subject line *(Length_subject)* | The number of characters in the subject line |
| | Length preheader *(Length_preheader)* | The number of characters in the preheader |
| | Personalization *(Personalization)* | Whether the email subject line and/or preheader has a personalized greeting (i.e., the recipient's name). Categories: No; Yes |
| | Digits *(HasDigits)* | Whether the email subject line and/or preheader has digits. Categories: No; Yes |
| | Punctuation *(Punctuation)* | The type of punctuation of the subject line. Categories: Affirmative; Exclamation; Interrogation |
| | Sender Recognizable *(Sender_recognize)* | Whether the email sender corresponds to the name of the financial institution that is sending the campaign, i.e., if the email sender is recognizable by the recipient. Categories: No; Yes |
| | Number of keywords *(Nr_keywords)* | The total number of keywords in the email subject line and preheader |
| Historical variables | Occurrence score $1^{st}$, $2^{nd}$, $3^{rd}$ and $4^{th}$ keyword *(OC_1stkey; OC_2ndkey; OC_3rdkey; OC_4thkey)* | Occurrence score of the first, second, third and fourth keyword of the subject line and preheader, counting from the left to the right |
| Target variable | Classification *(Classification)* | Categories: <br>•*Successful* – if the open rate is above the average; <br>•*Failure* – if the open rate is below the average; <br>The email open rate of a campaign is calculated by dividing the number of emails opened by the number of emails delivered. <br>Average Email Open Rate $\approx 16{,}81\%$ |

The occurrence score of a keyword is equal to the difference between the number of times the keyword was in the "Successful" set in past campaigns and the number of times it was in the "Failure" set [2]. This feature captures the consistency in performance of a keyword. In fact, a positive occurrence score indicates the keyword has a good opening performance, as it was mostly present in campaigns with an email open rate above the average.

## 4.3    Data Exploration

### Bivariate Analysis

This study aims to discover how marketers can improve the email open rate of their marketing campaigns. Hence, in this analysis the goal was to find the variables significantly associated with the open rate.

Since none of the numerical variables follows a Normal distribution, the proper test to analyze the correlation between pairs of numerical variables is the Spearman's rank correlation coefficient. The results of this test demonstrated that all the numerical variables are significantly correlated with the open rate, except the occurrence score of the first, third and fourth keyword (for a significance level of 5%). The variables more correlated with the open rate are the number of emails sent and the number of days the campaign was sent. Note that the length of the email subject line is negatively correlated with the open rate, as Chittenden and Rettie pointed out [7]. On the other hand, the length of the preheader and the number of keywords is slightly positively correlated. The dataset campaigns with a higher email open rate have a subject line with a number of characters between 33 and 42 and a preheader with a number of characters between 93 and 115. The occurrence score of the first and second keyword is positively correlated with the email open rate. This means that a better performance of the first and second keyword in past campaigns increases the open rate.

To finish, we performed the Kruskal-Wallis test to study the distribution of the open rate in the campaigns with and without a recognizable sender. We concluded that the dataset campaigns with a sender different than the company name tend to have a higher open rate (p-value approximately equal to 0). That statement is slightly controversial because, in general, using an email sender that is not recognizable by the recipient can negatively impact the open rate [4]. A possible reason that can justify this might be related to the bad sender's reputation of financial institutions in the market [1]. For those institutions, using their names as the sender can induce the recipient to not open the email.

### Text Visualization

The goal of this text analysis was to discover which keywords and keyword combinations are associated with a higher open rate. To derive the set of keyword combinations we used the *Knime* node *Term Neighborhood Extractor*, which extracts the first right neighbor of each keyword in the email subject line and preheader. Afterwards, the Frequency filter method was applied to remove the low-frequency terms, i.e., those that were present in less than two marketing campaigns. To finish, we used a keyword cloud to visualize the terms that were linked with higher email open rates. We could infer that some of the best keywords to use in the dataset campaigns are *rápido, comisión, crédito, préstamo, gratis, dinero* and *fácil*[5]. Additionally, some of the best keyword combinations are *{cambiar; banco}, {tarjeta; gratis}, {rápido; online}, {rápido; fácil}* and *{gratis; comisión}*[6].

## 5   Data Preparation

### 5.1   Data Transformation

Data transformation is the process of transforming data from one format to another, more suitable for applying Data Mining techniques. In this process we performed feature transformation, as many algorithms require the input features to be numerical.

The categorical variables, *HasDigits, Personalization* and *Sender_recognize*, only take two possible classes. Thus, for these variables the transformation made was to replace the observations belonging to the class *Yes* by the number 1 and the ones belonging to the class *No* by the number 0. For the categorical variables without any ordinal relationship between the categories, the One-Hot Encoding is one of the most used methods. According to this method, for each category of a variable a new column is created, where the value is 1 if for that observation the original feature assumes that value and 0 otherwise. We used this method for the variable *Punctuation*, that has three possible classes: *Affirmative, Exclamation* and *Interrogation*.

### 5.2   Feature Selection

The Feature Selection process has a huge impact in the performance of an algorithm. Hence, it is important to determine what are the most relevant features to the target variable. For that purpose, the feature selection experiment performed was to filter the input redundant features i.e., the features that are highly correlated. We identified these variables as being the pair of variables with a Spearman's rank correlation coefficient higher than 50%. Having redundant features does not add significant information to the existing set of features, as they carry similar information. Therefore, we can remove one of two highly correlated variables without losing important data. By reducing the set of features, the running time of the algorithm considerably decreases and, at the same time, the performance of the model increases [12].

## 6   Modelling

In this study, supervised classification algorithms were applied because the Data Mining problem in hand was to find to which set of classes (*Successful* or *Failure*) a new campaign belongs to, based on the training set containing past campaigns whose classification is already known. The classification algorithms used were the following: Decision Tree (C4.5) [14], Random Forest [5] and Gradient Tree Boosting [9]. The set

---

[5] Fast, commission, credit, loan, free, money and easy.

[6] {change; bank}, {card; free}, {fast; online}, {fast; easy} and {free; commission}.

of features that were used to train and test these algorithms were: *Sent; Sent_days; Nr_keywords; Length_subject; Length_preheader; Affirmative; Exclamation; Interrogation; HasDigits; Personalization; Sender_recognize; OC_1stkey; OC_2ndkey; OC_3rdkey; OC_4thkey* and *Classification*. To validate the performance of these models in new and unseen data, the 10-Fold Cross Validation method was used.

# 7 Evaluation

In this section we describe the experiments performed to select the best model for this classification problem, by using the 10-Fold Cross Validation method.

**Feature Selection Techniques*:**

1. No Feature Selection;
2. Filter the redundant variables[7]: variable *Sent_days, Nr_keywords* and *Exclamation*.

After comparing the results in Table 3, we concluded that the model that had the best performance was the Random Forest, when using the redundant feature selection technique. This model accurately predicted 82% of the observations, achieving approximately an AUC of 89% and a F-score of 71% (for the *Successful* class). The model achieved a very good precision and recall for the *Failure* class, of 83% and 93% respectively. The recall for the *Successful* class was slightly lower. This is probably justified by the unbalanced dataset, where 65,44% of the campaigns belong to the *Failure* class. The standard deviation of the 10-fold Cross Validation estimates of this model was approximately equal to 8,3%.

**Table 3.** Evaluation metrics.

| Technique* | Decision tree | | | Random forest | | | Gradient tree boosting | | |
|---|---|---|---|---|---|---|---|---|---|
| | AUC | F-s | Accuracy | AUC | F-s | Accuracy | AUC | F-s | Accuracy |
| 1 | 0,77 | 0,52 | 0,74 | 0,84 | 0,61 | 0,77 | 0,85 | 0,66 | 0,78 |
| 2 | 0,73 | 0,53 | 0,73 | 0,89 | 0,71 | 0,82 | 0,87 | 0,66 | 0,78 |

F-s: F-Score for the *Successful* class

The features that have the most significant impact on the classification are the number of emails sent, the occurrence score of the 4th, 3rd and 2nd keyword, the length of the preheader and the occurrence score of the 1st keyword (by decreasing order of weighted feature importance).

---

[7] This filter selection was performed inside each one of the ten Cross Validation loops.

# 8  Deployment

In this section, the insights gained from the previous analysis were validated in the company business through A/B testing. With this experiment, we could test small variations in the subject line of a campaign and find which version leads to a higher open rate. During the test period, half of the email recipients are randomly sent the control version (the actual version) and the other half receives simultaneously the treatment version (the new version being tested). After the test ends, the recipients will then receive the winning version, i.e. the version with a higher open rate. The elements of the subject line that we could test were the type of punctuation (with or without exclamation point; with or without question mark), the use of personalization (with or without a personalized greeting) and digits (with or without digits). These features were present in the Random Forest model developed, that correctly predicted the opening performance of 82% marketing campaigns.

To test the significance of the A/B test results, we used the Two Sample Z-Test with a significance level of 5%. The results regarding the use of personalization and the question mark in the subject line were statistically significant in increasing the open rate (p-value equal to 0 for both tests). The presence of digits and the exclamation point doesn't have a strong impact in the open rate, as the test results were not statistically significant (p-value equal to 0,42 and 0,35 respectively). Thus, we advise the company to include, if possible, personalization and the question mark in the email subject line to improve the open rate of the campaigns in study.

# 9  Conclusion

To our knowledge, this is the first publication that does a profound qualitative analysis of the key factors driving the opening behavior of financial marketing campaigns. These days, financial institutions are using Email Marketing as an important traffic source in their marketing strategy. Therefore, this study will be important in guiding financial marketers on how to improve the open rate of an email campaign. With that purpose, we developed a classification algorithm that can predict if a campaign will be labeled as *Successful* or *Failure*. A campaign is classified as *Successful* if it has an open rate above the average, else it is labeled as *Failure*. We tested three different classifiers – Decision Tree, Random Forest and Gradient Tree Boosting. The model that achieved the best performance was the Random Forest, when using the redundant filter selection technique. This model could accurately predict the performance of 82% campaigns, achieving an AUC of 89% and a F-score of 71% (for the *Successful* class). By using this model, marketers will have the chance to sooner correct potential problems in a campaign that could highly impact its revenue.

The features that revealed to be important to predict the opening performance of a campaign were the number of emails sent, the length of the preheader and the occurrence score of the keywords used in the subject line and preheader. As Balakrishnan and Parekh [2], our study acknowledges the importance of taking in account the historic performance of a keyword, in the past campaigns, when predicting the effectiveness of an email campaign. Concerning the email sender, we concluded that the dataset campaigns with a sender different than the name of the company tend to have a higher open rate. Note that we were not able to validate this in the deployment stage. Regarding the email subject line, we advise marketers to avoid using long subject lines since it can negatively impact the open rate. In addition, using a personalized greeting and the question mark in the subject line can significantly improve the open rate of the email marketing campaigns in study.

For future research, we consider important to also include features at the recipient level. For instance, the recipient location, device type, domain and time the email is sent and opened by the recipient. Lastly, it will be interesting to analyze the impact of email segmentation in the open rate of financial campaigns.

**Acknowledgements.** This work is financed by National Funds through the Portuguese funding agency, FCT - Fundação para a Ciência e a Tecnologia within project: UID/EEA/50014/2019.

# References

1. Afzal, H., Khan, M.A., ur Rehman, K., Ali, I., Wajahat, S.: Consumer's trust in the brand: can it be built through brand reputation, brand competence and brand predictability. Int. Bus. Res. **3**(1), 43 (2010)
2. Balakrishnan, R., Parekh, R.: Learning to predict subject-line opens for large-scale email marketing. In: 2014 IEEE International Conference on Big Data (Big Data), pp. 579–584. IEEE, October 2014
3. Berthold, M.R., et al.: KNIME: the konstanz information miner: version 2.0 and beyond. ACM SIGKDD Explor. Newslett. **11**(1), 26–31 (2009)
4. Biloš, A., Turkalj, D., Kelić, I.: Open-rate controlled experiment in e-mail marketing campaigns. Trziste/Market **28**(1), 93–109 (2016)
5. Breiman, L.: Random forests. Mach. Learn. **45**(1), 5–32 (2001)
6. Bonfrer, A., Drèze, X.: Real-time evaluation of email campaign performance. Mark. Sci. **28**(2), 251–263 (2009)
7. Chittenden, L., Rettie, R.: An evaluation of email marketing and factors affecting response. J. Target. Meas. Anal. Mark. **11**(3), 203–217 (2003)
8. Email marketing industry report. https://www.campaignmonitor.com/resources/guides/2018-email-marketing-industry-report/
9. Friedman, J.H.: Greedy function approximation: a gradient boosting machine. Ann. Stat. **29**, 1189–1232 (2001)
10. IBM Corp. Released 2016. IBM SPSS Statistics for Windows, Version 24.0. IBM Corp, Armonk, NY (2016)
11. Jaidka, K., Goyal, T., Chhaya, N.: Predicting email and article clickthroughs with domain-adaptive language models. In: Proceedings of the 10th ACM Conference on Web Science, pp. 177–184. ACM, May 2018
12. Koller, D., Sahami, M.: Toward optimal feature selection. In: Proceedings of the Thirteenth International Conference on International Conference on Machine Learning, pp. 284–292. ICML'96 (1996)
13. Luo, X., Nadanasabapathy, R., Zincir-Heywood, A.Nur, Gallant, K., Peduruge, J.: Predictive analysis on tracking emails for targeted marketing. In: Japkowicz, N., Matwin, S. (eds.) DS 2015. LNCS (LNAI), vol. 9356, pp. 116–130. Springer, Cham (2015). https://doi.org/10.1007/978-3-319-24282-8_11
14. Quinlan, J.R.: Induction of decision trees. Mach. Learn. **1**(1), 81–106 (1986)
15. Wirth, R., Hipp, J.: CRISP-DM: towards a standard process model for data mining. In: Proceedings of the Fourth International Conference on the Practical Applications of Knowledge Discovery and Data Mining, pp. 29–39. Citeseer, April 2000

# Enhancing BMI-Based Student Clustering by Considering Fitness as Key Attribute

Erik Dovgan[1]([envelope])([ORCID]), Bojan Leskošek[2], Gregor Jurak[2]([ORCID]), Gregor Starc[2]([ORCID]), Maroje Sorić[2]([ORCID]), and Mitja Luštrek[1]([ORCID])

[1] Department of Intelligent Systems,
Jožef Stefan Institute, Jamova cesta 39, 1000 Ljubljana, Slovenia
{erik.dovgan,mitja.lustrek}@ijs.si
[2] Faculty of Sport, University of Ljubljana,
Gortanova ulica 22, 1000 Ljubljana, Slovenia
{bojan.leskosek,gregor.jurak,gregor.starc,maroje.soric}@fsp.uni-lj.si

**Abstract.** The purpose of this study was to redefine health and fitness categories of students, which were defined based on body mass index (BMI). BMI enables identifying overweight and obese persons, however, it inappropriately classifies overweight-and-fit and normal-weight-and-non-fit persons. Such a classification is required when personalized advice on healthy life style and exercises is provided to students. To overcome this issue, we introduced a clustering-based approach that takes into account a fitness score of students. This approach identifies fit and not-fit students, and in combination with BMI, students that are overweight-and-fit and those that are normal-weight-and-non-fit. These results enable us to better target students with personalized advice based on their actual physical characteristics.

**Keywords:** Improving BMI-based classification · Fitness-based clustering · Multiobjective problem

## 1 Introduction

According to WHO, overweight and obesity have become urgent global health issues in recent decades [5]. Overweight and obese persons are classified according to the body mass index (BMI). This weight-to-height index enables defining categories of adolescents such as Overweight and Obese Adolescents (OOA) categories [1]. OOA defines four categories from the lowest to the highest BMI: underweight, normal weight, overweight, and obese. The BMI bounds for these categories are sex- and age-specific, and are typically given with sex-specific BMI-for-age charts.

This work is part of a project that has received funding from the European Union's Horizon 2020 research and innovation programme under grant agreement no. 727560.

© Springer Nature Switzerland AG 2019
P. Kralj Novak et al. (Eds.): DS 2019, LNAI 11828, pp. 155–165, 2019.
https://doi.org/10.1007/978-3-030-33778-0_13

The main advantage of the BMI index and the resulting categories is its simplicity to measure. More precisely, it requires only two easy-to-obtain measurements: body weight and height. Its simplicity also represents its drawback: BMI fails to identify persons that, for example, have high muscle mass. Although they are overweight according to BMI, they are fit and should be treated differently than overweight persons without high muscle mass. This is a key issue when providing personalized advice on healthy life style and exercises, e.g., to students in high school. For example, the advice for students with high BMI and high muscle mass should be significantly different than for those with high BMI only.

BMI in combination with OOA has been widely used to study the correlation between obesity and health conditions in the last decades. For example, various risk factors were analyzed with respect to the OOA categories [1]. In some cases, however, BMI is not enough for accurate prediction. For example, it was shown that the prevalence of excess adiposity is overestimated by BMI in blacks within the pediatric population [10], which mirrors our own observation that BMI is not always appropriate for health-related clustering.

There were also studies on the relation between BMI and fitness. For example, cardiovascular risk profile was investigated in Caucasian males with at least 3 h of sports activity per week and the results showed that the threshold for an optimal BMI concerning cardiovascular risk factors might be far below $25\,kg/m^2$ even if other lifestyle conditions are apparently optimal [7]. Heart failure mortality in men was studied in relation to cardiorespiratory fitness and BMI, and the results showed that the risk factor was significantly lower in fit compared with unfit men in normal and overweight body mass index but not in obese men [4].

The existing research shows that both BMI and fitness are important for assessing health status of persons and predicting health issues. In addition, it also shows that BMI and fitness score are two distinctive measurements: we cannot precisely predict one from the other, although some correlation exists. See, for example, Farrell et al. [4] who showed that there are unfit and normal weight persons, and those that are fit and obese. However, in contrast to BMI, there is no commonly used definition of fitness score. We propose to overcome this issue by considering a widely used test battery. This test battery is performed by students in Slovenian schools once a year and enables us to calculate an overall fitness score as well as access the main components of physical fitness (see Table 1). In contrast to related work, we do not predefine the clusters of fit and not fit persons, but we apply a multiobjective approach with three objectives to search for the best split into fit/non-fit clusters. In addition, the fitness score in combination with OOA categories enables the identification of persons that are overweight or obese but are fit, and those that have normal weight but are not fit. The resulting categories of students enable the teachers, parents and policy makers to create and provide personalized and better-targeted advice, recommendations and curricula.

The paper is further organized as follows. The fitness-based approach for clustering students is described in Sect. 2. Section 3 reports the experiments including the dataset and the results. Finally, Sect. 4 concludes the paper with ideas for future work.

**Table 1.** The physical fitness tests of the test battery. All the test measurements are in percentiles.

| Fitness test | Measurement |
|---|---|
| PTSF | Thickness of triceps skinfold |
| PAPT | Reaction time during arm plate tapping |
| PSBJ | Distance jumped during standing broad jump |
| POCB | Time to pass a polygon backwards and on all fours |
| PSU | Number of sit-ups in 60 s |
| PSR | Distance between fingertips and toes when standing and bending forward |
| PBAH | Time in a bent arm position while hanging from a bar |
| P60m | Time to run 60 m |
| P600m | Time to run 600 m |

## 2   A Fitness-Based Approach for Clustering Normal Weight, Overweight and Obese Students

There is no golden standard for deciding who is fit and who is not. The most straightforward approach to separate students who are fit from those who are not is to apply a threshold to the overall fitness score. However, it is not clear what this threshold should be, and since we have measurements of the main components of physical fitness available, we should consider whether they can be used to achieve a better separation. In our clustering, we explore these questions and finally propose an approach for separating the fit students from the non-fit.

### 2.1   Fitness Score

The fitness score is calculated by taking into account a set of physical fitness measurements. These measurements are obtained with the SLOfit test battery[1], i.e, a version of Eurofit Physical Fitness Test Battery [3], which is a set of physical fitness tests covering flexibility, speed, endurance, and strength. The selected set of measurements is shown in Table 1. For each measurement, a quantile (percentile) rank is calculated by taking into account sex and age. Utility functions then transform these ranks on a scale ranging from 0 to 100 points, where 0 is the worst possible score and 100 is the best. Finally, the points of all the measurements are summed up and the fitness score is determined as the quantile rank by taking into account reference population, sex and age.

### 2.2   Measuring Clustering Error

The fitness score enables the evaluation of clusters of students within a dataset: students within a cluster should have similar fitness score, while fitness score

---

[1] http://en.slofit.org/measurements/test-battery.

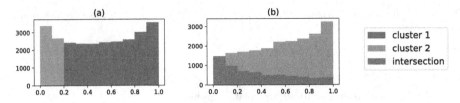

**Fig. 1.** Examples of clusters: (a) good clusters, i.e., there is no intersection between clusters; (b) bad clusters, i.e., intersection between the clusters is very high.

of various clusters should be different. To evaluate a pair of clusters, we firstly calculate the histograms of both clusters with respect to the fitness score. Next, we find the intersection between the histograms. The intersection represents the overlap between clusters, which ideally should be 0, since clusters should be disjunctive. Therefore, this intersection represents the error that is then normalized with respect to the size of both clusters. The resulting maximal error percentage between both clusters is then used as the amount of error with respect to the fitness score $(e_f)$. Examples of histograms of clusters and intersections between them are presented in Fig. 1: Good clusters with no overlap are shown in Fig. 1a, while Fig. 1b depicts bad clusters with high percentage of overlap.

The same error function can be also applied to percentile ranks of fitness components (i.e., physical fitness measurements), which can be interpreted as follows: we want to find clusters in which students have similar percentile ranks of fitness components, while the percentile ranks between clusters should differ. As a consequence, the performance of fit and non-fit students with respect to individual components should be different. The error measure based on percentile ranks of fitness components $(e_c)$ is thus calculated as the average of all the errors of individual fitness components.

Although the clusters of students with respect to the fitness score can significantly differ from the clusters based on OOA, it is reasonable to assume that the ratio between students with normal weight and those that are overweight or obese is similar to the ratio between fit and non-fit students. Note that the boundary between people with normal weight and those that are overweight or obese is to some degree arbitrary, and the same can be said for those who are fit or not. Therefore we assume the same ratio for the latter as for the former. As a consequence, the number of fat-and-fit students should be roughly the same as the number normal-weight-and-non-fit students. However, the exact numbers might differ, therefore we measure the error with respect to size difference $(e_s)$ as the normalized difference between the size of normal weight students and the size of fit students.

The proposed approach enables us to evaluate and compare various clustering algorithms that aim at clustering students into fit and non-fit clusters. The comparison is done in three-objective space, where the errors $(e_f, e_c, e_s)$ represent the dimensions, i.e., objectives, of this space: the error with respect to the total fitness $(e_f)$, the average error with respect to individual fitness components $(e_c)$, and the error with respect to the size $(e_s)$. Note that all the errors should be minimized.

### 2.3  Clustering Based on Fitness Score

Besides applying existing clustering algorithms to solve the problem of finding clusters of fit and non-fit students, we also propose the following algorithm. First, the fitness score is discretized equidistantly. Second, each discretized value is used as a limit as follows: all the students with lower fitness score are added to the first cluster, while the students with higher score are added to the second cluster. Each such pair of clusters is evaluated with respect to the error functions $(e_f, e_c, e_s)$. In comparison to other clustering algorithms, this approach has the advantage of being intuitive, easy to understand, and very effective. Its performance in comparison to other clustering algorithms is presented in the following section.

## 3  Experiments and Results

This section presents the dataset of students that were clustered, the clustering algorithms the were applied, and the obtained results with discussion.

### 3.1  Dataset of Physical Fitness Measurements

We evaluated our approach on a dataset of students from Slovenian schools, SLOfit[2]. More precisely, we only analyzed the data of high school students (ages 16–21). In addition, only the most recent year of measurements was used, i.e., 2018. The attributes for the clustering algorithms were percentile ranks of fitness components and are shown in Table 1. Moreover, only normal weight, overweight and obese students were selected. Note that the same approach can also be applied to underweight students, however, for the domain experts the most relevant division is between normal weight and overweight students. In total, 27,304 students were taken into account.

### 3.2  Clustering Algorithms

The clusters of fit and not-fit students were found with a set of clustering algorithms. Since the goal was to cluster in two clusters, only those algorithms that enabled defining the number of clusters were selected. However, several clustering algorithms have a high computational complexity, therefore, only a subset of data was clustered with those. In addition, some algorithms enabled creating a model on the subset and afterward cluster all the data with that model. The applied clustering algorithms and their characteristics are shown in Table 2. This table shows, for example, that spectral clustering has a high computational complexity, since only 5000 data could be clustered at once, and does not build

---

[2] http://www.slofit.org/.

**Table 2.** Evaluated clustering algorithms.

| Clustering algorithm | Clustered data | Cluster all data with a model | Randomly set parameters |
|---|---|---|---|
| OOA (Default, based on [5]) | All | Not needed | / |
| k-means [6] | All | Not needed | Random state |
| BIRCH [11] | 5000 | Yes | Threshold, Data sample |
| Spectral clustering [9] | 5000 | No | Random state, Data sample |
| Hierarchical clustering [8] | 5000 | No | Data sample |
| Fitness score (see Sect. 2.3) | All | Not needed | Fitness score bound |

a model to cluster the entire dataset after clustering the subset of data. On the contrary, k-means has a lower computational complexity since it was able to cluster the entire dataset at once. Consequently, it was not required to use subset of data and build a model to cluster all the data. BIRCH is something in between: it has a high computational complexity, therefore it could cluster only subset of data. However, it enables building a model on this subset of data, which was then used to cluster the entire dataset.

In our experiment, all of these algorithms were run 1000 times with randomly set parameter values (and randomly selected subset of data, if all the data could not be clustered due to algorithm's high computational complexity).

### 3.3   Results of the Clustering Algorithms

All the algorithms had to cluster the students into two clusters, i.e., students that are fit and those that are not fit. As described in Sect. 3.2, 1000 runs of each algorithm were performed, therefore the results of all the runs are presented. Each algorithm run was evaluated and is presented in terms of three objective/error functions $(e_f, e_c, e_s)$ as described in Sect. 2.3.

The results in three-dimensional objective space are shown in Fig. 2a. In addition, Figs. 2b–c show two additional perspective of the objective space: the first one focuses on the fitness score error, while the second one focuses on fitness components' and delta size errors. Since all three objectives are minimized, the optimal solution would be in (0, 0, 0), which is at the bottom left of all three figures.

These results show that the OOA clustering is not good with respect to the fitness score and fitness components' errors, since all the other algorithms are better in these two objectives. On the other hand, it is optimal with respect to the delta size error, which is true by definition, since delta size error measures the difference of sizes of the obtained clusters compared to the OOA clusters. In addition, k-means, spectral clustering and Fitness score clustering find the best splits with respect to the fitness components' error, while the Fitness score

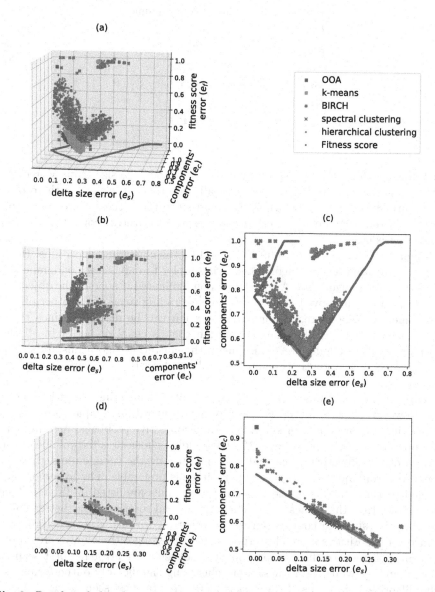

**Fig. 2.** Results of the clustering algorithms: (a) three-dimensional objective space; (b) focus on fitness score error; (c) projection on two objectives: fitness components' and delta size errors; (d) nondominated solutions in the three-dimensional objective space; (e) nondominated solutions projected on two objectives: fitness components' and delta size errors.

**Table 3.** Hypervolume of the clustering algorithms.

| Clustering algorithm | Hypervolume |
|---|---|
| OOA | 0.006 |
| k-means | 0.368 |
| BIRCH | 0.372 |
| Spectral clustering | 0.309 |
| Hierarchical clustering | 0.343 |
| Fitness score | 0.450 |

clustering also finds the best splits with respect to the delta size error. Moreover, the Fitness score clustering outperforms all the other algorithms with respect to the fitness score error, which is expected since the Fitness score clustering defines clusters that do not overlap with respect to fitness score (thus fitness score error is 0).

Figures 2a–c show all the solutions found by the clustering algorithms. However, when comparing the algorithms, it is simpler to only show nondominated solutions of each algorithm. A solution is nondominated if none of the objective functions can be improved in value without degrading some of the other objective values [2]. Therefore, a dominated solution can be discarded since there exists at least one (nondominated) solution that is equal or better in all the objectives. The nondominated solutions of the clustering algorithms are shown in Figs. 2d–e. These solutions confirm the above described comparison between the clustering algorithms.

Objective space enables us to compare results of clustering algorithms visually. However, a more appropriate approach for algorithm comparison consists of applying a unary operator suitable for multiobjective problems. A commonly used operator is the hypervolume [12]. Hypervolume measures the volume of the portion of the objective space that is dominated by the (nondominated) solutions. As a consequence, a higher hypervolume is preferable. The hypervolumes covered by the solutions of the clustering algorithms are listed in Table 3. This table shows that the Fitness score clustering found solutions that better cover the objective space in comparison to the other algorithms. Another argument in favor of the Fitness score clustering is that other algorithms only rarely outperform it in terms of fitness components and delta size error (as best seen in Figs. 2c and e), while the Fitness score clustering significantly outperforms the other algorithms in terms of the fitness score error (as best seen in Fig. 2b).

A solution found with the Fitness score clustering is presented in Fig. 3 in terms of distributions of BMI, OOA, fitness score and fitness components between the two clusters. This figure shows the clusters divided by fitness score 0.5, i.e., the division with the lowest fitness components' error.

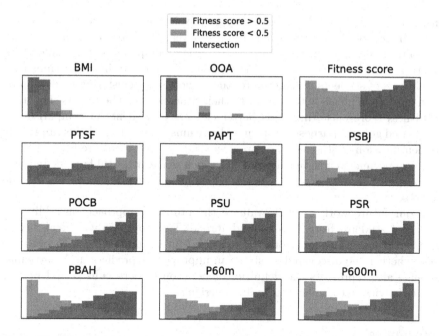

**Fig. 3.** Distribution of data with respect to BMI, OOA, fitness score and fitness components, which are additionally clustered with the Fitness score clustering in students with fitness score <0.5 and students with fitness score >0.5.

### 3.4  Discussion

The presented experiment has shown that the best clusters with respect to the three objectives are found by the Fitness score clustering. This can be seen in visual representation of the solutions in the objective space, and it is also confirmed by the hypervolumes obtained by the clustering algorithms. In addition, the Fitness score clustering enables finding clusters with the lowest (zero) delta size error, and with the lowest fitness components' error (the same fitness components' error is also achieved by the k-means algorithm). Even more, all the clusters of the Fitness score clustering have zero fitness score error, while none of the other clustering algorithms found clusters with zero fitness score error. Therefore, the Fitness score clustering performed the best among the tested algorithms.

## 4  Conclusion

This paper presented a new approach for identifying students that are overweight and fit, and those that have normal weight, but are not fit. This classification enhances the widely used BMI index that is suitable to classify students only as normal weight or overweight/obese. The presented approach introduces the fitness score calculated based on a set of physical fitness measurements performed in schools by all the students once a year. In addition, it also defines three objectives, i.e., (fitness score error, fitness components' error and delta size error), which are used to assess the quality of the clustering algorithms that find clusters of students. Furthermore, the Fitness score clustering is developed, which clusters students with respect to their fitness score. The results show that the Fitness score clustering finds better clusters of students in comparison to widely-used general-purpose clustering algorithms. The obtained clusters enable the identification of students that are overweight or obese but are fit, and those that have normal weight but are not fit, which makes it possible to define personalized and better targeted advice, recommendations and curricula for the students.

In our future work we will evaluate the proposed approach on additional datasets of students from Slovenia and abroad. This approach will be also combined with algorithms that predict students' future performance in order to assess whether the discovered clusters can improve this prediction. A particular challenge also represents the definition/generation of personalized and better-targeted advice, recommendations and curricula.

# References

1. Bacha, F., Saad, R., Gungor, N., Janosky, J., Arslanian, S.A.: Obesity, regional fat distribution, and syndrome X in obese black versus white adolescents: race differential in diabetogenic and atherogenic risk factors. J. Clin. Endocrinol. Metab. **88**, 2534–2540 (2003)
2. Deb, K., Pratap, A., Agrawal, S., Meyarivan, T.: A fast and elitist multiobjective genetic algorithm: NSGA-II. IEEE Trans. Evol. Comput. **6**(2), 182–197 (2002)
3. Eurofit: Eurofit Tests of Physical Fitness. Council of Europe, Strasbourg, 2 edn. (1993)
4. Farrell, S.W., Finley, C.E., Radford, N.B., Haskell, W.L.: Cardiorespiratory fitness, body mass index, and heart failure mortality in men. Circ. Hear. Fail. **6**(5), 898–905 (2013)
5. Kallioinen, M., Granheim, S.I.: Overweight and obesity in the western pacific region. Technical report, World Health Organization (2017)
6. Kanungo, T., Mount, D.M., Netanyahu, N.S., Piatko, C.D., Silverman, R., Wu, A.Y.: An efficient k-means clustering algorithm: analysis and implementation. IEEE Trans. Pattern Anal. Mach. Intell. **24**, 881–892 (2002)
7. Ortlepp, J.R., Metrikat, J., Albrecht, M., Maya-Pelzer, P., Pongratz, H., Hoffmann, R.: Relation of body mass index, physical fitness, and the cardiovascular risk profile in 3127 young normal weight men with an apparently optimal lifestyle. Int. J. Obes. **27**, 979–982 (2003)
8. Rokach, L., Maimon, O.: Clustering methods. In: Maimon, O., Rokach, L. (eds.) Data Mining and Knowledge Discovery Handbook, pp. 321–352. Springer, Boston (2005). https://doi.org/10.1007/0-387-25465-X_15
9. Shi, J., Malik, J.: Normalized cuts and image segmentation. IEEE Trans. Pattern Anal. Mach. Intell. **22**, 888–905 (2000)

10. Weber, D.R., Moore, R.H., Leonard, M.B., Zemel, B.S.: Fat and lean BMI reference curves in children and adolescents and their utility in identifying excess adiposity compared with BMI and percentage body fat. Am. J. Clin. Nutr. **98**(1), 49–56 (2013)
11. Zhang, T., Ramakrishnan, R., Livny, M.: Birch: an efficient data clustering method for very large databases. In: Proceedings of the 1996 ACM SIGMOD International Conference on Management of Data, pp. 103–114 (1996)
12. Zitzler, E., Thiele, L.: Multiobjective evolutionary algorithms: a comparative case study and the strength Pareto approach. IEEE Trans. Evol. Comput. **3**(4), 257–271 (1999)

# Deep Learning Does Not Generalize Well to Recognizing Cats and Dogs in Chinese Paintings

Qianqian Gu[1(✉)] and Ross King[2]

[1] The University of Manchester, Manchester M13 9PL, UK
qianqian.gu@manchester.ac.uk
[2] The Alan Turing Institute, London NW1 2DB, UK

**Abstract.** Although Deep Learning (DL) image analysis has made recent rapid advances, it still has limitations that indicate that its approach differs significantly from human vision, e.g. the requirement for large training sets, and adversarial attacks. Here we show that DL also differs in failing to generalize well to Traditional Chinese Paintings (TCPs). We developed a new DL object detection method A-RPN (Assembled Region Proposal Network), which concatenates low-level visual information, and high-level semantic knowledge to reduce coarseness in region-based object detection. A-RPN significantly outperforms YOLO2 and Faster R-CNN on natural images (P < 0.02). We applied YOLO2, Faster R-CNN and A-RPN to TCPs with a 12.9%, 13.2% and 13.4% drop in mAP compared to natural images. There was little or no difference in recognizing humans, but a large drop in mAP for cats and dogs (27% & 31%), and very large drop for horses (35.9%). The abstract nature of TCPs may be responsible for DL poor performance.

**Keywords:** Traditional Chinese Paintings · Computational aesthetics · Deep Learning · Object recognition · Machine learning

## 1 Introduction

One of the greatest mysteries in cognitive science is the architecture of human visual system and its virtuosity in object recognition. Human have the impressive ability to recognize visually presented objects with both high speed and accuracy. Attempts to replicate this ability have been made since the start of Artificial Intelligence (AI) research, but these have met with only limited success. Object recognition is traditionally one of the most intractable problems in AI [1].

Recently, with the advent of very large annotated image databases, and advances in Deep Learning (DL), have greatly improved AI object recognition [2]. However, despite many claims to the opposite, DL object recognition is still not nearly as good as in humans. Example of the weaknesses of DL are: the requirement for large training sets, and their susceptibility to adversarial attacks that do not confuse humans. It is hypothesized that the reason for this

© Springer Nature Switzerland AG 2019
P. Kralj Novak et al. (Eds.): DS 2019, LNAI 11828, pp. 166–175, 2019.
https://doi.org/10.1007/978-3-030-33778-0_14

is compiled background knowledge about the world encoded in the human visual system architecture [3,4]. This knowledge was learnt through millions of years of Darwinian evolution.

Visual art is present in all human societies, and dates back as long ago as the Paleolithic. Computational visual aesthetics investigates the relationship between art and computational science. There is a close relationship between the human visual system and the appreciation of art (Gombrich (2000) Art and Illusion, Princeton University Press;) *Computational Aesthetics* is the computational investigation of aesthetics. Computational aesthetics is a large subject [5–7], but relatively little work has been done at the interface between computational aesthetics and DL object recognition [8,9]. The limited evidence, based on Western art, points to significant differences in performance between natural image trained classifiers and painting-trained classifiers. To the best of our knowledge the relationship and differences between object recognition in natural images and art has not been investigated.

Traditional Chinese Painting (TCP) is one of the great art traditions of the World (206 BC - now) [10]. Its unique style, which dates back over 2000 years, is instantly recognizable to humans. Here, to explore the differences between human and DL object recognition, we investigate DL object recognition in natural images and TCPs.

## 2   Object Detection in Chinese Painting

Object detection is one of the most widely studied topics in image processing. It differs from the classical image classification problems where models classify images into a single category corresponding to their most salient object. Here we investigate transferring Deep Learning (DL) object detection models from natural image to TCPs.

There are two potential challenges in applying DL to detecting objects in Chinese Paintings. First, there does not exist an official well-structured Chinese Painting Image Database and associated Ontology. And existing work in this area [11] is limited. Second, the characteristics of TCP images are different from natural images as they are produced using specialized tools, materials and techniques.

### 2.1   Chinese Painting and TCPs Data

Chinese painting can be categorized into two major schools of styles: XieYi, or GongBi [12]. XieYi (freehand strokes, Fig. 2) paintings are characterized by exaggerated forms, with the painted objects abstract and sometimes without fully connected edges. GongBi (skilled brush or meticulous approach, Fig. 1) paintings are characterized by close attention to detail and fine brushwork, and the painted objects are more realistic. TCPs can be further classified as: figure painting, landscapes painting, or flower-and-bird painting; only figure paintings and flower-and-bird paintings depict objects as their main subject.

To form our TCPs object detection dataset we integrated two datasets: a self-collected dataset obtained from online open sources, and a high-resolution image

**Fig. 1.** Emperor Huizong Song's skilled brush flower-and-bird painting from Song Dynasty

**Fig. 2.** Beihong Xu's freehand strokes flower-and-bird painting

dataset provided by the Chinese Painting and Calligraphy Community. The sizes of these datasets are: 3,000 images, and 10,000 images respectively. The TCP objects dataset we used in this experiment contains 1,400 images in seven classes: human, horse, cow, bird, plant, cat, and dog. We manually annotated the data with class labels and segmentation knowledge (bounding box), as ground truth. (List of paintings we have used could be found: https://github.com/hiris1228/ TCP_object_detect.git).

## 2.2   Object Detection

Current DL object detection models generally identify various objects and location within one single image. A naive approach to solving this problem is to take different regions of interest from the image, and to classify the presence of the object within that region – usually using a ConvNets [13,14]. The R-CNN (Regions + CNN) method was developed to avoid the use of excessive number of regions in object detection. This relies on an external region proposal system based on a selective search algorithm [15–17]. The cost-reduced computational shared model R-FCN [18] attempts to balance translation-invariance in image classification, and translation translation-variance in object detection. The Mask R-CNN approach [19] applies FCN (Fully Convolutional Networks) [20] to each Region of Interest (RoI), and then applies classification and bounding box regression in parallel. Single Shot detectors like YOLO [21] and SSD (Single Shot MultiBox Detector [22]) and are able to obtain relatively good results.

## 3   Chinese Painting Object Detection Approach

The architecture of our TCPs Object Detector A-RPN is a VGG-16 [23] backbone base model as shown in Fig. 3. We adopted the popular two-stage object detection strategy of Faster R-CNN [17]. We generate a set of RoIs, and a classification model independent of selected RoIs. Comparing to ResNet [23], VGG-16 is not very 'deep' (ResNet has 152 layers), but its depth is appropriate given the limited TCP data.

The assembled Region Proposal Network (A-RPN) has three components: a general RPN that return RoIs and Intersection of Union (IoU) scores; a Chinese-Kitten RPN (CN-Kitten RPN), which concatenate both lower-level and high-level features to generate small object sensitive proposals, and refines the RoIs; and a Detection Network R-FCN that takes proposed RoIs, and classifies them into categories and background. All the sub-networks within the A-RPN are fully convolutional with respect to RoIs. All three components are initialized by a hierarchical feature map $M0$ through shared convolutional layers to form a pre-trained resembling VGG-16.

The region proposals (RoIs) processed in our A-RPN are divided into two parts. One comes from the generated RPN (RoIs-set1), the other is the small-scale region boxes generated by sliding window run on convCNK (RoIs-set2). The same NMS operation, with threshold 0.7, was applied to both general RPN and CN-Kitten RPN to reduce redundancy. This process returned two set of 2k proposals. Each of these regions has an Intersection of Union (IoU) score, which estimates the chance that the current RoI contains an object.

CN-Kitten RPN is an enhanced RPN that combines multi-layer features knowledge and RoI pooling layer to refine RoI proposals. This approach is designed to increase the model's sensitivity to small-scale object detection from knowledge of Feature Pyramid Networks [24]. With differing low-level visual information, the high-level convolutional features may be too coarse when we project our RoIs from the feature map to the original image. The aim is to better fit small objects, and to better utilize fine-grained features due to TCPs characteristics. We observed no significant drop in recall with small objects.

RoIs-set2 is generated from a $1 \times 1 \times 256$ fully-connected layer which is reduced from fine-grained feature map convCNK. convCNK is the multi-layer feature concatenation of conv5 and conv4 which is similar FCN-16s [20]. We apply an unsampling filter on conv5 to obtain transposed convolutional layer [25] conv5'. Then L2-normalize each layer per spatial location, and re-scales it with the same resolution as conv4 to obtain convCNK. Simultaneously, we applied a bilinear interpretation on convCNK, and obtained a semantic segmentation heat-map of the entire image for later refinement.

In order to obtain the final top 300 proposals as input of Detection Network R-FCN. A top rank voting model was applied to both RoI-set1 and RoI-set2. This ranks all anchors with their IoU scores from the highest to the lowest. It then projects all the selected anchors to the semantic segmentation heat-map, and computes an extra IoU score with the overlapped percentage. Then it looks to find whether there are existing anchors in its opposite set that have an overlapped rate higher than 0.7, if yes, it merges these two anchors by averaging their coordinates. The method repeats this procedure until it obtains proposals with the highest 300 pairs of IoU scores.

The detection network identifies and regresses the bounding box of regions likely to contain classes base on each proposed region. Unlike in the original Faster R-CNN [17], we apply Fast R-CNN [16] with R-FCN [18] as the detection network. As a member of FCN [20] family, it construct a set of position-sensitive

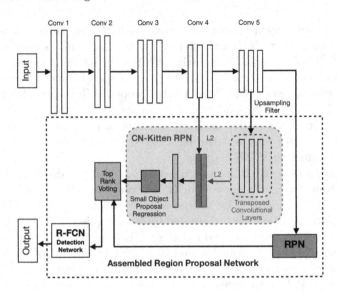

**Fig. 3.** Assembled Region Proposal Network (A-RPN) architecture.

score maps to incorporate translation variance. As there are no learning layers after the RoI layer, only a shared fully convolutional architecture, R-FCN is nearly cost-free, and even faster than the other pioneering DL image classification architectures [13, 15–17, 23] with a competitive mAP (mean Average Precision).

R-FCN uses a bank of specialized convolutional layers to encode as score maps position information with respect to a relative spatial position [18]. Our Detection network removes all FC layers, and computes all learn-able weight layers on the entire image to create a bank of position-sensitive score maps for C + 1 categories (C object categories + 1 background). Each set of score maps for one particular category represents a $k \times k$ spatial grid describing relative position information. Selective pooling only returns one score out of $k \times k$ on class prediction after performing average voting on these shared sets of score maps.

## 4    Experiment and Results

To investigate the differences between DL and human image analysis we applied DL to images that humans can easily interpret: natural image and Traditional Chinese Paintings (TCPs). We compared our A-RPN's outputs with two popular DL object detection models. We ran six experiments: Single Shot detectors on natural images, YOLO(N), and on Chinese paintings, YOLO(P); Faster R-CNN on natural images, Faster R-CNN(N), and on Chinese paintings, Faster R-CNN(P); A-RPN on natural images, A-RPN(N), and on Chinese paintings, A-RPN(P). The natural image dataset that we used was PASCAL VOC2007 trainval. Both datasets contains the same seven classes. We kept one-third of the data as test set, and divided the remaining data into 90% training set and 10% validation set. The results are shown in Table 1.

**Table 1.** The mAP Comparison for Flower-and-Bird and Figure Classification. And A-RPN has proved that it outperforms the other two models.

| Methods | YOLO(N) | YOLO(P) | Faster RCNN(N) | Faster RCNN(P) | ARPN(N) | ARPN(P) |
|---------|---------|---------|----------------|----------------|---------|---------|
| Human | 82.1 | 76.3 | 81.1 | 79.5 | 83.2 | 82.4 |
| Horse | 71.8 | 28.3 | 75.9 | 39.7 | 78.3 | 42.4 |
| Cow | 75.4 | 69.1 | 76.4 | 74.3 | 78.7 | 80.5 |
| Bird | 81.1 | 75.2 | 79.3 | 77.7 | 76.8 | 70.0 |
| Plant | 32.0 | 33.8 | 36.2 | 42.9 | 43.5 | 50.0 |
| Cat | 78.7 | 65.1 | 81.3 | 52.0 | 82.7 | 55.5 |
| Dog | 79.2 | 61.6 | 82.4 | 53.8 | 83.6 | 52.2 |
| Overall | 71.37 | 58.48 | 73.22 | 59.98 | **75.25** | **61.85** |

Table 1 shows that all the DL models can achieve more than 70% mAP on natural image object detection. A-RPN achieved the best performance in classification, and has significantly higher mAP ($P < 0.02$) than YOLO2 and Faster R-CNN. (Using the McNemar test, the Z-model statistic against YOLO2 and Faster R-CNN are respectively $-3.7905$ and $-2.4188$). Object recognition performance significantly drops when applied to TCPs. The YOLO2 model has a mAP performance drop of 12.9% while Faster R-CNN drops 13.2% and A-RPN drops 13.4%. The statistical difference in performance between the three methods on TCP object recognition is not significant.

The success of object recognition varies greatly between object class. The performance on classes 'Human' and 'Bird' is stable and accurate for all methods. Performance on the class 'Plant' is stable but has low mAP in all models. Class 'Cow' works better in natural image dataset than TCPs, but the drop is relatively small, while A-RPN increases with 1.8% mAP. All three models confuse the 'Cat' and 'Dog' classes in TCPs. For the class 'Cat' the decreases in performance were 13.6%, 29.3%, and 27.2% for A-RPN. For the class 'Dog' the decreases in performance were 17.6%, 28.6%, and 31.4%. The largest drop in performance was for the class 'Horse' in TCPs. Only A-RPN could achieve more than 40% mAP, while their mAPs on Nature Image data are all above 70%.

Table 2 and Fig. 4 show that our region proposals network has limitations when allocating object bounding boxes. There are 5 classes (out of 7 in total) that failed to achieve 85% detecting rate of one particular type of object. Intersections of Unions (IoU) is the region of interest union with the ground truth bounding box. Our threshold was set as 0.5. But after verifying, we noticed that only 63% of the ground truth bounding boxes were detected with ratios above 0.5. 22% of the objects were not detected, especially in the plant class, and 15% of all bounding box which does not cover objects.

**Table 2.** Heat-Map for A-RPN classification. Miss-classification scenarios with error more than 15% have been highlighted in the table. The N/A column represents the scenario that – there is no bounding box detected over the current ground truth bounding box.

| Truth | Prediction | | | | | | | |
|---|---|---|---|---|---|---|---|---|
| | Human | Horse | Cow | Bird | Plant | Cat | Dog | N/A |
| Human | 0.82 | 0 | 0 | 0 | 0 | 0 | 0 | **0.18** |
| Horse | 0 | 0.35 | 0.03 | **0.18** | 0.02 | 0 | **0.27** | **0.15** |
| Cow | 0 | 0.09 | 0.72 | 0 | 0 | 0 | 0.01 | **0.18** |
| Bird | 0 | 0 | 0 | 0.78 | 0.01 | 0.03 | 0 | **0.18** |
| Plant | 0 | 0 | 0 | 0.04 | 0.38 | 0.01 | 0.01 | **0.56** |
| Cat | 0 | 0 | 0 | 0.02 | 0.01 | 0.51 | **0.38** | 0.08 |
| Dog | 0 | 0.01 | 0.09 | 0 | 0 | **0.29** | 0.51 | 0.10 |

# 5    Discussion and Future Works

Putting to one side the differences between human and DL image recognition, it is interesting to consider whether in principle images in TCPs are harder to recognize using DL. We have shown A-RPN performs slightly better than YOLO and Faster R-CNN, but its success in natural image cannot be transferred to TCPs. There are a number of possible reasons for the drop in performance of DL in TCPs. First, DL models require large training sets, but the number of TCPs is quite limited, and their usage generally involve licenses. Therefore the limited size of the data set we used potentially prevented the DL models being fully trained.

Another reason may be the initialization of the layer M0 feature map. When the A-RPN is initialized, we use a pre-trained CNN layer trained on natural images. Natural images differ in a number of ways from images in TCPs. They inherently include perspective, while most Chinese paintings do not. In TCPs objects are often depicted in a highly abstract manner, easily comprehended by the human, but quite different from natural images.

Ineffective feature formation may also be a factor. CNN models have their own texture representations [23, 26]. Whether the work was 'linear' (contour-led) or 'painterly' (reliant more on brushstrokes denoting light and shadow) [27] has been proven to have effect on prediction. The majority of Chinese paintings are only black and white. In CNN models white color is often treated as 'no color', with a probability of being an object of zero, and the 'black ink' textures are similar in every single class.

We hypothesize that the abstract nature of TCPs may fundamentally restrict the ability of DL systems to recognize objects in TCPs. Objects in TCPs do not have fully connected edges. When CNN edge detection filters are applied on non-edge pixels in low-level layers, the resulting matrices will be filled with really small numbers or even zeros. In CNN texture representation strategy, the

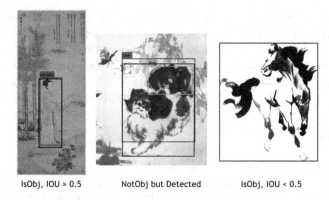

IsObj, IOU > 0.5          NotObj but Detected          IsObj, IOU < 0.5

**Fig. 4.** Bounding Box Detection Cases: Left, object detected with IoU score greater than 0.5; Middle, object detected with IoU less than threshold; Right, object not detected.

smaller the edge detection matrix is, the lower the convolution value will be. And this situation gets worse during strided convolutions. Finally, CNNs weakens the contribution from ambiguous edge pixels, and further separates edges that should be treated as connected in TCPs. This complicates object detection task in TCPs, and is also the reason why we absorbed the concept from FPN [24] to assemble features from different convolutional layers to try to maximize the sensitivity on learning edge information.

Many principles of Chinese Painting are derived from Daoism [28]. For example in Daoism empty space is an important concept, and a symbol of the void or nothingness. The most important text in Daoism states: 'Having and not having arise together' (Laozi 2). Chinese Painting has also been influenced by Buddhism, which emphasizes that 'What is form that is emptiness, what is emptiness that is form' (Paramita Hridaya Sutra). These beliefs have led Chinese paintings to stress the concept of Designing White Space — if one's mind can reach there, there is no need for the touch of any brush and 'formless is the image grand' (Laozi 4). Similarly, an important canon of Chinese painting describes its rhythmic vitality as Qi, a metaphysical concept of cosmic power: with Qi empty space is not blank, it is alive, like air. This prominent characteristic of Chinese painting turns its treatment of empty space as solid space. Taking the horse in XieYi style in Fig. 2 as an example, the absence of clear edges depicting the outline of a horse feed the human imagination.

## 6    Conclusion

We have demonstrated that state-of-the-art object DL recognition methods perform substantially worse on TCPs compared to natural images. We argue that these results point to interesting and important differences between DL systems and the human visual system that require further investigation. For example:

what features of the human visual system that enable it to effortlessly recognize cats and horses in Chinese paintings, deal with the white space, etc.; What is the relationship between aesthetics and image recognition? We hope, and expect, that further research will investigate these important questions, and lead to computational systems that can recognize objects in Chinese paintings as well as humans, and perhaps also appreciate their beauty as greatly as humans.

# References

1. Grill-Spector, K., Kourtzi, Z., Kanwisher, N.: The lateral occipital complex and its role in object recognition. Vis. Res. **41**(10–11), 1409–1422 (2001)
2. Schrimpf, M., et al.: Brain-score: which artificial neural network for object recognition is most brain-like? BioRxiv, p. 407007 (2018)
3. Grill-Spector, K., Weiner, K.S.: The functional architecture of the ventral temporal cortex and its role in categorization. Nat. Rev. Neurosci. **15**(8), 536 (2014)
4. Cichy, R.M., Khosla, A., Pantazis, D., Torralba, A., Oliva, A.: Comparison of deep neural networks to spatio-temporal cortical dynamics of human visual object recognition reveals hierarchical correspondence. Sci. Rep. **6**, 27755 (2016)
5. Birkhoff, G.D.: Aesthetic Measure. Harvard University Press, Cambridge (1933)
6. Greenfield, G.: On the origins of the term computational aesthetics (2005)
7. Neumann, L., Sbert, M., Gooch, B., Purgathofer, W., et al.: Defining computational aesthetics. In: Computational Aesthetics in Graphics, Visualization and Imaging, pp. 13–18 (2005)
8. Gatys, L.A., Ecker, A.S., Bethge, M.: A neural algorithm of artistic style. arXiv preprint: arXiv:1508.06576 (2015)
9. Crowley, E.J., Zisserman, A.: The art of detection. In: Hua, G., Jégou, H. (eds.) ECCV 2016, Part I. LNCS, vol. 9913, pp. 721–737. Springer, Cham (2016). https://doi.org/10.1007/978-3-319-46604-0_50
10. Lanciotti, L.: The British museum book of Chinese art (1993)
11. Wu, X., Li, G., Liang, Y.: Modeling Chinese painting images based on ontology. In: 2013 International Conference on Information Technology and Applications, pp. 113–116. IEEE (2013)
12. Jiang, S., Huang, T.: Categorizing traditional Chinese painting images. In: Aizawa, K., Nakamura, Y., Satoh, S. (eds.) PCM 2004, Part I. LNCS, vol. 3331, pp. 1–8. Springer, Heidelberg (2004). https://doi.org/10.1007/978-3-540-30541-5_1
13. Krizhevsky, A., Sutskever, I., Hinton, G.E.: ImageNet classification with deep convolutional neural networks. In: Advances in Neural Information Processing Systems, pp. 1097–1105 (2012)
14. LeCun, Y., et al.: Backpropagation applied to handwritten zip code recognition. Neural Comput. **1**(4), 541–551 (1989)
15. Girshick, R., Donahue, J., Darrell, T., Malik, J.: Rich feature hierarchies for accurate object detection and semantic segmentation. In: Proceedings of the IEEE Conference on Computer Vision and Pattern Recognition (CVPR), pp. 580–587 (2014)
16. Girshick, R.: Fast R-CNN. In: The IEEE International Conference on Computer Vision (ICCV), December 2015
17. Ren, S., He, K., Girshick, R., Sun, J.: Faster R-CNN: towards real-time object detection with region proposal networks. In: Advances in Neural Information Processing Systems, pp. 91–99 (2015)

18. Dai, J., Li, Y., He, K., Sun, J.: R-FCN: object detection via region-based fully convolutional networks. In: Advances in Neural Information Processing Systems, pp. 379–387 (2016)
19. He, K., Gkioxari, G., Dollár, P., Girshick, R.: Mask R-CNN. In: Proceedings of the IEEE International Conference on Computer Vision (ICCV), pp. 2961–2969 (2017)
20. Long, J., Shelhamer, E., Darrell, T.: Fully convolutional networks for semantic segmentation. In: Proceedings of the IEEE Conference on Computer Vision and Pattern Recognition (CVPR), pp. 3431–3440 (2015)
21. Redmon, J., Divvala, S., Girshick, R., Farhadi, A.: You only look once: unified, real-time object detection. In: Proceedings of the IEEE Conference on Computer Vision and Pattern Recognition (CVPR), pp. 779–788 (2016)
22. Liu, W., et al.: SSD: Single Shot MultiBox Detector. In: Leibe, B., Matas, J., Sebe, N., Welling, M. (eds.) ECCV 2016, Part I. LNCS, vol. 9905, pp. 21–37. Springer, Cham (2016). https://doi.org/10.1007/978-3-319-46448-0_2
23. Simonyan, K., Zisserman, A.: Very deep convolutional networks for large-scale image recognition. arXiv preprint: arXiv:1409.1556 (2014)
24. Lin, T.-Y., Dollar, P., Girshick, R., He, K., Hariharan, B., Belongie, S.: Feature pyramid networks for object detection. In: The IEEE Conference on Computer Vision and Pattern Recognition (CVPR), July 2017
25. Zeiler, M.D., Taylor, G.W., Fergus, R., et al.: Adaptive deconvolutional networks for mid and high level feature learning. In: The IEEE International Conference on Computer Vision (ICCV), vol. 1, p. 6 (2011)
26. Lin, T.-Y., Maji, S.: Visualizing and understanding deep texture representations. In: Proceedings of the IEEE Conference on Computer Vision and Pattern Recognition (CVPR), pp. 2791–2799 (2016)
27. Elgammal, A., Liu, B., Kim, D., Elhoseiny, M., Mazzone, M.: The shape of art history in the eyes of the machine. In: Thirty-Second AAAI Conference on Artificial Intelligence (2018)
28. Shaw, M.: Buddhist and taoist influences on Chinese landscape painting. J. Hist. Ideas **49**(2), 183–206 (1988)

# Temporal Analysis of Adverse Weather Conditions Affecting Wheat Production in Finland

Vladimir Kuzmanovski[1,3]([✉]), Mika Sulkava[2], Taru Palosuo[2], and Jaakko Hollmén[3]

[1] Jožef Stefan Institute, Jamova cesta 39, 1000 Ljubljana, Slovenia
vladimir.kuzmanovski@ijs.si
[2] Natural Resources Institute Finland (Luke),
P.O. Box 2, 00791 Helsinki, Finland
{mika.sulkava,taru.palosuo}@luke.fi
[3] Department of Computer Science, Aalto University,
P.O. Box 15400, 00076 Aalto, Finland
{vladimir.kuzmanovski,jaakko.hollmen}@aalto.fi

**Abstract.** Growing conditions of agricultural crops are increasingly affected by global climate change. Not only the overall agro-climatic conditions are changing, but also climatic variability and the occurrence of extreme weather events are becoming more frequent. This will affect crop yields and impact food supply both locally and globally. Located in the north, with short growing seasons and long days, Finland is not an exception. Drought- and temperature-related adverse events have been identified as most harmful abiotic factors on the production. Farmers try to mitigate with a range of management options. However, they need to adapt them over time as the climate is changing.

This study aims to identify the most adverse weather events that affect the spring wheat production in Finland and to ascertain if there have been changes on the most harmful abiotic weather-related factors during the last decades. Adverse weather conditions studied include frequency and length of periods with exceptional snow, drought, intensive rainfall and extreme heat. This was studied by modeling the wheat production using the adverse weather events as predictors with different lengths of training period (consecutive number of years) using LASSO regression.

The results reveal clear shift from early season drought and periodical intensive rainfall to the adverse effects of frequent and long periods of extremely high temperatures during later development stages.

**Keywords:** Wheat production · Adverse weather event · Data analysis · Time series

## 1 Introduction

Finland is located on the northern edge of the world's agricultural area, which makes the production conditions special. The growing seasons are short, with

© Springer Nature Switzerland AG 2019
P. Kralj Novak et al. (Eds.): DS 2019, LNAI 11828, pp. 176–185, 2019.
https://doi.org/10.1007/978-3-030-33778-0_15

long days, and variations in weather, both temporally and spatially. Most harmful weather events for cultivation of spring cereals under Finnish conditions are identified to be drought- and elevated temperature-related adverse events [9], with south-western Finland being an area where crop yield formation is mostly prone to climate-induced (abiotic) stresses [12,15]. The trend is likely to develop under future climate projections. [10] used temperature- and rainfall-based weather indices to explain variation in spring wheat production in Finland. The results hedge about 38% of wheat yield risk, suggesting that marginal products of weather events varies significantly during physiological development.

With the current knowledge and a range of management options farmers can try to mitigate many of the harmful effects from adverse weather conditions. Such management options include (but not limited to) changing the timing of growing season, selection of cultivar types or irrigation [6]. It has been shown, for example, that farmers use their earlier experiences of weather events to decide on crop and cultivars to be sown [8]. However, more detailed information on the relative impacts of adverse weather conditions on yields is very important to be able to find cost-effective means for adaptation.

This study aims to discover the impact of adverse weather conditions on production of spring wheat in Finland by analyzing: (i) effects of adverse weather events throughout the physiological development of the plants; and (ii) shifts in adverse weather conditions affecting crop yields. The impact is examined over both temporal scales, physiological development (short scale) and global

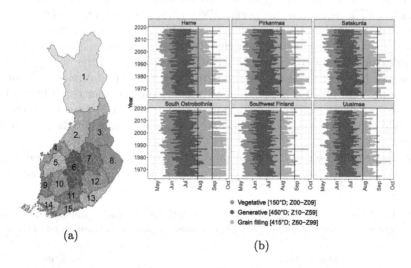

(a)

(b)

**Fig. 1.** (a) Map of ELY administrative regions in Finland. Following regions are considered in the study: *Uusimaa* (15), *Southwest Finland* (14), *South Ostrobothnia* (5), *Satakunta* (9), *Pirkanmaa* (10) and *Häme* (11); (b) Visualization of approximated spring wheat growth period and development stages across ELY regions, since 1965. Development stages are given with their length (in degree days) and corresponding Zadoks classification. Vertical lines mark the first day of August and September.

warming (long scale). Weather conditions are generalized over shorter and longer periods of recent history.

## 2    Material and Methods

### 2.1    Data

Data are collected from two sources: open-access databases of Natural Resource Institute Finland – Luke [5] and Finnish Meteorological Institute – FMI [2]. The former contains data on cultivation area, production and sowing time. The latter provides weather data such as daily temperature, precipitation and snow depth. The study is performed over 6 out of 15 ELY administrative regions representing the main cultivation area of spring wheat in Finland: *Uusimaa, Southwest Finland, South Ostrobothnia, Satakunta, Pirkanmaa* and *Häme* (Fig. 1a).

Data are transformed so that to fit the annual temporal scale of the defined problem. Each season, representing a single life cycle (plant growing period between sowing and harvesting) in a single region, is defined with three broader development stages: vegetative, generative and grain-filling, as defined in [7].

Development stages are expressed in thermal time, defined with degree days (°D). Review by [11] suggests a range of base temperature for calculating the degree days, but [10] recommend 5 °C as base in case of northern cultivars of spring wheat. Figure 1b visualizes length and Zadoks classification [16] of development stages, along with approximated duration of plants' life cycle across considered ELY regions, in the period 1965–2018. Each season is taken to start at the first day of the provided sowing period in corresponding region.

Adverse weather conditions are defined annually through a set of daily events with extreme weather condition. Single adverse event is in form of *frequency* of appearance (number of days within a period) or a length of a longest *streak* of such events (number of consecutive days with certain weather conditions). The literature emphasizes different definition of such events [4,11], out of which five are considered in this study (Table 1). The selection is bounded by the availability of data.

Set of descriptors is a Cartesian product of the defined development stages and the set of adverse weather events, given in the form of a frequency and

**Table 1.** List of events considered as annual adverse weather conditions. The last column gives the conditions in regard with variable names used throughout the study.

| Event | Description | Condition |
|---|---|---|
| Thin snow cover | Daily snow cover not exceeding 1 cm | $Snow\_depth \leq 1$ |
| Thick snow cover | Daily snow cover exceeding 4 cm | $Snow\_depth \geq 4$ |
| Drought | Total daily precipitation not exceeding 0.1 mm | $Precipitation \leq 0.1$ |
| Intensive rainfall | Total daily precipitation exceeding 40 mm | $Precipitation \geq 40$ |
| Extreme heat | Maximum daily temperature above 31 °C | $Max\_temp \geq 31$ |

length of a longest streak. Additionally, each development stage is given with number of days since: beginning of a year, and sowing date.

The complete dataset is defined with 36 descriptive variables over sample size of 318 instances (53 years of records in 6 ELY administrative regions).

## 2.2 Methodology

The study is performed with multivariate linear regression method using Least Absolute Shrinkage and Selection Operator – LASSO regression [13]. LASSO regression uses $L1$ regularization with penalties equal to absolute magnitude of coefficients. As such it builds sparse models that improves their interpretability. In addition, LASSO regression is indifferent to highly correlated descriptive variables, so it tends to pick one by ignoring the rest and setting their coefficient close to zero. Lasso penalty corresponds to Laplace prior by expecting more of coefficients to be close to zero, and a small subset of them greater than zero [3].

The selection of method for modeling the wheat production is justified by three main criteria: interpretability, variables vs. instances ratio, and parametric dimension of a model's run. Interpretability of models is of highest importance in this study as it comes along the need to understand the effect of adverse weather conditions over wheat production, by distinguishing effects of underlying adverse events over biologically processes and physiological development of plants.

High dimensionality of a variable's set in combination with a limited sample size can result in models with high bias. LASSO minimizes residual sum of squares by shrinking some coefficients to zero, which allows stable variable selection and avoids model's over-fitting [14].

Finally, LASSO implementation [3] requires setting up only one parameter - lambda ($\lambda$), a parameter for optimal $l_1$-norm regularization. It is optimized by performing a cross-validation over the training set and minimizing the $\lambda$ value that minimizes *mean squared error* – MSE.

## 2.3 Experimental Design

Experimental design defines the three-fold structure of the analysis in a workflow fashion: definition of short and long periods of training set, investigation of a model's robustness, and comprehensive analysis of effects from the adverse weather conditions.

Temporal analysis is performed using short and long memory, i.e., short and long periods of evidence in recent history. The analysis examines possible patterns from historical periods of 5 to 30 years. Intergovernmental Panel on Climate Change (IPCC) states that 30 years period is efficient for investigating the climate change and effects related to it [1]. However, due to the limited length of the time series of data on annual production of wheat, area and daily weather, the upper considered boundary is 30 years. Lower bound is set to be 5 years.

A selection of short and long periods is done in accordance to models' performance expressed with *Root Relative Squared Error* – *RRSE* and Pearson correlation coefficient. RRSE is an error metric that measures relative error of a model compared to the error of default model (average value):

$$RRSE = \sqrt{\frac{\sum_{i=1}^{n}(p_i - m_i)^2}{\sum_{i=1}^{n}(m_i - \bar{m})^2}}, \tag{1}$$

where $m_i$ is observed and $p_i$ predicted value, for $i \in [1, n]$. $\bar{m}$ is average observed value of $n$ observations (number of instances). In other words, it measures the fraction of unexplained variable and value range is $[0, \infty)$ with best value being the lower bound.

Periods with smaller number of years is set to randomly pick equal number of years for testing (24 years), as the longest possible period (30 years). Due to the stochastic properties of the testing set of years, the process is repeated 10 times and the average performance is reported.

Throughout the manuscript, the period of $n$ years is interchangeably referred to as *sliding window size* or *window size*, as well as *short* and *long window* for short and long historical periods, respectively. In fact, the period defined as a window represents a *training set* used for building models, while the consecutive year represents a *testing set*. All further elements of the experimental design adopt such definition and perform the tests with the technique of sliding window (Fig. 2), until the last year is tested, i.e., year 2018.

The robustness analysis includes comparison of performance of models built with a step-ahead predictions and a step-ahead simulation. Both utilizes the sliding window techniques over available years of the time series, for testing. The difference is in construction of training set. The prediction approach learns on collected data (original data on wheat production). The simulation approach constantly updates the training set by predictions from the previous year. The simulation approach allows predicting the effect of adverse weather conditions longer ahead. The length of such period depicts the robustness of a model build over certain window size.

Comprehensive analysis of variables selected in the process of building models for years from the testing set is the last part of the design. Due to the $\lambda$

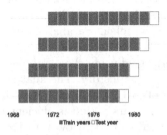

**Fig. 2.** Example of sliding window techniques applied throughout the experimental design. Window size represents the number of years in training set (green-filled tiles). The testing set (transparent tile) consists of one year in each trial. Trials are executed consequently, arranged from the bottom to the top of y-axis. (Color figure online)

parameter optimization using 10-fold cross validation, the process of building model for particular testing year is repeated 10 times. Weights of selected variables among all models built for particular testing year are averaged. Final selection is performed by picking out variables that appear in 8 out of 10 repeats and that have an average weight greater than $10^{-3}$.

## 3    Results and Discussion

The range of window sizes is visually examined in regard with median, intra-variance and outliers of models' performance, built over the given range of years.

The performance of models, built over different lengths of the sliding window ranging between 5 and 30, are given on Fig. 3a. As the window enlarges, the median performance in regard with both the error and correlation, increases and decreases, respectively. The variance of the performance among the models built with same window size is higher for very small and very high window sizes. The third criterion, appearance of outliers, shows significant decrease in regard with error and slight decrease in regard with correlation, as the window increases.

Accordingly, window sizes depicting lower variability, better median performance and fewer outliers are 12 and 23 years for short- and long-term conditions' aggregation, respectively. Built models will be referred to as *short-term models* and *long-term models*, respectively.

Average performance emphasizes that short-term models are more accurate compared to long-term (Table 2). However, considering the range of performance of individual models built over each year, higher stability and robustness is observed in long-term models compared to short-term (Fig. 3b). Such robustness is mainly adopted due to the encapsulated variety of events with higher

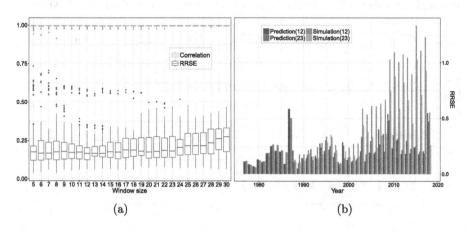

(a)    (b)

**Fig. 3.** (a) Performance of models (RRSE and Correlation coefficient) built over different window sizes varying from 5 to 30 years; (b) Predictive performance (RRSE) of the sliding-window models built with both step-ahead approaches (*prediction* and *simulation*) and window sizes of 12 and 23 years, for periods 1977–2018 and 1989–2018, respectively.

**Table 2.** Average performance of models built over short and long windows. Both window sizes are given with prediction (P) and simulation (S) approaches, as well as reference models built without weather conditions – given with *(ref)*

| Window size | RRSE (P) | Correlation (P) | RRSE (S) | Correlation (S) |
|---|---|---|---|---|
| Short (ref) | 0.191 | 0.995 | 0.506 | 0.995 |
| Short | 0.176 | 0.995 | 0.449 | 0.996 |
| Long (ref) | 0.237 | 0.995 | 0.426 | 0.995 |
| Long | 0.212 | 0.996 | 0.374 | 0.995 |

probability to appear in longer period of time. Consequently, the predictions are smoother when using the short-term models, while the long-term models sharpen the peaks (Fig. 4).

In addition, Table 2 shows performance of reference models, i.e. models built over data excluding variables that represent the adverse weather conditions. Although such models perform well, the difference of around 10% stresses out the impact of the adverse weather events. The impact is higher when long window is used and it is mostly reflected to the error measure. Considering the time-course of the predicted values, such improvement is mainly reflected to the magnitude of the residuals.

The robustness of models in regard with the step-ahead simulation is different compared to the task of prediction (Fig. 3b). Short-term term models fail to follow the performance of those with prediction approach and their performance deviates significantly after the year 2000 or 15 years of accumulation. The long-term models follow the prediction approach approximately until year 2005. Although they both depict good performance before year 2000 and 2005, it is

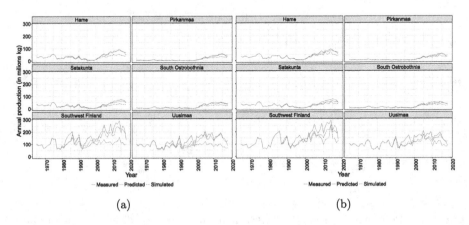

**Fig. 4.** Visualization of collected (red solid line), predicted (green dotted) and simulated (blue dashed) wheat production per ELY administrative centers, using models built with: **(a)** short window of 12 years; and **(b)** long window of 23 years. (Color figure online)

hard to conclude the exact range of stability of long-term simulations. Similar behaviour is observed in Fig. 4, where simulations are underestimating the real production of spring wheat in the years after 2005.

Examination of variables selected during models' building, reveals a change in abiotic weather-related factors with influence on wheat production in Finland, over the course of time (Fig. 5). A change in conditions can be analyzed from two dimensions: typomorphic and physiological. The former is a synonym for type of adverse weather event and examines whether certain type of event appears to affect the production. The latter, concentrates on the life cycle of wheat and depicts changes of effects observed in affected development stages.

*Drought* appears to be important in both short- and long-term models. Historically, it has been more harmful for the production, unlike recently. It causes a lot more damage in vegetative stage than in the later development. The longest streak is identified as more frequent source of negative effect, although frequency of dry days affects the production, as well.

*Intensive rainfall* appears in both short- and long-term models. Similar to drought, intensive rainfall used to have a large negative effect, during the vegetative stage. It is the case in both type of models where negative effect increases significantly until year 2010. Historically, the intensive rainfall is positive during generative and grain-filling stage, but without significant impact after the year 2005. The frequency of this event is predominant, thus the non-continuity of appearance does not have an effect.

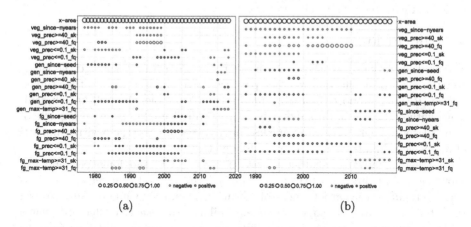

(a)                                  (b)

**Fig. 5.** Variables of models with windows size of: **(a)** 12 years; and **(b)** 23 years. Time is given on x-axis, while y-axis shows variables ordered by physiological stages of wheat development (from *vegetative* (upper), through *generative*, to *grain-filling* stage (lower)). Size of circles depicts variables' importance (weight), and color depicts their effect (positive or negative). Names are constructed of development stage, weather event and type of aggregation (frequency - "fq" or streak - "sk"). Exceptions are *x_area* that represents cultivated area, as well as *since-nyear* and *since-seed* – number of days since beginning of a year or seeding till beginning of a stage, respectively.

*Extreme temperature* shows most clear pattern of change on the long scale. Namely, historically it shows positive effect in the generative stage and less significant effect during the grain-filling stage. However, as approaching the recent years the effect shifted and posing a clear threat to the latest development stage. The changes are observable in both type of models with more clear evidence in the long-term models, with streaks being the main form of the adverse event.

Historically, *development delay* (postponed sowing) negatively influenced the overall production, unlike the length of the initial (vegetative) stage (*gen_since-seed*) that had positive affect on production. In recent years, postponed appearance of the grain-filling stage is beneficial for the overall production, a pattern visible in both type of models.

*Vegetative stage* is shown to be least affected by variety of conditions (adverse events), as both type of models stress out five lethal factors that affect the wheat development. In both cases, it turns being mostly vulnerable in the past especially under condition of frequent intensive rainfalls. Figure 5 emphasizes that the drought and, in particular, intensive rainfall had negative effect.

*Generative stage* has been historically stable in regard with negative effects from adverse conditions (Fig. 5a). Such stability diminish over time as the length of drought streaks increases, accompanied with extreme temperatures.

*Grain-filling stage* is shown to be constantly affected by variety of adverse conditions. However, the type of stress and their impact is changing along the course of time. Historically, both absence and intensive rainfall had positive impact. Nowadays, the maturity phase is highly affected by frequent and long periods of extremely high temperature.

## 4    Conclusion

The analysis revealed weather patterns that have importantly affected the yields and the way most effective weather patterns have changed during the studied period. Results support the earlier findings, for example, the well-known challenge with early-season droughts, which are also reflected in positive impacts of rainy seasons. There were also some less-obvious results, like the positive impacts of intensive rainfalls during the grain-filling stage.

A trend of adverse weather stress, in both short- and long-term models, shows clear shift throughout the time (Fig. 5). Historically, the adverse conditions negatively affect the earlier stages of wheat development, while it shifts toward the later stages over the course of time. Similarly, the negative effect originates as stress condition related to the precipitation and terminates as stress related to the elevated temperatures. Heat stress tolerance of cultivars are thus becoming increasingly important adaptation measure also for Northern production regions.

The selection of LASSO regression model class is justified by the consistent improvement of performance compared to the simpler reference models. Interpretation of models is easy for domain experts and simulation performance of the long-term models is rather close to step-ahead prediction for longer period, indicating that the models have incorporated many important weather-related factors affecting the production.

**Acknowledgment.** Authors acknowledge the Slovenian Ministry of Education, Science and Sport for funding the work through funding agreement C3330-17-529020.

# References

1. Allen, M., et al.: Technical summary: Global warming of 1.5 °C (2019). http://pure.iiasa.ac.at/id/eprint/15716/
2. FMI, F.M.I.: Daily weather data (2019). https://en.ilmatieteenlaitos.fi/download-observations
3. Friedman, J., Hastie, T., Tibshirani, R.: Regularization paths for generalized linear models via coordinate descent. J. Stat. Softw. **33**(1), 1–22 (2010)
4. Mäkinen, H., et al.: Sensitivity of European wheat to extreme weather. Field Crops Res. **222**, 209–217 (2018)
5. Natural Resource Institute Finland - LUKE: Agricultural statistics (2019). http://statdb.luke.fi/PXWeb/pxweb/en/LUKE
6. Olesen, J., et al.: Impacts and adaptation of European crop production systems to climate change. Eur. J. Agron. **34**(2), 96–112 (2011)
7. Peltonen-Sainio, P., Rajala, A., Seppälä, R.: Viljojen kehityksen ja kasvun ABC. Maa- ja elintarviketalous, Maa- ja elintarviketalouden tutkimuskeskus (2005). (in Finnish)
8. Peltonen-Sainio, P., Jauhiainen, L., Niemi, J., Hakala, K., Sipiläinen, T.: Do farmers rapidly adapt to past growing conditions by sowing different proportions of early and late maturing cereals and cultivars? Agric. Food Sci. **22**, 331–341 (2013)
9. Peltonen-Sainio, P., et al.: Harmfulness of weather events and the adaptive capacity of farmers at high latitudes of Europe. Clim. Res. **67**, 221–240 (2016)
10. Pietola, K., Myyrä, S., Jauhaianen, L.: Predicting the yield of spring wheat by weather indices in Finland: implications for designing weather index insurances. Agric. Food Sci. **20**(4), 269–286 (2011)
11. Porter, J.R., Gawith, M.: Temperatures and the growth and development of wheat: a review. Eur. J. Agron. **10**(1), 23–36 (1999)
12. Rötter, R.P., Höhn, J., Trnka, M., Fronzek, S., Carter, T.R., Kahiluoto, H.: Modelling shifts in agroclimate and crop cultivar response under climate change. Ecol. Evol. **3**(12), 4197–4214 (2013)
13. Tibshirani, R.: Regression shrinkage and selection via the LASSO. J. R. Stat. Soc. Ser. B (Methodol.) **58**(1), 267–288 (1996)
14. Tibshirani, R., Wainwright, M., Hastie, T.: Statistical Learning with Sparsity: The LASSO and Generalizations. Chapman and Hall/CRC, Boca Raton (2015)
15. Trnka, M., et al.: Adverse weather conditions for European wheat production will become more frequent with climate change. Nat. Clim. Change. **4**, 637 (2014)
16. Zadoks, J.C., Chang, T.T., Konzak, C.F.: A decimal code for the growth stages of cereals. Weed Res. **14**(6), 415–421 (1974)

# Predicting Thermal Power Consumption of the Mars Express Satellite with Data Stream Mining

Bozhidar Stevanoski[1], Dragi Kocev[2,3,4], Aljaž Osojnik[2,3], Ivica Dimitrovski[1], and Sašo Džeroski[2,3(✉)]

[1] Faculty of Computer Science and Engineering, Skopje, Macedonia
[2] Jožef Stefan International Postgraduate School, Ljubljana, Slovenia
[3] Department of Knowledge Technologies, Jožef Stefan Institute,
Ljubljana, Slovenia
Saso.Dzeroski@ijs.si
[4] Bias Variance Labs d.o.o., Ljubljana, Slovenia

**Abstract.** Orbiting Mars, the European Space Agency (ESA) operated spacecraft - Mars Express (MEX), provides extraordinary science data for the past 15 years. To continue the great contribution, MEX requires accurate power modeling, mainly to compensate for aging and battery degradation. The only unknown variable in the power budget is the power provided to the autonomous thermal subsystem, which in a challenging environment, keeps all equipment under its operating temperature. In this paper, we address the task of predicting the thermal power consumption (TPC) of MEX on all 33 thermal power lines, having available the stream of its telemetry data. Considering the problem definition, we face the task of multi-target regression, learning from data streams. To analyze such data streams, we use the incremental Structured Output Prediction tree (iSOUP-Tree) and the Adaptive Model Rules from High Speed Data Streams (AMRules) to model the power consumption. The evaluation aims to investigate the potential of the methods for learning from data streams for the task of predicting satellite power consumption and the influence of the time resolution of the measurements of thermal power consumption on the performance of the methods.

**Keywords:** Data streams · Multi-target regression · iSOUP-Trees · AMRules · Satellite · Thermal power consumption

## 1 Introduction

In June 2003, the Mars Express (MEX) spacecraft was launched from Earth, and after a six month cruise it arrived at Mars [5]. This mission of the European Space Agency (ESA) is ongoing up to the present time, and MEX is still in orbit around the planet Mars.

MEX's power source is electricity, either generated by its solar arrays, or alternatively (in the case of an eclipse), stored in its batteries. In addition to

© Springer Nature Switzerland AG 2019
P. Kralj Novak et al. (Eds.): DS 2019, LNAI 11828, pp. 186–201, 2019.
https://doi.org/10.1007/978-3-030-33778-0_16

powering platform units, electricity is used by the internal autonomous thermal subsystem, and only the remaining power can be allocated for science operations. Further considering the aging of the probe and the decaying capacity of the batteries, predicting the power consumption of the thermal subsystem allows for optimization of the science operations of the satellite.

Prediction of thermal power consumption (TPC) of MEX, is a crucial, but far from trivial task. The operating temperatures of the instruments and the on-board equipment vary from $-180\,°C$ for some, to room temperature for other. This spectrum of temperatures must be maintained in a challenging environment, where the side of an object illuminated by the Sun can reach temperatures more than $400\,°C$ higher than the unilluminated side. Even activating a radio transmitter results in a $28\,°C$ temperature increase of one side of MEX [14].

Initially, ESA used a manually constructed model for predicting TPC. However, it diverged from actual data year on year and an engineer's calibration was needed. Thus, a need arised for switching to a new approach, which would be able to automatically learn through experience.

Machine learning (ML) is the science that studies computer algorithms [15] with such abilities. In a very significant part of the research in machine learning, the experience is given as data in the form of a table. Each row in the table represents a data example and each column represents some feature (attribute).

A learning task can be categorized as *supervised* or *unsupervised*. The goal of the algorithms in the latter category is to find general descriptions of the examples, while the ones in the former category aim to predict the values for one or more target attributes for some data examples. If the values in the target attributes are nominal or numerical, the supervised learning task is said to be *classification* or *regression*, respectively. Another perspective for categorizing supervised learning tasks is based on the number of the target attributes. The tasks where one or more target attributes are present for the task of classification, are named *single-* and *multi-class* classification. The analogous terms for regression are *single-* and *multi-target* regression.

As devices that generate huge amounts of data are omnipresent, machine learning is facing increasing data complexity - not only in the number of target or descriptive columns, but also in the number of rows and the velocity at which they become available. In the extreme case, there are an infinite number of rows that are continuously arriving, whose storing for future knowledge extraction is obviously impossible. In this case, we are talking about data streams.

Learning from data streams (*data stream mining*) is a dynamic process. The implicit assumption made when working with finite amount of data examples (in the *classical* machine learning approach or in the *batch* setting) that all of the data is available before the learning process starts, is no longer valid for streams. Furthermore, since the data examples are arriving continuously and with high velocities, each can be processed at most once and in a limited amount of time.

The TPC prediction in MEX can be viewed as a task of data stream mining of its telemetry data, to produce real-time multi-target predictions of the electricity use on each of the 33 power lines of the heaters and coolers in its thermal subsystem. This field of multi-target learning from data streams has been given some attention, and has been researched only recently. Namely, there are two state-of-the-art methods in this field - the tree-based *incremental Structured Output Prediction tree* (iSOUP-Tree) [16], which is based on the *Fast Incremental Model Tree for Multi Target* (FIMT-MT) method [11] (the multi-target extension of *Fast Incremental Model Tree with Drift Detection* (FIMT-DD) [12]), and the rule-based *Adaptive Model Rules from High Speed Data Streams* (AMRules) [2].

In this paper, two research questions are investigated:

1. How do different methods for learning from multi-target data streams perform on the task of predicting satellite power consumption?
2. How does the time resolution, at which the measurements of thermal power consumption are considered, influence the performance of the methods?

The rest of the paper is organized as follows. In Sect. 2 MEX is discussed in greater detail, focusing on its telemetry data. Section 3 explains the used methods for data stream mining, while Sect. 4 discusses how they were employed for the task of TPC prediction. Section 5 presents the results, and finally Sect. 6 concludes the paper and gives directions for future work.

## 2    MEX Satellite and Its Power Consumption

### 2.1    Power Consumption of MEX

The 3D imagery of Mars that MEX has generated during the past 15 years has provided unprecedented information about the red planet. In order for MEX to continue providing valuable information, which would support ground exploration missions and other research, as well as to enable the proper function of MEX without breaking, twisting, deforming or failure of any equipment, careful power management is needed.

The available power, $\phi_{available}$, stored in the batteries or generated by the solar arrays, that is not consumed by the platform, $\phi_{platform}$, or by the thermal subsystem, $\phi_{thermal}$, can be used in science operations $\phi_{science}$.

$$\phi_{science} = \phi_{available} - \phi_{platform} - \phi_{thermal} \tag{1}$$

Two of the three terms in the right hand side of Eq. 1 are well known. The 200 thermistors in the spacecraft continually measure the temperatures around it and therefore enable the autonomous turning on or off of the electrical heaters, making the $\phi_{thermal}$ an unknown variable, difficult to predict.

In the initial, empirical, model to predict this variable, ESA has identified and incorporated key influencing factors, such as the distance of the spacecraft to the Sun and to Mars, the orbit phase and instrument and spacecraft operations [14]. However, the aging of the spacecraft has confronted this approach with many challenges. This motivated ESA to organize the Mars Express Power Challenge[1] and reach out to the machine learning community by releasing MEX data for four Martian years.

---

[1] https://kelvins.esa.int/mars-express-power-challenge/ [Last accessed: 12 June 2019].

## 2.2   MEX Power Challenge Data

The data released for the purpose of the MEX Power Challenge consists of (i) raw telemetry (context) data; and (ii) measurements of the electric current on the 33 thermal power lines (observation data). The time period covered in the data spans 4 Martian (or cca. 7.5 Earth) years, starting from the 22nd of August 2008 to the 1st of March 2016.

The context data consists of five components:

- **SAA** (Solar Aspect Data): timestamped data expressing the angles of the Sun-MEX line with the axes of MEX's coordinate system and with the panels' normal line
- **DMOP** (Detailed Mission Operations Plans): data about the execution of different subsystems' commands at a specific time, such as ON/OFF commands of radio communications or of the science instruments
- **FTL** (Flight dynamics TimeLine): timestamped data regarding the pointing and action commands that impact MEX's position, such as pointing the spacecraft towards Earth or Mars
- **EVTF** (Miscellaneous events): data about more events and their timestamp, such as the time intervals during which MEX was in Mars's shadow
- **LTDATA** (Long Term Data): timestamped long term data including the Sun-Mars distance and the value of the solar constant on Mars

The observational data represents the measurements of the electrical current/power on all 33 thermal power lines, recorded once or twice per minute. Predicting the average hourly power/current was the competition's goal.

## 2.3   Data Pre-processing and Feature Engineering

Considering the different time resolutions of the different components of the context data and the unstructured format of some data entries, the raw data cannot be used directly without pre-processing. The data pre-processing and feature engineering taken in this paper follows the approach taken in the winning solution of the MEX Power Challenge [4].

During this step, the time resolution is matched for all features. Namely, for a resolution $\Delta t$, the data is divided into intervals $[t_i, t_{i+1})$ with length $\Delta t$. The competition required the prediction of the average TPC over one-hour intervals. Since the context data allows for finer time resolutions, in this paper smaller values of $\Delta t$, i.e., $\Delta t \in \{5, 10, 15, 30, 60\}$ minutes, are considered.

A common issue in data science tasks is the handling missing data, which occurs in this task as well. Both the context and the observation data have missing values. In our work, examples with missing observation data for time periods longer than 10 min are removed. If the missing values for the context data span over a period shorter than 10 min, the values are linearly interpolated, and left intact otherwise.

The following subsections describe the feature categories constructed and later used by the data stream mining methods.

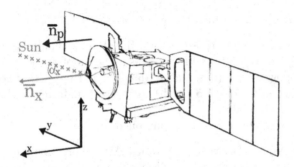

**Fig. 1.** Illustration of MEX and its coordinate system [4].

**Energy Influx Features.** As mentioned above, the solar energy irradiating a side of the spacecraft may increase its temperature by 400 °C. Furthermore, the energy collected by the solar panels is directly proportional with the generated energy. Hence, one of the feature classes represents the energy incident to the solar panels and to each of the six sides of MEX's cuboid in the interval $[t_i, t_{i+1})$.

Formally, the energy $E_S^i$ of the surface $S$ in time interval $[t_i, t_{i+1})$ is defined as

$$E_S^i = \int_{t_i}^{t_{i+1}} A \times max\{cos(\alpha(t)), 0\} \times c(t) \times U(t)dt \qquad (2)$$

where $A$, $cos(\alpha(t))$ and $c(t)$ denote the area of the surface $S$, angle between the Sun-MEX line and the normal $\boldsymbol{n}$ of the surface, and the solar coefficient at time $t$, respectively. A visual illustration of these notations is presented in Fig. 1. Since the data stream mining methods used in the paper are invariant to monotonic transformations of the features, $A$ is taken to be constant (1). $U(t)$ is the approximation of the Sun visible from the spacecraft at time $t$, referred to as the *umbra coefficient*, represented as a simple piecewise function:

$$U(t) = \begin{cases} 0 & \text{if MEX is in umbra} \\ 0.5 & \text{if MEX is in penumbra} \\ 1 & \text{otherwise} \end{cases} \qquad (3)$$

Although Eq. 2 defines the energy as an integral, the integral is not exactly computed. Instead, it is approximated by using the trapezoid-rule.

**Historical Energy Influx Features.** Since temperature is not a fast-changing variable, if a surface received a lot of solar energy and hence heated during the previous time interval, this will affect the current energy state. Therefore, for each surface $S$, a historical feature is defined as:

$$H_S^i = \sum_{j=1}^{H} E_S^{i-(j-1)} \qquad (4)$$

**Table 1.** Values of $H$ and the corresponding time spans for different time resolutions

| $\Delta t$ | Values of H | Time spans |
|---|---|---|
| 5 | {1, 3, 6, 13, 25} | {5, 15, 30, 65, 125} |
| 10 | {1, 2, 3, 6, 13} | {10, 20, 30, 60, 130} |
| 15 | {1, 2, 3, 4, 9} | {15, 30, 45, 60, 135} |
| 30 | {1, 2, 3, 4, 5} | {30, 60, 90, 120, 150} |
| 60 | {1, 2, 3} | {60, 120, 180} |

where $H$ is the total count of historical time intervals considered.

The variable $H$ is dependent on the time resolution. Intuitively, for $H = 25$ for example, the energy from 25 h ago, in the case of $\Delta t = 60$ min, has little to no impact in the current time, while the energy from about 2 h, when $\Delta t = 5$, has a large one. Given the time resolution, the values for $H$ used are given in Table 1.

**DMOP Features.** The raw DMOP data consists of a log of commands, whose names have been obfuscated, issued to MEX's subsystems. These commands can concern flight dynamics events or events that contain information about the subsystem and the executed command.

Assuming a delay between triggering a command and its thermal effect, the features constructed represent "time since last activation" of a specific subsystem command. The value of such a feature at time $t$ corresponding to event $k$ is:

$$f_k^i = \begin{cases} 0 & \text{if k is activated at } t_i \\ min\{f_k^{i-1} + \Delta t, \theta\} & \text{otherwise} \end{cases} \tag{5}$$

where $\theta$ (set to 1 day) regulates the diminishing importance of $f_k^i$ with time.

Such features are constructed for each flying dynamic event, each subsystem - command pair, and each subsystem when multiple commands are issued to it. Also binary indicator features for each subsystem and flying dynamic event are included, where $f_k^i = 1$ only if the subsystem was triggered within time step $t_i$, and $f_k^i = 0$ otherwise.

**FTL Features.** From the raw FTL data, containing logs of pointing events and their times, where simultaneously occurring events are possible, new features are constructed which display the proportion of the time in the interval $[t_i, t_{i+1})$ during which the event is in progress. The range of possible values these features can have is $[0, 1]$ - 0 if the event never happened, and 1 if it was active during the whole interval. Since these events typically last longer that the time unit considered, most values are the extremes of the domain range - either 0 or 1.

## 2.4   Final Dataset Overview

An overview of the datasets obtained for the different time resolutions is given in Table 2. Note that coarser time resolutions yield smaller datasets.

**Table 2.** Final datasets' properties

| $\Delta t$ [min] | Number of examples | Number of features | Memory size [MB] |
|---|---|---|---|
| 5 | 784773 | 462 | 2077 |
| 10 | 392474 | 462 | 1041 |
| 15 | 261697 | 462 | 699 |
| 30 | 130900 | 462 | 352 |
| 60 | 65493 | 448 | 167 |

# 3   Learning for Multi-target Prediction on Data Streams

## 3.1   Overview of Classical Batch Setting and Data Streams

The explosive growth of data generated and collected every day has brought data streams into the spotlight. Data streams are an algorithmic abstraction for continuously arriving sequences of examples, possibly infinitely many examples, at high velocities. The task of learning from such sequences poses challenges not present in the batch setting, where the entire dataset is available at the start of the learning process. For example, in data stream mining, real- or near-real time response is of crucial importance.

In addition to the fact that the dataset is not complete before the learning process begins, the theoretical infinite number of data examples makes it impossible for data stream mining methods to store them as they arrive, since there exists no infinitely large memory storage. Instead, each example is processed once, at the time of arrival, and later discarded. Besides memory efficiency, the high velocity of the stream imposes a time efficiency requirement even for the one-time processing of the example, in order to be ready to process the next one.

The underlying distribution of the data examples in the batch setting is assumed to be constant, and thus any learned generalization is applicable for future examples. This statement, however, does not hold in the data stream case. The temporal dimension there implies the possibility of distribution change, which is known as *concept drift*.

## 3.2   Multi-target Prediction in the Batch and Data Stream Settings

There are two general types of approaches to multi-target prediction - *local* and *global*. The former category builds models for each single primitive target, and later combines their prediction into a multi-target one. The global approach builds only one global model, for all targets.

Multi-target prediction has been researched extensively for the batch setting. The approaches taken for multi-target regression and classification include tree-based [6,8,20], rule-based [1,7], kernel-based [13,21] and instance-based [17,22] methods. In the current literature, little attention is given to multi-target prediction for data streams, as compared to the batch case. There are some methods for classification [18,19], while for regression there are rule-based [2] and tree-based [11,16] methods (the former is mainly based on the latter). Since our problem, i.e., predicting MEX's TPC, is a multi-target regression task, only the methods that fall under this category are discussed in greater detail.

**Massive Online Analysis (MOA).** The Java-based Massive Online Analysis (MOA) open-source framework includes a collection of data stream mining methods for classification, regression, multi-label classification, multi-target regression, clustering and concept drift detection [3]. The authors of iSOUP-Tree and AMRules have published the implementations of their methods in MOA.

MOA is similar and related to the WEKA project. However, MOA is designed to scale to more demanding problems. This work uses MOA and the original implementations of iSOUP-Tree and AMRules that MOA contains.

**iSOUP-Tree.** The iSOUP-Tree method is an instance-incremental method, meaning it takes exactly one instance to update the current generalization model. Initially, when no instance has been processed, the iSOUP-Tree is just an empty leaf node.

Once enough examples have been processed (but not stored directly) in a leaf node, a check is made to examine if there is a significant statistical support to split it. All possible binary splits, $A \leq c$ or $A = n$ for some numerical $c$ or nominal value $n$ of the attribute $A$, are calculated and evaluated using multi-target *intra-cluster variance reduction* (ICVR) as a heuristic function. Formally, the ICVR evaluates a split candidate $S$ as

$$ICVR(S) = \frac{1}{M} \sum_{j=1}^{M} \frac{1}{Var^j(S)} \left( Var^j(S) - \frac{|S_\top|}{|S|} Var^j(S_\top) - \frac{|S_\perp|}{|S|} Var^j(S_\perp) \right)$$

(6)

where $M$ is the total number of target variables (indexed by $j$), $S_\top$ and $S_\perp$ are the post-split sets of accumulated examples for which the evaluated split test is true or false respectively, $S$ is their union and $Var^j$ denotes the variance of the $j^{th}$ target attribute.

ICVR represents the homogeneity gain on the target values if split $S$ is chosen. Hence, the candidates with higher ICVR values are more desirable. It is important to note that the data examples of $S$, $S_\top$ or $S_\perp$ are *not* stored in memory. Instead, only the statistics necessary for computing the variances are recorded.

According to the heuristic in Eq. (6), the best candidate split $h_1$ is selected, as well as the second-best $h_2$. Next, the following sequence is constructed

$$\cdots \frac{h_2(k)}{h_1(k)}, \frac{h_2(k+1)}{h_1(k+1)}, \frac{h_2(k+2)}{h_1(k+2)} \cdots \tag{7}$$

where $k$ denotes the number of accumulated examples considered.

Let $X_k$ be a random variable denoting the ratio $\frac{h_2(k)}{h_1(k)}$, and $x_k$ be one sample of it. Then the observed average can be computed as $\bar{x} = \frac{1}{k}(x_1 + x_2 + \cdots + x_{|S|})$, which is a sample from the random variable $\bar{X} = \frac{1}{k}(X_1 + X_2 + \cdots + X_{|S|})$. The Hoeffding bound [10] is then applied to make an $(\epsilon, \delta)$-approximation, using the standard notation of $E[X]$ to denote the expected value of the random variable $X$. The Hoeffding bound is of the following form:

$$P(|\bar{X} - E[\bar{X}]| > \epsilon) \leq 2e^{-2|S|\epsilon^2} =: \delta \tag{8}$$

The value $\delta$ is a parameter to the iSOUP-Tree method named *splitting confidence*. The value $\epsilon$ can be formulated as an expression of $\delta$ and $|S|$.

Plugging $\bar{x}$ as observation of $\bar{X}$ in Eq. 8, one gets $E[\bar{X}] \in [\bar{x} - \epsilon, \bar{x} + \epsilon]$ with probability $1 - \delta$, i.e., if $\bar{x} + \epsilon < 1$ then $E[\bar{X}] < 1$ implying $\frac{h_2}{h_1} < 1$ (with probability $1 - \delta$), or in other words, there exists a significant support to take the currently best candidate and split the leaf node. In the case when $\bar{x} + \epsilon \geq 1$, the leaf waits for more examples.

This condition is checked only when enough examples have accumulated in the leaf. The check is made whenever the leaf has accumulated a number of examples which is a multiple of the parameter $GP$, which stands for *grace period*.

In order to overcome a drawback of the Hoeffding bound, which occurs when the values of the two best heuristics are close to each other, iSOUP-Tree introduces a new parameter - $\tau$. This so-called *tie breaking threshold* $\tau$ determines the minimal value $\epsilon$ can have before the leaf is split. The underlying data structure used to compute the statistics is the extended binary search tree (E-BST), also used by [12].

Each leaf makes a prediction by using an adaptive multi-target model, consisting of a multi-target perceptron and a multi-target mean predictor. The perceptron updates its weights by a backpropagation rule with a given *learning rate*. When a leaf is constructed, its learning rate is set to the parameter $\eta_0$, named *initial learning rate*. After each incoming example, the learning rate $\eta$ is updated by using the rule

$$\eta = \frac{\eta_0}{1 + n\eta_\Delta} \tag{9}$$

where $n$ is the number of recorded values and $\eta_\Delta$ is a parameter called *learning rate decay factor*. Finally, a prediction is made by using the perceptron or the mean regressor, depending on which one has the lower *fading mean absolute error* (fMAE) for that target:

$$fMAE^j(e_n) = \frac{\sum_{i=1}^{n} 0.95^{n-i} |\hat{y}_i^{\,j} - y_i^j|}{\sum_{i=1}^{n} 0.95^{n-i}} \tag{10}$$

where $e_n$, $\hat{y}_i{}^j$ and $y_i^j$ are the $n^{th}$ observed example, predicted and real values of the $j^{th}$ target for the $i^{th}$ example.

**AMRules.** This algorithm is a representative of rule-based approaches to multi-target prediction on data streams, which build rule sets (RS). Initially, AMRules starts with an empty RS and a default rule $\{\} \rightarrow \mathcal{L}$, where $\mathcal{L}$, initialized to $NULL$, is a modified version of the data structure E-BST used in iSOUP-Tree for storing statistics, which limits the maximum number of splitting points to a predefined value. This modification reduces the memory consumption as well as speeds up the split selection procedure [9].

When a new data example arrives, AMRules checks if some rule in the RS covers it, i.e., if all of the literals on the left hand side of the rule for that example are true. Target values are utilized to update the statistics of the rule. The Page-Hinkley (PG) change detection test is used to discover a concept drift. It considers a cumulative variable $m_T$:

$$m_T = \sum_{t=1}^{T} (x_t - \bar{x}_T - \alpha) \tag{11}$$

where $x_t$ is a previously observed value, $\bar{x}_T = \frac{1}{T}\sum_{t=1}^{T} x_t$, and $\alpha$ corresponds to the magnitude of allowed changes. Also, the minimum value of this variable is computed $M_T = \min_{t=1,\ldots,T}(m_t)$. When the difference between these values $PH_T = m_T - M_T$ is larger than the value of the parameter $\lambda$, concept drift is signaled. If change is detected, the rule is removed from the RS.

If a rule is not removed, it is considered for expansion. Here, again a grace period parameter is used. The expansion procedure is almost identical to the leaf node splitting of iSOUP-Tree discussed, using the Hoeffding bound, with the same heuristic function along with $\tau$ threshold and $\delta$ confidence parameters. The rule expansion is a process where the hypothetical candidate split is added to the literals on the rule's left hand side. As a special case, expanding the default rule means adding a new one in RS, with the extended literals.

The prediction and model building strategies depend on whether the rules are ordered or unordered. In the former case, only the first rule that covers the example is removed, expanded or used in prediction. The latter one enables all the rules that cover an example to have same treatment, independent of their order, and the final prediction is made as the aggregation (mean) of individual predictions.

The rules learned by the AMRules method generate predictions in a similar manner as the leaves in iSOUP-Tree. They use an adaptive strategy, choosing between a perceptron's and a mean regressor's prediction. AMRules and iSOUP-Tree differ is the learning rate, which in AMRules is a constant.

# 4   Experimental Design

## 4.1   Methods' Parameter Values

Both iSOUP-Tree and AMRules have been implemented and published into the open source framework MOA. Their authors recommend values for their parameters, which are implemented as default parameter values in MOA. Those values are again reused in this paper, e.g., we use the ordered version of AMRules.

Because of the discussed similarities between the two approaches, many of their parameters overlap. Table 3 provides an overview of the parameters, including their values. It specifies which parameters are unique to each methods and which are shared.

## 4.2   Evaluation Procedure and Evaluation Measures

In batch prediction, two phases, clearly separated in time, are present - first the model is trained, and only when this phase finishes, the evaluation can start. On the other hand, in data stream setting, since examples never stop arriving, the training and evaluation must be interleaved. Additionally, the streams' evaluation posses a new challenge - to assess how the models perform *over time*. Two main approaches are present:

- *Holdout:* Each incoming example is firstly used for testing. Once the model makes the prediction for its target, it is stored in a buffer, and only after it is filled, all stored examples are used for training.
- *Prequential:* This approach also uses the newly received example for testing, but once that is done, the model proceeds to training, without waiting for other examples. In other words, this approach is a special case of the holdout approach, where the buffer size is 1.

Since in holdout evaluation, the model is not updating for each example, towards the end of the buffer, it is getting "stale". Thus, in this work, the prequential approach is taken.

**Table 3.** The values of the parameters of iSOUP-Tree and AMRules

| Designation | Description | Value | Method |
|---|---|---|---|
| GP | Grace period | 200 | Both |
| $\delta$ | Split confidence | $10^{-7}$ | Both |
| $\tau$ | Tie breaking threshold | 0.05 | Both |
| $\eta_0$ | Initial learning rate | 0.2 | iSOUP-Tree |
| $\eta_\Delta$ | Learning rate decay | 0.001 | iSOUP-Tree |
| $\eta$ | Learning rate | 0.01 | AMRules |
| $\alpha$ | Magnitude of allowed changes | 0.05 | AMRules |
| $\lambda$ | Concept drift threshold | 50 | AMRules |

A prediction is generated for each example as it arrives. In order to evaluate model performance, as in batch setting, evaluation measures are needed. In the single target scenario, there exist wide-spread evaluation metrics adopted from statistics. When multiple target attributes are present, one possibility is to treat them individually, and produce as many evaluation scores as there are targets. However, this can be cumbersome, especially when there are more than just a few targets, as in the case of MEX's TPC where we have 33 targets.

A common approach in multi-target evaluation is to average the individual singe-target scores. This is also the one taken in this paper. In particular, the *average relative mean absolute error* $(\overline{RMAE})$ [16] is reported over a window with length $n$ (here, $n = 1000$), and calculated as follows:

$$\overline{RMAE} = \frac{1}{M} \sum_{j=1}^{M} \frac{\sum_{i=1}^{n} |y_i^j - \hat{y}_i^j|}{\sum_{i=1}^{n} |y_i^j - \bar{y}^j(i)|} \tag{12}$$

where $y_i^j$ and $\hat{y}_i^j$ are the real and predicted values of the target $j$ for the data example $i$ by the evaluated model, respectively, while the $\bar{y}^j(i)$ is the prediction by the mean regressor.

If the model evaluated is the mean regressor for each target value, then (12) would yield a score of 1, since the nominator and denominator would be the same expression. For any other model, its performance is compared with the mean regressor as a baseline, such that if the $\overline{RMAE}$ score is below 1, the evaluated model outperforms the baseline mean regressor. Lower values for $\overline{RMAE}$ scores are desired, where the perfect model has a score of 0.

# 5    Results and Discussion

In this Section, the empirical evaluation results are presented and discussed. They are given in the form of graphs in order to better capture and visualize the performance of the two methods over time. The results are presented in two parts - one for each of the two posed research questions.

## 5.1    Method Comparison on the Task of TPC on MEX

Figure 2 contains five graphs, one for each time resolution, comparing the performance of iSOUP-Tree and AMRules over time. At all time resolutions, concept drift as an important aspect of data stream mining is visible. Focusing on the last time period, AMRules score lower $\overline{RMAE}$ than iSOUP-Tree. AMRules detects and handles concept drift and is able to adapt to changes, even when changes occur after a long period without changes. On the other hand, iSOUP-Tree does not address change detection and adaptation explicitly: As seen in Fig. 2, this makes it vulnerable to concept drift, which is more likely to occur when learning over a long period of time.

If the last time period is left out of the analysis, iSOUP-Tree clearly outperforms AMRules. This is the case for all time resolutions, except for the 60 min one. For the 60 min resolution, the $\overline{RMAEs}$ of the two methods are very close.

Overall, AMRules has more stable error throughout time. iSOUP-Tree on the other hand, models MEX's TPC significantly better, but suffers from sensitivity to changes of the underlying distribution. Finally, it is important to note that both algorithms outperform the mean regressor, as both have $\overline{RMAE} < 1$.

## 5.2   The Influence of Time Resolution on Predictive Performance

To address the second research question, Fig. 3 compares the performance on the tasks of predicting the TPC of the MEX at different time resolutions, for each of the two methods. The lowest overall error is achieved at medium time resolutions, i.e., at 10 and 15 min, for both AMRules and iSOUP-Tree. For the finest resolution of 5 min, the performance is in the middle, while predicting at the coarsest resolution of 60 min results in the highest error through the majority of the time period.

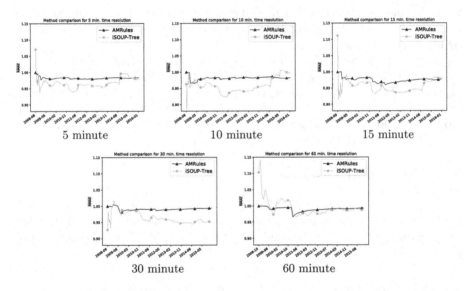

**Fig. 2.** A comparison of the performance of iSOUP-Tree and AMRules on the task of predicting the TPC of the MEX at different time resolutions.

At the 30 min resolution, on the other hand, there is a difference in behaviour between the two methods. In the last time period covered by the data, iSOUP-Tree performs the best. In contrast, for this resolution and time period, AMRules performs worst. Learning at coarser time resolutions yields less accurate models, since some of the fine-grained detail of the data are lost at these resolution. At very high resolutions, there is too much detail and noise can be confused for real signal more easily. Figure 3 shows that a medium resolution (10 to 15 min) is the most appropriate for the task at hand.

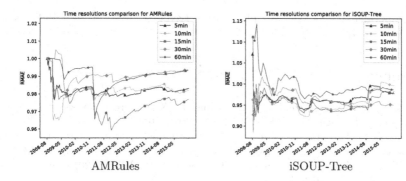

**Fig. 3.** The performance of each of AMRules and iSOUP-Trees at different time resolutions.

## 6   Summary, Conclusions and Further Work

In this paper, we first presented an overview of the MEX (Mars Express) spacecraft and its TPC (thermal power consumption). After a thorough discussion of the MEX telemetry data used to predict its TPC, we described the data preprocessing and feature extraction process. We continued with an introduction to data stream mining and multi-target prediction, focusing on their combination (multi-target regression on data streams) and two methods (AMRules and iSOUP-Tree) that address this task.

The central part of the paper addresses the details of using the two methods for multi-target regression on data streams to solve the problem of predicting TPC for the MEX. After clarifying the experimental design, the results are presented and discussed in the context of our research questions. More specifically, we compare the two methods as well as the performances of each method at different time resolutions.

Regarding the first question of how the two methods compare to each other, we note that iSOUP-Tree outperforms AMRules for most of the time. However, iSOUP-Tree does have a weakness, namely handling concept drift. At finer time resolutions, it performs slightly worse than AMRules at the end of the time period covered by the data released by ESA.

Regarding the second question, our results imply that medium time resolutions might be the best to consider when predicting the TPC of the MEX. Namely, at coarser time resolutions, some of the fine-grained detail of the data are lost. At very high resolutions, too much detail can contribute to confusing noise and signal.

Our current research agenda includes an extension of the work done in the paper by investigating how the iSOUP-Tree and AMRules methods perform on multi-target versus single-target versions of the task of predicting TPC for MEX. Although the methods considered outperform the mean regressor, we would also like to consider parameter optimization, to further boost their modeling capabilities. Finally, iSOUP-Tree is capable of learning tree ensembles, which can greatly improve performance - We thus plan to use tree ensembles in iSOUP-Trees for the task at hand.

# References

1. Aho, T., Ženko, B., Džeroski, S., Elomaa, T.: Multi-target regression with rule ensembles. J. Mach. Learn. Res. **13**, 2367–2407 (2012)
2. Almeida, E., Ferreira, C., Gama, J.: Adaptive model rules from data streams. In: Blockeel, H., Kersting, K., Nijssen, S., Železný, F. (eds.) ECML PKDD 2013, Part I. LNCS (LNAI), vol. 8188, pp. 480–492. Springer, Heidelberg (2013). https://doi.org/10.1007/978-3-642-40988-2_31
3. Bifet, A., Holmes, G., Kirkby, R., Pfahringer, B.: MOA: Massive Online Analysis. J. Mach. Learn. Res. **11**, 1601–1604 (2010)
4. Breskvar, M., et al.: Predicting thermal power consumption of the Mars Express satellite with machine learning. In: 6th International Conference on Space Mission Challenges for Information Technology, pp. 88–93. IEEE (2017)
5. Chicarro, A., Martin, P., Trautner, R.: The Mars express mission: an overview. In: Mars Express: The Scientific Payload, ESA SP 1240, pp. 3–13. European Space Agency, Publications Division (2004)
6. Clare, A., King, R.D.: Knowledge discovery in multi-label phenotype data. In: De Raedt, L., Siebes, A. (eds.) PKDD 2001. LNCS (LNAI), vol. 2168, pp. 42–53. Springer, Heidelberg (2001). https://doi.org/10.1007/3-540-44794-6_4
7. De Comité, F., Gilleron, R., Tommasi, M.: Learning multi-label alternating decision trees from texts and data. In: Perner, P., Rosenfeld, A. (eds.) MLDM 2003. LNCS, vol. 2734, pp. 35–49. Springer, Heidelberg (2003). https://doi.org/10.1007/3-540-45065-3_4
8. De'Ath, G.: Multivariate regression trees: a new technique for modeling species-environment relationships. Ecology **83**(4), 1105–1117 (2002)
9. Duarte, J., Gama, J., Bifet, A.: Adaptive model rules from high-speed data streams. ACM Trans. Knowl. Discov. Data **10**(3), 30 (2016)
10. Hoeffding, W.: Probability inequalities for sums of bounded random variables. J. Am. Stat. Assoc. **58**(301), 13–30 (1963)
11. Ikonomovska, E., Gama, J., Džeroski, S.: Incremental multi-target model trees for data streams. In: ACM Symposium on Applied Computing, pp. 988–993. ACM (2011)
12. Ikonomovska, E., Gama, J., Džeroski, S.: Learning model trees from evolving data streams. Data Min. Knowl. Discov. **23**(1), 128–168 (2011)
13. Khemchandani, R., Chandra, S., et al.: Twin support vector machines for pattern classification. IEEE Trans. Pattern Anal. Mach. Intell. **29**(5), 905–910 (2007)
14. Lucas, L., Boumghar, R.: Machine learning for spacecraft operations support - The Mars Express power challenge. In: International Conference on Space Mission Challenges for Information Technology, pp. 82–87. IEEE (2017)
15. Mitchell, T.: Machine Learning. McGraw Hill, Boston (1997)
16. Osojnik, A., Panov, P., Džeroski, S.: Tree-based methods for online multi-target regression. J. Intell. Inf. Syst. **50**(2), 315–339 (2018)
17. Pugelj, M., Džeroski, S.: Predicting structured outputs k-Nearest neighbours method. In: Elomaa, T., Hollmén, J., Mannila, H. (eds.) DS 2011. LNCS (LNAI), vol. 6926, pp. 262–276. Springer, Heidelberg (2011). https://doi.org/10.1007/978-3-642-24477-3_22

18. Shi, Z., Wen, Y., Feng, C., Zhao, H.: Drift detection for multi-label data streams based on label grouping and entropy. In: International Conference on Data Mining Workshops, pp. 724–731. IEEE (2014)
19. Spyromitros-Xioufis, E., Spiliopoulou, M., Tsoumakas, G., Vlahavas, I.: Dealing with concept drift and class imbalance in multi-label stream classification. In: 22nd International Joint Conference on Artificial Intelligence, pp. 1583–1588. AAAI (2011)
20. Struyf, J., Džeroski, S.: Constraint based induction of multi-objective regression trees. In: Bonchi, F., Boulicaut, J.-F. (eds.) KDID 2005. LNCS, vol. 3933, pp. 222–233. Springer, Heidelberg (2006). https://doi.org/10.1007/11733492_13
21. Vazquez, E., Walter, E.: Multi-output suppport vector regression. IFAC Proc. Vol. **36**(16), 1783–1788 (2003)
22. Zhang, M.L., Zhou, Z.H.: A k-nearest neighbor based algorithm for multi-label classification. In: International Conference on Granular Computing, pp. 718–721. IEEE (2005)

# Data and Knowledge Representation

# Parameter-Less Tensor Co-clustering

Elena Battaglia$^{(\boxtimes)}$ and Ruggero G. Pensa

Department of Computer Science, University of Turin, Turin, Italy
{elena.battaglia,ruggero.pensa}@unito.it

**Abstract.** Tensors co-clustering has been proven useful in many applications, due to its ability of coping with high-dimensional data and sparsity. However, setting up a co-clustering algorithm properly requires the specification of the desired number of clusters for each mode as input parameters. This choice is already difficult in relatively easy settings, like flat clustering on data matrices, but on tensors it could be even more frustrating. To face this issue, we propose a tensor co-clustering algorithm that does not require the number of desired co-clusters as input, as it optimizes an objective function based on a measure of association across discrete random variables (called Goodman and Kruskal's $\tau$) that is not affected by their cardinality. The effectiveness of our algorithm is shown on both synthetic and real-world datasets, also in comparison with state-of-the-art co-clustering methods based on tensor factorization.

**Keywords:** Clustering · Higher-order data · Unsupervised learning

## 1 Introduction

Tensors are widely used mathematical objects that well represent complex information such as social networks [12], heterogenous information networks [8,25], time-evolving data [1], behavioral patterns [11], and multi-lingual text corpora [17]. From the algebraic point of view, they can be seen as multidimensional generalizations of matrices and, as such, they can be processed with mathematical and computational methods that generalize those usually employed to analyze data matrices (e.g., non-negative factorization [21], singular value decomposition [26], clustering and co-clustering [2,24]). Clustering, in particular, is by far one of the most popular unsupervised machine learning techniques since it allows analysts to obtain an overview of the intrinsic similarity structures of the data with relatively little background knowledge about them. However, with the availability of high-dimensional heterogenous data, co-clustering has gained popularity, since it provides a simultaneous partitioning of each mode (rows and columns of the matrix, in the two-dimensional case). In practice, it copes with the curse of dimensionality problem by performing clustering on the main dimension (data objects or instances) while applying dimensionality reduction on the other dimension (features). Despite its proven usefulness, the correct application of tensor co-clustering is limited by the fact that it requires the specification of a congruent number of clusters for each mode, while, in realistic analysis scenarios,

© Springer Nature Switzerland AG 2019
P. Kralj Novak et al. (Eds.): DS 2019, LNAI 11828, pp. 205–219, 2019.
https://doi.org/10.1007/978-3-030-33778-0_17

the actual number of clusters is unknown. Furthermore, matrix/tensor clustering is often based on a preliminary tensor factorization step that, in its turn, requires further input parameters (e.g., the number of latent factors within each mode). As a consequence, it is merely impossible to explore all combinations of parameter values in order to identify the best clustering results.

The main reason for this problem is that most clustering algorithms (and tensor factorization approaches) optimize objective functions that strongly depend on the number of clusters. Hence, two solutions with two different numbers of clusters can not be compared directly. Although this reduces considerably the size of the search space, it prevents the discovery of a better partitioning once a wrong number of clusters is selected. In this paper, we address this limitation by proposing a tensor co-clustering algorithm that optimizes an objective function (a $n$-mode extension of an association measure called Goodman-Kruskal's $\tau$ [9]) whose local optima do not depend on the number of clusters. Additionally, we use an optimization schema that improves such objective function after each iteration. Consequently, our co-clustering approach can be also considered as an example of anytime algorithm, i.e., it can return a valid co-clustering even if it is interrupted before convergence is reached. We show experimentally that our algorithm provides accurate clustering results in each mode of the tensor. Compared with state-of-the-art techniques that require the desired number of clusters in each mode as input parameters, it achieves similar or better results. Additionally, it is also effective in clustering real-world datasets.

In summary, the main contributions of this paper are as follows: (i) we define an objective function for $n$-mode tensor co-clustering, based on Goodman-Kruskal's $\tau$ association measure, which does not require the number of clusters as input parameter (Sect. 3); (ii) we propose a stochastic optimization algorithm that improves the objective function after each iteration and supports the rapid convergence towards a local optimum (Sect. 4); (iii) we show the effectiveness of our metohd experimentally on both synthetic and real-world data, also in comparison with state-of-the-art competitors (Sect. 5).

## 2    Related Work

Analyzing multi-way data (or $n$-way tensors) has attracted a lot of attention due to their intrinsic complexity and richness. Hence, to deal with this complexity, in the last decade, many ad-hoc methods and extension of 2-way matrix methods have been proposed, many of which are tensor decomposition models and algorithms [16].

The problem of clustering and co-clustering of higher-order data has also been extensively addressed. Co-clustering has been developed as a matrix method and studied in many different application contexts including text mining [6,19], gene expression analysis [5] and graph mining [4] and has been naturally extended to tensors for its ability of handling high-dimensional data well. In [2], the authors perform clustering using a relation graph model that describes all the known relations between the modes of a tensor. Their tensor clustering formulation captures the maximal information in the relation graph by exploiting a family of loss function known as Bregman divergences. Instead, the authors of [28], use tensor-based latent factor analysis to address co-clustering in the context of web usage mining. Their algorithm is executed via the well-known multi-way decomposition algorithm called CANDECOMP/PARAFAC [10]. Papalex-akis *et al.* formulate co-clustering as a constrained multi-linear decomposition with sparse latent factors [18]. They propose a basic multi-way co-clustering algorithm exploiting multi-linearity using Lasso-type coordinate updates. Zhang *et al.* propose an extension of the tri-factor non-negative matrix factorization model [7] to a tensor decomposition model performing adaptive dimensionality reduction by integrating the subspace identification and the clustering process into a single process [27]. Finally, in [24], the authors introduce a spectral co-clustering method based on a new random walk model for nonnegative square tensors.

Differently from all these approaches, our tensor co-clustering algorithm is not based on any factorization model. Instead, it optimizes an extension of a measure of association whose effectiveness has been proven in matrix (2-way) co-clustering [15], and that naturally helps discover the correct number of clusters in tensor with arbitrary shape and density.

## 3   An Association Measure for Tensor Co-clustering

In this section, we introduce the objective function we optimize in our tensor co-clustering algorithm (presented in the next section). It consists in an association measure, called Goodman and Kruskal's $\tau$ [9], that evaluates the dependence between two discrete variables and has been used to evaluate the quality of 2-way co-clustering [20]. We generalize its definition to a $n$-mode tensor setting.

### 3.1   Goodman and Kruskal $\tau$ and Its Generalization

Goodman and Kruskal's $\tau$ [9] is an association measure that estimates the strength of the link between two discrete variables $X$ and $Y$ according to the proportional reduction of the error in predicting one of them knowing the other. In more details, let $x_1, \ldots, x_m$ be the values that variable $X$ can assume, with probability $p_X(1), \ldots, p_X(m)$ and let $y_1, \ldots, y_n$ be the possible values $Y$ can assume, with probability $p_Y(1), \ldots, p_Y(n)$. The error in predicting $X$ can be evaluated as the probability that two different observations from the marginal distribution of $X$ fall in different categories:

$$e_X = \sum_{i=1}^{m} p_X(i)(1 - p_X(i)) = 1 - \sum_{i=1}^{m} p_X(i)^2.$$

Similarly, the error in predicting $X$ knowing that $Y$ has value $y_j$ is

$$e_{X|Y=y_j} = \sum_{i=1}^{m} p_{X|Y=y_j}(i|j)(1 - p_{X|Y=y_j}(i|j)) = 1 - \sum_{i=1}^{m} p_{X|Y=y_j}(i|j)^2$$

and the expected value of the error in predicting $X$ knowing $Y$ is

$$\mathbb{E}[e_{X|Y}] = \sum_{j=1}^{n} e_{X|Y=y_j} p_Y(j)$$

$$= \sum_{j=1}^{n} (1 - \sum_{i=1}^{m} p_{X|Y=y_j}(i|j)^2) p_Y(j) = 1 - \sum_{i=1}^{m} \sum_{j=1}^{n} \frac{p_{X,Y}(i,j)^2}{p_Y(j)}.$$

Then the Goodman and Kruskall $\tau_{X|Y}$ measure of association is defined as

$$\tau_{X|Y} = \frac{e_X - \mathbb{E}[e_{X|Y}]}{e_X} = \frac{\sum_{i=1}^{m} \sum_{j=1}^{n} \frac{p_{X,Y}(i,j)^2}{p_Y(j)} - \sum_{i=1}^{m} p_X(i)^2}{1 - \sum_{i=1}^{m} p_X(i)^2}.$$

Conversely, the proportional reduction of the error in predicting $Y$ while $X$ is known is

$$\tau_{Y|X} = \frac{e_Y - \mathbb{E}[e_{Y|X}]}{e_Y} = \frac{\sum_{i=1}^{n} \sum_{j=1}^{m} \frac{p_{X,Y}(i,j)^2}{p_X(i)} - \sum_{j=1}^{n} p_Y(j)^2}{1 - \sum_{j=1}^{n} p_Y(j)^2}.$$

In order to use this measure for the evaluation of a tensor co-clustering, we need to extend it so that $\tau$ can evaluate the association of $n$ distinct discrete variables. Let $X_1, \ldots, X_n$ be discrete variables such that $X_i$ can assume $m_i$ distinct values (for simplicity, we will denote the possible values as $1, \ldots, m_i$), for $i = 1, \ldots, n$. Let $p_{X_i}(k)$ be the probability that $X_i = k$, for $k = 1, \ldots, m_i$, for $i = 1, \ldots, n$. Reasoning as in the two-dimensional case, we can define the reduction in the error in predicting $X_i$ while $(X_j)_{j \neq i}$ are all known as

$$\tau_{X_i} = \tau_{X_i|(X_j)_{j \neq i}} = \frac{e_{X_i} - \mathbb{E}[e_{X_i|(X_j)_{j \neq i}}]}{e_{X_i}}$$

$$= \frac{\sum_{k_1=1}^{m_1} \cdots \sum_{k_n=1}^{m_n} \frac{p_{X_1,\ldots,X_n}(k_1,\ldots k_n)^2}{P(X_j)_{j \neq i}((k_j)_{j \neq i})} - \sum_{k_i=1}^{m_i} p_{X_i}(k_i)^2}{1 - \sum_{k_i=1}^{m_i} p_{X_i}(k_i)^2}, \tag{1}$$

for all $i \leq n$. When $n = 2$, the measure coincides with Goodman-Kruskal's $\tau$.

Notice that, in the $n$-dimensional case as well as in the 2-dimensional case, the error in predicting $X_i$ knowing the value of the other variables is always positive and smaller or equal to the error in predicting $X_i$ without any knowledge about the other variables. It follows that $\tau_{X_i}$ takes values between $[0, 1]$. It will be 0 if knowledge of prediction of the other variables is of no help in predicting $X_i$, while it will be 1 if knowledge of the values assumed by variables $(X_j)_{j \neq i}$ completely specifies $X_i$.

### 3.2   Tensor Co-clustering with Goodman-Kruskal's $\tau$

Let $\mathcal{X} \in \mathbb{R}_+^{m_1 \times \cdots \times m_n}$ be a tensor with $n$ modes and non-negative values. Let us denote with $x_{k_1 \ldots k_n}$ the generic element of $\mathcal{X}$, where $k_i = 1, \ldots, m_i$ for each mode

$i = 1, \ldots, n$. A co-clustering $\mathcal{P}$ of $\mathcal{X}$ is a collection of $n$ partitions $\{\mathcal{P}_i\}_{i=1,\ldots,n}$, where $\mathcal{P}_i = \cup_{j=1}^{c_i} C_j^i$ is a partition of the elements on the $i$-th mode of $\mathcal{X}$ in $c_i$ groups, with $c_i \leq m_i$ for each $i = 1, \ldots, n$. Each co-clustering $\mathcal{P}$ can be associated to a tensor $\mathcal{T}^{\mathcal{P}} \in \mathbb{R}_+^{c_1 \times \cdots \times c_n}$, whose generic element is

$$
t_{i_1 \ldots i_n} = \sum_{k_1 \in C_{i_1}^1} \sum_{k_2 \in C_{i_2}^2} \cdots \sum_{k_n \in C_{i_n}^n} x_{k_1 \ldots k_n}. \tag{2}
$$

Consider now $n$ discrete variables $X_1, \ldots, X_n$, where each $X_i$ takes values in $\{C_1^i, \ldots C_{c_i}^i\}$. We can look at $\mathcal{T}^{\mathcal{P}}$ as the contingency $n$-modal table that empirically estimates the joint distribution of $X_1, \ldots, X_n$: the entry $t_{k_1 \ldots k_n}$ is the frequency of the event $(\{X_1 = C_{k_1}^1\} \cap \cdots \cap \{X_n = C_{k_n}^n\})$ and the frequency of $X_i = C_k^i$ is the marginal frequency obtained by summing all entries $t_{k_1 \ldots k_{i-1} k k_{i+1} \ldots k_n}$, with $k_1, \ldots, k_{i-1}, k_{i+1}, \ldots, k_n$ varying trough all possible values and the $i$-th index $k_i$ fixed to $k$. In the same way, we can compute the frequency of the event $(\{X_i = C_k^i\} \cap \{X_j = C_h^j\})$ as the sum of all elements $t_{k_1 \ldots k_n}$ of $\mathcal{T}^{\mathcal{P}}$ having $k_i = k$ and $k_j = h$. More in general, we can compute the marginal joint frequency of $d < n$ variables as the sum of all the entries of $\mathcal{T}^{\mathcal{P}}$ having the indices corresponding to the $d$ variables fixed to the values we are considering. For instance, given $\mathcal{T}^{\mathcal{P}} \in \mathbb{R}_+^{4 \times 3 \times 5 \times 2}$, the empirical frequency of the event $(\{X_1 = 3\} \cap \{X_3 = 4\})$ is

$$
t_{(3,4)}^{(1,3)} = \sum_{k_2=1}^{3} \sum_{k_4=1}^{2} t_{3,k_2,4,k_4}.
$$

From now on, we will use the newly introduced notation $t_{\mathbf{w}}^{\mathbf{v}}$ to denote the sum of all elements of a tensor having the modes in the upper vector $\mathbf{v}$ (in the example $(1, 3)$) fixed to the values of the lower vector $\mathbf{w}$ (in the example $(3,4)$). A formal definition of the scalar $t_{\mathbf{w}}^{\mathbf{v}}$ can result clunky: given a tensor $\mathcal{T} \in \mathbb{R}_+^{m_1 \times \cdots \times m_n}$ and two vectors $\mathbf{v}, \mathbf{w} \in \mathbb{R}_+{}^d$, with dimension $d \leq n$, such that $v_j \leq n$, $v_i < v_j$ if $i < j$ and $w_i \leq m_{v_i}$ for each $i, j = 1, \ldots, d$, we will use the following notation

$$
t_{\mathbf{w}}^{\mathbf{v}} = \sum_{k_{\bar{v}_1}=1}^{m_{\bar{v}_1}} \cdots \sum_{k_{\bar{v}_r}=1}^{m_{\bar{v}_r}} t_{e_1 \ldots e_n}
$$

where $\bar{\mathbf{v}}$ is the vector of dimension $r = n - d$ containing all the integers $i \leq n$ that are not in $\mathbf{v}$ and $e_i = w_i$ if $i \in \mathbf{v}$ while $e_i = k_i$ otherwise.

Summarizing, given a tensor $\mathcal{X}$ with $n$ modes and a co-clustering $\mathcal{P}$ over $\mathcal{X}$, we obtain a tensor $\mathcal{T}^{\mathcal{P}}$ that represents the empirical frequency of $n$ discrete variables $X_1, \ldots, X_n$ each of them with $c_i$ possible values (where $c_i$ is the number of clusters in the partition on the $i$-th mode of $\mathcal{X}$). Therefore, we can derive from $\mathcal{T}^{\mathcal{P}}$ the probability distributions of variables $X_1, \ldots, X_n$ and substitute them in Eq. 1: in this way we associate to each co-clustering $\mathcal{P}$ over $\mathcal{X}$ a vector $\tau^{\mathcal{P}} = (\tau_{X_1}^{\mathcal{P}}, \ldots, \tau_{X_n}^{\mathcal{P}})$ that can be used to evaluate the quality of the co-clustering.

In particular, for any $i, j \leq n$ and any $k_i = 1, \ldots, c_i$:

$$p_{X_1 \ldots X_n}(k_1, \ldots, k_n) = \frac{t_{k_1 \ldots k_n}}{T}, \quad p_{X_i}(k_1) = \frac{t^{(i)}_{(k_i)}}{T}, \quad p_{(X_j)_{j \neq i}}((k_j)_{j \neq i}) = \frac{t^{(j)_{j \neq i}}_{(k_j)_{j \neq i}}}{T},$$

where $T$ is the sum of all entries of $\mathcal{T}^{\mathcal{P}}$. It follows that

$$\tau^{\mathcal{P}}_{X_i} = \frac{\sum_{k_1=1}^{c_1} \cdots \sum_{k_n=1}^{c_n} \frac{t^2_{k_1 \ldots k_n}}{t^{(j)_{j \neq i}}_{(k_j)_{j \neq i}} \cdot T} - \sum_{k_i=1}^{c_i} \frac{\left(t^{(i)}_{(k_i)}\right)^2}{T^2}}{1 - \sum_{k_i=1}^{c_i} \frac{\left(t^{(i)}_{(k_i)}\right)^2}{T^2}} \tag{3}$$

for each $i = 1, \ldots, n$.

Suppose now we have two different partitions $\mathcal{P}$ and $\mathcal{Q}$ on the same tensor $\mathcal{X}$, corresponding to two different vectors $\tau^{\mathcal{P}}, \tau^{\mathcal{Q}} \in [0, 1]^n$. There is no obvious order relation in $[0, 1]^n$, so it is not immediately clear which one between $\tau^{\mathcal{P}}$ and $\tau^{\mathcal{Q}}$ is "better" than the other. In [15], the authors introduce a partial-order over $\mathbb{R}^n$ and exploit the notion of Pareto-dominance relation. Hence, their algorithm solves a multi-objective optimization problem. Instead, we propose another approach to compare partitions, based on a scalarization function, that maps the set of the partitions into $\mathbb{R}$ and then uses the natural order in $\mathbb{R}$ to compare partitions. In particular, we opt for the function $f$ that maps each partition $\mathcal{P}$ into a weighted sum $f(\mathcal{P}) = \sum_{i=1}^{n} w_i \tau^{\mathcal{P}}_{X_i}$, with fixed $w_i > 0$ such that $\sum_{i=1}^{n} w_i = 1$.

In this paper we will fix the weights $w_i = \frac{1}{n}$, for all $i = 1, \ldots, n$. We choose those values because we consider all modes equally important. Anyway, if other configurations of $\{w_i\}$ are used, the substance of the algorithm we will present in the following section does not change.

## 4    A Stochastic Local Search Approach to Tensor Co-clustering

Our co-clustering approach can be formulated as a maximization problem: given a tensor $\mathcal{X}$ with $n$ modes and dimension $m_i$ on mode $i$, an optimal co-clustering $\mathcal{P}$ for $\mathcal{X}$ is one that maximizes $f(\mathcal{P}) = \sum_{i=1}^{n} \tau^{\mathcal{P}}_{X_i}$. Since we do not fix the number of clusters, the space of possible solutions is huge (for example, given a very small tensor of dimension $10 \times 10 \times 10$, the number of possible partitions is $1.56 \times 10^{16}$): it is clear that a systematic exploration of all possible solutions is not feasible for a generic tensor $\mathcal{X}$. For this reason we propose a stochastic local search approach to solve the maximization problem.

### 4.1    Tensor Co-clustering Algorithm

Algorithm 1 provides a sketch of our tensor co-clustering algorithm, called $\tau TCC$. At each iteration $i$, it considers one mode by one, sequentially, and tries to improve the partition on that mode: fixed the $k$-th mode, the algorithm randomly

---

**Algorithm 1:** $\tau TCC(\mathcal{X}, N_{iter})$

---

**Input:** $\mathcal{X}$ tensor with $n$ modes, $N_{iter}$
**Result:** $\mathcal{P}_1, \ldots, \mathcal{P}_n$

1 Initialize $\mathcal{P}_1, \ldots, \mathcal{P}_n$ with discrete partitions;
2 $i \leftarrow 0$;
3 $T \leftarrow \mathcal{X}$;
4 $\max_\tau \leftarrow \sum_{j=1}^n \tau_{X_i}(T)$;
5 **while** $i \leq N_{iter}$ **do**
6     **for** $k = 1$ *to* $n$ **do**
7         Randomly choose $C_b^k$ in $\mathcal{P}_k$;
8         Randomly choose $o$ in $C_b^k$;
9         $c_k \leftarrow |\mathcal{P}_k \cup \emptyset|$;
10         $\max_\tau^e \leftarrow max_{e \in \{1, \ldots, c_k\}, e \neq b} \sum_{j=1}^n \tau_{X_j}(T^e)$ // see section 4.2;
11         $e \leftarrow argmax_{e \in \{1, \ldots, c_k\}, e \neq b} \sum_{j=1}^n \tau_{X_j}(T^e)$;
12         **if** $max_\tau^e > max_\tau$ **then**
13             $T \leftarrow T^e$;
14             $max_\tau \leftarrow max_\tau^e$;
15         **end**
16     **end**
17     $i \leftarrow i + 1$;
18 **end**

---

selects one cluster $C_b^k$ and one element $o \in C_b^k$. Then it tries to move $o$ in every other cluster $C_e^k$, with $e \neq b$, and in the empty cluster $C_e^k = \emptyset$: among them, it selects the one that optimizes the objective function. When all the $n$ modes have been considered, the $i$-th iteration of the algorithm is concluded. These operations are repeated until a stopping condition is met; although this condition can be a convergence criterion of $\tau$, for simplicity, we fix the maximum number of iterations by $N_{iter}$ in our algorithm. At the end of each iteration, one of the following possible moves has been done on mode $k$:

- an object $o$ has been moved from cluster $C_b^k$ to a pre-existing cluster $C_e^k$: in this case the final number of clusters on mode $k$ remains $c_k$ if $C_b^k$ is non-empty after the move. If $C_b^k$ is empty after the move, it will be deleted and the final number of clusters will be $c_k - 1$;
- an object $o$ has been moved from cluster $C_b^k$ to a new cluster $C_e^k = \emptyset$: the final number of clusters on mode $k$ will be $c_k + 1$ (the useless case when $o$ is moved from $C_b^k = \{o\}$ to $C_e^k = \emptyset$ is not considered);
- no move has been performed, thus the number of clusters remains $c_k$.

Thus, during the iterative process, the updating procedure is able to increase or decrease the number of clusters at any time. This is due to the fact that, contrary to other measures, such as the loss in mutual information [6], $\tau$ measure has an upper limit which does not depend on the numbers of co-clusters and thus enables comparison of co-clustering solutions of different cardinalities.

The proposed algorithm has the desirable property of increasing (or at least not worsening) the objective function after each iteration, i.e. $\sum_{i=1}^{n} \tau_{X_i}$ gets closer to the optimal value of the objective function. Notice, however, that it is not guaranteed that the global optimum will be reached. In fact, at each step $i$, the algorithm only allows to move from a partition $\mathcal{P}^{(i)}$ to a neighboring one, i.e., a partition obtainable by moving a single element from a cluster to another. It is not guaranteed that there is a path of neighboring partitions that connects $\mathcal{P}^{(i)}$ with an optimal partition $\mathcal{P}' \in O_\chi^f$. It is possible, conversely, that the algorithm comes to a partition with no neighboring solutions improving the objective function. In this case the algorithm ends in a local optimum.

Algorithm 1 modifies, at each iteration, every partition $\mathcal{P}_i$ by evaluating function $\tau_{X_i}^{\mathcal{P}}$. The computational complexity of this function is in $O(m_1 \cdot m_2 \cdot \ldots \cdot m_n)$. Moreover, during each iteration, for each mode these operations are performed for each cluster (including the empty cluster). Thus, in the worst case, the overall complexity of each iteration is in $O\left((\max_i m_i) \cdot (m_1 \cdot m_2 \cdot \ldots \cdot m_n)\right)$ for each mode. In the next section, we present an optimized version of the algorithm that reduces the overall time complexity.

## 4.2 Optimized Computation of $\tau$

In steps 10–11 of Algorithm 1, fixed a mode $k$, the following quantities are computed:

$$\max_{e \in \{1,\ldots,c_k\}, e \neq b} \sum_{j=1}^{n} \tau_{X_j}(T^e) \quad \text{and} \quad \underset{e \in \{1,\ldots,c_k\}, e \neq b}{\mathrm{argmax}} \sum_{j=1}^{n} \tau_{X_j}(T^e)$$

where $c_k$ is the number of clusters on mode $k$ (including the empty set) and $T^e$ is the contingency tensor associated to co-clustering $\mathcal{P}^e$ obtained by moving an object $o$ from cluster $C_b^k$ to cluster $C_e^k$ in partition $\mathcal{P}_k$, for each $e \in \{1,\ldots,c_k\}, e \neq b$.

A way to compute these quantities is to fix an arrival cluster $C_e^k$, move $o$ in $C_e^k$ obtaining a new partition $\mathcal{P}_k^e$, compute the contingency tensor associated to that partition (using Eq. 2), compute vector $\tau^e$ associated to tensor $T^e$ (using Eq. 3) and finally compute $\sum_{j=1}^{n} \tau_{X_j}(T^e)$. By repeating these steps for every $e \in \{1,\ldots,c_k\}, e \neq b$, we obtain a vector $\mathbf{v} = (\sum_{j=1}^{n} \tau_{X_j}(T^e))_{e \in \{1,\ldots,c_k\}, e \neq b}$ of dimension $c_k$ and we can compute $\max \mathbf{v}$ and $\mathrm{argmax}\, \mathbf{v}$. In order to obtain $\mathbf{v}$ in a more efficient way, we can reduce the amount of calculations by only computing the variation of $\tau^e$ from one step to another. We take advantage of the fact that a large part in the $\tau$ formula remains the same when moving a single element from a cluster to another. Hence, an important part of the computation of $\tau$ can be saved.

Imagine that $o$ has been selected in cluster $C_b^1$ and that we want to move it in cluster $C_e^1$ (for simplicity we consider $o$ on the first mode, but all the computations below are analogous on any other mode $k$). Object $o$ is a row on the first mode (let's say the $j$-th row) of tensor $\mathcal{X}$ and so $o$ can be expressed as a tensor $\mathcal{M} \in \mathbb{R}_+^{m_2 \times \cdots \times m_n}$ with $n-1$ modes, which generic entry is

$\mu_{k_2...k_n} = x_{jk_2...k_n}$. We will denote with $M$ the sum of all elements of $\mathcal{M}$. Let $\mathcal{T}$ and $\tau(\mathcal{T})$ be the tensor and the measure associated to the initial co-clustering and $\mathcal{S}$ and $\tau(\mathcal{S})$ the tensor and the measure associated to the final co-clustering obtained after the move. Tensor $\mathcal{S}$ differs from $\mathcal{T}$ only in those entries having index $k_1 \in \{b, e\}$. In particular, for each $k_i = 1, \ldots, c_i$ and $i = 2, \ldots, n$:

$$
\begin{aligned}
s_{bk_2...k_n} &= t_{bk_2...k_n} - \mu_{k_2...k_n} \\
s_{ek_2...k_n} &= t_{ek_2...k_n} + \mu_{k_2...k_n} \\
s_{k_1k_2...k_n} &= t_{k_1k_2...k_n}, \ if \ k_1 \notin \{b, e\}.
\end{aligned}
$$

Replacing these values in Eq. 1, we can compute the variation of $\tau_{X_1}$ moving object $o$ from cluster $C_b^1$ to cluster $C_e^1$ as:

$$\Delta\tau_{X_1}(\mathcal{T}, o, b, e, k = 1) = \tau_{X_1}(\mathcal{T}) - \tau_{X_1}(\mathcal{S})$$

$$= \frac{\Gamma_1 \left[ \frac{2M}{T^2}(M + t_{(b)}^{(1)} - t_{(e)}^{(1)}) \right] - \Omega_1 \left[ \frac{2}{T} \sum_{k_2,...,k_n} \frac{\mu_{k_2...k_n}(\mu_{k_2...k_n} + t_{ek_2...k_n} - t_{bk_2...k_n})}{t_{(k_2...k_n)}^{(2...n)}} \right]}{\Omega_1^2 - \Omega_1 \left[ \frac{2M}{T^2}(M + t_{(b)}^{(1)} - t_{(e)}^{(1)}) \right]}.$$

where $\Omega_1 = 1 - \sum_{k_1} \frac{\left(t_{(k_1)}^{(1)}\right)^2}{T^2}$ and $\Gamma_1 = 1 - \sum_{k_1,...,k_n} \frac{t_{k_1...k_n}^2}{T \cdot t_{(k_2...k_n)}^{(2...n)}}$ only depend on $\mathcal{T}$ and then can be computed once (before choosing $b$ and $e$). Thanks to this approach, instead of computing $m_i$ times $\tau_{X_i}$ with complexity $O(m_1 \cdot m_2 \cdot \ldots \cdot m_n)$, we compute $\Delta\tau_{X_i}(\mathcal{T}, o, b, e, k = i)$ with a complexity in $O(m_1 \cdot m_2 \cdot \ldots \cdot m_{i-1} \cdot m_{i+1} \cdot \ldots \cdot m_n)$ in the worst case with the discrete partition. Computing $\Gamma_i$ is in $O(m_1 \cdot m_2 \cdot \ldots \cdot m_n)$ and $\Omega_i$ in $O(m_i)$ and is done only once for each mode in each iteration.

In a similar way, we can compute the variation of $\tau_{X_j}$ for any $j \neq 1$:

$$\Delta\tau_{X_j}(\mathcal{T}, o, b, e, k = 1) = \tau_{X_j}(\mathcal{T}) - \tau_{X_j}(\mathcal{S})$$

$$= \frac{1}{\Omega_j T} \sum_{k_2...k_n} \left( \frac{t_{ek_2...k_n}^2}{t_{(k_i)_{i \neq j}}^{(i)_{i \neq j}}} - \frac{(t_{ek_2...k_n} + \mu_{k_2...k_n})^2}{t_{(k_i)_{i \neq j}, k_1 = e}^{(i)_{i \neq j}} + \mu_{(k_i)_{i \neq j}}^{(i)_{i \neq j-1}}} \right.$$

$$\left. + \frac{t_{bk_2...k_n}^2}{t_{(k_i)_{i \neq j}, k_1 = b}^{(i)_{i \neq j}}} - \frac{(t_{bk_2...k_n} - \mu_{k_2...k_n})^2}{t_{(k_i)_{i \neq j}, k_1 = b}^{(i)_{i \neq j}} - \mu_{(k_i)_{i \neq j}}^{(i)_{i \neq j-1}}} \right)$$

where $\Omega_j = 1 - \sum_{k_j} \frac{\left(t_{(k_j)}^{(j)}\right)^2}{T^2}$ only depends on $\mathcal{T}$ and can be computed once for all $e$. Consequently, instead of computing $m_j$ times $\tau_{X_j}$ in Algorithm 1 with a complexity in $O(m_1 \cdot m_2 \cdot \ldots \cdot m_n)$, we compute $\Delta\tau_{X_j}(\mathcal{T}, o, b, e, k = i)$ with a complexity in $O(m_1 \cdot m_2 \cdot \ldots \cdot m_{i-1} \cdot m_{i+1} \cdot \ldots \cdot m_n)$ in the worst case with the discrete partition. Computing $\Omega_j$ is in $O(m_j)$ and is done only once for each mode in each iteration.

Hence, when we have to decide in which cluster $C_e^k$ it is better to move object $o$, instead of computing vector $(\sum_{j=1}^n \tau_{X_j}(\mathcal{T}^e))_{e \in \{1,...,c_k\}, e \neq b}$ and its maximum, we can equivalently compute vector $\Delta\tau = (\sum_{j=1}^n \Delta\tau_{X_j}(\mathcal{T}, o, e, k))_{e \in \{1,...,c_k\}, e \neq b}$

and its minimum. In this way we reduce the amount of computations to be executed for each mode at each iteration of the algorithm from a complexity in $O((\max_i m_i) \cdot m_1 \cdot m_2 \cdot \ldots \cdot m_n)$ to $O(m_1 \cdot m_2 \cdot \ldots \cdot m_n)$.

Based on the above considerations, for a generic square tensor with $n$ modes, each consisting of $m$ dimensions, the overall complexity is in $O(\mathcal{I}n \cdot m^n)$, where $\mathcal{I}$ is the number of iterations (instead of $O(\mathcal{I}n \cdot m^{n+1})$).

## 5    Experiments

In this section, we evaluate the performance of our tensor co-clustering algorithm through experiments. We first apply the algorithm to synthetic data and then we show the results on a real-world dataset. To assess the quality of the clustering performances, we consider two measures commonly used in the clustering literature: normalized mutual information (NMI) [22] and adjusted rand index (ARI) [14]. We compare our results with those of other state-of-the-art co-clustering algorithms, based on CP [10] and Tucker [23] decomposition. nnCP is the non-negative CP decomposition. It can be used to co-cluster a tensor, as done in [28], by assigning each element in each mode to the cluster corresponding to the latent factor with highest value. The algorithm requires as input the number $r$ of latent factors of the decomposition: we set $r = \max(c_1, c_2, c_3)$, where $c_1, c_2$ and $c_3$ are the true numbers of classes on the three modes of the tensor. nnCP+kmeans combines CP with a post-processing phase in which $k$-means is applied on each of the latent factor matrices. Here, we set the rank $r$ to $\max(c_1, c_2, c_3) + 1$ and the number $k_i$ of clusters in each dimension equal to the real number of classes (according to our experiments, this is the choice that maximizes the performances of the algorithm). Similarly, nnTucker is the non-negative Tucker decomposition (here we set the ranks of the core tensor equal to $(c_1, c_2, c_3)$), while nnT+kmeans combines Tucker decomposition with $k$-means on the latent factor matrices [3,13]. Finally, SparseCP is a CP decomposition with non-negative sparse latent factors [18]. We set the rank $r$ of the decomposition equal to the maximum number of classes on the three modes of the tensor. It also requires one parameter $\lambda_i$ for each mode of the tensor: for the choice of their values we follow the instructions suggested in the original paper. All experiments are performed on a server equipped with 2 Intel Xeon E5-2643 quad-core CPU's, 128 GB RAM, running Arch Linux (kernel release: 4.19.14)[1].

### 5.1    Experiments on Synthetic Data

The synthetic data we use to assess the quality of the clustering performance are boolean tensors with three modes, created as follows. We fix the dimensions $m_1, m_2, m_3$ of the tensor and the number of embedded clusters $c_1, c_2, c_3$ on each

---

[1] The source code of our algorithm and all data used in this paper are available at: https://github.com/elenabattaglia/tensor_cc.

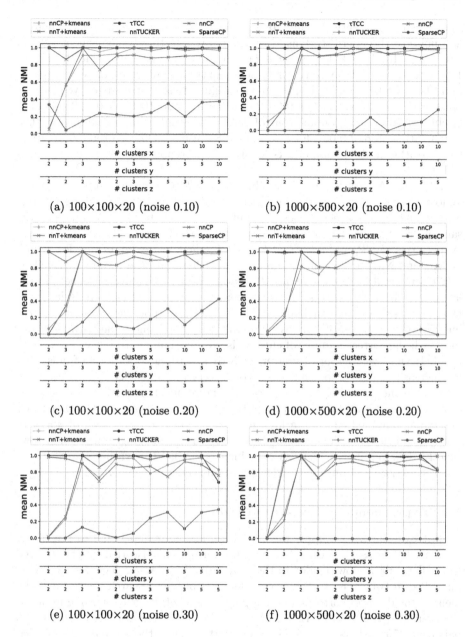

**Fig. 1.** Mean NMI on the three modes varying the number of embedded clusters on synthetic tensors with different sizes and levels of noise.

of the three modes. Then, we first construct a block tensor of dimensions $m_1 \times m_2 \times m_3$ with $c_1 \times c_2 \times c_3$ blocks. The blocks are created so that there are "perfect" clusters in each mode, i.e., all rows on each mode belonging to the same cluster are identical, while rows in different clusters are different. Then we add noise to the "perfect" tensor, by randomly selecting some element $t_{k_1 k_2 k_3}$, with $k_i \in \{1, \ldots, m_i\}$, for each $i \in \{1, 2, 3\}$, and changing its value (from 0 to 1 or vice versa). The amount of noise is controlled by a parameter $\epsilon \in [0, 1]$, indicating the fraction of elements of the original tensor we change. We generate tensors of different size ($100 \times 100 \times 20, 1000 \times 100 \times 20, 1000 \times 500 \times 20$), number of clusters (different combinations of 2, 3, 5, 10) and values of noise ($\epsilon = 0.05$ to 0.3 with a step of 0.05), for a total of 198 tensors[2]. On each tensor, we apply the algorithm and its competitors five times and report the mean of the results in Fig. 1.

NMI and ARI of the resulting clusters of $\tau TCC$ remain stably over 0.9 in almost all experiments and in the vast majority of cases the resulting clusters exactly match the correct classes (we omit the results in terms of ARI here for the sake of brevity, but they are similar to NMI ones). In particular, $\tau TCC$ always outperforms $nnCP$, $nnTucker$ and $SparseCP$ (the latter exhibits very low values of ARI and NMI for asymmetric tensors); furthermore, the results achieved by $\tau TCC$ are similar to those of $nnCP+kmeans$ and $nnT+kmeans$. Generally the latter get "better" clusters in cases where the number of clusters is large. In fact, in these cases it can happen that $\tau TCC$ does not identifies the correct number of clusters in all modes (we don't have the same issue with $k$-means, for which the correct number of clusters is given as input). To better investigate this behavior, we compute the average NMI according to all level of noise and for increasing number of embedded co-clusters (obtained as $c_1 \cdot c_2 \cdot c_3$). The results are shown in Fig. 2. In general, the noise and the number of embedded co-clusters do not affect the quality of the results to a great extent, although we observe a combined effect of a high number of co-cluster and level of noise. In this case, identifying the embedded co-clusters is challenging, unless one knows exactly their number, which, as explained beforehand, is rather unrealistic in the vast majority of unsupervised application scenarios.

## 5.2  Experiments on Real-World Data

As last experiment, we apply our algorithm and its competitors to the "four-area" DBLP dataset[3]. It is a bibliographic information network dataset extracted from DBLP data, downloaded in the year 2008. The dataset includes all papers published in twenty representative conferences of four research areas. Each element of the data set corresponds to a paper and contains the following information: authors, venue and terms in the title. The original dataset contains 14376 papers, 14475 authors and 13571 terms. Part of the authors (4057) are labelled in four classes, roughly corresponding to the four research areas. We select only these authors and their papers and perform some pre-processing step on the terms (stemming, stop-words removal). We obtain a dataset with 14328 papers, from which we create a ($6044 \times 4057 \times 20$)-dimensional tensor, highly sparse

---

[2] Here we report only the results of two representative tensors and three noise level.

[3] http://web.cs.ucla.edu/~yzsun/data/DBLP_four_area.zip.

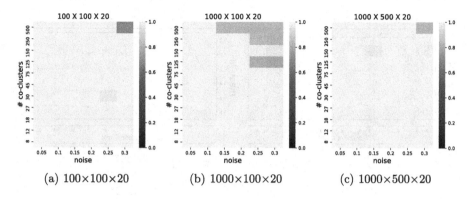

(a) 100×100×20          (b) 1000×100×20          (c) 1000×500×20

**Fig. 2.** Average NMI on the three modes varying the overall number of embedded co-clusters and the level of noise.

**Table 1.** Results of the co-clustering algorithms on "four-area" DBLP dataset. NMI, ARI and number of clusters identified are computed for the authors mode.

| Algorithm | NMI | ARI | # clusters |
|---|---|---|---|
| $\tau$TCC | 0.75 ± 0.01 | 0.80 ± 0.02 | 9 |
| nnTucker | **0.78** ± 0.00 | **0.84** ± 0.00 | 4 |
| nnCP | 0.74 ± 0.00 | 0.80 ± 0.00 | 4 |
| SparseCP | 0.00 ± 0.00 | 0.00 ± 0.00 | 1 |
| nnCP+kmeans | 0.24 ± 0.01 | 0.08 ± 0.00 | 4 |
| nnT+kmeans | 0.25 ± 0.01 | 0.06 ± 0.00 | 4 |

(99.98% of entries are equal to zero); the generic entry $t_{ijk}$ of the tensor counts the number of times term $i$ was used by author $j$ in conference $k$.

Table 1 shows that the best results are those of the non-negative Tucker decomposition where the number of latent factors is set to 4 (the correct number of embedded clusters). Observe, however, that in standard unsupervised settings, the number of "naturally" embedded clusters is unknown. Hence, by fixing the number of latent factors equal to the real number of natural clusters we are facilitating our competitors; if we modify the number of latent factors (see Figure 3(a)), the results get worse: this means that, if we don't specify the correct number of clusters on the author mode but we set an upper bound, the results of the Tucker based co-clustering algorithm become lower than those of $\tau$TCC. Also *nnCP* shows a similar behavior, but with slightly worse results (see Figure 3(b)). Note that $\tau$TCC achieves the second best performance, even if the number of clusters identified is higher than the correct number of classes (9 instead of 4): indeed, 4042 objects are correctly divided into four large groups and only 15 elements are assigned to 5 very small clusters, since they probably are candidate outliers. The ability of our algorithm to also identify outliers automatically will be investigated as future work.

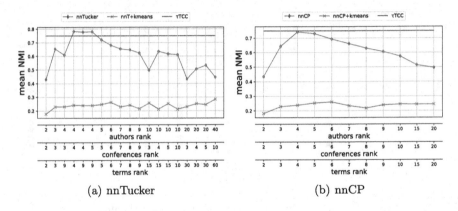

(a) nnTucker                    (b) nnCP

**Fig. 3.** Variation of nnTucker/nnCP results w.r.t. the rank of the decomposition.

## 6    Conclusions

The majority of tensor co-clustering algorithms optimizes objective functions that strongly depend on the number of co-clusters. This limits the correct application of such algorithms in realistic unsupervised scenarios. To address this limitation, we have introduced a new co-clustering algorithm specifically designed for tensors that does not require the desired number of clusters as input. Our experimental validation has shown that our approach is competitive with state-of-the-art methods that, however, can not work properly without specifying a correct number of clusters for each mode of the tensor. As future work, we will further investigate the ability of our method to identify candidate outliers as small clusters in the data.

## References

1. Araujo, M., Ribeiro, P.M.P., Faloutsos, C.: Tensorcast: forecasting time-evolving networks with contextual information. In: Proceedings of IJCAI 2018, pp. 5199–5203 (2018)
2. Banerjee, A., Basu, S., Merugu, S.: Multi-way clustering on relation graphs. In: Proceedings of SIAM SDM 2007, pp. 145–156 (2007)
3. Cao, X., Wei, X., Han, Y., Lin, D.: Robust face clustering via tensor decomposition. IEEE Trans. Cybern. 45(11), 2546–2557 (2015)
4. Chakrabarti, D., Papadimitriou, S., Modha, D.S., Faloutsos, C.: Fully automatic cross-associations. In: Proceedings of ACM SIGKDD 2004, pp. 79–88 (2004)
5. Cho, H., Dhillon, I.S., Guan, Y., Sra, S.: Minimum sum-squared residue co-clustering of gene expression data. In: Proceedings of SIAM SDM 2004, pp. 114–125 (2004)
6. Dhillon, I.S., Mallela, S., Modha, D.S.: Information-theoretic co-clustering. In: Proceedings of ACM SIGKDD 2003, pp. 89–98 (2003)
7. Ding, C.H.Q., Li, T., Peng, W., Park, H.: Orthogonal nonnegative matrix t-factorizations for clustering. In: Proceedings of ACM SIGKDD 2006, pp. 126–135 (2006)

8. Ermis, B., Acar, E., Cemgil, A.T.: Link prediction in heterogeneous data via generalized coupled tensor factorization. Data Min. Knowl. Discov. **29**(1), 203–236 (2015)
9. Goodman, L.A., Kruskal, W.H.: Measures of association for cross classification. J. Am. Stat. Assoc. **49**, 732–764 (1954)
10. Harshman, R.A.: Foundation of the parafac procedure: models and conditions for an" explanatory" multimodal factor analysis. UCLA Working Papers in Phonetics **16**, 1–84 (1970)
11. He, J., Li, X., Liao, L., Wang, M.: Inferring continuous latent preference on transition intervals for next point-of-interest recommendation. In: Proceesings of ECML PKDD 2018, pp. 741–756 (2018)
12. Hong, M., Jung, J.J.: Multi-sided recommendation based on social tensor factorization. Inf. Sci. **447**, 140–156 (2018)
13. Huang, H., Ding, C.H.Q., Luo, D., Li, T.: Simultaneous tensor subspace selection and clustering: the equivalence of high order svd and k-means clustering. In: Proceedings of the 14th ACM SIGKDD, pp. 327–335 (2008)
14. Hubert, L., Arabie, P.: Comparing partitions. J. Classif. **2**(1), 193–218 (1985)
15. Ienco, D., Robardet, C., Pensa, R.G., Meo, R.: Parameter-less co-clustering for star-structured heterogeneous data. Data Min. Knowl. Discov. **26**(2), 217–254 (2013)
16. Kolda, T.G., Bader, B.W.: Tensor decompositions and applications. SIAM Rev. **51**(3), 455–500 (2009)
17. Papalexakis, E.E., Dogruöz, A.S.: Understanding multilingual social networks in online immigrant communities. In: Proceedings of MWA 2015 (co-located with WWW 2015), pp. 865–870 (2015)
18. Papalexakis, E.E., Sidiropoulos, N.D., Bro, R.: From K-means to higher-way co-clustering: multilinear decomposition with sparse latent factors. IEEE Trans. Signal Process. **61**(2), 493–506 (2013)
19. Pensa, R.G., Ienco, D., Meo, R.: Hierarchical co-clustering: off-line and incremental approaches. Data Min. Knowl. Discov. **28**(1), 31–64 (2014)
20. Robardet, C., Feschet, F.: Efficient local search in conceptual clustering. In: Proceedings DS 2001, pp. 323–335 (2001)
21. Shashua, A., Hazan, T.: Non-negative tensor factorization with applications to statistics and computer vision. In: Proceedings of (ICML 2005), pp. 792–799 (2005)
22. Strehl, A., Ghosh, J.: Cluster ensembles - a knowledge reuse framework for combining multiple partitions. J. Mach. Learn. Res. **3**, 583–617 (2002)
23. Tucker, L.R.: Some mathematical notes on three-mode factor analysis. Psychometrika **31**, 279–311 (1966)
24. Wu, T., Benson, A.R., Gleich, D.F.: General tensor spectral co-clustering for higher-order data. In: Proceedings of NIPS 2016, 2559–2567 (2016)
25. Yu, K., He, L., Yu, P.S., Zhang, W., Liu, Y.: Coupled tensor decomposition for user clustering in mobile internet traffic interaction pattern. IEEE Access **7**, 18113–18124 (2019)
26. Zhang, T., Golub, G.H.: Rank-one approximation to high order tensors. SIAM J. Matrix Anal. Appl. **23**(2), 534–550 (2001)
27. Zhang, Z., Li, T., Ding, C.H.Q.: Non-negative tri-factor tensor decomposition with applications. Knowl. Inf. Syst. **34**(2), 243–265 (2013)
28. Zhou, Q., Xu, G., Zong, Y.: Web co-clustering of usage network using tensor decomposition. In: Proceedings of ECBS 2009, pp. 311–314 (2009)

# Deep Triplet-Driven Semi-supervised Embedding Clustering

Dino Ienco[1]([×])[ID] and Ruggero G. Pensa[2]([×])[ID]

[1] IRSTEA, UMR TETIS, LIRMM, Univiversity of Montpellier, Montpellier, France
dino.ienco@irstea.fr
[2] Department of Computer Science, University of Turin, Turin, Italy
ruggero.pensa@unito.it

**Abstract.** In most real world scenarios, experts dispose of limited background knowledge that they can exploit for guiding the analysis process. In this context, semi-supervised clustering can be employed to leverage such knowledge and enable the discovery of clusters that meet the analysts' expectations. To this end, we propose a semi-supervised deep embedding clustering algorithm that exploits triplet constraints as background knowledge within the whole learning process. The latter consists in a two-stage approach where, initially, a low-dimensional data embedding is computed and, successively, cluster assignment is refined via the introduction of an auxiliary target distribution. Our algorithm is evaluated on real-world benchmarks in comparison with state-of-the-art unsupervised and semi-supervised clustering methods. Experimental results highlight the quality of the proposed framework as well as the added value of the new learnt data representation.

## 1 Introduction

Clustering is by far one of the most popular machine learning task among computer scientists, machine learning specialists and statisticians. Although it is conceived to work in fully unsupervised scenarios, very often, its application in real-world domains is supported by the availability of some, scarce, background knowledge. Unfortunately, producing or extracting such background knowledge (in terms of available class labels or constraints) is a time consuming and expensive task. Hence, the amount of available background knowledge is not sufficient for driving a supervised task. Still, it can be helpful in guiding a semi-supervised learning process.

The aim of semi-supervised clustering is to take advantage of the few available side information to guide the clustering process towards a partitioning that takes into account both the natural distribution of the data and the expectations of the domain experts. One of the most popular class of semi-supervised clustering algorithms exploit the so-called pairwise constraints: the clustering process is driven by a set of *must-link* (or similarity) and *cannot-link* (or dissimilarity) pairs modeling the fact that two data examples involved in any of these constraints should belong to the same cluster (must-link) or not (cannot-link).

© Springer Nature Switzerland AG 2019
P. Kralj Novak et al. (Eds.): DS 2019, LNAI 11828, pp. 220–234, 2019.
https://doi.org/10.1007/978-3-030-33778-0_18

Such constraints are successively exploited to either learning a distance metric [7,12,14] or forcing constraints during the clustering process [23], although the most effective methods usually combine both strategies [3,4,19].

However, all these strategies suffer from the same two problems: (i) two examples involved in a cannot-link constraint may actually be assigned to the wrong clusters and still satisfy the constraint; (ii) when constraints are generated from the labeled portion of the training set (a common practice in semi-supervised learning), and the class is rather loose (e.g., multiple clusters co-exist within the same class), the must-link constraints would mislead the clustering algorithm resulting in poor partitioning results. To address this issue, an alternative form of supervision has been proposed: given three data examples $x_a$, $x_p$ and $x_n$, one may impose that $x_a$ (called reference or anchor example) is closer to $x_p$ (called positive example) than to $x_n$ (called negative example). Such relative comparisons form the so-called triplet constraints [15].

In this paper, we propose $Ts2DEC$ (Triplet Semi-Supervised Deep Embedding Clustering). $Ts2DEC$ is a deep embedding-based clustering framework that leverages triplet constraints to inject supervision in the learning process. The framework consists of a two-stage approach: (i) an autoencoder extracts a low-dimensional representation (embedding) of the original data and (ii) an initial cluster assignment is refined via the introduction of an auxiliary target distribution [25]. Both stages are guided by the knowledge supplied by triplet constraints.

By means of an extensive experimental study conducted on several real-world datasets, we show that our approach outperforms state-of-the-art semi-supervised clustering methods no matter how much supervision is considered.

## 2  Related Work

Early semi-supervised approaches used pairwise (e.g., must-link and cannot-link) constraints to learn a metric space before applying standard clustering [14], or to drive the clustering process directly [23]. In [23], a simple adaptation of k-means that enforces must-link and cannot-link constraints during the clustering process is described. [2] proposes a constrained clustering approach that leverages labeled data during the initialization and clustering steps. Instead, [4] integrates both constraint-based and metric-based approaches in a k-means-like algorithm. Davis et al., propose an information-theoretic approach to learning a Mahalanobis distance function [7]. They leverage a Bregman optimization algorithm [1] to minimize the differential relative entropy between two multivariate Gaussians under constraints on the distance function. This approach has been recently extended by Nogueira et al., who combine distance metric learning and cluster-level constraints [19]. Zhu et al. present a pairwise similarity framework to perform an effective constraint diffusion handling noisy constraints as well [28].

In recent years, the advances in the deep learning field have also fostered new research in semi-supervised clustering. For instance, in [11], the author propose a semi-supervised clustering algorithm that directly exploits labels, instead of pairwise constraints. Their algorithm generates an ensemble of multiresolution semi-supervised autoencoders. The final partitioning is obtained by applying k-means on the new data representation obtained by stacking together all the different low-dimensional embeddings. In a very recent work [21], a semi-supervised extension to Deep Embedded Clustering (DEC [25]) is proposed. DEC learns a low-dimensional representation via autoencoder and, successively, it gradually refines clusters with an auxiliary target distribution derived from the current softcluster assignment. Its semi-supervised extension [21], called Semi-supervised Deep Embedded Clustering (SDEC) makes use of pairwise constraints in the cluster refinement stage. Therefore, the learned feature space is such that examples involved in a must-link (resp. cannot-link) constraint are forced to be close (resp. far away) from each other.

Our approach is also based on DEC, but, contrary to [21], it exploits triplet constraints introducing the background knowledge at the different stages of the process: during the embeddings generation and during the clustering refinement. We remind that, the expressiveness of triplet constraints has already demonstrated to be effective in the constrained clustering task [15].

## 3    Triplet Semi-supervised Deep Embedding Clustering

In this section, we introduce our semi-supervised clustering approach, called $Ts2DEC$ (Triplet semi-supervised Deep Embedding Clustering). The goal is to group together a set of examples $X = \{x_i\}_{i=1}^N$ into $C$ clusters given some background knowledge in terms of constraints.

To this purpose, we model our problem using neural networks. In a nutshell, given $ML = \{(x_j, x_l)\}$ (resp. $CL = \{(x_j, x_l)\}$) the set of must-link (resp. cannot-link) constraints, first we derive triplet constraints from these two sets. A triplet constraint is defined as a tuple $(x_a, x_p, x_n)$ where $x_a$ is the anchor example and $x_p$ (resp. $x_n$) is the positive (resp. negative) example with the associated semantic that $x_a$ and $x_p$ (resp. $x_a$ and $x_n$) belong (resp. do not belong) to the same cluster. Furthermore, due to transitivity, we also have that $x_p$ and $x_n$ do not belong to the same cluster. Successively, due to the exponential number of triplets we can generate, we adopt a simple and practical strategy to sample a subset of such triplets. We remind that triplet selection is an hard task and some research works are investigating how to smartly sample useful and informative subsets of triplet constraints [26]. It is out of the scope of this work supplying a method that competes with such strategies. On the other hand, we set up an easy and ready to use approach that well fits our scenario. Once the set of triplet constraints are chosen, we inject such background information into a deep-learning based clustering algorithm [9,26,27].

More in detail, we integrate the semi-supervision during: (i) the data embedding generation, by alternating unsupervised and semi-supervised optimization of the network parameters and (ii) the clustering refinement stage when cluster assignment hardening loss [18] is employed. Figure 1(a) and 1(b) provide a general overview of the embedding generation and clustering refinement stage, respectively. For each stage, we depict with the rose color and the dotted line

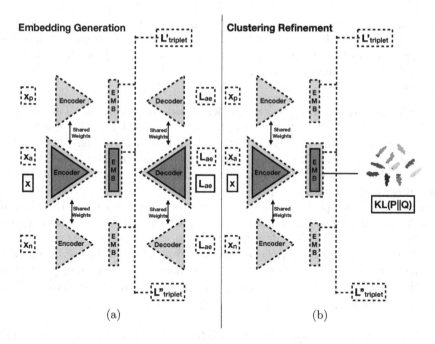

**Fig. 1.** General Overview of $Ts2DEC$: (a) Embedding Generation and (b) Clustering Refinement. We depict with the rose color and the dotted line the components related to the semi-supervised optimization (working on triplet constraints) while we depict with the blue color and the solid line the fully unsupervised components (working on the whole set of data $X$).

the components related to the semi-supervised optimization (working on triplet constraints) while we depict with the blue color and the solid line the fully unsupervised components (working on the whole set of data $X$). In the following, we provide the details of all the algorithmic steps of our approach.

**Triplets Generation Strategy.** The first preliminary step of $Ts2DEC$ is the generation of a set $T$ of triplet constraints from the set $ML$ and $CL$ of must-link and cannot-link constraints. To achieve this goal, first, we compute the transitive closure from both sets [6], then, we leverage it to generate all possible triplet constraints $(x_a, x_p, x_n)$. Generating triplets in such a way can produce a huge number of constraints. Consequently, we limit the number of triplet constraints by adopting the following strategy: for each pair $(x_a, x_p)$, we randomly sample a subset of possible examples that can play the role of negative examples $(x_n)$. Here, we give more importance to background knowledge that groups together similar examples (positive information) than information that forces examples to be clustered apart (negative information). We adopt this strategy because, during our experimental evaluations, we have empirically observed that positive information seems more effective in stretching the representation manifold thus respecting the given background knowledge. In our experiments, we sample 30% of all possible negative examples for each pair $(x_a, x_p)$ in order to obtain a reasonable trade off between performances and computational cost. The obtained set of triplet constraints is denoted by $T$.

**Embedding Generation with Background Knowledge.** The core of $Ts2DEC$ involves a first stage in which semi-supervised embedding representations are generated by means of autoencoder neural networks. This stage is depicted in Fig. 1(a). Autoencoders [16] are a particular kind of feed-forward neural network commonly employed to generate low-dimensional representation of the original data by setting up a reconstruction task. The autoencoder network is composed by two parts: (i) an encoder network that transforms the original data $X$ into an embedding representation (EMB) and (ii) a decoder network that reconstructs the original data from the embedding representation. Furthermore, the autoencoder network is layered and symmetric and the last layer of the encoder part is generally referred as bottleneck layer. The commonly adopted loss function optimized by an autoencoder network is the mean squared error between the original data and the reconstructed one:

$$L_{ae} = \frac{1}{|X|} \sum_{x_i \in X} ||x_i - dec(enc(x_i, \Theta_1), \Theta_2)||_2^2 \tag{1}$$

where $enc(z, \Theta_1)$ is the encoder network with parameters $\Theta_1$, while $dec(\cdot, \Theta_2)$ is the decoder network that reconstructs the data, with parameters $\Theta_2$.

For the encoder network, similarly to what proposed in [25], we adopt a feed-forward neural network with four layers (resp. 500, 500, 2000, 10 neurons per layer). The activation function associated to the first three (hidden) layers is the Rectifier Linear Unit (ReLU) while, for the last (bottleneck) layer, a simple linear activation function is employed [25]. The decoder is symmetrically derived from the encoder reversing the hidden layers.

A semi-supervised autoencoder [8,20] (denoted as SSAE), instead, is a multitask network that, in addition to the reconstruction task via its autoencoder structure, also deals with a discrimination task (mainly classification) leveraging the embedded representation. Conversely to most previous works on semi-supervised autoencoders [8,11,20] where the SSAE exploits labeled data to perform classification as supervised task, here, we design a SSAE that, associated to the reconstruction task, exploits the set $T$ of triplet constraints to generate the low-dimensional data embeddings EMB.

The set $T$ of triplets is used to learn a triplet network which consists of three different encoders/decoders with shared weights (highlighted in rose color and dotted line in Fig. 1(a)). In addition to the standard reconstruction loss, the specific loss function (triplet loss) optimized by the model is defined as follows:

$$L'_{triplet} = \sum_{(x_a, x_p, x_n) \in T} [d(x_a, x_p) - d(x_a, x_n) + \alpha]_+ \tag{2}$$

with

$$d(b,c) = ||norm_{L_2}(enc(b, \Theta_1)) - norm_{L_2}(enc(c, \Theta_1))||_2^2 \qquad (3)$$

where $T$ is the set of triplet constraints, $[x]_+ = max(0, x)$ is the hinge loss, $||x||_2^2$ is the squared $L_2$ norm of $x$, $enc(x, \Theta_1)$ is the encoder network, with weights parameters $\Theta_1$, applied on an example $x$, $norm_{L_2}$ is a function that performs the $L_2$ normalization of the output of the encoder and $\alpha$ is the margin hyperparameter usually involved in distance-based loss function to stretch the representation space [26]. We consider $\alpha$ equal to 1.0 since distances are derived by $L_2$ normalization.

Additionally, we can observe that, due to the transitivity relation among the examples in the triplet tuple, we can also define a second triplet loss function:

$$L_{triplet}'' = \sum_{(x_a, x_p, x_n) \in T} [d(x_a, x_p) - d(x_p, x_n) + \alpha]_+ \qquad (4)$$

where the second term of the hinge loss, this time, consider the relationship between the $x_p$ and $x_n$ examples. In the rest of the paper, $L_{triplet}'$ and $L_{triplet}''$ are exploited to introduce semi-supervision in the clustering process and we use the notation $L_{triplet}$ to indicate the sum of the two triplet loss functions: $L_{triplet} = (L_{triplet}' + L_{triplet}'')$.

The overall architecture of our semi-supervised autoencoder involves the optimization of $L_{triplet}$ loss as well as the simultaneous reconstruction of the examples concerned by the constraints. Given $T$, the set of triplet constraints, the loss function optimized by the $SSAE$ is as follows:

$$L_{ssae} = \frac{1}{|T|} \left( \left[ \sum_{t \in T} \sum_{x_i \in t} ||x_i - dec(enc(x_i, \Theta_1), \Theta_2)||_2^2 \right] + \lambda L_{triplet} \right) \qquad (5)$$

where $t = (x_a, x_p, x_n)$ is a generic triplet, $\lambda$ is a hyperparameter that controls the importance of the triplet loss term. Such loss function optimizes the parameters $\Theta_1$ and $\Theta_2$ so as to optimize the data reconstruction as well as to meet the constraint relationships expressed by the background knowledge. In $L_{ssae}$, the reconstruction term is considered with the aim of regularizing the action of the $L_{triplet}$ loss. Therefore, we obtain embeddings that meet the requirements expressed by the constraints as well as with the main reconstruction task.

We underline that, in our context, the embedding generation process involves two different stages: the first one implies the optimization of the autoencoder loss on the full set of data $X$ while, the second one regards the optimization of the semi-supervised autoencoder loss considering only the set of examples $X_t$ ($X_t = \{x_i \in t | t \in T\}$) covered by the triplet constraint set. Algorithm 1 reports the joint optimization procedure we employ to learn the weight parameters $\Theta_1, \Theta_2$. In a generic epoch, the procedure optimizes: (i) the unsupervised loss associated to data reconstruction on the set of data $X$ (line 3–4) and, (ii) both reconstruction and triplet losses ($L_{ssae}$) considering the set of data involved in the set $T$ (line 5–6). The learning of parameters is achieved via a gradient descent based approach using mini-batches. Finally, the data embeddings are generated considering the $\Theta_1$ parameters associated to the encoder network $enc(\cdot, \Theta_1)$.

---

**Algorithm 1.** Semi-supervised autoencoder optimization

---

**Require:** $X, T, N\_EPOCHS$

**Ensure:** $\Theta_1, \Theta_2$.

1: $i = 0$
2: **while** $i < N\_EPOCHS$ **do**
3:     Update $\Theta_1$ and $\Theta_2$ by descending the gradient:
4:     $\nabla_{\Theta_1,\Theta_2} \frac{1}{|X|} \sum_{x_i \in X} ||x_i - dec(enc(x_i, \Theta_1), \Theta_2)||_2^2$
5:     Update $\Theta_1, \Theta_2$ by descending the gradient:
6:     $\nabla_{\Theta_1,\Theta_2} \frac{1}{|T|} \left( \left[ \sum_{t \in T} \sum_{x_i \in t} ||x_i - dec(enc(x_i, \Theta_1), \Theta_2)||_2^2 \right] + \lambda L_{triplet}(T) \right)$
7:     $i = i + 1$
8: **end while**
9: **return** $\Theta_1, \Theta_2$

---

**Clustering Refinement with Background Knowledge.** Once the embedding representation produced by the SSAE is obtained, the final stage consists in a clustering refinement step via cluster assignment hardening [18, 25] as depicted in Fig. 1(b). Here, we iterate between computing an auxiliary target distribution and minimizing the Kullback-Leibler (KL) divergence with respect to it. More in detail, as depicted in Fig. 1(b), we discard the decoder part of the previous model ($\Theta_2$ parameters) but we still allow modifications of encoder parameters $\Theta_1$. Given the initial cluster centroids $\{c_j\}_{j=1}^{|C|}$, the cluster assignment hardening technique tries to improve the partitioning using an unsupervised algorithm that alternates between two steps: (i) compute a soft assignment between the embeddings and the cluster centroids and (ii) update the embedded data representation and refine the cluster centroids by learning from current high confidence assignments leveraging an auxiliary target distribution. The process is repeated until convergence is achieved or a certain number of iterations is executed. To generate the clustering centroids we use K-Means on the embeddings produced by the encoder network.

To compute the soft assignment, as commonly done in deep embedding clustering approaches, we exploit the Student's t-distribution as a kernel to measure the similarity [17]:

$$q_{ij} = \frac{(1 + ||EMB_i - c_j||^2)^{-1}}{\sum_{l=1}^{|C|} (1 + ||EMB_i - c_l||^2)^{-1}} \tag{6}$$

where $EMB_i$ is the embedded representation of the $i-th$ example obtained via $enc(x_i, \Theta_1)$, $c_j$ (resp. $c_l$) is the cluster centroid of the $j-th$ (resp. $l-th$) cluster, and $q_{ij}$ is the soft assignment between example $x_i$ and cluster $c_j$.

Once the soft assignments are computed, they are iteratively refined by learning from their high-confidence assignments with the help of an auxiliary target distribution. The target distribution is defined as:

$$p_{ij} = \frac{q_{ij}^2 / \sum_i q_{ij}}{\sum_l (q_{il}^2 / \sum_i q_{ij})} \tag{7}$$

Such distribution forces the assignment to have stricter probabilities (closer to 0 or 1) by squaring the original distribution and then normalizing it [18].

To match the soft-assignment $q$ with the auxiliary target distribution $p$, we employ the Kullback-Leibler (KL) divergence as loss function to evaluate the distance between the two probability distributions. The KL divergence is computed between the soft assignment $q_i$ and the auxiliary distribution $p_i$: $KL(P||Q) = \sum_i p_i \cdot \log \frac{p_i}{q_i}$.

Furthermore, we integrate the semi-supervision supplied by the background knowledge in this step as well, by adding the information carried out by the triplet constraints to the overall loss function:

$$L_{sscr} = KL(P||Q) + \lambda L_{triplet} \tag{8}$$

The resulting loss function considers the auxiliary target distribution together with the triplet constraints when upgrading the parameters of the encoder ($\Theta_1$). Hence, this last step has also an influence on the way embeddings are computed. As before, $\lambda$ is an hyperparameter controlling the importance of the triplet loss term and it is the same in the two steps of our framework. To optimize such semi-supervised loss $L_{sscr}$ we adopt a similar strategy to what proposed in Algorithm 1. Finally, once convergence is reached, each example is assigned to the cluster that maximizes its assignment score: $cluster(x_i) = argmax_j q_{ij}$.

# 4 Experiments

In this section, we assess the effectiveness of $Ts2DEC$ on several real world datasets comparing its behavior w.r.t. competitors. Then, we consider the impact of the different components of $Ts2DEC$ by means of an ablation study. Finally, we provide a visual inspection of the representation learnt by our strategy.

**Competitors.** For the quantitative evaluation, we compare the performances of $Ts2DEC$ with those obtained by different unsupervised and semi-supervised competing algorithms. The former are employed as baselines to understand the gain related to the introduction of weak supervision; the latter consist of fair state-of-the-art competitors that are more closely related to the task at hand. As regards the unsupervised approaches, we consider *K-Means* and DEC [25], a recent deep learning unsupervised clustering approach (the unsupervised clustering algorithm $Ts2DEC$ is built upon).

As semi-supervised clustering algorithms, we consider the following competitors: a semi-supervised variant of *K-Means*, named *MPCKmeans* [4]; a recent constrained spectral clustering method [5], called *Spectral*; two very recent deep learning based methods named *MSAEClust* [11] and *SDEC* [21]. *MPCKmeans* combines metric-learning and pairwise constraint processing to exploit the supplied supervision as much as possible. *Spectral* captures constrained clustering as a generalized eigenvalue problem via graph Laplacians. [11] employs an ensemble of semi-supervised autoencoders to learn embedding representations that fit the data as well as the background knowledge and that are finally used to perform clustering. Finally, *SDEC* is a direct extension of *DEC* that integrates the pairwise constraints in the clustering refinement stage, by adding an extra term to the cluster assignment hardening loss.

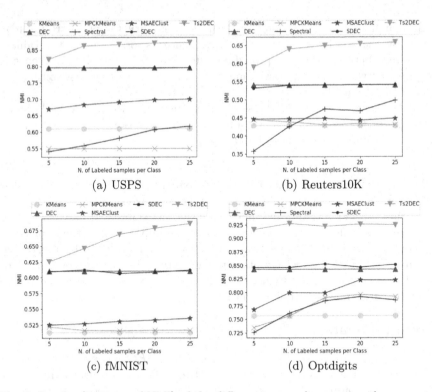

**Fig. 2.** Results (in terms of NMI) of the different approaches varying the amount of labeled samples per class on: (a) *USPS*, (b) *Reuters10K*, (c) *fMNIST* and (d) *Optdigits* benchmarks.

**Experimental Settings and Datasets.** To measure the clustering performances of all the methods, we use the Normalized Mutual Information (NMI) [22] as well the Adjusted Rand Index (ARI) [10]. Both NMI and ARI take their maximum value when the clustering partition completely matches the original one, i.e., the partition induced by the available class labels. The NMI measure ranges between $[0, 1]$ while the ARI index ranges between $[-1, 1]$. Both evaluation metrics can be considered as an indicator of the purity of the clustering result. For each dataset, both measures are computed considering the whole set of examples, including the ones on which the constraints are defined. We analyze the behavior of the different methods according to increasing levels of supervision. More in detail, we simulate the supervision in term of constraints, by selecting a number of labeled examples per class and, successively, inducing the corresponding full set of constraints. We vary such amount of labeled examples per class between 5 and 25 with a step of 5. Due to the randomness of the sample selection process and the non deterministic nature of the clustering algorithms, we repeat the sample selection step 5 times for each number of per-class labels and, successively, we repeat the clustering process 10 times. For *Ts2DEC* we derive the corresponding triplet constraints as explained in Sect. 3.

Finally, for each level of supervision, we report the average values of NMI and ARI. For all the methods, the number of clusters is equal to the number of classes.

*Ts2DEC* is implemented via the *Tensorflow* python library and the implementation is available online[1]. Model parameters are learnt using the Adam optimizer [13] with a learning rate equal to $1 \times 10^{-3}$ for the autoencoder (reconstruction and triplet loss functions) and we use Stochastic Gradient Descent with learning rate equal to $1 \times 10^{-2}$ for the Clustering Refinement stage (KL loss function) as done in *DEC* [25] and *SDEC* [21]. We set the value of $\lambda$ equal to $1 \times 10^{-3}$, a batch size of 256 and a number of epochs equal to 50 for the semi-supervised autoencoder. For the refinement clustering stage, we iterate the procedure for 20 000 batch iterations as done in *DEC* [25] and *SDEC* [21]. For all the competitors, we use publicly available implementations. For *SDEC*, the source code was kindly provided by the authors of the related paper. Experiments are carried out on a workstation equipped with an Intel(R) Xeon(R) W-2133, 3.6Ghz CPU, with 64Gb of RAM and one GTX1080 Ti GPU. To evaluate the behavior of all the competing approaches the experiments are performed on four publicly available datasets: (1) *USPS* is a handwritten digit recognition benchmark (10 classes) containing 9 298 grayscale images with size $16 \times 16$ pixels and provided by the United States Postal Service. (2) *fMNIST* is a dataset of Zalando's article images (shirt, sneakers, bags, etc.) consisting of 70 000 examples. Each example is a $28 \times 28$ grayscale image, associated with a label from 10 classes. It serves as a more complex drop-in replacement for the original MNIST benchmark [24]. (3) *Reuters10k* is an archive of English news stories labeled with a category tree that contains 810 000 textual documents. Following [25], we used 4 root categories: corporate/industrial, government/social, markets and economics as labels and excluded all documents with multiple labels. We randomly sampled a subset of 10 000 examples and computed TF-IDF features on the 2 000 most frequent words. (4) *Optdigits* is a dataset of the UCI repository involving optical recognition of handwritten digits. It contains 5 620 examples described by 64 feature each.

**Quantitative Evaluation.** Figures 2 and 3 report the performances of the different approaches on the four benchmarks in terms of NMI and ARI, respectively. Notice that *Spectral* was not able to process the *fMNIST* benchmark due to the fact that the original implementation cannot handle a dataset with 70 000 examples. We observe that both NMI and ARI depict a similar situation. At first look, we note that *Ts2DEC* outperforms all the competing approaches regarding any amount of supervision for all the four benchmarks. In addition, the graphs generally show that the margin gained by *Ts2DEC* increases with the amount of available supervision. This behavior is particularly evident in *USPS*, *fMNIST* and *Reuters10k*. Considering *Optdigits*, we observe an improvement between the supervision value 5 and 10 while, later on, *Ts2DEC* remains stable according to NMI and it slightly increases according to ARI. This is not the case for all

---

[1]  https://gitlab.irstea.fr/dino.ienco/ts2dec.

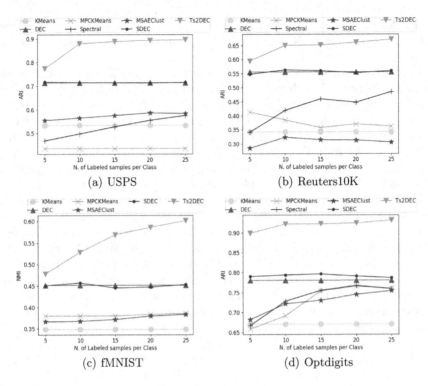

**Fig. 3.** Results (in terms of ARI) of the different approaches varying the amount of labeled samples per class on: (a) *USPS*, (b) *Reuters10K*, (c) *fMNIST* and (d) *Optdigits* benchmarks.

the other semi-supervised competitors. For instance, considering the *fMNIST* benchmark, we note that all competitors remain almost stable while varying the amount of supervision, underlying the fact that they are unable to exploit increasing amount of background knowledge properly. Unexpectedly, we observe that one of the best competitor is DEC, which is completely unsupervised. More surprisingly, *SDEC* performs similarly to its unsupervised counterpart.

**Ablation and Parameter Analysis.** In this section, we study the impact of the different components of $Ts2DEC$ that involve supervision, as well as the sensitivity of our method to hyperparameter $\lambda$. To do this, we fix the amount of supervision by considering 10 labeled examples from each class. For the first study, we derive two variants of our method: (i) $Ts2DEC_{v1}$ which considers background knowledge only to generate embeddings via semi-supervised autoencoder, and (ii) $Ts2DEC_{v2}$ which considers background knowledge only during the clustering refinement stage. Table 1 reports the results of this study in terms of NMI. We note that the best performances are obtained when semi-supervision is injected at both stages of our process. Furthermore, we observe that $Ts2DEC_{v1}$

**Table 1.** Impact of the different components of $Ts2DEC$ considering the NMI measure.

| Dataset | $Ts2DEC$ | $Ts2DEC_{v1}$ | $Ts2DEC_{v2}$ |
|---------|----------|---------------|---------------|
| *USPS* | **0.86** ± 0.02 | **0.86** ± 0.02 | 0.82 ± 0.03 |
| *fMNIST* | **0.65** ± 0.01 | 0.64 ± 0.01 | 0.64 ± 0.02 |
| *Reuters10k* | **0.64** ± 0.03 | **0.64** ± 0.03 | 0.61 ± 0.03 |
| *Optdigits* | **0.92** ± 0.01 | 0.91 ± 0.01 | 0.91 ± 0.02 |

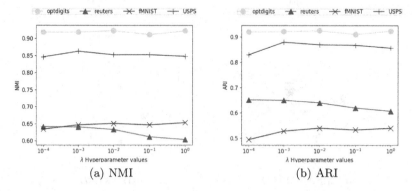

(a) NMI          (b) ARI

**Fig. 4.** Sensitivity analysis of the $\lambda$ hyperparameter: NMI (a) and ARI (b) are reported for increasing weight of semisupervision.

consistently achieves slightly better results than $Ts2DEC_{v2}$ in terms of NMI. The results of the sensitivity analysis are given in Fig. 4. In details, we let the hyperparameter $\lambda$ varies in the range $\{10^{-4}, 10^{-3}, 10^{-2}, 10^{-1}, 10^{0}\}$. At first look, *USPS*, *fMNIST* and *Optdigits* exhibit a similar behavior. When $\lambda$ is too small ($10^{-4}$), supervision is not that effective while, starting from $\lambda$ equal to $10^{-3}$, we observe that $Ts2DEC$ achieves stable performances and becomes insensitive to such parameter. On the other hand, for the *Reuters10k* dataset, we note that the performances slightly decrease when $\lambda$ increases. At a deeper inspection, we observe that raising the value of $\lambda$ results in an increase of the standard deviation associated to the average value plotted in Fig. 4. We remind that this benchmark is characterized by a high-dimensional feature space (2 000 features) and, the encoder/decoder architecture (inherited from the DEC method) is unable to compress the original data properly and, simultaneously, incorporate the supervision. This may explain the increasing instability and reduced performances when augmenting the importance of the semi-supervision.

**Visual Inspection.** Here, we visually analyze the embedding generated by our approach on the *Optdigits* benchmark. To this end, we visually compare the embeddings derived by $Ts2DEC$ with the embeddings generated by the other deep learning competitors considering an amount of labeled examples per class equal to 10 (i), and by increasing the amount of background knowledge from

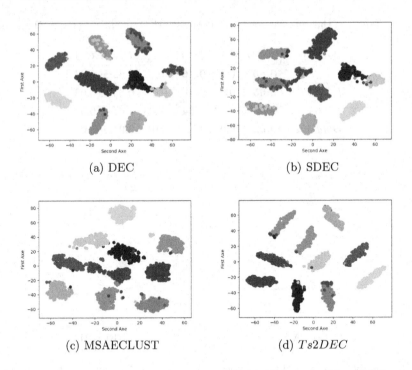

(a) DEC                    (b) SDEC

(c) MSAECLUST              (d) $Ts2DEC$

**Fig. 5.** Visual inspection of different embeddings processed by $TSNE$.

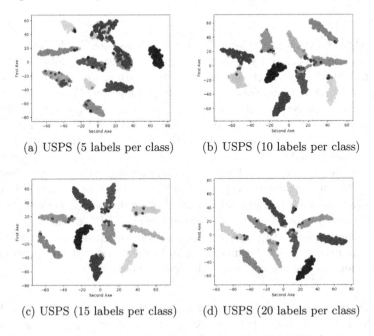

(a) USPS (5 labels per class)    (b) USPS (10 labels per class)

(c) USPS (15 labels per class)   (d) USPS (20 labels per class)

**Fig. 6.** Visual inspection of the embedding generated by $Ts2DEC$ (and processed by $TSNE$) for increasing amounts of background knowledge.

5 to 20 labels per class (ii). To obtain the two dimensional representations, we apply the $t$-distributed stochastic neighbor embedding ($TSNE$) approach [17]. For this evaluation we consider 300 instances per class. In the former evaluation (Fig. 5), we clearly note that the visual representation induced by $Ts2DEC$ provides a better separation among examples belonging to different classes and, simultaneously, locates examples belonging to the same class close to each other. The latter experiment (Fig. 5) shows the ability of $Ts2DEC$ to modify the data manifold exploiting the increasing amount of background knowledge. We observe that clear differences exist between the embeddings learnt when 5 (Fig. 6(a)) and 15 labeled examples (Fig. 6(c)) per class are considered, the latter providing significant better class separation than the former.

## 5   Conclusion

We have presented $Ts2DEC$, a new semi-supervised deep embedding clustering technique that integrates background knowledge as triplet constraints. More precisely, $Ts2DEC$ integrates the background knowledge at two stages: (i) during the data embedding generation and (ii) during the clustering refinement. Extensive evaluations on real-world benchmarks have shown that $Ts2DEC$ outperforms state-of-the-art competitors w.r.t different amount of background knowledge.

## References

1. Banerjee, A., Merugu, S., Dhillon, I.S., Ghosh, J.: Clustering with bregman divergences. J. Mach. Learn. Res. **6**, 1705–1749 (2005)
2. Basu, S., Banerjee, A., Mooney, R.J.: Semi-supervised clustering by seeding. In: ICML, pp. 27–34 (2002)
3. Basu, S., Bilenko, M., Mooney, R.J.: A probabilistic framework for semi-supervised clustering. In: KDD, pp. 59–68 (2004)
4. Bilenko, M., Basu, S., Mooney, R.J.: Integrating constraints and metric learning in semi-supervised clustering. In: ICML, pp. 81–88 (2004)
5. Cucuringu, M., Koutis, I., Chawla, S., Miller, G.L., Peng, R.: Simple and scalable constrained clustering: a generalized spectral method. In: AISTATS, pp. 445–454 (2016)
6. Davidson, I., Ravi, S.S.: Intractability and clustering with constraints. In: ICML, pp. 201–208 (2007)
7. Davis, J.V., Kulis, B., Jain, P., Sra, S., Dhillon, I.S.: Information-theoretic metric learning. In: ICML, pp. 209–216 (2007)
8. Haiyan, W., Haomin, Y., Xueming, L., Haijun, R.: Semi-supervised autoencoder: a joint approach of representation and classification. In: CICN, pp. 1424–1430 (2015)
9. Harwood, B., Kumar B.G., Carneiro, G., Reid, I.D., Drummond, T.: Smart mining for deep metric learning. In: ICCV, pp. 2840–2848 (2017)
10. Hubert, L., Arabie, P.: Comparing partitions. J. Classif. **2**(1), 193–218 (1985)
11. Ienco, D., Pensa, R.G.: Semi-supervised clustering with multiresolution autoencoders. In: IJCNN, pp. 1–8 (2018)

12. Kalintha, W., Ono, S., Numao, M., Fukui, K.: Kernelized evolutionary distance metric learning for semi-supervised clustering. In: AAAI, pp. 4945–4946 (2017)
13. Kingma, D.P., Ba, J.: Adam: a method for stochastic optimization. CoRR abs/1412.6980 (2014), http://arxiv.org/abs/1412.6980
14. Klein, D., Kamvar, S.D., Manning, C.D.: From instance-level constraints to space-level constraints: making the most of prior knowledge in data clustering. In: ICML, pp. 307–314 (2002)
15. Kumar, N., Kummamuru, K.: Semisupervised clustering with metric learning using relative comparisons. IEEE Trans. Knowl. Data Eng. **20**(4), 496–503 (2008)
16. LeCun, Y., Bengio, Y., Hinton, G.: Deep learning. Nature 521, 436, May 2015. https://doi.org/10.1038/nature14539
17. van der Maaten, L., Hinton, G.E.: Visualizing high-dimensional data using t-sne. JMLR **9**, 2579–2605 (2008)
18. Min, E., Guo, X., Liu, Q., Zhang, G., Cui, J., Long, J.: A survey of clustering with deep learning: from the perspective of network architecture. IEEE Access **6**, 39501–39514 (2018)
19. Nogueira, B.M., Tomas, Y.K.B., Marcacini, R.M.: Integrating distance metric learning and cluster-level constraints in semi-supervised clustering. In: IJCNN, pp. 4118–4125 (2017)
20. Rasmus, A., Berglund, M., Honkala, M., Valpola, H., Raiko, T.: Semi-supervised learning with ladder networks. In: NIPS, pp. 3546–3554 (2015)
21. Ren, Y., Hu, K., Dai, X., Pan, L., Hoi, S.C.H., Xu, Z.: Semi-supervised deep embedded clustering. Neurocomputing **325**, 121–130 (2019)
22. Strehl, A., Ghosh, J.: Cluster ensembles – a knowledge reuse framework for combining multiple partitions. J. Mach. Learn. Res. **3**, 583–617 (2002)
23. Wagstaff, K., Cardie, C., Rogers, S., Schrödl, S.: Constrained k-means clustering with background knowledge. In: ICML, pp. 577–584 (2001)
24. Xiao, H., Rasul, K., Vollgraf, R.: Fashion-mnist: a novel image dataset for benchmarking machine learning algorithms. CoRR abs/1708.07747 (2017)
25. Xie, J., Girshick, R.B., Farhadi, A.: Unsupervised deep embedding for clustering analysis. In: ICML, pp. 478–487 (2016)
26. Yu, B., Liu, T., Gong, M., Ding, C., Tao, D.: Correcting the triplet selection bias for triplet loss. In: ECCV, pp. 71–86 (2018)
27. Zhao, Y., Jin, Z., Qi, G., Lu, H., Hua, X.: An adversarial approach to hard triplet generation. In: ECCV, pp. 508–524 (2018)
28. Zhu, X., Loy, C.C., Gong, S.: Constrained clustering with imperfect oracles. IEEE Trans. Neural Networks Learn. Syst. **27**(6), 1345–1357 (2016)

# Neurodegenerative Disease Data Ontology

Ana Kostovska[1,2], Ilin Tolovski[1,2], Fatima Maikore[3], the Alzheimer's Disease
Neuroimaging Initiative, Larisa Soldatova[3], and Panče Panov[1,2(✉)]

[1] Department of Knowledge Technologies, Jožef Stefan Institute, Ljubljana, Slovenia
{ana.kostovska,ilin.tolovski,pance.panov}@ijs.si
[2] Jožef Stefan International Postgraduate School, Ljubljana, Slovenia
[3] Department of Computing, Goldsmiths University of London, London, UK
{fmaik001,l.soldatova}@gold.ac.uk

**Abstract.** In this paper, we report on the ontology for the representation of brain diseases data - NDDO. The proposed ontology facilitates
semantic annotation of datasets containing neurodegenerative diagnostic data (i.e. clinical, imaging, biomarker, etc.) and disease progression
data collected on patients by the hospitals. Rich semantic annotation of
datasets is essential for efficient support of data mining, for example for
the identification of suitable algorithms for data analytics, text mining,
and reasoning over distributed data and knowledge sources. To address
the data analytics perspective, we reused and extended our previous
work on ontology of data types (OntoDT) and ontology of core data
mining entities (OntoDM-core) to represent specific domain datatypes
that occur in the domain datasets. We demonstrate the utility of NDDO
on two use cases: semantic annotation of datasets, and incorporating
information about clinical procedures used to produce neurodegenerative data.

**Keywords:** Neurodegenerative diseases datasets · Ontology

## 1 Introduction

Data mining (DM) of medical data is one of the most important interdisciplinary
research areas pertinent to healthcare. It is a complex task that requires expertise
in DM and machine learning (ML), data ethics and relevant medical domains.
Such an expertise is not easily available. One way of dealing with the shortage
of the necessary expertise is to decrease the machine learning knowledge bar-
rier by providing more intelligent ML and DM tools. For example, an intelligent

---

the Alzheimer's Disease Neuroimaging Initiative—Data used in preparation of this
article were obtained from the Alzheimer's Disease Neuroimaging Initiative (ADNI)
database (adni.loni.usc.edu). As such, the investigators within the ADNI contributed
to the design and implementation of ADNI and/or provided data but did not participate in analysis or writing of this report. A complete listing of ADNI investigators
can be found at: http://adni.loni.usc.edu/wp-content/uploads/how_to_apply/ADNI_
Acknowledgement_List.pdf.

© Springer Nature Switzerland AG 2019
P. Kralj Novak et al. (Eds.): DS 2019, LNAI 11828, pp. 235–245, 2019.
https://doi.org/10.1007/978-3-030-33778-0_19

DM system should have the capability of automatically identifying for a given analysis task the most suitable DM and ML methods and available data, and invoking a pre-defined workflow or pipeline for constructing high-fidelity models. Then, domain practitioners, who are usually non-machine learning experts, will be able to construct accurate models for their tasks with minimal efforts and in accordance with the best practices. Having formal logical descriptions of datasets, and DM and ML tasks are essential for enabling such a system [9,10].

Formal knowledge representations define logical links between different domains, enabling data and knowledge interoperability. Here, we are focusing on the integration of the areas of data mining and brain diseases, more specifically neurodegenerative diseases. Neurodegenerative disease is a term that includes a range of conditions which primarily affect the neurons in the human brain. When neurons become damaged or die they cannot be replaced by the body. Neurodegenerative diseases are incurable conditions that result in progressive degeneration and consequently death of nerve cells. This leads to problems with movement, or mental functioning. Examples of neurodegenerative diseases include Parkinson's, Alzheimer's, and Huntington's diseases.

The results reported in this paper are inspired by the needs of the Human Brain Project (HBP)[1]. Recently HBP has identified a need to "develop a broad set of ontologies, covering wide range of brain diseases and types of data"[2]. Such ontologies should take into account the structure of the available hospital data and existing, well accepted ontologies.

In this paper, we report on the development of a specific ontology for neurodegenerative brain diseases named NDDO (Neurodegenerative Disease Data). NDDO is fully interoperable with Ontology of Core Data Mining Entities (OntoDM-core) [9], and many other relevant domain-specific knowledge representations (e.g., [6,11]). It supports a framework for reasoning over data on patients with neurodegenerative diseases and can assist in DM tasks. NDDO is consistent with the structure of available hospital data and with two well-known studies concerning neurodegenerative diseases: Alzheimer's Disease Neuroimaging Initiative (ADNI) and Parkinson's Progression Markers Initiative (PPMI).

## 2    NDDO: Neurodegenerative Disease Data Ontology

We designed the Neurodegenerative Disease Data Ontology (NDDO) in accordance with best practices in ontology engineering to provide a generic, compliant with existing standards, and easily extendable framework for the representation of brain diseases data. More specifically, we adhered to the OBO Foundry principles [12] to ensure interoperability with existing standards for brain diseases.

NDDO was developed to provide a comprehensive semantic model for describing patient data pertinent to two neurodegenerative diseases: Alzheimer's and Parkinson's Disease. The current NDDO model can be used as a template for the

---

[1] URL: https://www.humanbrainproject.eu/.

[2] See the HBP Call for Expression of Interest on comprehensive ontologies for brain diseases at https://www.humanbrainproject.eu/en/collaborate/open-calls/.

representation of other brain diseases. For modeling the core entities and processes in the ontology, we used descriptions of procedures from two documented medical initiatives as a guideline, i.e., the Alzheimer's Disease Neuroimaging Initiative[3] [1–3], and Parkinson's Progression Markers Initiative[4] [13,14].

The ontology version, reported in this paper, has 1731 classes, 638 of which are imported from external ontologies. NDDO has 100 object, 75 annotation, and 11 data properties. The level of expressivity is SROIQ(D). It is developed in the OWL2 ontology language, using the Protégé[5] environment. The ontology project and all of its resources are publicly available and regularly updated via a Git repository[6] as well as on BioPortal[7] [15], with a permanent URL at http://www.purl.org/nddo. It is published under the CC-BY 4.0. license.

NDDO's main focus is on formal representation of data collected in neurodegenerative studies. This includes representation of entities such as the study participants, their visits of physicians, different assessments conducted during that visits and their results, and the diagnosis. Figure 1 shows the structure of NDDO with the four core entities: *study participant, visit, health care process assay,* and *diagnosis.* Depending on the study, a *study participant* can be a person that is directly examined during an assessment, i.e. a *study subject,* or a person close to the study subject, i.e., *study partner,* that gives a report for the impact of their behaviour on the people around them (see Fig. 1a). The latter participates only in the ADNI study, since the study subjects can sometimes be unaware of what impact their behaviour has on people around them. The *study participant* is a part of a larger group, i.e., *cohort.* For the ADNI study the cohorts are determined solely from the subject's diagnosis, while in the PPMI study they are determined based on additional criteria, such as, genetic mutations, sleeping disorders and the Parkinson's disease history in the subject's family.

Each *study participant* participates in several visits (see Fig. 1b). The class *visit* is one of the central classes in the ontology, since it links the participants with the assessments. It is defined by the *visit specification* and occurs at a *site* which has a special *site ID.* Each *visit* consists of several *health care process assays.* The result of the visit is a *diagnosis* which is formed based on the scores from the *health care process assays.* The class *health care process assay* has three subclasses: *clinical assessment, biomarker assessment,* and *imaging assessment* (see Fig. 1c). Since there are two types of study participants, we differ between clinical assessments of study subject, study partner, or clinical assessment for both. The last two are conducted only in the ADNI study. Each of the assessments have their own output described by a score. Furthermore, the score of each assessment is connected with its corresponding datatype. This is done by reusing and extending classes from the OntoDT ontology.

---

[3] ADNI webpage: http://adni.loni.usc.edu/.

[4] PPMI webpage: http://www.ppmi-info.org/.

[5] URL: Protégé web page: http://protege.stanford.edu.

[6] NDDO Git URL: http://source.ijs.si/ppanov/nddo/tree/master/Development.

[7] NDDO at BioPortal: https://bioportal.bioontology.org/ontologies/NDDO.

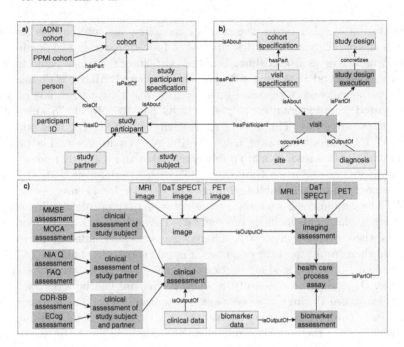

**Fig. 1.** NDDO core entities: (a) study participant; (b) visit; (c) health care process assay. The blue rectangles are continuant classes, while green rectangles are process classes. (Color figure online)

## 3   Use Cases

NDDO is specifically designed to be easily extendable to support different use cases. Adherence to the OBO Foundry design principles and NDDO's modular architecture enables seamless integration with other ontologies when extended representations are necessary for new use cases and DM workflows. For example, an Ontology for Clinical Laboratory Standard Operating Procedures (OCL-SOP) [7] can be seamlessly integrated with NDDO to capture information about clinical procedures pertinent to ADNI and PPMI domains. Such full integration is possible because both representations were designed following the same principles. In the same way, we can integrate NDDO with the ontology of datatypes (OntoDT) [10] and provide semantic annotation of datasets originating from neurodegenerative studies (such as ADNI and PPMI) as well as with ontology of data mining (OntoDM) [9] to provide means of constructing (semi)automatic analysis workflows. In this paper, we briefly describe two example use cases.

### 3.1   Use Case: Semantic Annotation of Datasets

In this use case, we show how the NDDO ontology can be used for semantic annotation of datasets containing data on patients with neurodegenerative diseases.

We integrated NDDO with the OntoDT ontology of datatypes and OntoDM ontology of data mining to semantically annotate neurodegenerative datasets. Rich semantic annotation of datasets enables semantic search of datasets suitable for a particular DM task. It also enables intelligent data mining tools to suggest adequate for the dataset at hand analysis methods. The considered datasets were previously used in data analysis tasks: clustering task [5] and predictive modeling task [8]. Datasets contain data on patients with neurodegenerative diseases (Alzheimer's and Parkinson's diseases) from two initiatives (ADNI and PPMI).

We reused classes from OntoDM-core and OntoDT for the annotation of the dataset features, e.g. data types. From OntoDM-core, we reused the class that contains the specification of a dataset (*OntoDM-core: dataset specification*) and the class for the representation of data mining tasks (*OntoDM-core: data mining task*). Both classes are linked with *OntoDT: datatype* class via `has-part` relation (see Fig. 2a). We generated the semantic annotations using the Apache Jena library[8]. This was done in semi-automatic fashion, since we need to manually map the dataset features to the adequate classes defined in NDDO. The annotations are expressed as RDF facts. Ontology-based annotation of datasets enables reasoners to infer new knowledge based on the asserted facts.

**Example of ADNI Dataset Used for Clustering.** The annotation schema that we propose for the semantic annotation of datasets is governed by the data mining task at hand. For example in clustering, the task of grouping a set of objects in such a way that objects in the same group (or cluster) are more similar to each other than to those in other groups (clusters). On the other hand, predictive modeling involves prediction of a single property or a set of properties of a given object from (feature-based) descriptions of the object.

Gamberger et al. [5] performed cluster analysis on an instance of the ADNI dataset intending to generate homogeneous clusters of male and female Alzheimer's disease patients. The dataset they used contains information about 317 female and 342 male patients and has 243 clinical and biological descriptors.

In Fig. 2b, we show the specific classes and relations required for the annotation of datasets used in cluster analysis. First, we reused the *OntoDM-core: feature-based unlabeled dataset class*, which is a subclass of the more general *OntoDM-core: dataset specification* class. For the representation of the clustering task, we reused *OntoDM-core: batch clustering task*. The two classes are connected with the corresponding datatype class *OntoDT: feature-based unlabeled data*. This class contains one field component (*OntoDT: descriptive record of primitives field component*). The class describes the datatypes of the descriptive features, which in this case can only be of primitive datatype.

**Example of PPMI Dataset for Predictive Modeling.** Mileski et al. [8] used datasets from the PPMI database to predict the motor impairment assessment scores by utilising the scores of regions of interest (ROIs) from the fMRI imaging assessment and the DaT scans. The data mining task they were solving was multi-target regression (MTR).

---

[8] https://jena.apache.org/.

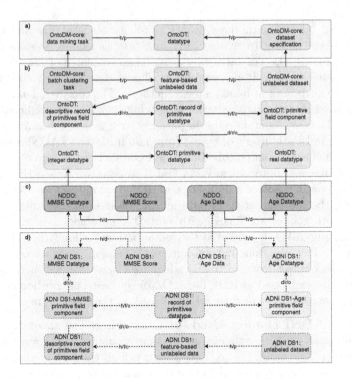

**Fig. 2.** Semantic annotation schema for the ADNI dataset [5]: (a) top level classes;
(b) specific classes and relations used in cluster analysis; (c) specific NDDO datatype
classes and (d) example annotations of two descriptive features.

In order to semantically annotate the PPMI dataset used in the study, we
followed the same principles as in the ADNI study (see Fig. 3). The difference
is mainly in the representation of the datatype information due to the different structure of the dataset in the MTR learning scenario. Namely, opposed to
having just descriptive features, which was the case in the previous example, in
MTR we predict multiple numeric values simultaneously.

From OntoDM-core, we reused the classes representing the MTR task and
MTR dataset specification and connected them with the relevant datatype class
from OntoDT (*OntoDT: feature-based completely labeled data with record of
numeric ordered primitive output*). This class has two field components. The first
one describes the datatypes of the descriptive features, which are of primitive
datatype. The latter describes the datatypes of the features on the target side.
In the MTR learning setting each target feature is described with the numeric
datatype. The instantiation of the ontology classes was done in the same way as
in ADNI use case (see Fig. 3d).

Semantic annotations of datasets used for ML not only guide the selection
of adequate algorithms, but also promote explicitness of such selections, greater
clarity and reusability.

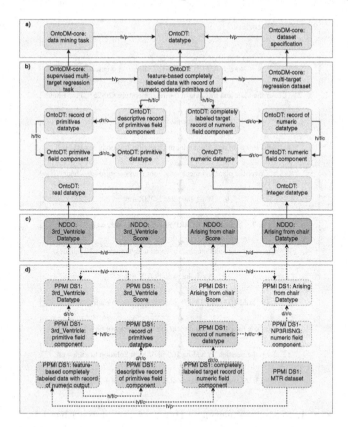

**Fig. 3.** Semantic annotation schema for the PPMI dataset used in [8]: (a) top level classes (b) specific classes and relations for annotation of datasets used in the MTR task; (c) NDDO datatype classes and (d) example annotations of two descriptive features.

## 3.2   Use Case: Clinical Laboratory Procedures

Clinical laboratory procedures are essential for ADNI and PPMI domains. Such procedures are usually captured in a form of manuals expressed in natural language (e.g. [1–3,13,14]). Unfortunately, such representations suffer from inherit ambiguities of human languages. This may lead to different interpretations by different agents, and consequently - to different implementations and outcomes. Moreover, it complicates their computational processing and analysis, e.g. it is difficult to compare procedures expressed in natural language, to identify gaps, and to check them for logical consistency and completeness.

In our previous work, we have developed a translation engine for translating and disambiguating laboratory protocols from natural language to a standardized machine processable format [4]. The translation engine parses natural language laboratory protocols using the terminology in OCL-SOP (an Ontology for Clinical Laboratory Standard Operating Procedures). OCL-SOP formally

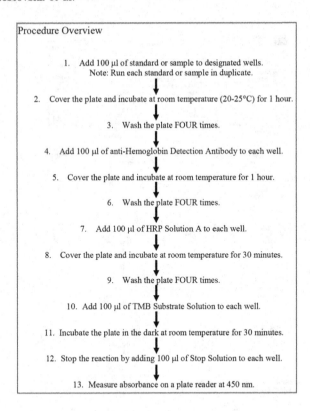

**Fig. 4.** Procedure overview: Human Hemoglobin Elisa Kit.

represents key entities relevant to clinical laboratory procedures [7]. The translation engine identifies key entities from a given protocol and matches them to relevant terms in OCL-SOP, to a provided list of chemical entities and to a list of equipment, and then generates a protocol in a machine processable form. Laboratory protocols in the OCL-SOP form are suitable for processing by software and robotic agents. They are also more accurate, explicit and easier to understand. We developed a dedicated mobile application for displaying such processed protocols in a user friendly format to the laboratory practitioners.

In order to adequately represent ADNI and PPMI procedures described in the respective study procedure manuals, we integrated OCL-SOP with NDDO. We imported from NDDO such terms as *data item, clinical finding,* and *laboratory finding.* We linked the imported class *laboratory finding* with the OCL-SOP data action branch using the `has-specified-output` object property.

Since the alignment of OCL-SOP with NDDO, now ADNI and PPMI procedures can be annotated with OCL-SOP terms, standardized, disambiguated and displayed in a user-friendly application. As an example, using our translation engine we processed the protocol for testing the presence and levels of hemoglobin in cerebrospinal fluid (CSF). The PPMI analysis of hemoglobin in

| Statement | Action | Entity | min Temperature | max Temperature | Volume | Equipment | Period | Condition |
|---|---|---|---|---|---|---|---|---|
| Add 100 μl of standard or sample to designated wells. | add | | | | 100 μl: | wells: | | |
| Note: Run each standard or sample in duplicate. | | | | | | | | |
| Cover the plate and incubate at room temperature (20-25°C) for 1 hour. | cover | | | | | plate: | | |
| Incubate at room temperature (20-25°C) for 1 hour. | incubate | | 18°C:UO:0000027 | 18°C:UO:0000027 | | | 1 hour | at room temperature: |
| Wash the plate FOUR times. | wash | | | | | | | |
| Add 100 μl of anti-Hemoglobin Detection Antibody to each well. | add | | | | 100 μl: | well: | | |
| Cover the plate and incubate at room temperature for 1 hour. | cover | | | | | plate: | | |
| Incubate at room temperature for 1 hour. | incubate | | 18°C:UO:0000027 | 18°C:UO:0000027 | | | 1 hour | at room temperature: |
| Wash the plate FOUR times. | wash | | | | | | | |
| Add 100 μl of HRP Solution A to each well. | add | HRP Solution A: | | | 100 μl: | well: | | |
| Cover the plate and incubate at room temperature for 30 minutes. | cover | | | | | plate: | | |
| Incubate at room temperature for 30 minutes. | incubate | | 18°C:UO:0000027 | 18°C:UO:0000027 | | | 30 minutes | at room temperature: |
| Wash the plate FOUR times. | wash | | | | | | | |
| Add 100 μl of TMB Substrate Solution to each well. | add | TMB Substrate Solution: | | | 100 μl: | well: | | |
| Incubate the plate in the dark at room temperature for 30 minutes. | incubate | | 18°C:UO:0000027 | 18°C:UO:0000027 | | plate: | 30 minutes | in the dark::at room temperature |
| Stop the reaction by adding 100 μl of Stop Solution to each well. | | | | | | | | |
| Measure absorbance on a plate reader at 450 nm. | measure | | | | | | | |

**Fig. 5.** Output of the translation engine (a segment).

CSF samples procedure calls for the use of an ELISA assay obtained from Bethyl Laboratories[9]. Figure 4 shows a segment of the Human Hemoglobin Elisa Kit protocol in natural language. The output protocol in a machine readable format is shown in Fig. 5. Explicit linking of datasets and procedures used to produce those datasets is important for data and knowledge integration and processing.

# 4    Conclusions and Future Work

We developed NDDO, an ontology for the representation of data on patients with neurodegenerative data, required for the hospital data integration and reasoning. NDDO was specifically designed to be easily extendable to support various applications. Depending on the application, a suitable representation can be 'plugged-in' to NDDO.

In this paper, we have demonstrated the utility of NDDO on two use cases. The first use case demonstrates the interoperability of NDDO with OntoDM and OntoDT to provide semantically rich annotations for neurodegenerative datasets and advise on data mining tasks. The second use case shows the integration of NDDO with OCL-SOP for the representation and processing of clinical procedures. This enables a linkage of neurodegenerative datasets with procedures that were used to obtain them. Having rich semantic representations of neurodegenerative data will enable more efficient data mining and knowledge discovery. For example, information about the procedures should be used for the integration of datasets to ensure that all the analysed data were collected following similar procedures. Otherwise, data mining results can be meaningless.

**Acknowledgements.** The authors would like to acknowledge the support of the Slovenian Research Agency through the project J2-9230 and the Public Scholarship, Development, Disability and Maintenance Fund of the Republic of Slovenia through its scholarship program.

---

[9] https://www.bethyl.com/product/pdf/E88-134-180409.pdf.

Data collection and sharing for this project was funded by the Alzheimer's Disease Neuroimaging Initiative (ADNI) (National Institutes of Health Grant U01 AG024904) and DOD ADNI (Department of Defense award number W81XWH-12-2-0012). ADNI is funded by the National Institute on Aging, the National Institute of Biomedical Imaging and Bioengineering, and through generous contributions from the following: AbbVie, Alzheimer's Association; Alzheimer's Drug Discovery Foundation; Araclon Biotech; BioClinica, Inc.; Biogen; Bristol-Myers Squibb Company; CereSpir, Inc.; Cogstate; Eisai Inc.; Elan Pharmaceuticals, Inc.; Eli Lilly and Company; EuroImmun; F. Hoffmann-La Roche Ltd and its affiliated company Genentech, Inc.; Fujirebio; GE Healthcare; IXICO Ltd.; Janssen Alzheimer Immunotherapy Research & Development, LLC.; Johnson & Johnson Pharmaceutical Research & Development LLC.; Lumosity; Lundbeck; Merck & Co., Inc.; Meso Scale Diagnostics, LLC.; NeuroRx Research; Neurotrack Technologies; Novartis Pharmaceuticals Corporation; Pfizer Inc.; Piramal Imaging; Servier; Takeda Pharmaceutical Company; and Transition Therapeutics. The Canadian Institutes of Health Research is providing funds to support ADNI clinical sites in Canada. Private sector contributions are facilitated by the Foundation for the National Institutes of Health (www.fnih.org). The grantee organization is the Northern California Institute for Research and Education, and the study is coordinated by the Alzheimer's Therapeutic Research Institute at the University of Southern California. ADNI data are disseminated by the Laboratory for Neuro Imaging at the University of Southern California.

Data used in the preparation of this article were obtained from the Parkinson's Progression Markers Initiative (PPMI) database (www.ppmi-info.org/data). For up-to-date information on the study, visit www.ppmi-info.org. PPMI–a public-private partnership–is funded by the Michael J. Fox Foundation for Parkinson's Research and funding partners, including AbbVie, Allergan, Avid, Biogen, BioLegend, Bristol-Myers Squibb, GE Healthcare, Genentech, GlaxoSmithKline, Lilly, Lundbeck, Merck, Meso Scale Discovery, Pfizer Inc., Piramal Imaging, F. Hoffmann-La Roche AG, Sanofi Genzyme, Servier, Takeda, Teva, UCB, and Global Capital.

# References

1. Alzheimer's Disease Neuroimaging Initiative: ADNI 1 Procedures Manual
2. Alzheimer's Disease Neuroimaging Initiative: ADNI 2 Procedures Manual
3. Alzheimer's Disease Neuroimaging Initiative: ADNI GO Procedures Manual
4. Maikore, F.S., Haddi, E., Soldatova, L.: A framework for it support of clinical laboratory standards. Int. J. Priv. Health Inf. Manage. **6**(2), 13–25 (2018)
5. Gamberger, D., et al.: Clusters of male and female Alzheimer's disease patients in the Alzheimer's Disease Neuroimaging Initiative (ADNI) database. Brain Inform. **3**(3), 169–179 (2016)
6. Jensen, M., et al.: The neurological disease ontology. J. Biomed. Seman. **4**(1), 42 (2013)
7. Maikore, F., et al.: An ontology for clinical laboratory standard operating procedures. In: Proceedings of the Joint Ontology Workshops 2017 (2017)
8. Mileski, V., et al.: Multi-dimensional analysis of PPMI data. In: Proceedings of the 8th IPS Students' Conference 2016, Ljubljana, Slovenia, pp. 175–178 (2016)
9. Panov, P., Soldatova, L., Džeroski, S.: Ontology of core data mining entities. Data Min. Knowl. Disc. **28**(5–6), 1222–1265 (2014)

10. Panov, P., Soldatova, L.N., Džeroski, S.: Generic ontology of datatypes. Inf. Sci. **329**, 900–920 (2016)
11. Scheuermann, R.H., Ceusters, W., Smith, B.: Toward an ontological treatment of disease and diagnosis. Summit Transl. Bioinform. **2009**, 116 (2009)
12. Smith, B., et al.: The OBO foundry: coordinated evolution of ontologies to support biomedical data integration. Nat. Biotechnol. **25**(11), 1251 (2007)
13. The Parkinson's Progression Markers Initiative: PPMI Case Report Forms
14. The Parkinson's Progression Markers Initiative: PPMI Study Protocol
15. Whetzel, P.L., et al.: Bioportal: enhanced functionality via new web services from the national center for biomedical ontology to access and use ontologies in software applications. Nucleic Acids Res. **39**(Suppl. 2), W541–W545 (2011)

# Embedding to Reference t-SNE Space Addresses Batch Effects in Single-Cell Classification

Pavlin G. Poličar[1]($\boxtimes$), Martin Stražar[1], and Blaž Zupan[1,2]

[1] University of Ljubljana, 1000 Ljubljana, Slovenia
{pavlin.policar,martin.strazar,blaz.zupan}@fri.uni-lj.si
[2] Baylor College of Medicine, Houston, TX 77030, USA

**Abstract.** Dimensionality reduction techniques, such as t-SNE, can construct informative visualizations of high-dimensional data. When working with multiple data sets, a straightforward application of these methods often fails; instead of revealing underlying classes, the resulting visualizations expose data set-specific clusters. To circumvent these batch effects, we propose an embedding procedure that uses a t-SNE visualization constructed on a reference data set as a scaffold for embedding new data points. Each data instance in the secondary data is embedded independently, and does not change the reference embedding. This prevents any interactions between instances in the secondary data and implicitly mitigates batch effects. We demonstrate the utility of this approach by analyzing six recently published single-cell gene expression data sets with up to tens of thousands of cells and thousands of genes. The batch effects in our studies are particularly strong as the data comes from different institutions and was obtained using different experimental protocols. The visualizations constructed by our proposed approach are cleared of batch effects, and the cells from secondary data sets correctly co-cluster with cells of the same type from the primary data.

**Keywords:** Batch effects · Embedding · t-SNE · Visualization · Single-cell transcriptomics · Data integration · Domain adaptation

## 1 Introduction

Two-dimensional embeddings and their visualizations may assist in the analysis and interpretation of high-dimensional data. Intuitively, two data instances should be co-located in the resulting visualization if their multi-dimensional profiles are similar. For this task, non-linear embedding techniques such as t-distributed stochastic neighbor embedding (t-SNE) [1] or uniform manifold approximation and projection [2] have recently complemented traditional data transformation and embedding approaches such as principal component analysis (PCA) and multi-dimensional scaling [3,4]. While useful for visualizing data from a single coherent source, these methods may encounter problems with multiple

© Springer Nature Switzerland AG 2019
P. Kralj Novak et al. (Eds.): DS 2019, LNAI 11828, pp. 246–260, 2019.
https://doi.org/10.1007/978-3-030-33778-0_20

data sources. Here, when performing dimensionality reduction on a merged data set, the resulting visualizations would typically reveal source-specific clusters instead of grouping data instances of the same class, regardless of data sources. This source-specific confounding is often referred to as *domain shift* [5], *covariate shift* [6] or *data set shift* [7]. In bioinformatics, the domain-specific differences are more commonly referred to as *batch effects* [8–10].

Massive, multi-variate biological data sets often suffer from these source-specific biases. The focus of this work is single-cell genomics, a domain that was selected due to high biomedical relevance and abundance of recently published data. Single-cell RNA sequencing (scRNA-seq) data sets are the result of isolating RNA molecules from individual cells, which serve as an estimate of the expression of cell's genes. The studies can exceed thousands of cells and tens of thousands of genes, and typically start with cell type analysis. Here, it is expected that cells of the same type would cluster together in two-dimensional data visualization [10]. For instance, Fig. 1a shows t-SNE embedded data from mouse brain cells originating from the visual cortex [11] and the hypothalamus [12]. The figure reveals distinct clusters but also separates the data from the two brain regions. These two regions share the same cell types and—contrary to the depiction in Fig. 1a—we would expect the data points from the two studies to overlap. Batch effects similarly prohibit the utility of t-SNE in the exploration of pancreatic cells in Fig. 1b, which renders the data from a pancreatic cell atlas [13] and similarly-typed cells from diabetic patients [14]. Just like with data from brain cells, pancreatic cells cluster primarily by data source, again resulting in a visualization driven by batch effects.

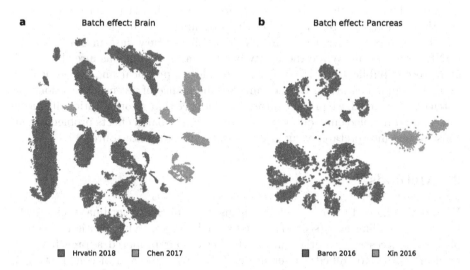

**Fig. 1.** Batch effects are a driving factor of variation between the data sets. We depict a t-SNE visualization of two pairs of data sets. In each pair, the data sets share cell types, so we would expect cells from the reference data (blue) to mix with the cells in a secondary data sets (orange). Instead, t-SNE clusters data according to the data source. (Color figure online)

Current solutions to embedding the data from various data sources address the batch effect problems up-front. The data is typically preprocessed and transformed such that the batch effects are explicitly removed. Recently proposed procedures for batch effect removal include canonical correlation analysis [8] and mutual nearest-neighbors [9,10]. In these works, batch effects are deemed removed when cells from different sources exhibit good mixing in a t-SNE visualization. The elimination of batch effects may require aggressive data preprocessing which may blur the boundaries between cell types. Another problem is also the inclusion of any new data, for which the entire analysis pipeline must be rerun, usually resulting in a different embedding layout and clusters that have little resemblance to original visualization and thus require reinterpretation.

We propose a direct solution of rendering t-SNE visualizations to address batch effects. Our approach treats one of the data sets as a *reference* and embeds the cells from another, *secondary data set* to a reference-defined low-dimensional space. We construct a t-SNE embedding using the reference data set, which is then used as a scaffold to embed the secondary data. The key idea underpinning our approach is that secondary data points are embedded independently of one another.

Independent embedding of each secondary datum causes the clustering landscape to depend only on the reference scaffold, thus removing data source-driven variation. In other words, when including new data, the scaffold inferred from the reference data set is kept unchanged and defines a "gravitational field", independently driving the embedding of each new instance. For example, in Fig. 2, the cells from the visual cortex define the scaffold (Fig. 2a) into which we embed the cells from the hypothalamus (Fig. 2b). Unlike in their joint t-SNE visualization (Fig. 1a), the hypothalamic cells are dispersed across the entire embedding space and their cell type correctly matches the prevailing type in reference clusters.

The proposed solution implements a mapping of new data into an existing t-SNE visualization. While the utility of such an algorithm was already hinted at in recent publication [15], we here provide its practical and theoretically-grounded implementation. Considering the abundance of recent publications on batch effect removal, we present surprising evidence that a computationally more direct and principled embedding procedure solves the batch effects problem when constructing interpretable visualizations from different data sources.

## 2   Methods

We describe an end-to-end pipeline that uses fixed t-SNE coordinates as a scaffold for embedding new (secondary) data, enabling joint visualization of multiple data sources while mitigating batch effects. Our proposed approach starts by using t-SNE to embed a reference data set, with the aim of constructing a two-dimensional visualization to facilitate interpretation and cluster classification. Then, the placement of each new sample is optimized independently via the t-SNE loss function. Independent treatment of each data instance from a secondary data set disregards any interactions present in that data set, and prevents

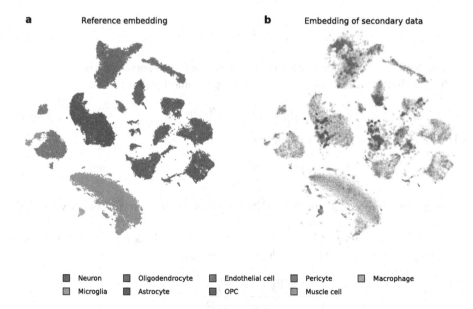

**a** Reference embedding
**b** Embedding of secondary data

| | Neuron | | Oligodendrocyte | | Endothelial cell | | Pericyte | | Macrophage |
| | Microglia | | Astrocyte | | OPC | | Muscle cell | | |

**Fig. 2.** A two-dimensional embedding of a reference containing brain cells (**a**) and the corresponding mapping of secondary data containing hypothalamic cells. (**b**) The majority of hypothalamic cells were mapped to their corresponding reference cluster. For instance, astrocyte cells marked with red on the right were mapped to an oval cluster of same-typed cells denoted with the same color in the visualization on the left. (Color figure online)

the formation of clusters that would be specific to the secondary data. Below, we start with a summary of t-SNE and its extensions (Sect. 2.1), introducing the relevant notation, upon which we base our secondary data embedding approach (Sect. 2.2).

## 2.1   Data Embedding by T-SNE and Its Extensions

Local, non-linear dimensionality reduction by t-SNE is performed as follows. Given a multi-dimensional data set $\mathbf{X} = \{\mathbf{x}_1, \mathbf{x}_2, \ldots, \mathbf{x}_N\} \in \mathbb{R}^D$ where $N$ is the number of data points in the reference data set, t-SNE aims to find a low dimensional embedding $\mathbf{Y} = \{\mathbf{y}_1, \mathbf{y}_2, \ldots, \mathbf{y}_N\} \in \mathbb{R}^d$ where $d \ll D$, such that if points $\mathbf{x}_i$ and $\mathbf{x}_j$ are close in the multi-dimensional space, their corresponding embeddings $\mathbf{y}_i$ and $\mathbf{y}_j$ are also close. Since t-SNE is primarily used as a visualization tool, $d$ is typically set to two. The similarity between two data points in t-SNE is defined as:

$$p_{j|i} = \frac{\exp\left(-\frac{1}{2}\mathcal{D}(\mathbf{x}_i, \mathbf{x}_j)/\sigma_i^2\right)}{\sum_{k \neq i} \exp\left(-\frac{1}{2}\mathcal{D}(\mathbf{x}_i, \mathbf{x}_k)/\sigma_i^2\right)}, \quad p_{i|i} = 0 \tag{1}$$

where $\mathcal{D}$ is a distance measure. This is then symmetrized to

$$p_{ij} = \frac{p_{j|i} + p_{i|j}}{2N}. \tag{2}$$

The bandwidth of each Gaussian kernel $\sigma_i$ is selected such that the perplexity of the distribution matches a user-specified parameter value

$$\text{Perplexity} = 2^{H(P_i)} \tag{3}$$

where $H(P_i)$ is the Shannon entropy of $P_i$,

$$H(P_i) = -\sum_i p_{j|i} \log_2(p_{j|i}). \tag{4}$$

Different bandwidths $\sigma_i$ enable t-SNE to adapt to the varying density of the data in the multi-dimensional space.

The similarity between points $\mathbf{y}_i$ and $\mathbf{y}_j$ in the embedding space is defined using the $t$-distribution with one degree of freedom

$$q_{ij} = \frac{\left(1 + ||\mathbf{y}_i - \mathbf{y}_j||^2\right)^{-1}}{\sum_{k \neq l} \left(1 + ||\mathbf{y}_k - \mathbf{y}_l||^2\right)^{-1}}, \quad q_{ii} = 0. \tag{5}$$

The t-SNE method finds an embedding $\mathbf{Y}$ that minimizes the Kullback-Leibler (KL) divergence between $\mathbf{P}$ and $\mathbf{Q}$,

$$C = \text{KL}(\mathbf{P} \,||\, \mathbf{Q}) = \sum_{ij} p_{ij} \log \frac{p_{ij}}{q_{ij}}. \tag{6}$$

The time complexity needed to evaluate the similarities in Eq. 5 is $\mathcal{O}(N^2)$, making its application impractical for large data sets. We adopt a recent approach for low-rank approximation of gradients based on polynomial interpolation which reduces its time complexity to $\mathcal{O}(N)$. This approximation enables the visualization of massive data sets, possibly containing millions of data points [16].

The resulting embeddings substantially depend on the value of the perplexity parameter. Perplexity can be interpreted as the number of neighbors for which the distances in the embedding space are preserved. Small values of perplexity result in tightly-packed clusters of points and effectively ignore the long-range interactions between clusters. Larger values may result in a more globally consistent visualizations—preserving distances on a large scale and organizing clusters in a more meaningful way—but can lead to merging small clusters and thus obscuring local aspects of the data [15].

The trade-off between the local organization and global consistency may be achieved by replacing the Gaussian kernels in Eq. 1 with a mixture of Gaussians of varying bandwidths [17]. Multi-scale kernels are defined as

$$p_{j|i} \propto \frac{1}{L} \sum_{l=1}^{L} \exp\left(-\frac{1}{2}\mathcal{D}(\mathbf{x}_i, \mathbf{x}_j)/\sigma_{i,l}^2\right), \quad p_{i|i} = 0 \tag{7}$$

where $L$ is the number of mixture components. The bandwidths $\sigma_{i,l}$ are selected in the same manner as in Eq. 1, but with a different value of perplexity for each $l$. In our experiments, we used a mixture of two Gaussian kernels with perplexity values of 50 and 500. A similar formulation of multi-scale kernels was proposed in [15], and we found the resulting embeddings are visually very similar to those obtained with the approach described above (not shown for brevity).

## 2.2  Adding New Data Points to Reference Embedding

Our algorithm, which embeds new data points to a reference embedding, consists of estimating similarities between each new point and the reference data and optimizing the position of each new data point in the embedding space. Unlike parametric models such as principal component analysis or autoencoders, t-SNE does not define an explicit mapping to the embedding space, and embeddings need to be found through loss function optimization.

The position of a new data point in embedding space is initialized to the median reference embedding position of its $k$ nearest neighbors. While we found the algorithm to be robust to choices of $k$, we use $k = 10$ in our experiments.

We adapt the standard t-SNE formulation from Eqs. 1 and 5 with

$$p_{j|i} = \frac{\exp\left(-\frac{1}{2}\mathcal{D}(\mathbf{x}_i, \mathbf{v}_j)/\sigma_i^2\right)}{\sum_i \exp\left(-\frac{1}{2}\mathcal{D}(\mathbf{x}_i, \mathbf{v}_j)/\sigma_i^2\right)}, \tag{8}$$

$$q_{j|i} = \frac{\left(1 + ||\mathbf{y}_i - \mathbf{w}_j||^2\right)^{-1}}{\sum_i \left(1 + ||\mathbf{y}_i - \mathbf{w}_j||^2\right)^{-1}}, \tag{9}$$

where $\mathbf{V} = \{\mathbf{v}_1, \mathbf{v}_2, \ldots, \mathbf{v}_M\} \in \mathbb{R}^D$ where $M$ is the number of samples in the secondary data set and $\mathbf{W} = \{\mathbf{w}_1, \mathbf{w}_2, \ldots, \mathbf{w}_M\} \in \mathbb{R}^d$. Additionally, we omit the symmetrization step in Eq. 2. This enables new points to be inserted into the embedding independently of one another. The gradients of $\mathbf{w}_j$ with respect to the loss (Eq. 6) are:

$$\frac{\partial C}{\partial \mathbf{w}_j} = 2 \sum_i \left(p_{j|i} - q_{j|i}\right)\left(\mathbf{y}_i - \mathbf{w}_j\right)\left(1 + ||\mathbf{y}_i - \mathbf{w}_j||^2\right)^{-1} \tag{10}$$

In the optimization step, we refine point positions using batch gradient descent. We use an adaptive learning rate scheme with momentum to speed up the convergence, as proposed by Jacobs [18,19]. We run gradient descent with momentum $\alpha$ of 0.8 for 250 iterations, where the optimization converged in all our experiments. The time complexity needed to evaluate the gradients in Eq. 10 is $\mathcal{O}(N \cdot M)$, however, by adapting the same polynomial interpolation based approximation, this is reduced to $\mathcal{O}(\max\{N, M\})$.

Special care must be taken to reduce the learning rate $\eta$ as the default value in most implementations ($\eta = 200$) may cause points to "shoot off" from the reference embedding. This phenomenon is caused due to the embedding to a previously defined t-SNE space, where the distances between data points and corresponding gradients of the optimization function may be quite large. When running standard t-SNE, points are initialized and scaled to have variance 0.0001. The resulting gradients tend to be very small during the initial phase, resulting in stable convergence. When embedding new samples, the span of the embedding is much larger, resulting in substantially larger gradients, and the default learning rate causes points to move very far from the reference embedding. In our experiments, we found that decreasing the learning rate to $\eta \sim 0.1$ produces stable solutions. This is especially important when using the interpolation-based approximation, which places a grid of interpolation points over the embedding space, where the number of grid points is determined by the span of the embedding. Clearly, if even one point "shoots off" far from the embedding, the number of required grid points may grow dramatically, increasing the runtime substantially. The reduced learning rate suppresses this issue, and does not slow the convergence because of the adaptive learning rate scheme, provided the optimization is run for a sufficient number of steps.

## 3    Experiments and Discussion

We apply the proposed approach to t-SNE visualizations of single-cell data. Data in this realm include a variety of cells from specific tissues and are characterized through gene expression. In our experiments, we considered several recently published data sets where cells were annotated with the cell type. Our aim was to construct t-SNE visualizations where similarly-typed cells would cluster together, despite systematic differences between data sources. To that end, we focus on comparing different ways of using t-SNE rather than differences to embeddings like PCA or MDS, which have been substantially covered before [1,4]. Below, we list the data sets, describe single-cell specific data preprocessing procedures, and display the resulting data visualizations. Finally, we discuss the success of the proposed approach in alleviating the batch effects.

### 3.1    Data

We use three pairs of reference and secondary single-cell data sets originating from different organisms and tissues. The data in each pair were chosen so that the majority of cell types from the secondary data set were included in the reference set (Table 1). The cells in the data sets originate from the following three tissues:

**Mouse Brain.** The data set from Hrvatin *et al.* [11] contains cells from the visual cortex exploring transcriptional changes after exposure to light. This was used as a reference for the data from Chen *et al.* [12], containing cells from the mouse hypothalamus and their reaction to food deprivation. From the secondary data, we removed cells with no corresponding types in the reference: tanycytes, ependymal, epithelial, and unlabelled cells.

**Human Pancreas.** Baron *et al.* [13] created an atlas of pancreatic cell types. We used this set as a reference for data from Xin *et al.* [14], who examined transcriptional differences between healthy and type 2 diabetic patients.

**Mouse Retina.** Macosko *et al.* [20] created an atlas of mouse retinal cell types. We used this as a reference for the data from Shekhar *et al.* [21], who built an atlas for retinal bipolar cells.

**Table 1.** Data sets used in our experiments. The first data set in each pair (Hrvatin *et al.*, Baron *et al.*, and Macoscko *et al.*) was used as a reference. We relied on the quality control and annotations from the original publication and report the number of cell types after preprocessinng. The cell annotations were made consistent to annotations from the Cell Ontology [22]. Notice that different RNA sequencing protocols were used to estimate gene expressions.

| Study | Organism/Tissue | Protocol | Cells | Cell Types | Sparsity (%) |
|-------|-----------------|----------|-------|------------|--------------|
| Hrvatin *et al.* | mouse brain | inDrop | 48,266 | 9 | 94 |
| Chen *et al.* | | Drop-seq | 14,437 | 6 | 93 |
| Baron *et al.* | human pancreas | inDrop | 8,569 | 9 | 91 |
| Xin *et al.* | | SMARTer | 1,492 | 4 | 86 |
| Macosko *et al.* | mouse retina | Drop-seq | 44,808 | 12 | 97 |
| Shekhar *et al.* | | Drop-seq | 27,499 | 5 | 96 |

## 3.2  Single-Cell Data Preprocessing Pipeline

Due to the specific nature of single-cell data, additional steps must be taken to properly apply t-SNE. We use a standard single-cell preprocessing pipeline, consisting of the selection of 3,000 representative genes (see Sect. 3.3), library size normalization, log-transformation, standardization, and PCA-based representation that retains 50 principal components [10,23]. To obtain the reference embedding, we apply multi-scale t-SNE using PCA initialization [15]. Due to high-dimensionality of the preprocessed input data we use cosine distance to estimate similarities between reference data points [24]. When adding new data points from the secondary data set to the reference embedding, we select 1,000 genes present in both data sets and use the cosine similarity to estimate the similarities between the secondary data item and reference data points. We note that similarities are computed using the raw count matrices. The preprocessing stages are detailed in accompanying Python notebooks (Sect. 3.5).

## 3.3  Gene Selection

Single-cell data sets suffer from high levels of technical noise and low capture efficiency, resulting in sparse expression matrices [25]. To address this problem, we use a specialized feature-selection method, which exploits the mean-dropout relationship of expression counts as recently proposed by Kobak and Berens [15]. Here, genes with higher than expected dropout rate are regarded as potential markers for cell subpopulations and are retained in the data.

Given an expression matrix $\mathbf{X} \in \mathbb{R}^{N \times G}$ where $N$ is the number of samples and $G$ is the number of genes in the data set, we compute the fraction of cells where a gene $g$ was not expressed

$$d_g = \frac{1}{N} \sum_i I\left(X_{ig} = 0\right) \tag{11}$$

The mean $\log_2$ expression of gene $g$ considers only cells expressing $g$:

$$m_g = \langle \log_2 X_{ig} \mid X_{ig} > 0 \rangle. \tag{12}$$

All genes expressed in less than ten cells are discarded. In order to select a specific number of $\hat{G}$ genes, we use a binary search to find a value $b$ such that

$$\sum_g I\left(d_g > \exp\left[-(m_g - b)\right] + 0.02\right) = \hat{G}. \tag{13}$$

## 3.4    Results and Discussion

Figures 2, 3, and 4 show the embeddings of the reference data sets and their corresponding embeddings of the secondary data sets. In all the figures, the cells from the secondary data sets were positioned in the cluster of same-typed reference cells, providing strong evidence of the success of our approach. There are some deviations to these observations; for instance, in Fig. 2 several oligodendrocyte precursor cells (OPCs) were mapped to oligodendrocytes. This may be due to differences in annotation criteria by different authors, or due to inherent similarities of these types of cells. Examples of such erroneous placements can be found in other figures as well, but are uncommon and constitute less then 5% of the cells (less than 5% in brain, 1% in pancreas and 2% in retina secondary data).

Notice that we could simulate the split between reference and secondary data sets using one data set only and perform cross-validation, however this type of experiment would not incorporate batch effects. We want to remind the reader that handling batch effects were central to our endeavour and that the disregard of this effect could lead to overly-optimistic results and data visualizations strikingly different from ours. For example, compare the visualizations from Figs. 1a and 2b, or Figs. 1b and 3b.

We use a number of additional, recently proposed modifications to enhance the t-SNE visualization of the reference data set. One important extension is the use of multi-scale similarities, which, in addition to finding local structure, also ensure consistent organization of the resulting clusters. For illustration, consider visualizations with standard and multi-scale t-SNE in Fig. 5. Notice, for instance, that in multi-scale t-SNE (Fig. 5b) the clusters with neuronal cells are clumped together, while their placement in standard t-SNE is arbitrary (Fig. 5a).

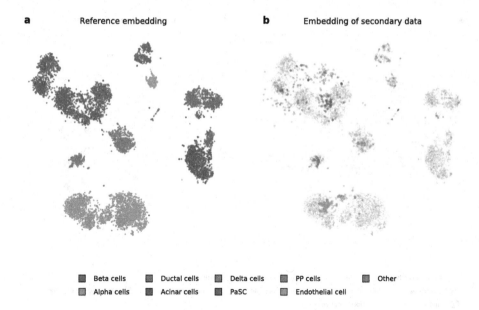

**a**  Reference embedding          **b**  Embedding of secondary data

Beta cells    Ductal cells    Delta cells    PP cells    Other
Alpha cells    Acinar cells    PaSC    Endothelial cell

**Fig. 3.** Embedding of pancreatic cells from Baron *et al.* [13] and cells from the same tissue from Xin *et al.* [14]. Just like in Fig. 2, the vast majority of the cells from the secondary data set were correctly mapped to the same-typed cluster of reference cells.

We also observed the important role of gene selection in crafting the reference embedding spaces. We found that when selecting an insufficient number of genes, the resulting visualizations display overly-fragmented clusters. When the selection is too broad and includes lowly expressed genes, the subclusters tend to overlap. These effects can all be attributed to sparseness of the data sets and may be intrinsic to single-cell data. In our studies, we found that selection of 3,000 genes yields most informative visualizations (Fig. 6).

In principle, our theoretically-grounded embedding of secondary data into the scaffold defined by the reference embedding could be simplified with the application of the nearest neighbors-based procedure. For example, while describing a set of tricks for t-SNE, Kobak and Berens [15] proposed positioning new points into a known embedding by placing them in the median position of their 10 nearest neighbors, where the neighborhood was estimated in the original data space. Notice that we use this approach as well, but only for the initialization of positions of new data instances that are subject to further optimization. Despite both nearest-neighbors search and t-SNE optimization can be computed in linear time, the former dominates the runtime (mouse retina example; 44,808 reference, 26,830 secondary cells, 9 min NN-search, 13 s optimization).

We demonstrate a case where nearest neighbor-based positioning is insufficient in Fig. 7. It may yield clumped visualizations where the optimal positioning using the t-SNE loss function is much more dispersed and rightfully shows a more considerable variation in the secondary data. Some points may also fall

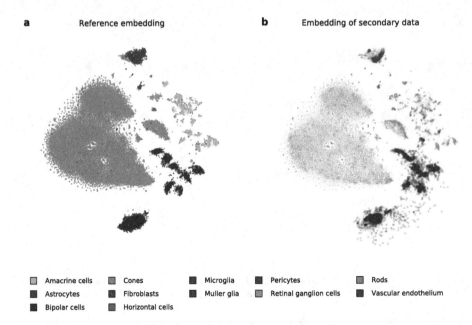

**Fig. 4.** An embedding of a large reference of retinal cells from Macosco *et al.* [20] **(a)** and mapping of cells from a smaller study that focuses on bipolar cells from Shekhar *et al.* [21]. **(b)** We use colors consistent with the study by Macosko *et al.*

into the empty regions, while after optimization they typically move closer to same-typed groups. While the nearest neighbor-based approach always places new data points to their most similar cluster, the proposed t-SNE approach can identify extreme outliers and place them far from the reference cluster.

Our approach assumes that all cell types from the secondary data set are present in the reference. The method would fail to reveal novel cell types in the secondary data set, possibly positioning them arbitrarily close to unrelated clusters. Procedures such as scmap were recently proposed to cope with such cases and identify the cells whose type is new and not included in the reference [26].

Our procedure is, therefore, asymmetrical in the choice of reference and secondary data set. In practice, however, newly produced *secondary* data would be embedded into previously-prepared reference landscapes. Large collections of data *e.g.* the Human Cell Atlas initiative [27] make it possible to scale up our approach to wider sets of cell types.

## 3.5   Implementation

The procedures described in this paper are provided as Python notebooks that are, together with the data, available in an open repository[1]. All experiments were run using **openTSNE**, our open and extensible t-SNE library for Python [28].

---

[1] https://github.com/biolab/tsne-embedding.

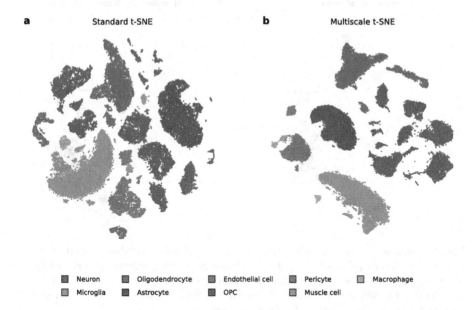

**Fig. 5.** A comparison of standard and multi-scale t-SNE on data from the mouse visual cortex [11]. **(a)** Standard t-SNE places clusters arbitrarily. **(b)** Augmenting t-SNE with multi-scale similarities and using proper initialization provides a more meaningful layout of the clusters. Neuronal types occupy one region of the space. Oligodendrocyte precursor cells (OPCs) are mainly progenitors to oligodendrocytes, but may also differentiate into neurons or astrocytes.

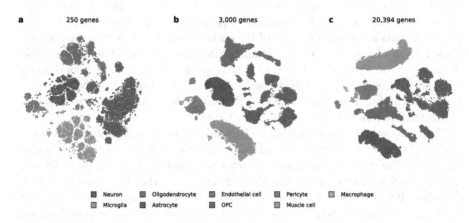

**Fig. 6.** Gene selection plays an important role when constructing the reference embedding. **(a)** Using too few genes results in fragmented clusters. **(b)** Using an intermediate number of genes reveals clustering mostly consistent with cell annotations. **(c)** Including all the genes may lead to under-clustering of the more specialized cell types.

**a**    Median initialization              **b**    t-SNE embedding of secondary data

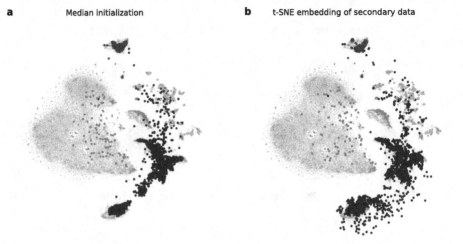

**Fig. 7.** Comparison of data placement using the nearest neighbors approach from Kobak and Berens [15] and the optimized placement using our algorithm. (a) Data points are placed to the median position of their 10 nearest neighbors in the reference set. (b) Point positions are optimized, revealing a different, more dispersed placement that better reflects the variety of cells in the secondary data set.

## 4    Conclusion

Almost all recent publications of single-cell studies begin with a two-dimensional visualization of the data that reveals cellular diversity. While many dimensionality reduction techniques are available, different variants of t-SNE are most often used to produce such visualizations. Single-cell studies enable the exploration of biological mechanisms at a cellular level, and their publications in the past couple of years are abundant. One of the central tasks in single-cell studies is the classification of new cells based on findings from previous studies. Such transfer of knowledge is often difficult due to batch effects present in data from different sources. Addressing batch effects by adapting and extending t-SNE, the prevailing method used to present single-cell data in two-dimensional visualization, motivated the research presented in this paper.

The proposed approach uses a t-SNE embedding as a scaffold for the positioning of new cells within the visualization, and possibly for aiding in their classification. The three case studies incorporating pairs of data sets from different domains but with similar classifications demonstrate that our proposed procedure can effectively deal with batch effects to construct visualizations that correctly map secondary data sets onto an embedding of the data from an independent study that possibly uses different experimental protocol. While we focused here on reference visualizations constructed using t-SNE, this approach can be applied using any existing two-dimensional visualization.

**Acknowledgements.** This work was supported by the Slovenian Research Agency Program Grant P2-0209, and by the BioPharm.SI project supported from European Regional Development Fund and the Slovenian Ministry of Education, Science and Sport. We would also like to thank Dmitry Kobak for many helpful discussions on t-SNE.

# References

1. van der Maaten, L., Hinton, G.: Visualizing data using t-SNE. J. Mach. Learn. Res. **9**(Nov), 2579–2605 (2008)
2. McInnes, L., Healy, L., Melville, L.: UMAP: uniform manifold approximation and projection for dimension reduction. ArXiv e-prints, February 2018
3. Wattenberg, M., Viégas, F., Johnson, I.: How to use t-SNE effectively. Distill **1**(10), e2 (2016)
4. Becht, E., et al.: Dimensionality reduction for visualizing single-cell data using UMAP. Nat. Biotechnol. **37**(1), 38 (2019)
5. Gopalan, R., Li, R., Chellappa, R.: Domain adaptation for object recognition: an unsupervised approach. In: 2011 International Conference on Computer Vision, pp. 999–1006. IEEE (2011)
6. Bickel, S., Brückner, M., Scheffer, T.: Discriminative learning under covariate shift. J. Mach. Learn. Res. **10**(Sep), 2137–2155 (2009)
7. Quionero-Candela, J., Sugiyama, M., Schwaighofer, A., Lawrence, N.D.: Dataset Shift in Machine Learning. The MIT Press, Cambridge (2009)
8. Butler, A., Hoffman, P., Smibert, P., Papalexi, E., Satija, R.: Integrating single-cell transcriptomic data across different conditions, technologies, and species. Nat. Biotechnol. **36**(5), 411 (2018)
9. Haghverdi, L., Lun, A.T.L., Morgan, M.D., Marioni, J.C.: Batch effects in single-cell RNA-sequencing data are corrected by matching mutual nearest neighbors. Nat. Biotechnol. **36**(5), 421 (2018)
10. Stuart, T., et al.: Comprehensive Integration of Single-Cell Data. Cell **177**(7), 1888–1902 (2019)
11. Hrvatin, S., et al.: Single-cell analysis of experience-dependent transcriptomic states in the mouse visual cortex. Nat. Neurosci. **21**(1), 120 (2018)
12. Chen, R., Xiaoji, W., Jiang, L., Zhang, Y.: Single-cell RNA-seq reveals hypothalamic cell diversity. Cell Rep. **18**(13), 3227–3241 (2017)
13. Baron, M., et al.: A single-cell transcriptomic map of the human and mouse pancreas reveals inter-and intra-cell population structure. Cell Syst. **3**(4), 346–360 (2016)
14. Xin, Y., et al.: RNA sequencing of single human islet cells reveals type 2 diabetes genes. Cell Metab. **24**(4), 608–615 (2016)
15. Kobak, D., Berens, P.: The art of using t-SNE for single-cell transcriptomics. bioRxiv, p. 453449 (2018)
16. Linderman, G.C., Rachh, M., Hoskins, J.G., Steinerberger, S., Kluger, Y.: Fast interpolation-based t-SNE for improved visualization of single-cell RNA-seq data. Nat. Methods **16**(3), 243 (2019)
17. Lee, J.A., Peluffo-Ordóñez, D.H., Verleysen, M.: Multi-scale similarities in stochastic neighbour embedding: reducing dimensionality while preserving both local and global structure. Neurocomputing **169**, 246–261 (2015)
18. Jacobs, R.A.: Increased rates of convergence through learning rate adaptation. Neural Networks **1**(4), 295–307 (1988)

19. van der Maaten, L.: Accelerating t-SNE using tree-based algorithms. J. Mach. Learn. Res. **15**(1), 3221–3245 (2014)
20. Macosko, E.Z., et al.: Highly parallel genome-wide expression profiling of individual cells using nanoliter droplets. Cell **161**(5), 1202–1214 (2015)
21. Shekhar, K., et al.: Comprehensive classification of retinal bipolar neurons by single-cell transcriptomics. Cell **166**(5), 1308–1323 (2016)
22. Bard, J., Rhee, S.Y., Ashburner, M.: An ontology for cell types. Genome Biol. **6**(2), R21 (2005)
23. Wolf, F.A., Angerer, P., Theis, F.J.: SCANPY: large-scale single-cell gene expression data analysis. Genome Biol. **19**(1), 15 (2018)
24. Domingos, P.M.: A few useful things to know about machine learning. Commun. ACM **55**(10), 78–87 (2012)
25. Islam, S., et al.: Quantitative single-cell RNA-seq with unique molecular identifiers. Nat. Methods **11**(2), 163 (2014)
26. Kiselev, V.Y., Yiu, A., Hemberg, M.: scmap: projection of single-cell RNA-seq data across data sets. Nat. Methods **15**(5), 359 (2018)
27. Rozenblatt-Rosen, O., Stubbington, M.J.T., Regev, A., Teichmann, S.A.: The Human Cell Atlas: from vision to reality. Nat. News **550**(7677), 451 (2017)
28. Poličar, P.G., Stražar, M., Zupan, B.: openTSNE: a modular Python library for t-SNE dimensionality reduction and embedding. bioRxiv (2019)

# Symbolic Graph Embedding
# Using Frequent Pattern Mining

Blaž Škrlj[1,2], Nada Lavrač[1,2,3], and Jan Kralj[2(✉)]

[1] Jožef Stefan International Postgraduate School, Ljubljana, Slovenia
[2] Jožef Stefan Institute, Ljubljana, Slovenia
`jan.kralj@ijs.si`
[3] University of Nova Gorica, Nova Gorica, Slovenia

**Abstract.** Relational data mining is becoming ubiquitous in many fields of study. It offers insights into behaviour of complex, real-world systems which cannot be modeled directly using propositional learning. We propose Symbolic Graph Embedding (SGE), an algorithm aimed to learn symbolic node representations. Built on the ideas from the field of inductive logic programming, SGE first samples a given node's neighborhood and interprets it as a transaction database, which is used for frequent pattern mining to identify logical conjuncts of items that co-occur frequently in a given context. Such patterns are in this work used as features to represent individual nodes, yielding interpretable, symbolic node embeddings. The proposed SGE approach on a venue classification task outperforms shallow node embedding methods such as DeepWalk, and performs similarly to metapath2vec, a black-box representation learner that can exploit node and edge types in a given graph. The proposed SGE approach performs especially well when small amounts of data are used for learning, scales to graphs with millions of nodes and edges, and can be run on an of-the-shelf laptop.

**Keywords:** Graphs · Machine learning · Relational data mining ·
Symbolic learning · Embedding

## 1 Introduction

Many contemporary databases are comprised of vast, linked and annotated data, which can be hard to exploit for various modeling purposes. In this work, we explore how learning from heterogeneous graphs (i.e. heterogeneous information networks with different types of nodes and edges) can be conducted using the ideas from the fields of symbolic relational learning and inductive logic programming [16], as well as contemporary representation learning on graphs [2].

Relational datasets have been considered in machine learning since the early 1990s, where tools such as Aleph [27] have been widely used for relational data analysis. However, recent advancements in deep learning, a field of subsymbolic machine learning, which allows for learning from relational data in the form of graphs, was shown as useful for many contemporary relational learning tasks

© Springer Nature Switzerland AG 2019
P. Kralj Novak et al. (Eds.): DS 2019, LNAI 11828, pp. 261–275, 2019.
https://doi.org/10.1007/978-3-030-33778-0_21

at scale, including recommendation, anomaly detection and similar [25]. The state-of-the-art methodology exploits the notion of *node embeddings*—nodes, represented using real-valued vectors. As such, node embeddings can be simply used with propositional learners such as e.g., logistic regression or neural networks. The node embeddings, however, directly offer little to no insight into connectivity patterns relevant for representing individual nodes.

In this work we demonstrate that symbolic pattern mining can be used for learning *symbolic node embeddings* in heterogeneous information graphs. The main contributions of this work include:

1. An efficient graph sampler which samples based on a distribution of lengths of random walks, implemented in Numba, offering 15x faster sampling than a Python-native implementation, scaling to graphs with millions of nodes and edges on an of-the-shelf laptop.
2. Symbolic graph embedding (SGE), a symbolic representation learner that is explainable and achieves state-of-the-art performance for the task of node classification.
3. Evidence that symbolic node embeddings can perform comparably to black-box node embeddings, whilst requiring *less* space and data.

This paper is structured as follows. We first discuss the related work (Sect. 2), followed by the description of the proposed approach (Sect. 3), its computational and spatial complexity (Sect. 4), and its empirical evaluation (Sect. 5). We finally discuss the obtained results and potential further work (Sect. 6).

## 2    Related Work

Symbolic representation learning has already been considered in the early 1990s in the inductive learning community, when addressing multi-relational learning problems through the so-called *propositionalization* approach [16]. The goal of propositionalization is to transform multi-relational data into real-valued vectors describing the individual training instances, that are a part of a relational data structure. The values of the vectors are obtained by evaluating a relational feature (e.g., a conjunct of conditions) as true (value 1) or false (value 0). For example, if all conditions of a conjunct are true, the relational feature is evaluated as true, resulting in value 1, and gets value 0 otherwise. We next discuss the approaches which were most influential for this work. The in-house developed Wordification [22] explores how relational databases can be unfolded into bags of relational words, used in the same manner as done in the area of natural language processing via Bag-of-words-based representations. Wordification, albeit very fast, can be spatially expensive, and was designed for SQL-based datasets. Our work was also inspired by the recently introduced HINMINE methodology [14], where Personalized PageRank vectors were used as the propositionalization mechanism. Here, each node is described via its probability to visit any other node, thus, a node of a netwrk is described using a distribution over the remainder of the nodes. Further, propositionalization has recently been explored in combination with artificial neural networks [8], and as a building block of deep relational machines [6].

Frequent pattern mining is widely used for identifying interesting patterns in real-world transaction databases. Extension of this paradigm to graphs was already explored [12], using the Apriori algorithm [1] for the pattern mining. In this work we rely on the efficient FP-Growth [4] algorithm, which employs more structured counting compared to Apriori using fp-trees as the data structure. Frequent pattern mining is commonly used to identify logical patterns which appear above a certain e.g., frequency threshold. Efficiently mining for such patterns remains a lively research area on its own, and can be scaled to large computing clusters [11].

The proposed work also explores how a given graph can be sampled, as well as embedded efficiently. Many contemporary node representation learning methods, such as node2vec [9], DeepWalk [23], PTE [28] and metapath2vec [7], exploit such ideas in combination with e.g., the skip-gram model in order to obtain low-dimensional embeddings of nodes. Out of the aforementioned methods, only metapath2vec was adapted specifically to operate on *heterogeneous information networks*, i.e. graphs with additional information on node and edge types. It samples pre-defined meta paths, yielding type-aware node representations which serve better for classifying e.g., different research venues to topics.

Heterogeneous (non-attributed) graphs are often formalized as RDF triplets. Relevant methods, which explore how such triplets can be embedded are considered in [5], as well as in [24]. The latter introduced RDF2vec, a methodology for direct transformation of a RDF database to the space of real-valued entity embeddings. Understanding how such graphs can be efficiently sampled, as well as embedded into low-dimensional, real-valued vectors is a challenging problem on its own.

## 3 Proposed SGE Algorithm

In this section we describe Symbolic Graph Embedding (SGE), a new algorithm for symbolic node embedding. The algorithm is summarized in Fig. 1. The algorithm consists of two basic steps. First, for each node in an input graph, the neighborhood of the node is sampled (Sect. 3.2). Next, the patterns, emerging from the walks around a given node are transformed into a set of features whose values describe the node (Sect. 3.3). In this section, we first introduce some basic definitions and then explain both steps of SGE in more detail.

### 3.1 Overview and Definitions

We first define the notions of a graph as used in this work.

**Definition 1 (Graph).** *A graph is a tuple $G = (N, E)$, where $N$ is a set of nodes and $E$ is a set of edges. The elements of $E$ can be either subset $N$ of size 2 (e.g., $\{n_1, n_2\} \subseteq N$), in which case, we say the graph is undirected. Alternatively, $E$ can consist of ordered pairs of elements from $N$ (e.g., $(n_1, n_2) \in N \times N$) – in this case, the graph is directed.*

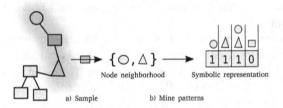

**Fig. 1.** Schematic represenatation of SGE. The red square's neighborhood (red highlight) is first used to construct symbolic features, forming a propositional graph representation (part a of the figure). The presence of various symbolic patterns (FP) is recorded and used to determine feature vectors for individual nodes (part b of the figure). The obtained representation can be used for subsequent data analysis tasks such as classification or visualization. (Color figure online)

In this work we focus on directed graphs, yet the proposed methodology can also be extended to undirected ones. In this work we also use the notion of a *walk*.

**Definition 2 (Walk).** *Given a directed graph $G = (N, E)$, a walk is any sequence of nodes $n_1, n_2 \ldots, n_k \in N$ so that each pair $n_i, n_{i+1}$ of consecutive nodes is connected by an edge, i.e. $(n_i, n_{i+k}) \in E$.*

Finally, we define the notion of node embedding as used throughout this work.

**Definition 3 (Symbolic node embedding).** *Given a directed graph $G = (N, E)$, a d-dimensional node embedding of graph $G$ is a matrix $\mathcal{M}$ in a vector space $\mathbb{R}^{|N| \times d}$, i.e. $\mathcal{M} \in \mathbb{R}^{|N| \times d}$. Such embedding is considered* symbolic, *when each column represents a symbolic expression, which, when evaluated against a given node's neighborhood information, returns a real number representing a given node.*

We first discuss the proposed neighborhood sampling routine, followed by the description of pattern learning as used in this work.

## 3.2    Sampling Node Neighborhoods

Sampling a given node's local and global neighborhoods offers insights into connectivity patterns of the node with respect to the rest of the graph. Many contemporary methods resort to node neighborhood sampling for obtaining the node co-occurrence information. In this work we propose a simple sampling scheme which produces a series of graph walks. The walks can be further used for learning tasks—in this work, we use them to produce node representations. Building on recent research ideas [17,18], the proposed scheme consists of two steps: selection of walk sampling distribution and sampling. We first discuss the notion of distribution-based sampling, followed by the implemented sampling scheme.

**Distribution-Based Sampling.** Our algorithm is based on the assumption that when learning a representation of a given node, nodes at various distances from the considered node are relevant. In order to use the information on neighborhood nodes, we sample several random walks starting at each node of the graph. Because real-world graphs are diverse, it is unlikely that the same sampling scheme would suffice for arbitrary graphs. To account for such uncertainty, we introduce the notion of *walk distribution vector*, a vector describing how many walks of a certain length shall be sampled. Let $w \in \mathbb{R}^s$ denote a vector of length $s$ (a parameter of the approach). The $i$-th value of the vector corresponds to the proportion of walks of length $i$ that are to be sampled. Note that the longest walk that can occur is of length $s$. For example consider the following vector $w$ of length $s = 4$, $w = [0.2, 0, 0.5, 0.3]$. Assuming we sample e.g, 100 random walks, 20 walks will be of length one, zero of length 2, 50 of length three and 30 of length four. As $w$ represents a probability distribution of different walk lengths to be sampled, $\sum_{i=1}^{s} w_i = 1$ must hold.

Having defined the formalism for describing the number of walks of different lengths, we have yet to describe the following two aspects in order to fully formalize the proposed sampling scheme: how to parametrize $w$ and walk efficiently?

**How to Parametrize $w$.** We next discuss the considered initialization of the probability vector $w$. We attempt to model such vector by assuming a prior *walk length* distribution, from which we first sample $\phi$ samples—these samples represent different random walk *lengths*. In this work, we consider Uniform walk length distribution, where the $i$-th element of vector $w$ is defined as $\frac{1}{s}$, where $s$ represents the length of $w$ (maximum walk length).

The considered variant of graph sampling procedure does not take into account node or edge types. One of the purposes of this work is to explore whether such naïve sampling—when combined with symbolic learning—achieves good performance. The rationale for not exploring how to incorporate node and edge types is thus twofold: First, we explore whether symbolic learning, as discussed in the next section, detects heterogeneous node patterns on its own,

---

**Algorithm 1:** Order-aware random walker.

**Data:** A graph $G = (N, E)$
**Parameters :** Starting node $n_i$, walk length $s$
1  $c \leftarrow n_i$;
2  $\delta \leftarrow 0$;
3  $\mathcal{W} \leftarrow$ multiset;
4  **for** $\alpha \in [1 \ldots s]$ **do**
5     $o := \text{Uniform}(N_G(c))$ ;           ▷ Select a random node.
6     $\mathcal{W} \leftarrow \mathcal{W} \cup (o, \alpha)$;             ▷ Store visited node.
7     $c \leftarrow o$;
8  **end**
**Result:** A random walk $\mathcal{W}$

as the node representations are discrete and could, as such, provide such information. Further, as exact node information is kept intact, and each node can be mapped to its type, the node types are implicitly incorporated. Next, we believe that by selecting the appropriate prior walk distribution $w$, node types can be to some extent taken into account (yet this claim depends largely on a given graph's topology).

**How to Walk Efficiently.** An example random walker, which produces walks of a given length (used in this work) is formalized in Algorithm 1. Here, we denote with $N_G(n_i)$ the neighbors of the $i$-th node. The Uniform($N_G(c)$) represents a randomly picked neighbor of a given node $c$, where picking each neighbor is equiprobable. We mark such picked node with $o$. Note that the walker is essentially a probabilistic depth-first search. The algorithm returns a list of tuples where each tuple $(o, \alpha)$ contains both the visited node $o$ and the step $\alpha$ at which the node was visited. On line 6, we append to a current walk a tuple, comprised of a certain node and the overall walk length, it is a part of. Note that such inclusion of node IDs is suitable in a learning setting, where e.g., part of the graph's labels are not known and are to be predicted using the remainder of the graph. Even though inclusion of such information might seem redundant, we would like to remind the reader that the presented algorithm represents only a single random walker for a single walk length. In reality, multiple walkers yielding walks of different lengths are simulated, since including such positional (walk length) information can be beneficial for the subsequent representation learning step. The proposed Algorithm 1 represents a simple random walker. In practice, thousands of random walks are considered. As discussed, their lengths are distributed according to $w$. In theory, one could learn the optimal $w$ by using e.g., stochastic optimization, yet we explore a different, computationally more feasible approach for obtaining a given $w$. We next discuss the notion of symbolic pattern mining and final formalization of the proposed SGE algorithm.

### 3.3 Symbolic Pattern Mining

In the previous section we discussed how a given node's neighborhood can be efficiently sampled. In this section we first discuss the general idea behind forming node representations, followed by a description of the frequent pattern mining algorithms employed.

**Forming Node Representations.** Algorithm 1 outputs a multiset comprised of nodes, represented by (node id, walk order) tuples. Such multisets are in the following discussion considered as *itemsets*, as this is the terminology used in [4]. In the next step of SGE, we use the itemsets to obtain individual node representations. We first give an outline of this step, and provide additional details in the next section. The set of all nodes is considered as a transaction database, and the itemsets comprised of node id and walk order are used to identify frequent patterns (of tuples). Best patterns, selected based on their frequency of occurrence, are used as *features*. The way of determining the best patterns is approach specific, and is discussed in the following paragraphs. Feature values are determined based on the pattern identification method, and are either real-, natural- or binary-valued. Intuitively, they represent the presence of a given node pattern in a given node's neighborhood.

**Frequent Pattern Mining.** We next discuss the frequent pattern mining approaches explored as part of SGE. The described approaches constitute the *findPatterns* method discussed in the next section. For each node, a multiset of (node ID, walk length) tuples is obtained. In each of the described approaches, the result is a transformation of the set of multisets, describing the network nodes, into a set of feature vectors describing these nodes.

**Relational BoW**. This paradigm leverages the Bag-of-words (BOW) constructors widely known in natural language processing [31]. Here, the tuples forming the itemsets, output by Algorithm 1, are considered as words. Thus, each word is comprised of a node and the order of a random walk in which that node was identified as connected to the node for which the representation is being constructed.

For the purpose of BOW construction, we consider the multiset of (node ID, walk length) tuples, generated by random walks that start at node $n$. This multiset is viewed as a *"node document"*, consisting of individual words—i.e. the tuples contained in the multiset. The number of total features, $d$, is a parameter of the SGE algorithm. We consider the following variations of this paradigm for transforming each node "document" $\mathfrak{T}_n$ into one feature vector of size $d$:

- **Binary**. In a binary conversion, the values of the vector represent the presence or absence of a given tuple $k$-gram (a combination of $k$ tuples; $k$ is a free parameter of SGE) in the set of random walks associated with a given node document. The features, represented by such tuple $k$-grams, can have values of either 0 or 1.
- **TF**. Here, counts of a given $k$-gram $t$ in a given node document $\mathfrak{T}_n$ are used as feature values ($TF_{t,\mathfrak{T}_n}$). The values are integers. Note that $TF_{t,\mathfrak{T}_n}$ represents the multiplicity of a given tuple $k$-gram in the multiset (node document).
- **TF-IDF**. Here, TF-IDF weighting scheme is employed to weight the values of individual features. The obtained values are real numbers. Given a tuple $k$-gram $t$ and the transaction database $\mathfrak{T}$, it is computed as:

$$\text{TF-IDF}(t,n) = (1 + \log \text{TF}_{t,\mathfrak{T}_n}) \cdot \log \frac{|N|}{\mathfrak{T}_t},$$

where $TF_{t,\mathfrak{T}_N}$ is the number of $t$'s occurrences in a given document of node $n$ and $\mathfrak{T}_t$ is the overall occurrence of this $k$-gram in the whole transaction database.

**FP-Growth**. This well known variant of association rule learning [4] constructs a specialized data structure termed fp-tree, which is used to count combinations of tuples of different lengths. It is more efficient than the well known Apriori algorithm [3, 21].

For the purpose of FP-growth, the obtained multiset $\mathfrak{T}$ is viewed as a set of *itemsets* - a transaction database. For each itemset, only the set of unique tuples is considered as the input while their multiplicity is ignored. The FP-Growth algorithm next considers such non-redundant transaction databse $\mathfrak{T}$ to identify frequent combinations of tuples, similarly to the TF and TF-IDF schemes described above. The free parameter we consider in this work is *support*, which controls how frequent tuple combinations shall be considered. Similarly to TF and TF-IDF schemes, once the representative tuple combinations are obtained, they are considered as features, whose values are determined based on their presence in a given node document, and are binary (0 = not present, 1 = present). Note that some of these features may correspond to the features generated by TF-IDF, however, in the case of FP-growth, we allow sizes of tuple combinations to be arbitrary, rather than fixed to $k$. Also unlike the BOW approaches, the dimension of the constructed feature vectors constructed is not fixed but is controlled implicitly by varying the value of the *support* parameter.

**SGE Formulation.** The formulation of the whole approach is given in Algorithm 2. Here, first the sampling vector $w$ is constructed. Next, the vector is traversed. The $i$-th component of vector $w$ represents the number of walks of length $i$ that will be simulated. For each component of $w$ cell, a series of

---

**Algorithm 2:** Symbolic graph embedding.

**Data:** A graph $G = (N, E)$
**Parameters :** Number of walk samples $\nu$, sampling distribution $\eta$, pattern
finder $r$, embedding dimensionality $d$, starting node $n_i$
**Result:** Symbolic node embedding $M$

1  $\tau :=$ generateSamplingVector($\eta, \nu$);
2  $\mathfrak{T} \leftarrow$ multiset;
3  **for** $o \in \tau$ **do**
4  $\quad \alpha \leftarrow o$'s index ;                        ▷ Walk length.
5  $\quad \mathcal{D} \leftarrow \{\}$;
6  $\quad$ **for** $k \in [1 \ldots o]$ **do**
7  $\quad\quad \mid \mathcal{D} \leftarrow \mathcal{D} \cup$ Walk($G, n_i, \alpha$) ;        ▷ Sample with Algorithm 1.
8  $\quad$ **end**
9  $\quad \mathfrak{T} \leftarrow \mathfrak{T} \cup \mathcal{D}$ ;                      ▷ Update walk object.
10 **end**
11 $\mathcal{P} \leftarrow$ findPatterns($\mathfrak{T}, r$) ;              ▷ Find patterns.
12 $\mathcal{M} \leftarrow$ representNodes($\mathfrak{T}, \mathcal{P}, r, d$) ;        ▷ Represent nodes.
13 **return** $\mathcal{M}$;

---

random walks (lines 6–8) is simulated, which produces sequences of nodes that are used to fill a node-level walk container $\mathcal{D}$. Thus, $\mathcal{D}$, once filled, consists of $o$ sets representing individual random walks of length $\alpha$. The walks are added into a single multiset, prior to being stored into the global transaction structure $\mathfrak{T}$. Once $w$ is traversed, frequent patterns are found (line 11), where the transaction structure $\mathfrak{T}$ comprised of all node-level walks is used as the input. The findPatterns method in line 11 can be any method that takes a transaction database as input, the considered ones are discussed in the following section. The top most frequent $d$ patterns are used as features, and represent the columns (dimensions) of the final representation $\mathcal{M}$. Here, the representNodes method (line 12) fills the values according to the considered weighting scheme (part of $r$)[1].

## 4    Computational and Spatial Complexity

In this section we discuss the computational aspects of the proposed approach. We split this section into two main parts, where we first discuss the complexity of the sampling, followed by the pattern mining part.

The time complexity of the proposed sampling strategy depends on the number of simulated walks and the walk lengths. The complexity of a single walk is linear with respect to the length of the walk. If we define the average walk length as $\bar{l}$, and the number of all samples as $\nu$, the spatial complexity, required to store all walks amounts to $\mathcal{O}(|N| \cdot \nu \cdot \bar{l})$. As the complexity of a single random walk is linear with respect to the length of the walk, the considered sampling's time complexity amounts to $\mathcal{O}(\nu \cdot \bar{l})$ for a single node. The proposed approach is also linear with respect to the number of nodes both in space and time.

The computational complexity of pattern mining varies based on the algorithm used for this step. The considered FP-Growth's complexity is linear with respect to the number of transactions, whereas its spatial complexity is, due to efficient counting employed, similarly efficient and does not explode as for example with the Apriori family of algorithms. The result of the pattern mining step is a $|N| \times d$ matrix, where $d$ is the number of patterns considered as features. Compared to e.g., metapath2vec and other shallow graph embedding methods, which yield a dense matrix, this matrix is *sparse*, and potentially requires orders of magnitude less space for the same $d$[2]. As storing large dense matrices can be spatially demanding, the proposed sparse feature representation requires less space, especially if high-dimensional embeddings are considered (the black-box methods commonly yield dense representation matrices). The difference arises especially for very large datasets, where dense node representations can become a spatial bottleneck. We observe that $\approx 10\%$ of elements are non-zero, indicating that storing the feature space as a sparse matrix results in smaller time complexity. Worst case spatial complexity of storing the embedding, however, is for both types of methods $\mathcal{O}(|N| \cdot d)$.

---

[1] Note that this method takes as input random walk samples for *all* nodes.

[2] In practice, however, larger dimensions are needed to represent the set of nodes well by using symbolic representations.

## 5    Empirical Evaluation

In this section we present the evaluation setting, where we demonstrate the performance of the proposed Symbolic Graph Embedding approach. We follow closely the evaluation introduced by metapath2vec, where the representation is first obtained, and next used for the classification task, where logistic regression is used as a classifier of choice. We test the performance on a heterogeneous information graph, comprised of authors, papers and venues[3]. The task is to classify venues into one of eight possible topics. The dataset was first used for evaluation of metapath2vec, hence we refer to the original results when comparing with the proposed approach. The considered graph consists of 2,766,148 nodes and 2,503,628 edges, where the 133 venues are to be classified into correct classes. We compare SGE against previously reported performances [7] of DeepWalk [23], LINE [29], PTE [28], metapath2vec and metapath2vec++ [7]. All methods are considered state-of-the-art for black-box node representation learning. The PTE and two variations of metapath2vec can take into account different (typed) paths during sampling.

We tested the following SGE variants. For pattern learning, we varied the TF-IDF, BoW and TF-, as well as the FP-Growth methods. The parameter search space used to obtain the results was as follows. The number of features = [500,1000,1500,2000,3000], considered vectorizers = ["TF-IDF","TF","FP-growth","Binary"], relation order (relevant for TF-based vectorizers—the highest $k$-gram order considered) = [2, 3, 4], walks of lengths = [2, 3, 5, 10], and number of walk samples = [1000, 10000] were considered. The support parameter of the FP-Growth parameter was varied in the range [3, 5, 8]. The Uniform walk length distribution was used. In addition to the proposed graph sampling (FS), a simple breadth-first search (BFS) that explores neighborhood of order two was also tested. We report the best performing learners' scores based on the type of the vectorizer and the sampling distribution. Ten repetitions of ten-fold, stratified cross validation is used, the resulting micro and macro F1 scores are averaged to obtain the final performance estimate. We report the performance of logistic regression classifier when varying the percentage of training data.

### 5.1    Results

In this section we discuss in detail the results for the node classification task. The results in Table 1 are presented in terms of micro and macro F1 scores, with respect to training set percentage. We visualize the performance of the compared representations in Fig. 2.

The first observation is that shallow node embedding methods, e.g., node2vec and LINE, do not perform as well as the best performing SGE variants (Binary with Uniform sampling). Further, we can observe that best performing SGE also outperforms metapath2vec and metapath2vec++, indicating that symbolic representations can (at least for this particular dataset) offer sufficient node

---

[3] Accessible at https://ericdongyx.github.io/metapath2vec/m2v.html.

**Table 1.** Numeric results of the proposed SGE approach compared to the state-of-the-art approaches, presented in terms of micro and macro F1 scores, with respect to training set percentage. Best performing approaches are highlighted in green.

| Method / Percentage | 10% | 20% | 30% | 40% | 50% | 60% | 70% | 80% | 90% |
|---|---|---|---|---|---|---|---|---|---|
| Macro-F1 | | | | | | | | | |
| DeepWalk/node2vec | 0.140 | 0.191 | 0.280 | 0.343 | 0.391 | 0.442 | 0.478 | 0.496 | 0.446 |
| LINE (1st+2nd) | 0.463 | 0.701 | 0.847 | 0.895 | 0.920 | 0.931 | 0.947 | 0.941 | 0.947 |
| PTE | 0.170 | 0.654 | 0.830 | 0.894 | 0.921 | 0.935 | 0.951 | 0.953 | 0.949 |
| metapath2vec | 0.525 | 0.803 | 0.897 | 0.940 | 0.953 | 0.953 | **0.970** | **0.968** | **0.967** |
| metapath2vec++ | 0.544 | 0.805 | 0.900 | **0.947** | **0.958** | **0.956** | 0.968 | 0.953 | 0.950 |
| SGE (binary + FS) | **0.815** | **0.883** | **0.918** | 0.919 | 0.922 | 0.937 | 0.942 | 0.931 | 0.950 |
| SGE (TF + FS) | 0.716 | 0.769 | 0.826 | 0.853 | 0.875 | 0.886 | 0.906 | 0.914 | 0.919 |
| SGE (TF-IDF + FS) | 0.364 | 0.114 | 0.121 | 0.03 | 0.354 | 0.353 | 0.363 | 0.148 | 0.146 |
| SGE (FP-growth + FS) | 0.523 | 0.684 | 0.712 | 0.771 | 0.785 | 0.815 | 0.801 | 0.816 | 0.838 |
| SGE (Binary + BFS) | 0.396 | 0.589 | 0.685 | 0.710 | 0.702 | 0.772 | 0.792 | 0.759 | 0.778 |
| SGE (TF + BFS) | 0.054 | 0.058 | 0.070 | 0.087 | 0.091 | 0.083 | 0.091 | 0.088 | 0.091 |
| SGE (TF-IDF + BFS) | 0.360 | 0.090 | 0.113 | 0.047 | 0.324 | 0.321 | 0.325 | 0.122 | 0.122 |
| SGE (FP-growth + BFS) | 0.400 | 0.547 | 0.586 | 0.591 | 0.594 | 0.646 | 0.587 | 0.609 | 0.553 |
| Micro-F1 | | | | | | | | | |
| DeepWalk/node2vec | 0.214 | 0.249 | 0.327 | 0.379 | 0.409 | 0.463 | 0.498 | 0.526 | 0.529 |
| LINE (1st+2nd) | 0.517 | 0.716 | 0.846 | 0.895 | 0.920 | 0.933 | 0.950 | 0.956 | 0.957 |
| PTE | 0.427 | 0.688 | 0.837 | 0.895 | 0.924 | 0.935 | 0.955 | 0.967 | 0.957 |
| metapath2vec | 0.598 | 0.833 | 0.901 | 0.940 | 0.952 | 0.954 | **0.973** | **0.982** | **0.986** |
| metapath2vec++ | 0.619 | 0.834 | 0.903 | **0.946** | **0.958** | **0.957** | 0.970 | 0.974 | 0.979 |
| SGE (Binary + FS) | **0.815** | **0.880** | **0.918** | 0.918 | 0.921 | 0.935 | 0.940 | 0.933 | 0.964 |
| SGE (TF + FS) | 0.718 | 0.771 | 0.824 | 0.850 | 0.872 | 0.881 | 0.900 | 0.911 | 0.921 |
| SGE (TF-IDF + FS) | 0.477 | 0.231 | 0.245 | 0.138 | 0.518 | 0.522 | 0.528 | 0.289 | 0.279 |
| SGE (FP-growth + FS) | 0.515 | 0.655 | 0.700 | 0.758 | 0.775 | 0.807 | 0.800 | 0.815 | 0.864 |
| SGE (Binary + BFS) | 0.388 | 0.557 | 0.657 | 0.692 | 0.678 | 0.750 | 0.785 | 0.759 | 0.821 |
| SGE (TF + BFS) | 0.148 | 0.150 | 0.155 | 0.168 | 0.175 | 0.167 | 0.172 | 0.159 | 0.193 |
| SGE (TF-IDF + BFS) | 0.461 | 0.195 | 0.226 | 0.144 | 0.469 | 0.467 | 0.478 | 0.252 | 0.250 |
| SGE (FP-growth + BFS) | 0.381 | 0.512 | 0.553 | 0.558 | 0.563 | 0.619 | 0.577 | 0.600 | 0.607 |

description. The best performing SGE variant was the simplest one, with simple binary features obtained via fast sampling. Here, 10,000 walks were sampled and feature matrix of dimension 3000 was considered along with up to three-gram patterns. Finally, we visualized the embeddings by projecting them to 2D using the UMAP algorithm [19]. The resulting visualization, shown in Fig. 3, shows that the obtained symbolic node embeddings maintain the class structure of the data.

## 5.2 Implementation Details and Reproducibility

In this section we discuss the details of the proposed SGE. The main part of the implementation is Python-based, where Numpy [30] and Scipy [13] libraries were used for efficient processing. The Py3plex library[4] was used to parse the heterogeneous graph used as input [26]. The final graph was returned as a MultiDiGraph object compatible with NetworkX [10]. The TF, TF-IDF and Binary vectorizers implementations from the Scikit-learn library [20] were used. As the main bottleneck we recognized the graph sampling, which we further implemented using the Numba [15] framework for production of compiled code from native Python. After re-implementing the walk sampling part in Numba, we achieved approximately

---

[4] https://github.com/SkBlaz/Py3plex.

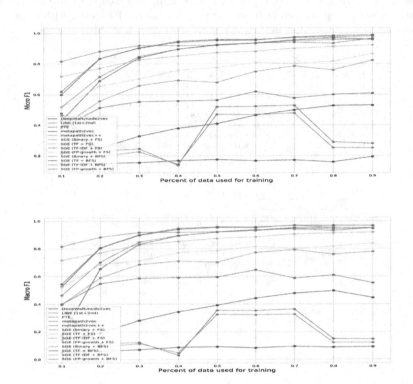

**Fig. 2.** Micro and macro F1 performance with respect to train percentages.

**Fig. 3.** UMAP projection of the best performing SGE embedding into 2D. Colors represent different types of venues (the class to be predicted). It can be observed that the obtained embeddings maintain the class-dependent structure, even though they were constructed in a completely unsupervised manner. The visualization was obtained using UMAP's default parameters.

15x speedup, which was enough to consider up to 10,000 walk samples of the large benchmark graph used in this work[5].

# 6   Discussion and Conclusions

In this work we compared the down-stream learning performance of symbolic features, obtained by sampling a given node's neighborhood, to the performance of black-box learners. Testing the approaches on the venue classification task, we find that Symbolic Graph Embedding offers similar performance on a large, real-world graph comprised of millions of nodes and edges. The proposed method outperforms the state-of-the-art shallow embeddings by up to ≈65%, and heterogeneous graph embeddings by up to ≈27% when only small percentages of the representation are used for learning (e.g., 10%). The method performs comparably to metapath2vec and metapath2vec++ when the whole embedding is considered for learning. One of the most apparent results is the well performing Binary + Uniform SGE, which indicates that simply checking the presence of relational features potentially offers enough descriptive power for successful classification. This result indicates that certain graph patterns emerge as important, where their presence or absence in a given node's neighborhood can serve as relevant for classification. The TF-IDF-based SGE variants performed the worst, indicating that more complex weighting schemes are not as applicable as in the other areas of text mining. We believe the proposed methodology could be further compared with RDF2vec and similar triplet embedding methods. The obtained symbolic embeddings were also explored qualitatively, where UMAP projection to 2D was leveraged to inspect whether the SGE symbolic node representations group according to their assigned classes. Such grouping indicates potential quality of the embedding, as venues of similar topics should be clustered together in the latent space.

# References

1. Agrawal, R., Srikant, R., et al.: Fast algorithms for mining association rules. In: Proceedings of 20th International Conference on Very Large Data Bases, VLDB, vol. 1215, pp. 487–499 (1994)
2. Bengio, Y., Courville, A., Vincent, P.: Representation learning: a review and new perspectives. IEEE Trans. Pattern Anal. Mach. Intell. **35**(8), 1798–1828 (2013)
3. Borgelt, C.: Efficient implementations of apriori and eclat. In: FIMI 2003: Proceedings of the IEEE ICDM Workshop on Frequent Itemset Mining Implementations (2003)
4. Borgelt, C.: An implementation of the FP-growth algorithm. In: Proceedings of the 1st International Workshop on Open Source Data Mining: Frequent Pattern Mining Implementations, pp. 1–5. ACM (2005)
5. Cochez, M., Ristoski, P., Ponzetto, S.P., Paulheim, H.: Global RDF vector space embeddings. ISWC 2017. LNCS, vol. 10587, pp. 190–207. Springer, Cham (2017). https://doi.org/10.1007/978-3-319-68288-4_12

---

[5] The code repository is available at https://github.com/SkBlaz/SGE.

6. Dash, T., Srinivasan, A., Vig, L., Orhobor, O.I., King, R.D.: Large-scale assessment of deep relational machines. In: Riguzzi, F., Bellodi, E., Zese, R. (eds.) ILP 2018. LNCS (LNAI), vol. 11105, pp. 22–37. Springer, Cham (2018). https://doi.org/10. 1007/978-3-319-99960-9_2

7. Dong, Y., Chawla, N.V., Swami, A.: metapath2vec: scalable representation learning for heterogeneous networks. In: Proceedings of the 23rd ACM SIGKDD International Conference on Knowledge Discovery and Data Mining, pp. 135–144. ACM (2017)

8. França, M.V., Zaverucha, G., Garcez, A.S.D.: Fast relational learning using bottom clause propositionalization with artificial neural networks. Mach. Learn. **94**(1), 81–104 (2014)

9. Grover, A., Leskovec, J.: node2vec: Scalable feature learning for networks. In: Proceedings of the 22nd ACM SIGKDD International Conference on Knowledge Discovery and Data Mining, pp. 855–864. ACM (2016)

10. Hagberg, A., Swart, P., Chult, D.S.: Exploring network structure, dynamics, and function using NetworkX. In: Proceedings of the 7th Python in Science Conference (SciPy), January 2008

11. Han, J., Cheng, H., Xin, D., Yan, X.: Frequent pattern mining: current status and future directions. Data Min. Knowl. Discov. **15**(1), 55–86 (2007)

12. Inokuchi, A., Washio, T., Motoda, H.: An apriori-based algorithm for mining frequent substructures from graph data. In: Zighed, D.A., Komorowski, J., Żytkow, J. (eds.) PKDD 2000. LNCS (LNAI), vol. 1910, pp. 13–23. Springer, Heidelberg (2000). https://doi.org/10.1007/3-540-45372-5_2

13. Jones, E., Oliphant, T., Peterson, P., et al.: SciPy: Open source scientific tools for Python (2001). http://www.scipy.org/

14. Kralj, J., Robnik-Šikonja, M., Lavrač, N.: HINMINE: heterogeneous information network mining with information retrieval heuristics. J. Intell. Inf. Syst. **50**(1), 29–61 (2018)

15. Lam, S.K., Pitrou, A., Seibert, S.: Numba: A LLVM-based python JIT compiler. In: Proceedings of the Second Workshop on the LLVM Compiler Infrastructure in HPC, p. 7. ACM (2015)

16. Lavrač, N., Džeroski, S.: Inductive Logic Programming: Techniques and Applications. Ellis Horwood, New York (1994)

17. Leskovec, J., Faloutsos, C.: Sampling from large graphs. In: Proceedings of the 12th ACM SIGKDD International Conference on Knowledge Discovery and Data Mining, pp. 631–636. ACM (2006)

18. Maiya, A.S., Berger-Wolf, T.Y.: Sampling community structure. In: Proceedings of the 19th International Conference on World Wide Web, pp. 701–710. ACM (2010)

19. McInnes, L., Healy, J., Saul, N., Grossberger, L.: UMAP: uniform manifold approximation and projection. J. Open Source Softw. **3**(29), 861 (2018)

20. Pedregosa, F., et al.: Scikit-learn: machine learning in python. J. Mach. Learn. Res. **12**(Oct), 2825–2830 (2011)

21. Perego, R., Orlando, S., Palmerini, P.: Enhancing the *Apriori* algorithm for frequent set counting. In: Kambayashi, Y., Winiwarter, W., Arikawa, M. (eds.) DaWaK 2001. LNCS, vol. 2114, pp. 71–82. Springer, Heidelberg (2001). https://doi.org/10.1007/3-540-44801-2_8

22. Perovšek, M., Vavpetič, A., Kranjc, J., Cestnik, B., Lavrač, N.: Wordification: propositionalization by unfolding relational data into bags of words. Expert Syst. Appl. **42**(17–18), 6442–6456 (2015)

23. Perozzi, B., Al-Rfou, R., Skiena, S.: Deepwalk: online learning of social representations. In: Proceedings of the 20th ACM SIGKDD International Conference on Knowledge Discovery and Data Mining, pp. 701–710. ACM (2014)
24. Ristoski, P., Paulheim, H.: RDF2Vec: RDF graph embeddings for data mining. In: Groth, P., et al. (eds.) ISWC 2016. LNCS, vol. 9981, pp. 498–514. Springer, Cham (2016). https://doi.org/10.1007/978-3-319-46523-4_30
25. Shi, C., Hu, B., Zhao, W.X., Philip, S.Y.: Heterogeneous information network embedding for recommendation. IEEE Trans. Knowl. Data Eng. **31**(2), 357–370 (2018)
26. Škrlj, B., Kralj, J., Lavrač, N.: Py3plex: a library for scalable multilayer network analysis and visualization. In: Aiello, L.M., Cherifi, C., Cherifi, H., Lambiotte, R., Lió, P., Rocha, L.M. (eds.) COMPLEX NETWORKS 2018. SCI, vol. 812, pp. 757–768. Springer, Cham (2019). https://doi.org/10.1007/978-3-030-05411-3_60
27. Srinivasan, A.: The Aleph Manual (2001)
28. Tang, J., Qu, M., Mei, Q.: Pte: predictive text embedding through large-scale heterogeneous text networks. In: Proceedings of the 21th ACM SIGKDD International Conference on Knowledge Discovery and Data Mining, pp. 1165–1174. ACM (2015)
29. Tang, J., Qu, M., Wang, M., Zhang, M., Yan, J., Mei, Q.: Line: large-scale information network embedding. In: Proceedings of the 24th International Conference on World Wide Web, pp. 1067–1077. International World Wide Web Conferences Steering Committee (2015)
30. Walt, S.V.D., Colbert, S.C., Varoquaux, G.: The NumPy array: a structure for efficient numerical computation. Comput. Sci. Eng. **13**(2), 22–30 (2011)
31. Zhang, Y., Jin, R., Zhou, Z.H.: Understanding bag-of-words model: a statistical framework. Int. J. Mach. Learn. Cybern. **1**(1–4), 43–52 (2010)

# Feature Importance

Feature Departure

# Feature Selection for Analogy-Based Learning to Rank

Mohsen Ahmadi Fahandar[✉] and Eyke Hüllermeier

Heinz Nixdorf Institute and Department of Computer Science,
Intelligent Systems and Machine Learning Group, Paderborn University,
Paderborn, Germany
ahmadim@mail.upb.de, eyke@upb.de

**Abstract.** Learning to rank based on principles of analogical reasoning has recently been proposed as a novel method in the realm of preference learning. Roughly speaking, the method proceeds from a regularity assumption as follows: Given objects $A$, $B$, $C$, $D$, if $A$ relates to $B$ as $C$ relates to $D$, and $A$ is preferred to $B$, then $C$ is presumably preferred to $D$. This assumption is formalized in terms of so-called analogical proportions, which operate on a feature representation of the objects. Consequently, a suitable feature representation is an important prerequisite for the success of analogy-based learning to rank. In this paper, we therefore address the problem of feature selection and adapt common feature selection techniques, including forward selection, correlation-based filter techniques, as well as Relief-based methods, to the case of analogical learning. The usefulness of these approaches is shown in experiments with synthetic and benchmark data.

**Keywords:** Feature selection · Leaning to rank · Analogical reasoning

## 1 Introduction

The idea of applying principles of analogical reasoning in preference learning [6] has recently attracted increasing attention in the machine learning literature [1,2,4]. For example, a method for so-called *object ranking* is proposed in [1]. This task consists of learning a ranking function that accepts any set of objects (typically though not necessarily represented as feature vectors) as input, and predicts a ranking in the form of a total strict order as output. To this end, the authors invoke the following inference pattern: If object $A$ relates to $B$ as $C$ relates to $D$, and $A$ is preferred to $B$, then $C$ is presumably preferred to $D$.

As an illustration, Fig. 1 shows objects represented as points in $\mathbb{R}^2$. In the left panel, the relationship between $A$ and $B$ is roughly the same as the relationship between $C$ and $D$, in case "relationship" is understood in the (geometric) sense of "relative location". Therefore, if $A$ is known to be preferred to $B$, we may suspect that $C$ is preferred to $D$ (suggesting that decreasing the value of $x_1$ and increasing the value of $y_1$ has a positive influence on preference).

© Springer Nature Switzerland AG 2019
P. Kralj Novak et al. (Eds.): DS 2019, LNAI 11828, pp. 279–289, 2019.
https://doi.org/10.1007/978-3-030-33778-0_22

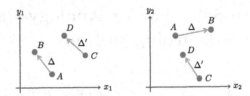

**Fig. 1.** Geometric illustration of analogical relationships: In the left picture, the four objects are in analogical relationship, because the "delta" between $A$ and $B$ is roughly the same as between $C$ and $D$. In the right picture, this is not the case.

Obviously, preferences will not always satisfy this form of analogical relationship (e.g., right panel of Fig. 1). Nevertheless, even if they do so only approximately, or the number of quadruples $(A, B, C, D)$ violating the relationship is sufficiently small, good overall predictions may still be produced, for example via averaging effects. In any case, as already suggested by Fig. 1, the feature representation of objects may have an important influence on how well the analogy assumption applies. Therefore, prior to applying analogical reasoning methods, it could be useful to embed the objects in a suitable space, so that the assumption of the above inference pattern holds true in that space. This is comparable, for example, to embedding objects in $\mathbb{R}^d$ such that the nearest neighbor rule with Euclidean distance yields good predictions in a classification task.

In this paper, by addressing the problem of *feature selection* in analogical inference, we make a first step in that direction. Feature selection, which has been studied quite thoroughly in machine learning [8], can be seen as a specific type of embedding, namely a projection of the data from the original feature space to a subspace. By ignoring irrelevant or noisy features and restricting to the most relevant dimensions, the performance can often be improved.

In the next section, we recall the problem of object ranking and the analogy-based approach to this problem put forward in [1]. In Sect. 3, we address the task of feature selection and adapt existing techniques to the case of analogy-based learning to rank. An empirical evaluation of the proposed methods is presented in Sect. 4, prior to concluding the paper in Sect. 5.

## 2    Analogy-Based Learning to Rank

Consider a reference set of objects $\mathcal{X}$, and assume each object $\boldsymbol{x} \in \mathcal{X}$ to be described in terms of a feature vector; thus, an object is a vector $\boldsymbol{x} = (x_1, \ldots, x_d) \in \mathbb{R}^d$ and $\mathcal{X} \subseteq \mathbb{R}^d$. The goal in *object ranking* is to learn a ranking function $\rho$ that accepts any (query) subset $Q = \{\boldsymbol{x}_1, \ldots, \boldsymbol{x}_n\} \subseteq \mathcal{X}$ of $n = |Q|$ objects as input. As output, the function produces a ranking $\pi$ represented by a bijection $\{1, \ldots, n\} \longrightarrow \{1, \ldots, n\}$, such that $\pi(k)$ is the position of the $k^{th}$ object $\boldsymbol{x}_k$. The ranking function $\rho$ is learned on a set of training data $\mathcal{D} = \{(Q_1, \pi_1), \ldots, (Q_M, \pi_M)\}$, where each ranking $\pi_\ell$ defines a total order of the set of objects $Q_\ell$. The predicted ranking is evaluated in terms of a suitable loss function or performance metric. A common choice is the *ranking loss*,

$d_{RL}(\pi, \hat{\pi})$, which is the number of incorrectly ranked pairs $(i, j)$, i.e., such that $\pi(i) < \pi(j)$ but $\hat{\pi}(i) > \hat{\pi}(j)$, divided by the number $n(n-1)/2$ of all pairs.

A new approach to object ranking was recently proposed on the basis of analogical reasoning [1]. This approach essentially builds on the following inference pattern: If object $a$ relates to object $b$ as $c$ relates to $d$, and knowing that $a$ is preferred to $b$, we (hypothetically) infer that $c$ is preferred to $d$. This principle is formalized using the concept of analogical proportion [13].

Consider four values $a, b, c, d$ from an attribute domain $\mathbb{X}$. The quadruple $(a, b, c, d)$ is said to be in analogical proportion, denoted by $a : b :: c : d$, if "$a$ relates to $b$ as $c$ relates to $d$". More specifically, the quadruple can be in analogy to a certain degree, denoted $v(a, b, c, d)$, which can be expressed as $E(\mathcal{R}(a, b), \mathcal{R}(c, d))$. Here, the relation $E$ denotes the "as" part of the above description, and $\mathcal{R}$ can be instantiated in different ways, depending on the underlying domain $\mathbb{X}$. An example is the arithmetic proportion

$$v(a, b, c, d) = \begin{cases} 1 - |(a - b) - (c - d)|, & \text{if } (a - b)(c - d) > 0, \\ 0, & \text{otherwise}, \end{cases} \tag{1}$$

on $\mathbb{X} = [0, 1]$. Finally, an analogical proportion on complete feature vectors, $v(\boldsymbol{a}, \boldsymbol{b}, \boldsymbol{c}, \boldsymbol{d})$, can be obtained by averaging the proportions $v(a_i, b_i, c_i, d_i)$ on the individual components.

With a measure of analogical proportion at hand, the object ranking task is tackled as follows: Consider any pair of query objects $\boldsymbol{x}_i, \boldsymbol{x}_j \in Q$. Every preference $\boldsymbol{z} \succ \boldsymbol{z}'$ (i.e., $\boldsymbol{z}$ is preferred to $\boldsymbol{z}'$) observed in the training data $\mathcal{D}$, such that $(\boldsymbol{z}, \boldsymbol{z}', \boldsymbol{x}_i, \boldsymbol{x}_j)$ are in analogical proportion, suggests that $\boldsymbol{x}_i \succ \boldsymbol{x}_j$. This principle is referred as *analogical transfer* of preferences, because the observed preference for $\boldsymbol{z}, \boldsymbol{z}'$ is (hypothetically) transferred to $\boldsymbol{x}_i, \boldsymbol{x}_j$. Accumulating all pieces of evidence that can be collected in favor of $\boldsymbol{x}_i \succ \boldsymbol{x}_j$ and, vice versa, the opposite preference $\boldsymbol{x}_j \succ \boldsymbol{x}_i$, an overall degree $p_{i,j}$ is derived for this pair of objects. The same is done for all other pairs in the query. Eventually, all these degrees are combined into an overall consensus ranking by using a suitable rank aggregation technique; see e.g., [3]. We refer to [1] for a detailed description of this method, which is called "analogy-based **learning to rank**" (able2rank).

## 3   Feature Selection

In this paper, we adapt common feature selection techniques for the task of learning to rank based on analogical reasoning, including filter methods based on correlation, forward selection as a well-known wrapper approach, and Relief-based algorithms. More specifically, we propose a strategy that can be seen as a combination of a filter and a wrapper approach:

(S1) Starting from $d$ original features $X_1, \ldots, X_d$, we first evaluate each feature $X_i$ in terms of a score $e(X_i)$, and sort all features in decreasing order of their evaluation. The result is a permutation $\sigma$ of $\{1, \ldots, d\}$, where $\sigma(k)$ is the index of the $k^{th}$ best feature.

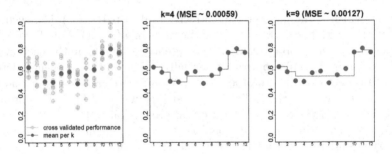

**Fig. 2.** Illustration of the procedure for determining the optimal number of features: Fit of an isotonic function for $k = k^*$ (middle) and for a suboptimal $k$ (right). X-axis shows $k$ values, and the y-axis is the rank loss estimations.

(S2) As candidates for feature subsets, we then only consider the "top-$k$" subsets $S_k = \{X_{\sigma(1)}, \ldots, X_{\sigma(k)}\}$, $k \in \{1, \ldots, d\}$, instead of the exponential number of all subsets. For each candidate $S_k$, we determine the ranking performance in a wrapper mode, i.e., by running the learning-to-rank algorithm with this subset of features, and finally pick the subset $S_{k^*}$ with the best performance.

Both the evaluation of individual features in step S1 and the evaluation of feature subsets in step S2 are done on the training data. In particular, to get an unbiased estimate of the performance, step S2 is done in a (nested) cross-validation mode, i.e., by splitting the training data again into two parts, running the algorithm with feature subset $S_k$ on the first part and estimating performance on the second part. Since different results will be obtained for different splits, the evaluation of a candidate subset $S_k$ is "noisy" and consists of $r$ performance degrees (estimations of the rank loss) $\hat{p}_{k,1}, \ldots, \hat{p}_{k,r}$, with $r$ the number of repetitions (which we assume to be fixed). The question, then, is how to choose the optimal subset $S_{k^*}$, or equivalently, the optimal number of features $k^*$. This problem can be seen as a specific type of *noisy function optimization*.

The simplest approach is to aggregate the losses $\hat{p}_{k,1}, \ldots, \hat{p}_{k,r}$ into a single estimate $\bar{p}_k$, for example the mean or the median, and determine $k^*$ on the basis of these estimates: $k^* = \operatorname{argmin}_{k \in \{1,\ldots,d\}} \bar{p}_k$. Yet, to increase the robustness of the selection, and to benefit from an averaging effect over the estimates for different values $k$, we choose $k^*$ based on how well a specific $k$ complies with the hypothesis of being the true optimum (i.e., that $k = k^*$). To this end, assuming that the true loss $p_k$ is minimal for $k$ and increases toward both sides (referred to as "umbrella ordering" in [7]), we fit a performance curve with these properties using isotonic regression.

More specifically, as proposed in [17], we find estimates $\hat{\bar{p}}_1, \ldots, \hat{\bar{p}}_d$ minimizing

$$s_k = \sum_{i=1}^{d} (\hat{\bar{p}}_i - \bar{p}_i)^2 ,$$

under the constraint $\hat{p}_1 \geq \cdots \geq \hat{p}_{k-1} \geq \hat{p}_k \leq \hat{p}_{k+1} \leq \cdots \leq \hat{p}_d$. The sum of squared deviations, $s_k$, can be seen as a measure of how well the data supports the hypothesis $k = k^*$ (see Fig. 2 for an illustration). Correspondingly, we obtain $s_k$ for all $k \in \{1, \ldots, d\}$ and eventually select $k^* = \mathrm{argmin}_{k \in \{1, \ldots, d\}} s_k$.

In what follows, we outline different methods we used to realize step S1 of our approach, i.e., to evaluate individual features $X_i$ and rank them from (presumably) most to least informative.

## 3.1  Correlation-Based Selection

The intuition behind the correlation-based approach is to assess the merit of each feature on the basis of the information it provides about the target variable. In the context of analogy, we propose the following measure, which is inspired by measures of correlation as commonly used in regression and classification.

Consider objects (feature vectors) $a, b, c, d$ in the training data, such that $a : b :: c : d$ and $a_i : b_i :: c_i : d_i$. Thus, since the objects as a whole are in analogical proportion, as well as the values for the $i^{th}$ feature $X_i$, one can say that the feature contributes to the analogy. Now, we consider the analogy as a positive example if the preferences on both sides are coherent, i.e., either $a \succ b$ and $c \succ d$, or $b \succ a$ and $d \succ c$; otherwise, the analogy is a negative example. To evaluate the attribute $X_i$, we count the number of positive examples (i.e., quadruples $a, b, c, d$ with the above properties) and negative examples in the training data, and take the difference between these counts.

## 3.2  Forward Selection

Sequential forward selection is a well-known wrapper method. Starting from the empty feature set $S = \emptyset$, it successively adds the feature that leads to the highest improvement. More specifically, forward selection tentatively adds each feature $X_j \notin S$, runs the learning algorithm with $S^{(j)} = S \cup \{X_j\}$, and then adopts the $S^{(j)}$ with the best performance.

Normally, forward selection has a stopping condition and terminates with the current feature subset $S$ if a further expansion does not seem to improve performance. Recall, however, that we are using this method for the purpose of feature ranking instead of selection, i.e., to realize step S1 of our general approach. Therefore, we run forward selection until all features are consumed. Obviously, this may become quite costly, since the total number of subsets to be evaluated is as large as $\frac{d(d+1)}{2}$.

## 3.3  Relief-Based Algorithms

The Relief family of algorithms is among the most successful approaches for feature weighting and selection. The original Relief algorithm, proposed by Kira and Rendell [10,11], is inspired by instance-based learning for classification [9]: It seeks to find a weighting of features such that nearest neighbor classification with weighted Euclidean distance leads to strong performance. To this end, the algorithm estimates feature weights iteratively. In each iteration, a *target instance* $x = (x_1, \ldots, x_d)$ is selected at random, and two nearest neighbors of this instance are found: a first one from the same class, termed near-hit (NH), and a second one from a different class, called near-miss (NM). The weight $w_i$ of the $i^{th}$ feature $X_i$ is then updated by $w_i - |x_i - \mathrm{NH}_i| + |x_i - \mathrm{NM}_i|$, where $\mathrm{NH}_i$ is the value of $X_i$ in the near-hit instance. After $m$ iterations, the weight vector $w = (w_1, \ldots, w_d)$ defines a relevance score for each feature.

Since then, several modifications and extensions of Relief have been proposed; we refer the reader to [18] for a thorough review of Relief-based algorithms. The most commonly used variant today is Relief-F [12], which enhances the original algorithm mainly in three aspects: it considers the $k$ nearest neighbors in the update rule (instead of only a single neighbor), it handles missing (incomplete) data, and can be applied to multi-class problems.

Another interesting variant is *Iterative Relief* (I-Relief). It overcomes certain difficulties of Relief with handling non-monotonic features [5] and, moreover, addresses two other important flaws of the original version: The nearest neighbors in the original feature space might not be the nearest neighbors in the weighted feature space, and the computation of feature weights is quite sensitive toward outliers in the data [15,16]. In I-Relief, all instances contribute to the weight update (to some degree, which is controlled by a sigmoidal function of the distance to the target instance). Moreover, as the name suggests, the algorithm is iterative and, in each iteration, computes distances between instances based on the weights from the previous iteration. The algorithm runs until the weight update is negligible or some other stopping condition is met.

We consider adaptations of Relief-F [12] and I-Relief [15,16] for feature selection in the context of analogy-based learning to rank, focusing on the analogical proportion (1), which is a map $v : [0,1]^4 \longrightarrow [0,1]$. To this end, we establish a connection between this problem and the problem of binary classification: Suppose that the (latent) utility of an object $x \in \mathcal{X}$ can be expressed in terms of a linear function $w^\top x$ specified by a weight vector $w$. A preference $x_i \succ x_j$ is then equivalent to $w^\top x_i > w^\top x_j$, and hence to $w^\top (x_i - x_j) > 0$. From the point of view of a linear classifier, which produces binary predictions (positive or negative) by thresholding scores at 0, $z_{i,j} = x_i - x_j$ can be seen as a positive example (with class $c_{i,j} = +1$) and $z_{j,i} = x_j - x_i$ as a negative example (with class $c_{j,i} = -1$). Thus, solving the learning-to-rank problem in the sense of learning the weight vector $w = (w_1, \ldots, w_d)$ could principally be reduced to training a linear classifier on positive and negative examples of that kind. The training data for this classifier, $\mathcal{D}_{bin}$, can be extracted from the original preference data $\mathcal{D}$ as described above.

With this view of learning to rank as binary classification, and the weight $w_i$ as a degree of relevance of the feature $X_i$, Relief-based algorithms can be applied on the data $\mathcal{D}_{bin}$ in a more or less straightforward way. Our main modification concerns the similarity function used for finding the nearest neighbors of the target instance in the transformed feature space, which we reasonably define as a function of our measure of analogical proportion: Given transformed feature vectors $z_1 = (z_{1,1}, \ldots, z_{1,d}), z_2 = (z_{2,1}, \ldots, z_{2,d}) \in [-1,1]^d$, their *weighted similarity* is defined as

$$\mathrm{sim}_w(z_1, z_2) = \frac{1}{d} \sum_{i=1}^{d} w_i \, v(z_{1,i}, z_{2,i}), \qquad (2)$$

where $v$ measures the analogical proportion between $z_1$ and $z_2$ (or, more precisely, the analogical proportion of the objects from which they have been derived):

$$v(a, b) = \begin{cases} 1 - |a - b|, & \text{if } ab > 0, \\ 0, & \text{otherwise.} \end{cases}$$

The weight vector in the case of Relief-F is simply the all-one vector. The input of our adapted Relief-F algorithm comprises the data $\mathcal{D}_{bin}$ and the number $K$ of nearest neighbors. What the algorithm is doing on the original preference data $\mathcal{D}$ can roughly be described as follows: Given objects $a, b, c, d$ with a high degree of analogy, and such that a feature $X_i$ contributes to this analogy, the weight of $X_i$ is increased if the preferences are coherent ($a \succ b$ and $c \succ d$, or $b \succ a$ and $d \succ c$), and decreased otherwise. Eventually, the algorithm will return a weight vector representing the merit of each feature for the task of analogy-based learning to rank. Again, these weights can be used to determine a ranking $\sigma$ of the features as required for step S2 of our approach[1].

Using I-Relief with the modified similarity measure (2), the weight vector $w$ derived in each iteration is used for computing the similarity in the subsequent iteration. In addition to the similarity function, we slightly modified the objective function to account for similarity instead of distance. Our adapted I-Relief algorithm accepts as input the data set $\mathcal{D}_{bin}$ and a kernel width $\gamma$ together with a stopping parameter $\theta$. As stated in [15,16], the convergence rate of I-Relief is fully controlled by the choice of $\gamma$. Once convergence is reached, the weight vector $w$ representing the score of each feature will be returned.

## 4   Experiments

We implemented the able2rank algorithm (with slight modification: (1) is used to account for our similarity measure (2)), including the pre-processing and normalization of the data, as described in [2]. The implementation of the feature selection methods follows the description in the previous section. For I-Relief, the stopping parameter was set to $\theta = 10^{-8}$, and the algorithm was run for kernel width $\gamma \in \{0.1, 0.3, 0.5, 1, 3, 5\}$. The final feature weights were then obtained by averaging the resulting feature weights per attribute. Likewise, in Relief-F, feature weights were averaged over parameters $K \in \{1, 5, 10, 15\}$. The performances of candidate subsets $S_k$ in step S2 of our approach (cf. Sect. 3) were estimated on the basis of a two-fold cross validation (repeated 5 times) on the training data.

We used the same data sets as in [1]: Bundesliga (B), Decathlon (D), Footballers (F), Hotels (H), University Rankings (U), Volleyball (V). In addition, we

---

[1] Actually, we do not produce a ranking of all features, but include only those features whose scores are positive.

**Table 1.** Results (real data) in terms of loss $d_{RL}$ on the test data.

| $D_{train} \rightarrow D_{test}$ | Full | Forward | Relief-F | I-Relief | Correlation |
|---|---|---|---|---|---|
| B1 → B2 | .031 ± .009(5) | .017 ± .006(4) | .012 ± .006(1) | .012 ± .006(1) | .012 ± .006(1) |
| B2 → B1 | .013 ± .000(1) | .013 ± .000(1) | .013 ± .000(1) | .014 ± .007(4) | .014 ± .007(4) |
| Average | .022 (3.00) | .015 (2.50) | .012 (1.00) | .013 (2.50) | .013 (2.50) |
| D1 → D2 | .063 ± .000(1) | .169 ± .001(5) | .063 ± .000(1) | .063 ± .000(1) | .063 ± .000(1) |
| D1 → D3 | .079 ± .003(1) | .139 ± .008(5) | .079 ± .003(1) | .079 ± .003(1) | .079 ± .003(1) |
| D1 → D4 | .111 ± .008(1) | .154 ± .006(5) | .111 ± .008(1) | .111 ± .008(1) | .111 ± .008(1) |
| D2 → D1 | .091 ± .001(1) | .150 ± .001(5) | .103 ± .001(4) | .091 ± .001(1) | .091 ± .001(1) |
| D2 → D3 | .083 ± .006(2) | .096 ± .006(5) | .055 ± .006(1) | .083 ± .006(2) | .083 ± .006(2) |
| D2 → D4 | .151 ± .004(2) | .186 ± .007(5) | .125 ± .007(1) | .151 ± .004(2) | .151 ± .004(2) |
| D3 → D1 | .140 ± .001(1) | .213 ± .001(4) | .227 ± .001(5) | .140 ± .001(1) | .197 ± .001(3) |
| D3 → D2 | .117 ± .001(1) | .165 ± .001(4) | .167 ± .001(5) | .117 ± .001(1) | .162 ± .000(3) |
| D3 → D4 | .165 ± .004(2) | .174 ± .005(4) | .163 ± .007(1) | .165 ± .004(2) | .179 ± .008(5) |
| D4 → D1 | .164 ± .001(1) | .191 ± .001(5) | .175 ± .001(4) | .173 ± .001(3) | .164 ± .001(1) |
| D4 → D2 | .156 ± .001(1) | .180 ± .001(4) | .194 ± .001(5) | .167 ± .001(3) | .156 ± .001(1) |
| D4 → D3 | .122 ± .010(2) | .117 ± .007(1) | .128 ± .005(4) | .147 ± .008(5) | .122 ± .010(2) |
| Average | .120 (1.33) | .161 (4.33) | .133 (2.75) | .124 (1.92) | .130 (1.92) |
| F3 → F4 | .162 ± .002(4) | .156 ± .002(3) | .139 ± .002(1) | .166 ± .003(5) | .143 ± .002(2) |
| F4 → F3 | .156 ± .002(3) | .154 ± .002(2) | .165 ± .002(5) | .156 ± .002(3) | .152 ± .003(1) |
| Average | .159 (3.50) | .155 (2.50) | .152 (3.00) | .161 (4.00) | .147 (1.50) |
| FB1 → FB2 | .043 ± .001(5) | .031 ± .001(4) | .011 ± .001(1) | .014 ± .001(2) | .014 ± .001(2) |
| FB2 → FB1 | .050 ± .001(4) | .052 ± .001(5) | .013 ± .001(3) | .012 ± .001(1) | .012 ± .001(1) |
| Average | .046 (4.50) | .041 (4.50) | .012 (2.00) | .013 (1.50) | .013 (1.50) |
| G1 → G2 | .073 ± .006(1) | .094 ± .015(5) | .083 ± .011(2) | .083 ± .011(2) | .083 ± .011(2) |
| G1 → G3 | .019 ± .003(1) | .023 ± .005(5) | .019 ± .006(1) | .019 ± .006(1) | .019 ± .006(1) |
| G2 → G1 | .047 ± .007(3) | .034 ± .008(1) | .043 ± .007(2) | .047 ± .007(3) | .047 ± .007(3) |
| G2 → G3 | .029 ± .002(2) | .023 ± .004(1) | .029 ± .002(2) | .029 ± .002(2) | .029 ± .002(2) |
| G3 → G1 | .048 ± .006(5) | .038 ± .006(1) | .041 ± .010(2) | .041 ± .010(2) | .041 ± .010(2) |
| G3 → G2 | .078 ± .006(2) | .074 ± .003(1) | .084 ± .016(3) | .084 ± .016(3) | .084 ± .016(3) |
| Average | .049 (2.33) | .048 (2.33) | .050 (2.00) | .050 (2.17) | .050 (2.17) |
| H1 → H2 | .060 ± .000(1) | .062 ± .000(3) | .065 ± .000(4) | .061 ± .000(2) | .067 ± .000(5) |
| H2 → H1 | .101 ± .001(5) | .100 ± .001(3) | .096 ± .001(2) | .100 ± .001(3) | .095 ± .000(1) |
| Average | .080 (3.00) | .081 (3.00) | .080 (3.00) | .080 (2.50) | .081 (3.00) |
| T1 → T2 | .029 ± .004(5) | .019 ± .005(4) | .007 ± .003(1) | .007 ± .003(1) | .007 ± .003(1) |
| T2 → T1 | .066 ± .005(5) | .054 ± .004(4) | .001 ± .001(2) | .002 ± .001(3) | .000 ± .000(1) |
| Average | .048 (5.00) | .036 (4.00) | .004 (1.50) | .004 (2.00) | .004 (1.00) |
| P1 → P2 | .021 ± .002(5) | .019 ± .001(4) | .006 ± .001(1) | .006 ± .001(1) | .006 ± .001(1) |
| P2 → P1 | .018 ± .001(5) | .012 ± .002(4) | .004 ± .001(1) | .004 ± .001(1) | .004 ± .001(1) |
| Average | .019 (5.00) | .015 (4.00) | .005 (1.00) | .005 (1.00) | .005 (1.00) |
| PK1 → PK2 | .131 ± .010(1) | .147 ± .013(3) | .156 ± .016(4) | .182 ± .024(5) | .131 ± .010(1) |
| PK1 → PK3 | .172 ± .018(1) | .187 ± .023(3) | .196 ± .024(4) | .215 ± .038(5) | .172 ± .018(1) |
| PK2 → PK1 | .133 ± .009(1) | .172 ± .019(5) | .152 ± .012(4) | .146 ± .010(3) | .133 ± .009(1) |
| PK2 → PK3 | .159 ± .019(1) | .202 ± .033(5) | .177 ± .023(4) | .164 ± .025(3) | .159 ± .019(1) |
| PK3 → PK1 | .183 ± .014(1) | .195 ± .015(3) | .223 ± .022(4) | .223 ± .022(4) | .183 ± .014(1) |
| PK3 → PK2 | .168 ± .014(1) | .183 ± .015(3) | .212 ± .022(4) | .212 ± .022(4) | .168 ± .014(1) |
| Average | .158 (1.00) | .181 (3.67) | .186 (4.00) | .190 (4.00) | .158 (1.00) |
| U1 → U2 | .078 ± .001(1) | .102 ± .001(5) | .092 ± .001(4) | .078 ± .001(1) | .078 ± .001(1) |
| U2 → U1 | .058 ± .001(1) | .066 ± .002(2) | .067 ± .001(3) | .077 ± .001(4) | .077 ± .001(4) |
| Average | .068 (1.00) | .084 (3.50) | .080 (3.50) | .077 (2.50) | .077 (2.50) |
| V1 → V2 | .030 ± .000(1) | .030 ± .000(1) | .077 ± .015(4) | .030 ± .000(1) | .077 ± .015(4) |
| V2 → V1 | .015 ± .000(1) | .030 ± .000(3) | .035 ± .007(4) | .035 ± .007(4) | .015 ± .000(1) |
| Average | .022 (1.00) | .030 (2.00) | .056 (4.00) | .032 (2.50) | .046 (2.50) |

**Table 2.** Results (synthetic data) in terms of loss $d_{RL}$ on the test data.

| $D_{train} \rightarrow D_{test}$ | Full | Forward | Relief-F | I-Relief | Correlation |
|---|---|---|---|---|---|
| AS1 → AS1 | .040 ± .008(5) | .036 ± .010(4) | .010 ± .004(1) | .010 ± .004(1) | .010 ± .004(1) |
| AS2 → AS2 | .036 ± .006(5) | .028 ± .009(4) | .007 ± .002(1) | .007 ± .002(1) | .007 ± .002(1) |
| AS3 → AS3 | .048 ± .015(5) | .044 ± .013(4) | .009 ± .003(1) | .009 ± .003(1) | .009 ± .003(1) |
| AS4 → AS4 | .037 ± .005(5) | .036 ± .005(4) | .007 ± .001(1) | .007 ± .001(1) | .007 ± .001(1) |
| Average | .040 (5.00) | .036 (4.00) | .008 (1.00) | .008 (1.00) | .008 (1.00) |

include FIFA WC goals statistics (G), Facebook Metrics (FB), NBA Teams and Players (T, P), and Poker (PK). All data sets together with a description, as well as the implementation of the presented algorithms, are publicly available[2].

In our experiments, predictions were produced for certain data set $D_{test}$ of the data, using other parts $D_{train}$ as training data; an experiment of that kind is denoted by $D_{train} \rightarrow D_{test}$. The averaged ranking loss together with the standard deviation of the conducted experiments (repeated 20 times) are summarized in Table 1, where the numbers in parentheses indicate the rank of the achieved score in the respective problem. Moreover, the table shows average ranks per problem domain. As can be seen, the relative performance of the feature selection methods depends on the domain, and there is no clear winner. Yet, feature selection as such turns out to be useful most of time, in the sense that improvements over the use of all features (Full) can be achieved.

That being said, on some of the problem domains, the gains through feature selection might not be as pronounced as one may expect. As one quite obvious explanation, let us mention that most of the features in these data sets are hand-picked, and hence supposedly relevant. A good example is the decathlon data (D), in which features correspond to performances in the 10 disciplines. Unsurprisingly, feature selection does not yield any improvements on this data, simply because all the 10 features are important.

Therefore, we additionally conducted experiments with synthetic data, which allows one to examine the effectiveness of the proposed feature selection algorithms in a controlled fashion. One example is a data set we called *Answer Sheets*[3]. In addition to 3 relevant attributes, this data is corrupted by noisy (irrelevant) features of different type: 2 binary features in $\{0, 1\}$, 2 nominal in $\{1, 2, 3, 4, 5\}$, and 3 integer features in $\{1, \ldots, Q\}$. This process gives rise to data sets AS1 ($Q = 50$) and AS2 ($Q = 100$). Further, we doubled the number of noisy features (of each type) to produce data sets AS3 ($Q = 50$) and AS4 ($Q = 100$).

As can be seen in Table 2, the gains through feature selection are more significant now. Indeed, the feature selection methods are producing almost perfect results, and upon closer inspection, it turns out that they are almost always selecting the right subset of relevant features. Similar results (omitted here due to space restrictions) were obtained for other synthetic data sets.

---

[2]  https://github.com/mahmadif/able2rank.
[3]  The description is available at https://github.com/mahmadif/able2rank.

# 5    Conclusion and Future Work

This paper elaborates on the problem of feature selection for analogy-based learning to rank by adapting common feature selection techniques including forward selection as a wrapper method, filter-techniques based on correlation, and Relief-based approaches. The efficacy of these techniques are evaluated on both synthetic and real-world data sets. The experimental results show that the studied feature selection methods serve their purpose.

Going beyond the mere inclusion or exclusion of features, we plan to elaborate on feature weighting and metric learning [19] for analogical preference learning in future work, as well as general data embedding techniques. Specifically interesting appears the idea of *co-embedding* [14], that is, the idea of embedding possibly different types of objects in a common (Euclidean) space. This may provide a means for realizing *transfer learning*, where source objects $A$ and $B$ are from one domain, and target objects $C$ and $D$ from a possibly different domain.

# References

1. Ahmadi Fahandar, M., Hüllermeier, E.: Learning to rank based on analogical reasoning. In: AAAI (2018)
2. Ahmadi Fahandar, M., Hüllermeier, E.: Analogy-based preference learning with kernels. In: Benzmüller, C., Stuckenschmidt, H. (eds.) KI 2019. LNCS (LNAI), vol. 11793, pp. 34–47. Springer, Cham (2019). https://doi.org/10.1007/978-3-030-30179-8_3
3. Ahmadi Fahandar, M., Hüllermeier, E., Couso, I.: Statistical inference for incomplete ranking data: the case of rank-dependent coarsening. In: ICML (2017)
4. Bounhas, M., Pirlot, M., Prade, H.: Predicting preferences by means of analogical proportions. In: ICCBR (2018)
5. Draper, B., Kaito, C., Bins, J.: Iterative relief. In: 2003 Conference on Computer Vision and Pattern Recognition Workshop (2003)
6. Fürnkranz, J., Hüllermeier, E.: Preference Learning. Springer, Heidelberg (2011). https://doi.org/10.1007/978-3-642-14125-6
7. Geng, Z., Shi, N.Z.: Algorithm AS 257: isotonic regression for umbrella orderings. J. R. Stat. Soc. Seri. C (Appl. Stat.) **39**(3), 397–402 (1990)
8. Guyon, I., Elisseeff, A.: An introduction to variable and feature selection. JMLR **3**, 1157–1182 (2003)
9. Keogh, E.: Instance-Based Learning, pp. 549–550. Springer, Boston (2010). https://doi.org/10.1007/978-0-387-30164-8_409
10. Kira, K., Rendell, L.A.: The feature selection problem: traditional methods and a new algorithm. In: AAAI (1992)
11. Kira, K., Rendell, L.A.: A practical approach to feature selection. In: Proceedings ML-92, 9th International Workshop on Machine Learning (1992)
12. Kononenko, I.: Estimating attributes: analysis and extensions of relief. In: ECML (1994)
13. Miclet, L., Prade, H.: Handling analogical proportions in classical logic and fuzzy logics settings. In: Sossai, C., Chemello, G. (eds.) ECSQARU 2009. LNCS (LNAI), vol. 5590, pp. 638–650. Springer, Heidelberg (2009). https://doi.org/10.1007/978-3-642-02906-6_55

14. Mirzazadeh, F., Guo, Y., Schuurmans, D.: Convex co-embedding. In: AAAI (2014)
15. Sun, Y.: Iterative relief for feature weighting: algorithms, theories, and applications. IEEE TPAMI **29**(6), 1035–1051 (2007)
16. Sun, Y., Li, J.: Iterative relief for feature weighting. In: ICML (2006)
17. Turner, T., Wollan, P.: Locating a maximum using isotonic regression. Comput. Stat. Data Anal. **25**(3), 305–320 (1997)
18. Urbanowicz, R., Meeker, M., LaCava, W., Olson, R., Moore, J.: Relief-based feature selection: introduction and review. J. Biomed. Inform. **85**, 189–203 (2017)
19. Weinberger, K.Q., Saul, L.K.: Distance metric learning for large margin nearest neighbor classification. JMLR **10**, 207–244 (2009)

# Ensemble-Based Feature Ranking
# for Semi-supervised Classification

Matej Petković[1,2(✉)], Sašo Džeroski[1,2], and Dragi Kocev[2]

[1] Jožef Stefan International Postgraduate School, Ljubljana, Slovenia
[2] Department of Knowledge Technologies, Jožef Stefan Institute, Ljubljana, Slovenia
{matej.petkovic,saso.dzeroski,dragi.kocev}@ijs.si
http://kt.ijs.si

**Abstract.** In this paper, we propose three feature ranking scores (Symbolic, Genie3, and Random Forest) for the task of semi-supervised classification. In this task, there are only a few labeled examples in a dataset and many unlabeled. This is a highly relevant task, since it is increasingly easy to obtain unlabeled examples, while obtaining labeled examples is often an expensive and tedious task. Each of the proposed feature ranking scores can be computed by using any of three approaches to learning predictive clustering tree ensembles (bagging, random forests, and extra trees). We extensively evaluate the proposed scores on 8 benchmark datasets. The evaluation finds the most suitable ensemble method for each of the scores, shows that taking into account unlabeled examples improves the quality of a feature ranking, and demonstrates that the proposed feature ranking scores outperform a state-of-the-art semi-supervised feature ranking method SEFR. Finally, we identify the best performing pair of a feature ranking score and an ensemble method.

**Keywords:** Semi-supervised learning · Feature ranking · Ensembles

## 1 Introduction

A center task in machine learning is predictive modeling concerned with learning a predictive model, from a given training dataset of values of features-target pairs $(\boldsymbol{x}, y)$, where $\boldsymbol{x} = (x_1, \ldots, x_F)$. The learned predictive models can then be used to predict target values for previously unseen values of features. In this paper, we focus on the classification task, where the domain $\mathcal{Y}$ of the target $y$ is a finite set of discrete values. The task at hand is called binary classification if $|\mathcal{Y}| = 2$, and multi-class classification if $|\mathcal{Y}| > 2$. In both cases, we refer to the target values of $y$ as labels or classes.

Typically, all the examples in the training set are labeled, i.e., have a known value of $y$. In that case, we say that the learning is supervised. However, determining which class an example belongs to, might be very expensive or take too much time in some domains (e.g., compound toxicity). Hence, there is a multitude of datasets with only a handful of labeled examples and many unlabeled

© Springer Nature Switzerland AG 2019
P. Kralj Novak et al. (Eds.): DS 2019, LNAI 11828, pp. 290–305, 2019.
https://doi.org/10.1007/978-3-030-33778-0_23

ones. To address this challenge, methods that can use the unlabeled data in the learning phase have been developed [6,14]. These semi-supervised learning (SSL) methods are applicable mostly when there are only a few labeled examples and plenty of unlabeled data, which makes applying the supervised learning hard.

Another prominent task in machine learning is feature ranking where, the goal is to discover to what extent each of the features $x_i$, $1 \leq i \leq F$ is relevant for the class $y(x)$. Formally, given a dataset $\mathscr{D}_{\mathrm{TRAIN}}$, the output of a feature ranking algorithm is a list of feature importance scores $importance(x_i)$, where a higher score corresponds to a higher relevance of the feature for the target values. The task of feature ranking is typically seen as a data preprocessing step. We can perform dimensionality reduction on the data to make the learning of a predictive model faster or even at all feasible. This is done by discarding the features that have lower importance than some threshold $\vartheta \in \mathbb{R}$. Lower dimensionality also results in models that are easier to understand – this is particularly important when a data scientist collaborates with a domain expert. Lately, much work is being done in the field of explainable artificial intelligence. In the case of black box models, such as neural networks and ensembles, feature ranking is the only way to at least partially explain the obtained predictions.

A task that is related to feature ranking is feature selection. The goal of feature selection is to identity a subset of the features that yield better (or at least the same) predictive performance when used to learn predictive models (as compared to learning a predictive model on the complete feature set). Note that this is a task different from feature ranking: the former looks for the best subset, while the latter focuses on ordering the features based on their relevance for the target. Notwithstanding, as mentioned above, feature ranking can be used to perform feature selection by applying a threshold on the importance scores.

We propose a method for SSL feature ranking, i.e., learning a feature ranking for datasets with a handful of labeled examples and many unlabeled examples. The proposed method is based on the ensemble learning paradigm [16]. It uses tree-based methods for semi-supervised learning of classification trees [14].

We perform an empirical evaluation of the proposed method on 8 benchmark multi-class classification datasets. In the empirical evaluation, we set out to investigate the influence of the combination of ensemble learning methods and feature ranking scores, the number of labeled examples, and the number of unlabeled examples. Moreover, we compare the performance of the variants of the proposed method to the performance of their fully supervised counterparts as well as to the performance of another competing method for semi-supervised feature ranking based on ensemble learning [1].

The remainder of this paper is organized as follows. In Sect. 2, we review the background and the related work. Next, in Sect. 3, we describe the proposed method for feature ranking for semi-supervised classification. Furthermore, we outline the design of the empirical evaluation of the proposed method in Sect. 4 and then we discuss the results of the evaluation in Sect. 5. Finally, we conclude and provide directions for further work in Sect. 6.

# 2    Background and Related Work

## 2.1    Related Work

There are three major groups of methods that address the SSL task. The simplest option is to *discard the unlabeled data*, and then use an existing algorithm for supervised learning. However, when the number of labeled examples is really low (e.g., 10 or 20 examples), this approach has severely limited success.

The second group of methods performs *self-training* [17] where an algorithm for supervised learning is first applied to the labeled examples in $\mathscr{D}_{\text{TRAIN}}$. Next, the resulting model is used to predict the target values of the unlabeled examples, and a heuristic score is used to assess the certainty/reliability of the predictions of the models. The examples for which the algorithm is the most certain in its predictions, keep their labels and the examples are added to the current set of labeled examples. The cycle is iteratively repeated until a stopping criterion is met (e.g., no more unlabelled examples or no more reliable predictions). At the end, a model is learned on the final set of labeled examples.

The last class of SSL methods are *algorithm-adaptation* methods, where an existing algorithm, e.g., for learning decision trees [3], is adapted so that it can also take into account the unlabeled examples. In the case of decision trees, this was done by adapting the heuristic that measures the impurity of the current dataset [14], so that not only the target $y$ is taken into account, but also the features $x_i$. Under the clustering assumption, i.e., the assumption that the class values correspond to well-defined clusters of data [6], the last two approaches are expected to be superior to the first solution.

Like supervised feature ranking methods, feature ranking methods for SSL belong to three major groups [18]: *filter* methods, where no predictive model is needed for feature ranking; *embedded* methods, where feature ranking is computed directly from a predictive model; and *wrapper* methods, where a predictive model is typically retrained more than once and the ranking is built iteratively.

Filter methods are typically the fastest, but can be myopic. Namely, they do not take into consideration possible feature interactions and have limited scope, e.g., the variance score [2] is applicable to datasets with numeric features only. A representative of embedded methods is the SEFR feature ranking [1] computed from an ensemble of decision trees: this is an SSL adaptation of the random forest ranking [4], where the trees are learned in a self-training fashion. A representative of the wrapper methods is the method for recursive feature elimination with support vector machines (SVMs) [20]. At each iteration, an SVM model is trained and thus a normal $w$ is obtained. Features $x_i$ for which the absolute value of components $|w_i|$ are the smallest are removed, and the procedure is iteratively repeated until a complete ranking is obtained. For other SSL feature ranking methods (which are mostly limited to feature selection of numeric features), we refer the reader to a recent survey [18].

## 2.2  Semi-supervised Predictive Clustering Trees

The proposed feature ranking method is based on ensembles of predictive clustering trees (PCTs) for classification. The PCT framework views a decision tree as a hierarchy of clusters, which are induced with the standard top-down induction process [5] described in Algorithm 1. The root of a PCT corresponds to a cluster containing all data, which is recursively partitioned into subclusters while moving down the tree. The leaves represent the clusters at the lowest level of the tree hierarchy and each leaf is labeled with its cluster's prototype (prediction). PCTs generalize decision trees and can be used for a variety of learning tasks, including clustering tasks, different types of structured output prediction tasks [3,12], as well as SSL tasks [14]. The generalization is based on appropriately adapting the heuristic for inducing PCTs and the prototype function to the given structured output prediction task.

Algorithm 1 is used for learning PCTs. Its input is a set of examples $E \subseteq \mathscr{D}_{\text{TRAIN}}$, and its output is a tree. The heuristic $h$ that is used for selecting the best test at a node is the weighted impurity reduction of the subsets of $E$ (lines 3 and 4), induced by the tests. By maximizing it (line 5 of Algorithm 2), the algorithm is guided towards small trees with good predictive performance. If no test reduces the impurity, a leaf is created and the prototype (e.g., average) of the instances belonging to that leaf is computed.

| **Algorithm 1.** PCT($E$) | **Algorithm 2.** BestTest($E$) |
|---|---|
| 1: $(t^*, h^*, \mathcal{P}^*) = \text{BestTest}(E)$ | 1: $(t^*, h^*, \mathcal{P}^*) = (none, 0, \emptyset)$ |
| 2: **if** $t^* \neq none$ **then** | 2: **for each** test $t$ **do** |
| 3:     **for each** $E_i \in \mathcal{P}^*$ **do** | 3:     $\mathcal{P} = $ partition induced by $t$ on $E$ |
| 4:         $tree_i = \text{PCT}(E_i)$ | 4:     $h = |E|impu(E) - \sum_{E_i \in \mathcal{P}} |E_i|impu(E_i)$ |
| 5:     **return** $Node(t^*, \bigcup_i \{tree_i\})$ | 5:     **if** $h > h^*$ **then** |
| 6: **else** | 6:         $(t^*, h^*, \mathcal{P}^*) = (t, h, \mathcal{P})$ |
| 7:     **return** $Leaf(Prototype(E))$ | 7: **return** $(t^*, h^*, \mathcal{P}^*)$ |

In the standard classification scenario, the impurity $impu(E)$ of a data subset $E$ is defined as the Gini impurity of the class $y$, i.e., $Gini(E, y) = 1 - \sum_c p_E^2(c)$, where $p_E(c)$ is the relative frequency of the class value $c$ in the subset $E$. In the semi-supervised scenario, where some or most of the target values are missing, the heuristic is adapted so that also impurity of the features is taken into account. To this end, we first define the normalized version of $Gini$, namely $Gini'(E, y) = Gini(E, y)/Gini(\mathscr{D}_{\text{TRAIN}}, y)$. Analogously, we introduce the impurity measure for numeric features as the normalized variance $Var'(E, x_i) = Var(E, x_i)/Var(\mathscr{D}_{\text{TRAIN}}, x_i)$. Finally, the impurity of a set $E$ is defined as $impu(E) = w \, Gini'(E, y) + (1 - w) \cdot \frac{1}{F} \sum_{i=1}^{F} impu(E, x_i)$, where the $impu(E, x_i)$ is calculated as $Gini'(E, x_i)$ if $x_i$ is nominal and as $Var'(E, x_i)$ if $x_i$ is numeric, and $F$ is the number of features and the influence of $Gini'(E, y)$ is controlled by the parameter $w \in [0, 1]$, whose optimal value is set by internal cross-validation. When computing the relative frequencies that are used in the

*Gini'* score, only the examples with known values of the target $y$ (or feature $x_i$) are taken into account. The prediction, i.e., the prototype in a leaf of a PCT, is defined as the majority class value in the leaf.

### 2.3  Ensembles of PCTs

An ensemble is a set of base predictive models constructed with a given algorithm. The prediction for a new example $x$ is made by combining the predictions of the models from the ensemble. In classification tasks, this is typically done by voting, where the $i$-th base model either votes for the class value it predicts, or computes the probabilities $p_i(c \mid x)$, for all class values $c$. In the first case, the prediction of the ensemble is the class with the most votes. In the second case, the prediction is the class with the highest sum $\sum_i p_i(c \mid x)$.

A necessary condition for an ensemble to be more accurate than its members, is that the members are accurate and diverse models [9], i.e., that they make different errors on new examples. However, we do not use the ensembles as predictive models. Rather, we use them as a basis for computing feature ranking scores. As it is evident from the scores' definitions in Sect. 3 (Eqs. (1), (2) and (3)), one can also compute the ranking from a single tree, but the variance of feature importances decreases when the number of trees is higher. There are several ways to introduce diversity among the PCTs in an ensemble. We describe and make use of three of them.

**Bagging and Random Forests.** Instead of being learned from the original dataset $\mathscr{D}_{\text{TRAIN}}$, each tree in the bagging/random forest ensemble is built from a different bootstrap replicate $\mathcal{B}$ of the dataset $\mathscr{D}_{\text{TRAIN}}$, called bag. The examples $\mathscr{D}_{\text{TRAIN}} \setminus \mathcal{B}$ are called out-of-bag examples (OOB). Additionally, the line 2 of the *BestTest* procedure (see Algorithm 2) is modified to change the feature set during learning by introducing randomization in the test selection. More precisely, at each node in a decision tree, a random subset of the features is taken, and the best test is selected from the splits defined by these features. The number of the retained features $F'$ is given as a function of the total number of features $F$, e.g., $\lceil \sqrt{F} \rceil$, $\lceil F/4 \rceil$, etc. We obtain the bagging procedure if we keep all the features, and the random forest procedure otherwise.

**Extra Tree Ensembles.** As in random forests, we consider $F'$ features in each node, but we do not evaluate all potential tests that the features could yield. Rather, we choose randomly only one test per feature. Among these $F'$ tests, we choose the best one. From the bias-variance point of view, the rationale behind the Extra-Trees method is that the explicit randomization of the cut-point and feature combined with ensemble averaging should be able to reduce variance more strongly than the weaker randomization schemes used by other methods [7]. Note that originally, Extra-Tree ensembles use no bootstrapping. However, we introduced it due to two main reasons: i) some preliminary experiments showed that it is beneficial to do so from the predictive power point of view, and ii) the Random Forest score (see Eq. (3)) requires OOB examples for its computation.

# 3    Feature Ranking Scores for SSL Classification

We first propose and describe the *Symbolic score*. Then, we proceed explaining the *Genie3* [11] and the *Random Forest* scores [4]. To avoid confusion, Random Forest score will be always in singular form and capitalized, whereas the ensemble method random forests will be in plural form and not capitalized. In the following, a tree is denoted by $\mathcal{T}$, whereas $\mathcal{N} \in \mathcal{T}$ denotes a node in a tree. Trees form an ensemble $\mathcal{E}$ of size $|\mathcal{E}|$. The set of all internal nodes of a tree $\mathcal{T}$ in which the feature $x_i$ appears as part of a test is denoted as $\mathcal{T}(x_i)$.

**Symbolic Score.** In the simplest version of the score, we count the occurrence of a given feature in the tests in the internal nodes of the trees. Since the features appearing closer to the root influence more examples and are intuitively more important, we define the importance of the feature $x_i$ as

$$importance_{\text{SYMB}}(x_i) = \frac{1}{|\mathcal{E}|} \sum_{\mathcal{T} \in \mathcal{E}} \sum_{\mathcal{N} \in \mathcal{T}(x_i)} |E(\mathcal{N})| / |\mathscr{D}_{\text{TRAIN}}|, \tag{1}$$

so that the appearances of the feature $x_i$ are weighted by the number of examples the corresponding node influences. The term $1/|\mathscr{D}_{\text{TRAIN}}|$ is just a scaling factor. This is a parameter-less version of the previously defined Symbolic score [16], where an appearance of a feature $x_i$ in node $\mathcal{N}$ was awarded $\alpha^{\text{depth}}(\mathcal{N})$, where the value of the $\alpha \in (0,1]$ had to be chosen by the user.

**Genie3 Score.** The main motivation for this score is that splitting the current subset $E \subseteq \mathscr{D}_{\text{TRAIN}}$, according to a test where an important feature appears, should result in high impurity reduction. The Genie3 importance of the feature $x_i$ is thus defined as

$$importance_{\text{GENIE3}}(x_i) = \frac{1}{|\mathcal{E}|} \sum_{\mathcal{T} \in \mathcal{E}} \sum_{\mathcal{N} \in \mathcal{T}(x_i)} |E(\mathcal{N})| h^*(\mathcal{N}), \tag{2}$$

where $E(\mathcal{N})$ is the set of examples that come to the node $\mathcal{N}$, and $h^*(\mathcal{N})$ is the value of the variance reduction function described in Algorithm 2. Greater emphasis is again put on the features higher in the tree, where $|E|$ is larger.

**Random Forest (RF) Score.** This score tests how much noising a given feature decreases the predictive performance of the trees in the ensemble. The greater the performance degradation, the more important the feature is.

Once a tree $\mathcal{T}$ is grown, the algorithm evaluates the performance of the tree by using the corresponding $\text{OOB}_{\mathcal{T}}$ examples. This results in the accuracy $a(\text{OOB}_{\mathcal{T}})$. Afterward, we randomly permute the values of feature $x_i$ in the set $\text{OOB}_{\mathcal{T}}$ and obtain the set $\text{OOB}_{\mathcal{T}}^i$ with the corresponding accuracy $a(\text{OOB}_{\mathcal{T}}^i)$.

The importance of the feature $x_i$ for the tree $\mathcal{T}$ is defined as the relative decrease of accuracy after noising. The final Random Forest score of the feature is the average of these values across all trees in the forest:

$$importance_{\mathrm{RF}}(x_i) = \frac{1}{|\mathcal{E}|} \sum_{\mathcal{T} \in \mathcal{E}} \frac{a(\mathrm{OOB}_{\mathcal{T}}) - a(\mathrm{OOB}_{\mathcal{T}}^i)}{a(\mathrm{OOB}_{\mathcal{T}})}. \tag{3}$$

Note that $a(\mathrm{OOB}_{\mathcal{T}}^i) = a(\mathrm{OOB}_{\mathcal{T}})$ if the feature $x_i$ does not appear in $\mathcal{T}$. This can speed up the computation of $importance_{\mathrm{RF}}$, but this feature ranking method is still the most time consuming. While the time complexity of the first two is negligible as compared to the one of growing the forest, this one has an additional linear factor: the number of examples in the training set.

## 4    Experimental Design

In this section, we define the experimental questions, briefly describe the datasets, define the evaluation procedure and describe which parameters of the algorithms were used in the experiments. With the empirical evaluation, we set out to answer the following four questions:

1. Which ensemble method suits a given feature ranking score the most?
2. Can our SSL-feature ranking scores make effective use of the unlabeled examples, especially when the number of labeled examples is small?
3. Do our SSL-feature ranking scores yield state-of-the-art feature rankings?
4. Which SSL-feature ranking score has the highest quality?

Before proceeding to the rest of the section, let us mention that the proposed SSL feature ranking scores are implemented in the CLUS software http://source.ijs.si/ktclus/clus-public, and that all the datasets, the results are available at http://source.ijs.si/mpetkovic/ssl-fr, together with our implementation of the competing SEFR method [1].

We use 8 benchmark classification datasets from various domains: medicine (*Arrhythmia*, *Dis*), science and technology (*Gasdrift*, *Pageblocks*, *Phishing*), gaming (*Chess*, *Tic-tac-toe*) and economy (*Bank*). The main properties of the datasets, i.e., the number of features, classes, and examples are given in Table 1.

Prior to performing any experiments, each dataset $\mathscr{D}$ was randomly split into $x = 10$ stratified folds which resulted in the test sets $\mathscr{D}_{\mathrm{TEST}i}$, $0 \leq i < x$. In contrast to the cross-validation in the standard classification scenario where $\mathscr{D}_{\mathrm{TRAIN}i} = \cup_{j \neq i} \mathscr{D}_{\mathrm{TEST}j}$ we first define the copy $\mathscr{D}_{\mathrm{TEST}i}^{\ell}$ of $\mathscr{D}_{\mathrm{TEST}i}$ in which we remove the labels (class values) of all but $\lfloor \ell/(x-1) \rfloor + r_i$ randomly selected examples, where $\lfloor \cdot \rfloor$ is the floor function, $r$ is the reminder of $\ell$ when divided by $x - 1$, and $r_i = 1$ if $i < r$ and 0 otherwise. This assures that every training set $\mathscr{D}_{\mathrm{TRAIN}i}^{\ell} = \cup_{j \neq i} \mathscr{D}_{\mathrm{TEST}i}^{\ell}$ contains precisely $\ell$ labeled examples. We make sure that the implication $\ell_1 \leq \ell_2 \Rightarrow$ *labeled examples of* $\mathscr{D}_{\mathrm{TRAIN}i}^{\ell_1}$ *are a subset of the labeled examples in* $\mathscr{D}_{\mathrm{TRAIN}i}^{\ell_2}$ holds.

**Table 1.** Basic properties of the datasets in the experiments: number of nominal and numeric features, number of examples, number of classes and the proportion of examples belonging to the majority class.

| Dataset | Nominal | Numeric | Examples | Classes | Majority class [%] |
|---------|---------|---------|----------|---------|--------------------|
| Arrhythmia [15] | 73 | 206 | 452 | 16 | 54 |
| Bank [15] | 9 | 7 | 4521 | 2 | 88 |
| Chess [15] | 36 | 0 | 3196 | 2 | 52 |
| Dis [8] | 22 | 6 | 3772 | 2 | 98 |
| Gasdrift [15] | 0 | 129 | 13910 | 6 | 22 |
| Pageblocks [15] | 0 | 10 | 5473 | 5 | 90 |
| Phishing [15] | 30 | 0 | 11055 | 2 | 56 |
| Tic-tac-toe [15] | 9 | 0 | 958 | 2 | 65 |

An SSL-ranking score is computed from $\mathscr{D}\text{TRAIN}_i^\ell$ and its standard-classification counterpart is computed on the $\mathscr{D}\text{TRAIN}_i^\ell$ with the unlabeled examples removed. Afterward, both rankings are evaluated on $\mathscr{D}\text{TEST}_i^\ell$. This is done by using the $k$NN algorithm with $k = 20$ where weighted version of the standard squared Euclidean distance is used. For two input vectors $\boldsymbol{x}^1$ in $\boldsymbol{x}^2$, the distance $d$ between them is defined as $d(\boldsymbol{x}^1, \boldsymbol{x}^2) = \sum_{i=1}^{F} w_i d_i^2(\boldsymbol{x}_i^1, \boldsymbol{x}_i^2)$, where $d_i$ is defined as the absolute difference of the feature values scaled to the $[0, 1]$-interval, if $x_i$ is numeric, and as $\mathbf{1}[\boldsymbol{x}_i^1 \neq \boldsymbol{x}_i^2]$ ($\mathbf{1}$ is the indicator function), if $x_i$ is nominal.

We first define $w_i = \max\{importance(x_i), 0\}$, since Random Forest ranking can award a feature a negative score. In the degenerated case when the resulting values all equal 0, we define $w_i = 1$, for all features $x_i$. The first step is necessary to ignore the features that are of lower importance than a randomly generated one would be. The second step is necessary to ensure $d$ is well-defined.

The evaluation through $k$NN was chosen because of three main reasons. First, this is a distance based method, hence, it can easily make use of the information contained in the feature importances in the learning phase. Second, $k$NN is simple: Its only parameter is the number of neighbors. In the prediction stage, the neighbors' contributions to the predicted value are equally weighted, so we do not introduce additional parameters that would influence the performance. Finally, if a feature ranking is meaningful, then when the feature importances are used as weights in the calculation of the distances, $k$NN should produce better predictions as compared to $k$NN without using these weights [19].

To assess the predictive performance, we first compute the sum $M$ of the $x$ confusion matrices $M_i$ that we obtain from the $x$-fold cross-validation, i.e., $M = \sum_{i=0}^{x-1} M_i$. We use the confusion matrix $M$ to compute accuracy, Cohen's $\kappa$, $F_1$ score and Matthew's correlation coefficient as evaluation measures. The latter three were used since they do not give misleading results in the case of skewed class distribution. Due to space limitations, we will report only accuracy and $F_1$ score (considering Cohen's $\kappa$, $F_1$ score and Matthew's correlation coefficient leads to the same conclusions as presented here). We compute 1-versus-other macro-averaged versions of the $F_1$ measure and Matthew's coefficient (to also consider multi-class classification problems). Note that defining the matrix $M$ is necessary, because averaging the scores over the folds is wrong, as the measures are not additive.

We parametrize the used methods as follows. The number of trees in the ensembles was set to 100. The number of features that are considered in each internal node was set to $\sqrt{F}$ for random forests and $F$ for extra trees [7]. The optimal value of the parameter $w$ for computing the ensembles of PCTs was selected by 5-fold internal cross-validation. The considered values were $w \in \{0, 0.1, 0.2, \ldots, 0.9, 1\}$. The possible numbers of labeled examples $\ell$ in the training datasets were $\ell \in \{50, 100, 200, 350, 500\}$ [13].

## 5   Results and Discussion

### 5.1   Most Appropriate Ensemble

We start our analysis by choosing the most appropriate ensemble for a given feature ranking score. To this end, we fix the score (Symbolic, Genie3 or Random Forest) and compute it from the three ensembles (random forest, bagging, and extra trees), for all values of the number of labeled examples $\ell$. When we evaluate these rankings on a given test set with a given evaluation measure $m$, we obtain a curve consisting of points $(\ell, m(\ell))$, for every score-ensemble pair. The final measure for the performance of an ensemble for a given score is the area under the corresponding curve, denoted by $auc$. When computing the $auc$, we assume that the points $\ell$ are equidistant, which effectively puts more weight on the lower values of $\ell$, where the SSL methods are the most applicable.

Thus, for every feature ranking score, dataset and evaluation measure, we obtain curves that belong to the three ensemble methods, and rank them with respect to the $auc$: the one with the highest $auc$ is assigned rank 1. We then average these ranks over the datasets and evaluation measures.

The summary of the comparison of the pairs (ensemble, score) is given in Table 2. We can see that the differences among different ensemble methods are not large which means that for a given score, all three ensemble methods are

**Table 2.** Average ranks of the feature rankings computed from a fixed feature ranking score and varying ensemble method: random forests (RF), bagging and extra trees (ET). The ranks are reported for SSL ensembles and supervised ensembles separately.

| Score | SSL ensembles | | | Supervised ensembles | | |
|---|---|---|---|---|---|---|
| | RF | Bagging | ET | RF | Bagging | ET |
| Symbolic | 1.86 | 2.07 | 2.07 | 1.93 | 1.95 | 2.11 |
| Genie3 | 2.00 | 1.93 | 2.07 | 1.91 | 1.75 | 2.34 |
| Random forest | 2.07 | 1.82 | 2.11 | 1.98 | 1.89 | 2.14 |

appropriate to some extent. This is also evident from the fact that no average rank - except maybe that of supervised extra trees coupled with Genie3 score (2.34) is close to the worst possible average rank of 3. However, note that ensembles of extra trees always have the worst average rank, hence we decide only between random forests and bagging. If we are interested only in the quality of a ranking, then the following choices are made: for Symbolic score, random forests are the most appropriate (in both the SSL and the supervised case), whereas for the other two scores (Genie3 and Random Forest) the bagging method performs best. We will stick to these choices throughout the rest of the paper. If the time complexity of inducing the ranking is also taken into account, then random forests would be preferred over bagging, at least in the case of the supervised version of the Symbolic score, where the difference in average ranks is only 0.02, but random forests rankings are computed $\sqrt{F}$-times faster.

**Table 3.** Differences between the $auc$ of SSL feature rankings and their supervised counterparts, measured in terms of accuracy and $F_1$ measure of 20NN classifier. If $\Delta > 0$, the SSL feature ranking outperforms the supervised one.

| Dataset | Symbolic | | Genie3 | | Random forest | |
|---|---|---|---|---|---|---|
| | $\Delta acc$ | $\Delta F_1$ | $\Delta acc$ | $\Delta F_1$ | $\Delta acc$ | $\Delta F_1$ |
| Arrhythmia | 0.015 | 0.046 | −0.010 | 0.012 | 0.004 | 0.031 |
| Bank | −0.030 | 0.058 | −0.028 | 0.060 | −0.039 | 0.057 |
| Chess | −0.153 | −0.168 | −0.152 | −0.166 | −0.091 | −0.107 |
| Dis | −0.017 | 0.095 | −0.022 | 0.113 | −0.011 | 0.117 |
| Gasdrift | −0.237 | −0.217 | −0.267 | −0.244 | −0.262 | −0.237 |
| Pageblocks | 0.017 | 0.202 | 0.018 | 0.228 | 0.017 | 0.190 |
| Phishing | −0.257 | −0.300 | −0.243 | −0.288 | −0.252 | −0.297 |
| Tic-tac-toe | 0.110 | 0.201 | 0.105 | 0.191 | 0.142 | 0.265 |

## 5.2   Can Unlabeled Data Improve a Feature Ranking?

Here, we compare the quality of the SSL feature rankings to their supervised counterparts in more depth. Since the latter use only labeled data, this shows whether SSL feature rankings can make effective use of unlabeled data. We draw the same curves as in the previous section and obtain three types of diagrams as shown in Fig. 1. The Tic-tac-toe dataset (Figs. 1a and b) is a representative dataset where SSL rankings outperform their supervised counterparts, for all considered numbers of labeled examples $\ell$. Note that for 50 labeled examples, the 20NN model that uses the supervised rankings only achieves a default accuracy (see Table 1), hence taking into account unlabeled examples clearly helps.

The next type of diagrams presented in Fig. 1c and d shows the curves for the Arrhythmia dataset. This dataset is the only one where the curves of the SSL and supervised rankings intersect: for the lower number of labeled examples

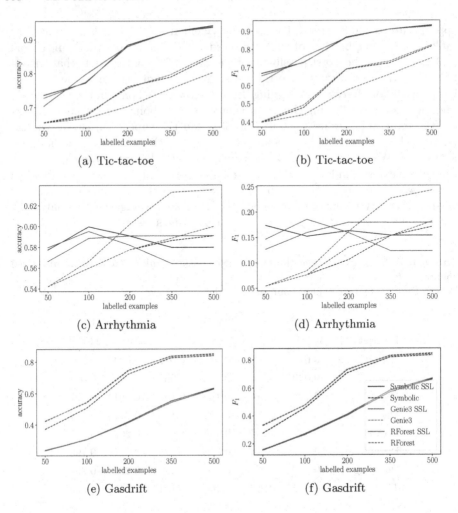

**Fig. 1.** Comparison of the SSL feature ranking with the supervised feature rankings on the Tic-tac-toe, Arrhythmia and Gasdrift datasets, in terms of accuracy and $F_1$ measure of the 20NN classifier. The legend in (f) applies to all subfigures.

(up to 200), the SSL feature rankings outperform their supervised counterparts, which again shows the usefulness of the unlabeled examples. Adding more labeled examples (at least 350), boosts the performance of supervised rankings more and they achieve better performance in this case. Sometimes, taking into account the unlabeled examples does not help, as shown in the diagrams of the last type whose representative, the Gasdrift dataset, is presented in Figs. 1e and f.

To explain this, we test the validity of the clustering assumption. Recall that, in general, SSL should be the most effective when the clusters in the data are in concordance with the values of the target variable $y$. To check this, we compute the $k$-means clustering of a dataset $\mathscr{D}$, where the parameter $k$ is set to the number of class values. We compare the resulting partition of the $\mathscr{D}$ to that induced by the class labels $y(\boldsymbol{x})$, in terms of the Adjusted Rand Index (ARI) [10], which equals 1 when the partitions are equal and 0 if the partitions are random. The whole procedure is repeated 10 times and in the end, we compute the median of the ARI scores.

It turns out that the median ARI score for the Gasdrift dataset is quite low $(2.19 \cdot 10^{-2})$, whereas the Tic-tac-toe dataset has high ARI score $(7.00 \cdot 10^{-1})$. For the other datasets, we do not show the graphs, but only the differences between the *auc* values between the SSL and the supervised version of every ranking score. They are given in the Table 3. We will focus mainly on the columns that belong to the $F_1$ measure results, since some of the datasets (Bank, Dis, Pageblocks) are very imbalanced. We see that results are consistent for all three feature ranking scores: SSL rankings outperform their supervised versions in 5 out of 8 cases. The remaining three include the datasets Gasdrift (discussed above), Phishing and Chess. The Phishing dataset has an ARI score of $-7.16 \cdot 10^{-5}$, which effectively means that class labels are randomly distributed among the clusters; Thus, the better performance of the supervised rankings comes as no surprise. The last one is the Chess dataset which has an ARI score of $2.20 \cdot 10^{-1}$.

### 5.3   Are the Proposed Methods State-of-the-Art?

To answer this question, we compare our SSL feature rankings to the SEFR ranking. We follow the structure of the previous section and first show the graphs for the datasets Tic-tac-toe, Arrhythmia and Gasdrift in Fig. 2.

Figure 2a and b depict the quality of the feature rankings computed from the Tic-tac-toe dataset, in terms of the accuracy and $F_1$ measure respectively. We observe that the differences between SSL and SEFR rankings are not as remarkable as those between the SSL and the supervised rankings. However, all SSL rankings still consistently outperform the SEFR ranking, except for the Random Forest ranking computed from 50 labeled examples.

Similar results are obtained for the Arrhythmia dataset as shown if Fig. 2c and d. Again, the differences are in favor of the SSL rankings - in this case, without exceptions. From these two graphs, we can also deduce that adding more labeled examples does not necessarily help: the quality of rankings (except for the Genie3 rankings, shown in green) does not monotonically increase and the $F_1$ scores of the 20NN classifiers that use the Random Forest, Symbolic and SERF importance scores computed from 500 labeled examples are even slightly lower than the $F_1$ scores of the same rankings computed from 50 labeled examples. The Gasdrift dataset is the one of the eight datasets for which the differences are the smallest, as shown in Fig. 2e and f. The quality of the rankings here increases as more labeled examples are provided. However, the differences equal approximately 0.2 percentage points, as evident from Table 4.

The same table also reveals that typically, the Symbolic, Genie3 and Random Forest rankings outperform the SEFR rankings, since the large majority of the $\Delta$ values in the table are positive. In terms of $F_1$ score, Random Forest and Symbolic rankings outperform the SEFR ranking in 7 out of 8 cases, and Genie3 ranking outperforms it in 6 out of 8 cases. The numbers for accuracy are similar, so we can rightly conclude that our rankings exhibit state-of-the-art performance.

## 5.4   Which Ranking Score Is the Best?

The answers to the previous two experimental questions are pretty clear. However, determining the best ranking is a bit harder. The average ranks (over all datasets and evaluation measures) of the Genie3, Random Forest and Symbolic scores are 1.90, 2.03 and 2.06 respectively, so Genie3 score performs best on average, but at the same time, it is ranked first only five times. Random Forest

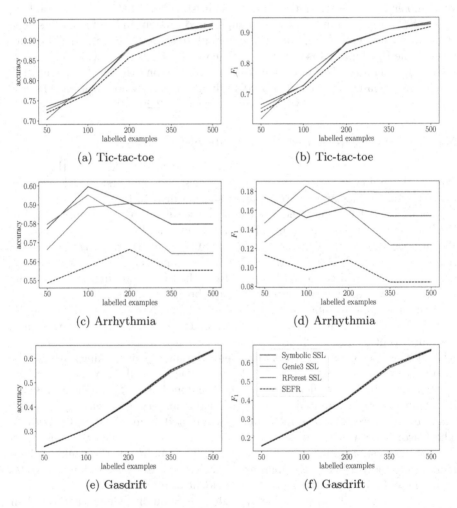

(a) Tic-tac-toe

(b) Tic-tac-toe

(c) Arrhythmia

(d) Arrhythmia

(e) Gasdrift

(f) Gasdrift

**Fig. 2.** Comparison of the SSL and SEFR feature rankings on the Tic-tac-toe, Arrhythmia and Gasdrift datasets, in terms of accuracy and $F_1$ measure of the 20NN classifier. The legend in (f) applies to all subfigures.

**Table 4.** Differences between the *auc* of SSL feature rankings and SEFR feature rankings, measured in terms of accuracy and $F_1$ measure, of 20NN classifier. If $\Delta > 0$, the SSL feature ranking outperforms the SEFR one.

| Dataset | Symbolic | | Genie3 | | Random forest | |
|---|---|---|---|---|---|---|
| | $\Delta acc$ | $\Delta F_1$ | $\Delta acc$ | $\Delta F_1$ | $\Delta acc$ | $\Delta F_1$ |
| Arrhythmia | $2.9 \cdot 10^{-2}$ | $6.1 \cdot 10^{-2}$ | $2.9 \cdot 10^{-2}$ | $7.1 \cdot 10^{-2}$ | $2.1 \cdot 10^{-2}$ | $5.4 \cdot 10^{-2}$ |
| Bank | $6.0 \cdot 10^{-3}$ | $6.9 \cdot 10^{-3}$ | $8.5 \cdot 10^{-3}$ | $1.1 \cdot 10^{-2}$ | $-3.4 \cdot 10^{-3}$ | $6.1 \cdot 10^{-3}$ |
| Chess | $-2.4 \cdot 10^{-2}$ | $-2.6 \cdot 10^{-2}$ | $-8.6 \cdot 10^{-4}$ | $-2.8 \cdot 10^{-3}$ | $6.3 \cdot 10^{-3}$ | $2.4 \cdot 10^{-3}$ |
| Dis | $-8.9 \cdot 10^{-3}$ | $1.3 \cdot 10^{-2}$ | $-1.4 \cdot 10^{-2}$ | $3.1 \cdot 10^{-2}$ | $-2.6 \cdot 10^{-3}$ | $3.5 \cdot 10^{-2}$ |
| Gasdrift | $2.5 \cdot 10^{-3}$ | $3.7 \cdot 10^{-3}$ | $-1.6 \cdot 10^{-3}$ | $-1.7 \cdot 10^{-3}$ | $2.6 \cdot 10^{-3}$ | $5.2 \cdot 10^{-3}$ |
| Pageblocks | $8.5 \cdot 10^{-4}$ | $1.2 \cdot 10^{-2}$ | $2.5 \cdot 10^{-3}$ | $4.0 \cdot 10^{-2}$ | $-8.0 \cdot 10^{-4}$ | $-5.1 \cdot 10^{-3}$ |
| Phishing | $1.4 \cdot 10^{-2}$ | $1.7 \cdot 10^{-2}$ | $7.9 \cdot 10^{-3}$ | $9.7 \cdot 10^{-3}$ | $5.4 \cdot 10^{-3}$ | $6.4 \cdot 10^{-3}$ |
| Tic-tac-toe | $1.7 \cdot 10^{-2}$ | $2.2 \cdot 10^{-2}$ | $1.5 \cdot 10^{-2}$ | $1.9 \cdot 10^{-2}$ | $1.8 \cdot 10^{-2}$ | $2.2 \cdot 10^{-2}$ |

score is ranked first most frequently (15 times), but is also ranked last most frequently (16 times). The distribution of the Symbolic ranking is the closest to the uniform one (it is ranked 1st, 2nd and 3rd 12-times, 11-times, and 9-times respectively) and has the worst average (2.06).

Since all the feature ranking scores offer state-of-the-art performance, we may conclude that they are approximately equally good, so we can again make a decision based on the second criterion: computational complexity. Recall that random forests were the most appropriate method for the Symbolic score, whereas bagging was the most suitable for Genie3 and Random Forest score, which means that Symbolic rankings are computed $\sqrt{F}$-times faster than the Genie3 rankings. The Random Forest rankings demand even more time than Genie3 rankings, due to the permutations of feature values and evaluating each tree of the ensemble on the corresponding out-of-bag examples multiple times, as explained in Sect. 3.

## 6   Conclusions

In this paper, we propose three feature ranking scores (Symbolic, Genie3 and Random Forest). Each can be computed from three different ensembles (bagging, random forests and extremely randomized trees) of predictive-clustering trees (PCTs), which were adapted to the semi-supervised classification task. We evaluate the obtained feature rankings on 8 benchmark classification datasets. We first determine the most suitable ensemble for each of the scores. For the Symbolic score, these are random forests, whereas for the Genie3 and Random Forest score, this is bagging.

Next, we show that using unlabeled data leads to improvements in the ranking since the proposed semi-supervised feature rankings mostly outperform their supervised analogs, which use only labeled data. We analyzed the datasets where this does not hold by checking the validity of the clustering assumption of SSL and show that most probably the assumption is not valid in these cases.

After that, we compare our feature ranking scores to the SEFR feature ranking method and empirically show that we consistently outperform this state-of-the-art baseline with every proposed feature ranking score. Finally, we compare the proposed scores among each other and conclude that they are all equally good. We suggest using the Symbolic score because it has the best (lowest) time complexity.

We will continue our work on this topic in three directions. We first plan to extend the number of the benchmark datasets and then to define and evaluate rankings that are obtained from gradient boosting ensembles. Next, we will extend the approach towards regression and more complex predictive modeling tasks including structured output prediction. Finally, we will work to circumvent the use of the internal cross-validation for determining the best value of $w$ in the PCT heuristic by computing the ARI score instead.

**Acknowledgements.** We acknowledge the financial support of the Slovenian Research Agency via the grants P2-0103 and a young researcher grant to MP. SD and DK acknowledge the support by the Slovenian Research Agency (via grants J2-9230, and N2-0056), and the European Commission (project LANDMARK and The Human Brain Project SGA2). The computational experiments presented here were executed on a computing infrastructure from the Slovenian Grid (SLING) initiative.

# References

1. Bellal, F., Elghazel, H., Aussem, A.: A semi-supervised feature ranking method with ensemble learning. Pattern Recognit. Lett. **33**(10), 1426–1433 (2012)
2. Bishop, C.M.: Neural Networks for Pattern Recognition. Oxford University Press, New York (1995). https://dl.acm.org/citation.cfm?id=525960
3. Blockeel, H.: Top-down Induction of First Order Logical Decision Trees. Ph.D. thesis, Katholieke Universiteit Leuven, Leuven, Belgium (1998)
4. Breiman, L.: Random forests. Mach. Learn. **45**(1), 5–32 (2001)
5. Breiman, L., Friedman, J., Olshen, R., Stone, C.J.: Classification and Regression Trees. Chapman and Hall/CRC, New York (1984)
6. Chapelle, O., Schölkopf, B., Zien, A.: Semi-supervised Learning. MIT Press, Cambridge (2010)
7. Geurts, P., Erns, D., Wehenkel, L.: Extremely randomized trees. Mach. Learn. **36**(1), 3–42 (2006)
8. Gijsbers, P.: OpenML repository (2017). https://www.openml.org/d/40713
9. Hansen, L.K., Salamon, P.: Neural network ensembles. IEEE Trans. Pattern Analy. Mach. Intell. **12**, 993–1001 (1990)
10. Hubert, L., Arabie, P.: Comparing partitions. J. Classif. **2**(1), 193–218 (1985)
11. Huynh-Thu, V.A., Irrthum, A., Wehenkel, L., Geurts, P.: Inferring regulatory networks from expression data using tree-based methods. PLoS One **5**(9), 1–10 (2010)
12. Kocev, D., Vens, C., Struyf, J., Džeroski, S.: Tree ensembles for predicting structured outputs. Pattern Recognit. **46**(3), 817–833 (2013)
13. Levatić, J.: Semi-supervised Learning for Structured Output Prediction. Ph.D. thesis, Jožef Stefan Postgraduate School, Ljubljana, Slovenia (2017)
14. Levatić, J., Ceci, M., Kocev, D., Džeroski, S.: Semi-supervised classification trees. J. Intell. Inf. Syst. **49**(3), 461–486 (2017)

15. Lichman, M.: UCI machine learning repository (2013). http://archive.ics.uci.edu/ml
16. Petković, M., Kocev, D., Džeroski, S.: Feature ranking for multi-target regression. Mach. Learn. J. (2019, accepted)
17. Raina, R., Battle, A., Lee, H., Packer, B., Ng, A.Y.: Self-taught learning: transfer learning from unlabeled data. In: 24th International Conference on Machine Learning, pp. 759–766. ACM (2007)
18. Sheikhpour, R., Sarram, M., Gharaghani, S., Chahooki, M.: A survey on semi-supervised feature selection methods. Pattern Recognit. **64**((C)), 141–185 (2017)
19. Wettschereck, D.: A Study of Distance Based Algorithms. Ph.D. thesis, Oregon State University, Corvallis, OR (1994)
20. Xu, Z., King, I., Lyu, M.R.T., Jin, R.: Discriminative semi-supervised feature selection via manifold regularization. Trans. Neural Netw. **21**(7), 1033–1047 (2010)

# Variance-Based Feature Importance
# in Neural Networks

Cláudio Rebelo de Sá[(⊠)] [iD]

Data Science Research Group, University of Twente, Enschede, Netherlands
c.f.pinhorebelodesa@utwente.nl

**Abstract.** This paper proposes a new method to measure the relative importance of features in Artificial Neural Networks (ANN) models. Its underlying principle assumes that the more important a feature is, the more the weights, connected to the respective input neuron, will change during the training of the model. To capture this behavior, a running variance of every weight connected to the input layer is measured during training. For that, an adaptation of Welford's online algorithm for computing the online variance is proposed. When the training is finished, for each input, the variances of the weights are combined with the final weights to obtain the measure of relative importance for each feature. This method was tested with shallow and deep neural network architectures on several well-known classification and regression problems. The results obtained confirm that this approach is making meaningful measurements. Moreover, results showed that the importance scores are highly correlated with the variable importance method from Random Forests (RF).

## 1 Introduction

Effectively measuring the relevance of features in Artificial Neural Networks (ANN) can foster its usage in new domains where some interpretability is required. Current studies show that there has been some effort to bring more interpretability to Artificial Neural Networks in the recent years [3,5]. However, despite the various approaches available in the literature, there is a lack of simple, and yet reliable, variable importance approaches for ANN.

The classic and most simple architecture of a feed-forward neural network is composed by one input layer, one or more hidden layers and one output layer. These layers are connected by weights and each is composed of a certain number of neurons. Each neuron in the input layer represents one independent variable, or feature, from the data. These neurons connect to the first hidden layer, which in turn connects either to the next hidden layer (and so on) or to the output layer. The neurons in this output layer represent the target variable.

During the training of ANN, the weights (which connect the neurons) are constantly being changed for better fitting the data. This process occurs for every batch of data, and it lasts until all the epochs are finished. Once this process is finished, it is natural to assume that, the higher the absolute values of the weights, the more important a variable would be [2]. However, on the other

© Springer Nature Switzerland AG 2019
P. Kralj Novak et al. (Eds.): DS 2019, LNAI 11828, pp. 306–315, 2019.
https://doi.org/10.1007/978-3-030-33778-0_24

hand, we also know that regularization techniques force the weights to become smaller (e.g., L1 and L2). Besides, the choice of the initialization of these weights can also interfere with their final absolute value.

In this paper, a very simple method is proposed to measure the relative feature importance (or variable importance) of ANN models. Its underlying principle assumes that the more important a feature is, the more the weights, connected to the respective input neuron, will change during the training of the model. This means that, it expects bigger changes in the weights connected to more relevant variables, independently from their absolute value. Under this assumption, by measuring the total variance of the weights connected to each input node, one should be able to measure its relative importance.

Since the weights are being changed at every batch and neural networks could easily take hundreds of epochs to be trained, it is not practical to store all the values for later computing its variance. For this reason, the running variance of each weight that is connected to the input layer is used instead. For that, an adapted version of Welford's online algorithm [9] is proposed. This algorithm updates the mean and variance of the weights, connected to the input layer, at an user defined step (e.g., per batch or per epoch).

Finally, the variances from all the weights connected to a feature are combined into a single value, which are then used for assessing their relative importance. This contrasts with most of the approaches available in the literature, because the variable importance is not measured on the absolute values of the weights of the network, but on their variance during the training.

Some well-known regression and classification datasets were used to test the proposed approach, which included one artificial dataset. Empirical results presented in this paper, using shallow and deep ANN, show that this approach holds promise and can be used to effectively assess the relative importance of variables in different datasets.

## 2  Variable Importance in FNN

Most approaches assess the feature importance based on the final weights of the trained neural networks [2,4,5]. One of the most well-known was proposed in 1991 by Garson [2] and it is still being used [3,8]. It basically consist in adding up the absolute values of the weights between each input node and the response variables. In other words, all the weights connecting a given input node, including the hidden layers, to a specific response variable will contribute to measure feature importance. Finally, the total score of the input nodes is scaled relatively to all other inputs.

However, we know that during the training phase, the weights of a neural network are being modified. These updates of the weight are repeated until the model reaches its final state. Therefore, we propose to measure the variance of these weights in order to get the relative variable importance based for ANN.

In [6], the authors observed that the difference between quartiles of the distributions seemed to be related with their relative importance. In this work, a similar approach is taken, however the focus is in the variance instead of the interquartile range.

Considering the size of current ANN and the high number of epochs to train them, storing the values of the all the weights to compute the variance would be computationally expensive. However, a method proposed by Welford in 1962 [9] (See Sect. 2.1) allows one to compute and update the variance as the measurements are given, one at a time. This has the advantage that, the values do not need to be saved to compute the variance in the end.

A simple adaptation of Welford's online variance, proposed here, makes this approach much simpler to implement. It is similar to the method in [9], except that this was adapted to deal with matrices. These matrices represent the weights connecting the input layer with the first hidden layer.

## 2.1    Welford's Online Variance

The variance of a sample of size $n$ is defined as:

$$S_n^2 = \frac{SS_n}{n-1} = \sum_{i=1}^{n} \frac{(x_i - \bar{x}_n)^2}{n-1} \tag{1}$$

where the corrected sum of squares $SS_n$ is:

$$SS_n = \sum_{i=1}^{n} (x_i - \bar{x}_n)^2 \tag{2}$$

and the mean, $\bar{x}_n$, is defined as:

$$\bar{x}_n = \sum_{i=1}^{n} \frac{x_i}{n} \tag{3}$$

However, we can write the corrected sum of squares $SS_n$ as:

$$SS_n = SS_{n-1} + \left(\frac{n-1}{n}\right)(x_n - \bar{x}_{n-1})^2$$

This way, if we replace this in Eq. 1, we can update the variance of a sample, originally with size $n-1$, by adding one more measurement to the sample, $x_n$ [9]. This can be represented as:

$$Var(x_n) = \frac{SS_{n-1}}{n-1} + \frac{(x_n - \bar{x}_{n-1})}{n} \tag{4}$$

where $n$ represents the total number of updates. This computes the online variance of the weights (also known as running variance).

## 2.2 Online Variance of the Weights

Let us define a dataset $D = \{\langle v_i \rangle\}$, $i = 1, \ldots, z$ with $z$ instances, where $v_i$ is a vector containing the values $v_i^j, j = 1, \ldots, m$ of $m$ independent variables, $\{A_1, \ldots, A_m\}$, describing instance $i$.

To represent the weights between layers in ANN, we define $w_{a,b}$ as the weight connecting node $a$ to node $b$. As mentioned before, $m$ represents the number of input variables and $q$ the number of neurons in the first hidden layer. Now we can represent the Variance-based feature Importance of Artificial Neural Networks (VIANN) score of the weights as:

$$\text{VIANN}(A_s) = \sum_{k=1}^{q} Var(w_{s,k}) \times |w_{s,k}| \tag{5}$$

where $t$ represents the total number of updates and $w_{s,k}$ the weights of the first hidden layer connected to the input $A_s$. This means that the final score will depend on both the final weights and the variance of the weights during the training.

We will use the variance as in Eq. 5, to score the importance of the features. The assumption is that, the more the $w_{a,b}$ varies in the training phase, the higher the relevance of the nodes $a$ to the prediction. When using VIANN we need to define at which steps of the training we update the variance. Several options can be considered, per iteration (after every batch), per epoch or with an user defined interval. For simplicity, in this work we update the variance of the weights at each epoch.

# 3 Experimental Setup and Results

In this section we explain how we setep the experiments and describe the architecture of the neural networks tested. We also present the results obtained by the proposed approach and compare with the most used algorithm to measure feature importance in Neural Networks, the Garson's algorithm. We also describe the datasets which were selected for this study.

## 3.1 Experimental Setup

In this experimental setup, many parameters could have been modified and studied. However, as proof of concept, we tried to make some simple and reasonable choices for the design of the architectures and its parameters.

Since we are testing both regression and classification tasks, the neural networks were built in such a way that the last layer (the output layer) can change to classification or regression mode. When the task is *classification*, the last layer will have the same number of neurons as the number of classes and an activation function *softmax*. The loss, in this case is *categorical cross-entropy*. On the other hand, when the task is regression, the last layer has one neuron and the activation is *linear*. The loss function in this case is the *Mean Squared Error*.

Besides that, two types of activation functions are tested in the remaining layers. One neural network has a *linear* activation function (NN1) and the other has the RELU (NN2). Both NN1 and NN2 have 3 hidden layer, one with 50 neurons, other with 100 and the last hidden layer with 50. Besides NN1 and NN2, we also tested the approach in a deeper neural network, DeepNN. The dimensions of the three neural networks are:

NN1: input, 50, 100, 50, output
NN2: input, 50, 100, 50, output
DeepNN: input, 500, 1024, 2048, 4096, 2048, 1024, 500, output

We note that the input and output layers are adapted according to the size of the number of features and number of classes per dataset, respectively.

In previous experiments, we observed that the final accuracy of the model, strongly affects the scoring of the most relevant features. This means that, if the accuracy is low, the importance scores tend to be misleading. For this reason, a simple procedure was used to find a more appropriate number of training epochs per dataset. An early stopping function monitored the validation accuracy in the classification datasets during training. When the validation accuracy is >95% the training stops. The maximum number of epochs was set to 1000 and the minimum to 5. In the regression datasets, the number of epochs was fixed to 100 epochs.

The optimizer used in the experiments was set to the default parameters of the Stochastic Gradient Descent (SGD) optimizer from the python package *Keras*. Due to the different dataset sizes, the batch size was adapted for each dataset. It was defined as the rounded number obtained from the division of the number of instances in the dataset by 7.

We compare the obtained scores of the variables with other approaches such as Random Forests (RF) [1] and Garson's algorithm (GA) [2][1]. We also measured the loss of the models when each feature was removed (replaced with zeros) which is the same as the Leave-One-Feature-Out (LOFO) approach. Then, the features, which after being removed, resulted in the highest loss, are considered more important. Finally, to compare the orders between the different techniques, we use the Kendall's tau correlation coefficient.

## 3.2   Datasets

In the experiments several classification and regression datasets from the *scikit learn* python package [7] were used (Table 1). These particular datasets were chosen to illustrate the effectiveness of this approach. All the features of the dataset were normalized to zero mean and standard deviation 1.

---

[1] All the results presented in this paper can be replicated using the python file in https://github.com/rebelosa/feature-importance-neural-networks.

**Table 1.** Datasets used in the experiments

| Name | #Features | #Instances | Type | #Classes |
|---|---|---|---|---|
| Breast cancer | 30 | 569 | Class | 2 |
| Digits | 64 | 1797 | Class | 10 |
| Iris | 4 | 150 | Class | 3 |
| Wine | 13 | 178 | Class | 3 |
| Boston | 12 | 506 | Regr. | - |
| Diabetes | 10 | 442 | Regr. | - |

### 3.3   Testing the Feature Importance

In this experimental part, we compare the ranking of the features obtained with VIANN, for NN1 and NN2. As previously mentioned, there are several approaches to measure the relative variable importance in datasets. Besides comparing with the RF variable importance measure, we also compare with the Leave-One-Feature-Out (LOFO) approach. By measuring the loss difference after removing each variable, we can sort the variables by the ones which have a greater impact.

The presented in Table 2 show the Kendall tau correlation between the different variable rankings when using the neural network with *linear* activation function, NN1. We can see that VIANN obtains a higher correlation than the Garson technique, as compared with the LOFO. Besides, the relative order of the variables seems to fluctuate much more when obtained with the Garson approach.

**Table 2.** Kendall tau correlation between the different variable importance techniques using the NN1 (linear activation function)

| Dataset | VIANN-LOFO | VIANN-RF | Garson-LOFO | Garson-VIANN |
|---|---|---|---|---|
| Breast cancer | **0.43** | 0.85 | 0.17 | 0.20 |
| Digits | **0.96** | 0.66 | 0.90 | 0.87 |
| Iris | **0.99** | 0.88 | 0.73 | 0.62 |
| Wine | **0.93** | 0.64 | 0.91 | 0.85 |
| Boston | **0.95** | 0.79 | 0.84 | 0.80 |
| Diabetes | **0.86** | 0.68 | 0.85 | 0.79 |

In Table 3 we observe a very similar behavior of the one observed in Table 2. In this case the neural network NN2 is exactly the same as before, except that it has a *RELU* activation function instead of *linear*. In general, we observe that the correlation VIANN-LOFO is worst when using the *RELU*, specially for the classification datasets.

**Table 3.** Kendall tau correlation between the variable importance technique VIANN and other approaches with the NN2 (RELU activation function)

| Dataset | VIANN-LOFO | VIANN-RF | Garson-LOFO | Garson-VIANN |
|---|---|---|---|---|
| Breast cancer | **0.78** | 0.76 | 0.43 | 0.61 |
| Digits | **0.94** | 0.76 | 0.92 | 0.84 |
| Iris | 0.92 | 0.87 | **0.97** | 0.81 |
| Wine | 0.88 | 0.50 | **0.95** | 0.93 |
| Boston | **0.96** | 0.81 | 0.76 | 0.60 |
| Diabetes | **0.98** | 0.90 | 0.64 | 0.60 |

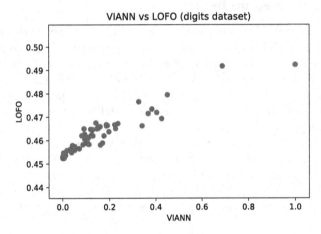

**Fig. 1.** Plot of the feature importance scores obtained with the NN2 using VIANN (x axis) and LOFO (y axis) in the *Digits* dataset.

In Fig. 1 we can see how the LOFO and VIANN relate in terms of feature importance scores. It seams that the variance of the weights combined with the final weights has a linear relation with the increase in the loss of the model.

### 3.4 Deep Neural Network

Finally, despite the small size of the dataset, we wanted to test the approach in the deep learning context. For this reason, we used a deeper network, which we refer as *DeepNN*. The results obtained are presented in Table 4. In this case, we observe that VIANN is still better at giving a more meaningful score of the variables. The exception is the *Wine* dataset where the correlation is not so high.

We observe in Fig. 2, how the scores of VIANN with the DeppNN are related with the variable importance scores of RF. Even though the two models have different biases, the importance of the variables is only slightly changed, which intuitively makes sense.

**Table 4.** Kendall tau correlation between the variable importance technique VIANN and other approaches with the DeepNN (RELU activation function)

| Dataset | VIANN-LOFO | VIANN-RF | Garson-LOFO | Garson-VIANN |
|---|---|---|---|---|
| Breast cancer | **0.60** | 0.45 | 0.22 | 0.37 |
| Digits | **0.83** | 0.80 | 0.60 | 0.46 |
| Iris | **0.90** | 0.98 | 0.73 | 0.49 |
| Wine | 0.41 | 0.43 | **0.74** | 0.51 |
| Boston | **0.76** | 0.86 | **0.76** | 0.79 |
| Diabetes | **0.86** | 0.88 | 0.64 | 0.80 |

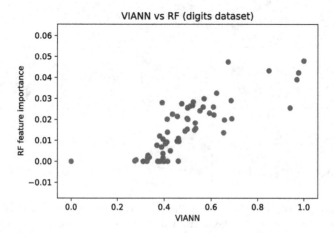

**Fig. 2.** Plot of the feature importance scores obtained with the DeepNN using VIANN (x axis) and RF (y axis) in the *Digits* dataset.

In general, we observe that the feature importance scores are better than the Garson's technique, which shows that VIANN has potential as a measure of importance of features in shallow and deep networks. Even considering that different features can be more or less important for different models, it does not seem likely that the features will have completely distinct relevance for each model. Therefore, we believe that VIANN is measuring some phenomenon that is closely related with the importance of the features.

## 3.5 Evolution of Weights During Training

Since the motivation was to use the variance to measure the feature importance, we wanted to understand if the behavior of the weights of the most relevant features was actually changing more than the others. Therefore, in this experiment we trained the NN2 and captured the weights in every iteration between the input layer and the first hidden layer during the training phase.

The result can be seen in Fig. 3, where the subplots are sorted by the relative importance of the respective inputs (higher to lower), obtained with the LOFO approach. One pattern that can be observed, is that, in fact, the weights seem to change more in the first inputs (e.g. input 12,0 and 9). On the other hand, the weights which affect less the loss, are mostly constant during the entire training. Besides that, the absolute value of the weights seems to also be higher in more important features. These observations support the motivation of this paper (Eq. 5).

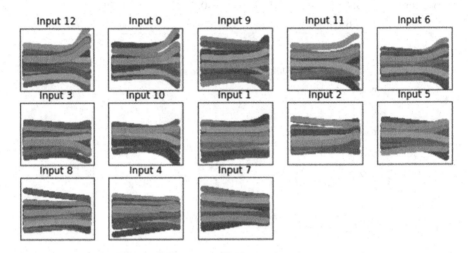

**Fig. 3.** Evolution of the weights (coloured lines) of the NN2, per iteration, between the input layer and the first hidden layer trained in the *Wine* dataset (x-axis: iterations; y-axis: weights)

## 4    Conclusions

In this work we compare the performance of a feature importance technique which is based on the variance of the weights during the training of neural networks. We compare our results with one of the most widely used variable importance techniques in ANN, Garson's algorithm. The results showed that this approach is more reliable in identifying the order of the variables which have a greater influence in the loss. In comparison to Garson's method it has the advantage that it does not require that the first and last hidden layers have the same number of neurons.

We also observed that, when the validation accuracy is low, the scores of the features can be misleading. Some tests, which are not reported in this paper, indicated that without proper regularization techniques, the variable importance scores also do not make sense.

Considering the results obtained, this approach holds promise to effectively measure the relevance of features (or even just any neuron). That is, the simplicity of VIANN makes it straightforward to measure the relevance of every node and not only the input layer. Moreover, it can be easily extended to other neural network layers, such as recurrent or convolutional.

Since this work is only a preliminary study, it does not provide a comprehensive overview of feature importance for ANN. However, as future work we would like to study if VIANN can be used to obtain the importance of the variables on distinct ANN architectures. A thorough experimental study should be made in the future to fully understand the advantages and limitations of this approach.

**Acknowledgments.** I gratefully acknowledge the support of NVIDIA Corporation with the donation of the Titan X Pascal GPU used for this research.

# References

1. Breiman, L.: Random forests. Mach. Learn. **45**(1), 5–32 (2001)
2. David Garson, G.: Interpreting neural-network connection weights. AI Expert **6**(4), 46–51 (1991)
3. Heaton, J., McElwee, S., Fraley, J.B., Cannady, J.: Early stabilizing feature importance for tensorflow deep neural networks. In: 2017 International Joint Conference on Neural Networks, IJCNN 2017, Anchorage, AK, USA, 14–19 May, 2017, pp. 4618–4624 (2017)
4. Martínez, A., Castellanos, J., Hernández, C., de Mingo López, L.F.: Study of weight importance in neural networks working with colineal variables in regression problems. In: Multiple Approaches to Intelligent Systems, 12th International Conference on Industrial and Engineering Applications of Artificial Intelligence and Expert Systems, IEA/AIE-99, Cairo, Egypt, May 31 – June 3, 1999, Proceedings, pp. 101–110 (1999)
5. Olden, J.D., Jackson, D.A.: Illuminating the "black box": a randomization approach for understanding variable contributions in artificial neural networks. Ecol. Model. **154**(1), 135–150 (2002)
6. Paliwal, M., Kumar, U.A.: Assessing the contribution of variables in feed forward neural network. Appl. Soft Comput. **11**(4), 3690–3696 (2011)
7. Pedregosa, F., et al.: Scikit-learn: machine learning in python. J. Mach. Learn. Res. **12**, 2825–2830 (2011)
8. Shavitt, I., Segal, E.: Regularization learning networks: deep learning for tabular datasets. In: Advances in Neural Information Processing Systems 31: Annual Conference on Neural Information Processing Systems 2018, NeurIPS 2018, 3–8 December 2018, Montréal, Canada, pp. 1386–1396 (2018)
9. Welford, B.P.: Note on a method for calculating corrected sums of squares and products. Technometrics **4**(3), 419–420 (1962)

# Interpretable Machine Learning

# A Density Estimation Approach for Detecting and Explaining Exceptional Values in Categorical Data

Fabrizio Angiulli, Fabio Fassetti, Luigi Palopoli, and Cristina Serrao[(✉)]

DIMES, University of Calabria, 87036 Rende, CS, Italy
{f.angiulli,f.fassetti,l.palopoli,c.serrao}@dimes.unical.it

**Abstract.** In this work we deal with the problem of detecting and explaining exceptional behaving values in categorical datasets. As a first main contribution we provide the notion of frequency occurrence which can be thought as a form of Kernel Density Estimation applied to the domain of frequency values. As a second contribution, we define an outlierness measure for categorical values that, leveraging the cdf of the density described above, decides if the frequency of a certain value is rare if compared to the frequencies associated with the other values. This measure is able to simultaneously identify two kinds of anomalies called *lower outliers* and *upper outliers*, namely exceptionally low or high frequent values. The experiments highlight that the method is scalable and able to identify anomalies of different nature from traditional techniques.

**Keywords:** Outliers · Categorical attributes · Outlier explanation

## 1 Introduction

Outlier detection is a well known discovery problem. Outliers arise due to mechanical faults, fraudulent behaviour, human errors, instrument error or simply through natural deviations in populations.

As outliers are interesting because they are suspected of not being generated by the same mechanisms as the rest of the data, it is important to justify why detected outliers are generated by some other mechanisms. However, the border between data normality and abnormality is often not clear cut, consequently, while some outlier detection methods assign to each object in the input data set a label of either "normal" or "outlier", in this paper we describe a method able to single out anomalous values occurring within the dataset.

We deal with categorical data and, specifically, we take the perspective of perceiving an attribute value as anomalous if its frequency occurrence is exceptionally typical or un-typical within the distribution of frequencies occurrences of any other attribute value. However, within the categorical scenario the process of comparing frequencies poses different challenges. Indeed, if we take the point of view that the data at hand is the result of a sampling procedure in which data values are associated with some pre-defined occurrence probabilities, then the fact that a certain categorical value is observed exactly $f$ times is a matter

© Springer Nature Switzerland AG 2019
P. Kralj Novak et al. (Eds.): DS 2019, LNAI 11828, pp. 319–334, 2019.
https://doi.org/10.1007/978-3-030-33778-0_25

of chance rather than being a hard property of that value. This has led us to the definition of the concept of *soft frequency occurrence* which, intuitively, consists in the *estimate of the density associated with frequency occurrences*. We obtain this measure by specializing the classical Kernel Density Estimation technique to the domain of frequency values.

As a second contribution, we leverage the cumulated frequency distribution of the above density estimate to decide if the frequency of a certain value is rare when compared to the frequencies associated with the other values. In particular, we are able to identify two kind of anomalies, namely *lower outliers* and *upper outliers*. A *lower outlier* is a value whose frequency is low while, typically, the dataset objects assume a few similar values, namely the frequencies of the other values are high. An *upper outlier* is a value whose frequency is high while, typically, the dataset objects assume almost distinct values, namely the frequencies of the other values are low. Both scenarios can be singled out by one unified outlierness measure; this peculiarity clearly differentiates our proposal from almost all the existing measures of outlierness.

Despite values can show exceptional behaviour with respect to the whole population, it should be noted that very often a value emerges as exceptional only when we restrict our attention to a subset of the whole population. Thus, our technique is designed to output the so-called *explanation-property pairs* $(E, p)$, where $E$, called *explanation*, denotes a condition used to determine the target subpopulation and $p$, called *property*, represents an attribute $p_a$ and a value $p_v$ such that $p_v$ is exceptionally frequent or infrequent within the subpopulation selected by the explanation $E$. This allows us to provide an interpretable explanation for the abnormal values discovered. The output of the algorithm corresponds to the so-called *explanation-property pairs*.

As a further difference, our technique is *knowledge-centric* as the search space we visit is formed by explanation-property pairs and the outliers we provide can be seen as a product of the knowledge mined. This is clearly different from traditional outlier detection approaches which are object-centric.

The rest of the work is organised as follows. Section 2 discusses work related with the present one. Section 3 introduces the frequency occurrence function. Section 4 describes the outlierness function for ranking categorical values. Section 5 describes experimental results.

## 2   Related Works

Categorical outlier detection has witnessed some interest from literature, however the identification of both *features* and *subpopulations* which characterise anomalies has not been studied in depth, in fact, there is little literature about detecting anomalous properties, and/or related outlier objects, equipped with explanations.

Moreover, to the best of our knowledge, no technique is able to natively detect *upper* outliers, while our measure of outlierness based on *density estimation* of frequency occurrences is completely novel. Most of the traditional outlier detection methods have been basically explored in the context of numerical data, among them there are distance-based methods [4], density-based methods [5], and many others [6].

Dealing with anomalies in categorical data is still evolving. The fundamental challenge in solving this problem in presence of categorical values is the difficulty in defining a suitable similarity measure over them. So, different strategies have been proposed to face with the problem above.

In [7] some methods are presented to *map categorical data on numerical data* together with a framework for categorical data analysis using a rich set of techniques that are usually applicable to continuous data. However, the effectiveness of these techniques is hardly related to the choice of the mapping. The ROAD algorithm [15], applies the classical *density-based approach* on categorical data. The algorithm exploits both densities and distances. In particular, the Hamming distance is employed to compute cluster-based outliers and the density, evaluated as the mean frequency of the values, to compute frequency-based outliers. Both measures present some limitations: as for Hamming distance, outliers with few exceptional attributes are not captured, as for densities, they are not compared with expected values and then attributes associated with distinct values, as primary keys, can affect results. Ruled-based methods borrow the concept of frequent item-set from association rule mining, as done in [12]. The algorithm uses the concept of Frequent Item-sets Mining (FIM): it first extracts patterns, or sets of items (categorical values), that co-occur frequently in the dataset and then assigns an outlier score to each data point based on the number of frequent sets it contains. Thus, outliers are likely to be points that contain relatively few frequent patterns (item-sets). According to this approach, however, outliers are points that do not contain frequent patterns and not points containing infrequent patterns and anomalies due to exceptionally high frequencies are not taken into account. *Information-theoretic measures*, such as the entropy, are used in [10] to measure the disorder of a dataset with the outliers removed: a dataset that contains many outliers supposedly has a great amount of mess, so removing outliers will lead to a dataset that is less disordered. However, this measure has to be intended as a global measure involving simultaneously all the attributes, so it is neither able to detect outliers in sub-populations nor to identify outliers characterised by one (or few) outlying attributes. Moreover, computing the measure on a fixed dataset is a hard problem. The methods proposed in [9] computes an outlier factor on the basis of the ratio between the probability of *co-occurrence of two sets of attributes* and the product between the probabilities of occurrence of the two sets taken separately. Thus, authors are interested in properties consisting in at least two attributes and do not address subpopulations. Recently [14] proposed a novel coupled unsupervised outlier detection method CBRW. It estimates the outlierness of each feature value which can either detect outliers directly or determine feature selection for subsequent outlier detection. The value is computed by *comparing the frequency of each value with the most frequent value (the mode)*. However, this is just a measure of deviation, explanations are not provided and, by definition, upper outliers cannot be detected.

A problem associated with outlier detection, but less explored in literature, is that of outlier explanation [2], which consists in finding features that can justify the outlierness of an object. [1,2] propose a technique for categorical and numerical domains respectively that, *given in input one single object known to be outlier*, provides features justifying its anomaly and subpopulations where its exceptionality is evident. A generalisation is proposed in [3] where a set, required to be small, of outliers is provided in input. The work [11] provides *intentional knowledge* by finding the subspaces that better explain why the object is an outlier, that are those where objects score the largest scores as distance-based outliers. However, the above measure is monotonic with respect to the subset inclusion relationship: if an object is an outlier in a subspace, then it will be an outlier in all its supersets. Although meaningful when dealing with homogeneous numerical attributes, this kind of monotonicity may not be in general suitable for categorical attributes, where often objects exhibit outlierness only in characterising subspaces. [8] use *spectral embeddings* to detect subspaces where anomalous objects achieve high outlier scores and normal objects keep the same distances from each other. [13] instead proposes a technique to explain outliers based on the building of a *binary classifier* to separate inliers from outliers based on subspace explorations. However, these methods are neither designed for categorical data nor to exploit subspaces to select subpopulations as is natural in our scenario.

## 3   Frequency Occurrence

In this section we give some preliminary definitions and introduce the notation employed throughout the paper.

A *dataset* $\mathcal{D}$ on a set of *categorical* attributes $\mathbf{A}$ is a set of objects $o$ assuming values on the attributes in $\mathbf{A}$. By $o[a]$ we denote the value of $o$ on the attribute $a \in \mathbf{A}$. $\mathcal{D}[a]$ denotes the multiset $\{o[a] \mid o \in \mathcal{D}\}$.

A *condition* $\mathcal{C}$ is a set of pairs $(a_i, v_i)$ where each $a_i$ is an attribute and each $v_i \in \mathcal{D}[a_i]$. A singleton condition is said to be *atomic*. By $\mathcal{D}_\mathcal{C}$ we denote the new dataset $\{o \in \mathcal{D} \mid o[a_i] = v_i, \forall (a_i, v_i) \in \mathcal{C}\}$.

**Definition 1 (Frequency distribution).** *A frequency distribution $\mathcal{H}$ is a multiset of the form $\mathcal{H} = \{f_1^{(1)}, \ldots, f_1^{(w_1)}, \ldots, f_n^{(1)}, \ldots, f_n^{(w_n)}\}$ where each $f_i^{(j)} \in \mathbb{N}$ is a distinct frequency, $f_i^{(j)} = f_i^{(k)} = f_i$ for each $1 \leq j, k \leq w_i$, and $w_i$ denotes the number of occurrences of the frequency $f_i$. By $N(\mathcal{H})$ (or simply $N$ whenever $\mathcal{H}$ is clear from the context) we denote $w_1 \cdot f_1 + \ldots + w_n \cdot f_n$.*

*For the sake of simplicity, we will refer to a frequency distribution as a set $\mathcal{H} = \{f_1, f_2, \ldots, f_n\}$ and to the number of occurrences $w_i$ of $f_i$ as $w(f_i)$. To ease the writing of expressions, we also assume that the dummy frequency $f_0 = 0$ with $w_0 = 0$ is always implicitly part of any frequency distribution.*

Given a multiset $V$, the frequency $f_v^V$ of the value $v \in V$ is the number of occurrences of $v$ in $V$.

The frequency distribution of the dataset $\mathcal{D}$ on the attribute $a$ is the multiset $\mathcal{H}_a^{\mathcal{D}} = \{f_v^{\mathcal{D}[a]} \mid v \in \mathcal{D}[a]\}$. Note that $N(\mathcal{H}_a^{\mathcal{D}}) = |\mathcal{D}|$.

**Theorem 1.** *Let $\mathcal{H} = \{f_1, \dots, f_n\}$ be a frequency distribution. Then, $n \leq \sqrt{N(\mathcal{H})}$.*

*Proof.* Since $N(\mathcal{H}) = w_1 \cdot f_1 + w_2 \cdot f_2 + \cdots + w_n \cdot f_n$, $n$ is maximized when $(i)$ $f_1 = 1$, $(ii)$ $\forall i, w_i = 1$, and $(iii)$ $\forall i > 1, f_{i+1} = f_i + 1$. Thus, the maximum $n$ is such that $1 + 2 + \cdots + n = N(\mathcal{H})$ and, since $1 + 2 + \cdots + n = \frac{n(n+1)}{2}$, it follows that $n \cdot (n + 1) = 2 \cdot N(\mathcal{H})$ and, then, that $n = O(\sqrt{N(\mathcal{H})})$.

From the above theorem, it immediately follows that the number of distinct frequencies in $\mathcal{H}^D$ is at most $\sqrt{|\mathcal{D}|}$.

Now we define the notion of *frequency occurrence* as a tool for quantifying how frequent is a certain frequency.

**Definition 2 (Hard frequency occurrence).** *Given a frequency distribution $\mathcal{H}$, the frequency occurrence $\mathcal{F}_{\mathcal{H}}(f_i)$ of $f_i$, also denoted by $\mathcal{F}(f_i)$ whenever $\mathcal{H}$ is clear from the context, is the product $w_i \cdot f_i$.*

The above definition allows us to associate with each distinct value in $\mathcal{D}[a]$ a score that is related not only to its frequency in the dataset but also to how many other values have its same frequency.

A major drawback of the previous definition is that close frequency values do not interact with each other and, as a consequence, small variations of the frequency distribution may cause sensible variations in the *frequency occurrence* values. E.g., consider the case in which the frequencies $f_i = 49, w_i = 1$ and $f_{i+1} = 51, w_{i+1} = 1$ are replaced with $f_i' = 50, w_i' = 2$. While in the former case $\mathcal{F}(f_i) = 49$ and $\mathcal{F}(f_{i+1}) = 51$, in the latter case we have that $\mathcal{F}(f_i') = 100$ that is about twice the frequency occurrence associated with $f_i$ and $f_{i+1}$. Intuitively, we do not desire a similar small variation in the frequency distribution to impact so largely on the outcome of the measure. Indeed, if we take the point of view that the data at hand is the result of a sampling procedure in which data values are associated with some pre-defined occurrence probabilities, then the fact that a certain categorical value is observed exactly $f$ times is a matter of chance, rather than being an hard property of that value.

Thus, now we refine the previous definition of frequency occurrence in order to cope with the scenario depicted above. Specifically, to overcome the mentioned drawback, we need to force close frequency values to influence each other in order to jointly contribute to the frequency occurrence value. With this aim, we inspired to Kernel Density Estimation (KDE) methods to design an ad-hoc density estimation procedure.

First of all, we point out that we are working in a discrete domain composed of frequency values, a peculiarity that differentiates it from the standard framework of KDE. We start by illustrating the proposed density estimation procedure.

A *(discrete) kernel function $K_{f_i}$ with parameter $f_i$* is a probability mass function having the property that $\sup_{f \geq 0} K_{f_i}(f) = K_{f_i}(f_i)$.

Given an interval $I = [f_l, f_u]$ of frequencies, a frequency $f_i$, and a kernel function $K$, the *volume* of $K_{f_i}$ in $I$, denoted as $V_I(K_{f_i})$, is given by $\sum_{f=f_l}^{f_u} K_{f_i}(f)$. The following expression

$$\mathcal{F}(f) = \sum_{\varphi \in I(f)} \left\{ \sum_{i=1}^{n} w_i \cdot f_i \cdot K_{f_i}(\varphi) \right\}.$$

where $I(f)$ represents an interval of frequencies centred in $f$, provides the density estimate of the *frequency occurrence* of the frequency $f$.

Since $K_{f_i}(\cdot)$ is a probability mass function, the frequency $f_i$ provides a contribution to the *frequency occurrence* of $f$ corresponding to the portion of the volume of $K_{f_i}$ which is contained in $I(f)$, that is $V_{I(f)}(K_{f_i})$. Hence, if the interval $I(f)$ contains the entire domain of $K_{f_i}$ then $f_i$ provides its maximal contribution $w_i \cdot f_i$. Frequencies $f_i$ whose domain do not intersect $I(f)$ do not contribute to the *frequency occurrence* of $f$ at all.

The above definition needs to properly calibrate the width $I(f)$ of the interval to be centred in $f$. To eliminate the dependence of the formulation from an arbitrary interval, we resort to the following alternative formulation in which frequencies $\varphi$ are not constrained to belong to the interval $I(f)$. However, since the generic kernel $K_{f_i}(\cdot)$ could be arbitrarily far from the frequency of interest $f$, now its contribution has to be properly weighted

$$\mathcal{F}(f) = \sum_{\varphi \geq 0} \left\{ \sum_{i=1}^{n} \left[ w_i \cdot f_i \cdot K_{f_i}(\varphi) \cdot \frac{Pr[X_{f_i} = f]}{Pr[X_{f_i} = f_i]} \right] \right\}.$$

Let $X_{f_i}$ denote the random variable distributed according to $K_{f_i}$ and, hence, having $f_i$ as the value that is most likely to be observed. The ratio $\frac{Pr[X_{f_i}=f]}{Pr[X_{f_i}=f_i]} \leq 1$ represents a weight factor for the kernel $K_{f_i}(\cdot)$ which is maximum, in that evaluates to 1, for $f = f_i$. Hence, the closer the kernel $K_{f_i}(\cdot)$ to the frequency of interest $f$, the larger its contribution to the *frequency occurrence* of $f$. Since the above probabilities can be directly obtained from the associated kernel, it can be rewritten as follows

$$\mathcal{F}(f) = \sum_{\varphi \geq 0} \left\{ \sum_{i=1}^{n} \left[ w_i \cdot f_i \cdot K_{f_i}(\varphi) \cdot \frac{K_{f_i}(f)}{K_{f_i}(f_i)} \right] \right\}. \tag{1}$$

Equation (1) can be rewritten as

$$\mathcal{F}(f) = \sum_{i=1}^{n} \left[ w_i \cdot f_i \cdot \frac{K_{f_i}(f)}{K_f(f)} \cdot \sum_{\varphi \geq 0} K_{f_i}(\varphi) \right].$$

Since $K_{f_i}(\cdot)$ is a probability mass function, the summation over the domain of all its values is equal to 1, thus, the above expression can be finally simplified in

$$\mathcal{F}(f) = \sum_{i=1}^{n} \left[ w_i \cdot f_i \cdot \frac{K_{f_i}(f)}{K_{f_i}(f_i)} \right]. \tag{2}$$

Since $\mathcal{F}$ represents a notion of density function associated with frequency occurrences, it is preferable that its volume evaluated in the frequencies $\mathcal{H} = \{f_1, \ldots, f_n\}$ evaluates to $N(\mathcal{H})$. This leads to the following final form of the *frequency occurrence* function.

**Definition 3 (Soft occurrence function).** *Given a frequency distribution $\mathcal{H}$, the frequency occurrence $\mathcal{F}_{\mathcal{H}}(f_i)$ of $f_i$, also denoted by $\mathcal{F}(f_i)$ whenever $\mathcal{H}$ is clear from the context, is given by the following expression*

$$\mathcal{F}(f) = \frac{N(\mathcal{H})}{N_{\mathcal{F}}(\mathcal{H})} \cdot \sum_{i=1}^{n} \left[ w_i \cdot f_i \cdot \widehat{K}_{f_i}(f) \right], \tag{3}$$

*where*

$$\widehat{K}_{f_i}(f) = \frac{K_{f_i}(f)}{K_{f_i}(f_i)} \quad and \quad N_{\mathcal{F}}(\mathcal{H}) = \sum_{j=1}^{n} \left\{ \sum_{i=1}^{n} \left[ w_i \cdot f_i \cdot \widehat{K}_{f_i}(f_j) \right] \right\}.$$

**Theorem 2.** *Let $\mathcal{D}$ be a dataset and let $\mathcal{H}^{\mathcal{D}} = \{f_1, \ldots, f_n\}$. Then, the cost of computing the set of frequency occurrences $\{\mathcal{F}_{\mathcal{H}^{\mathcal{D}}}(f_1), \ldots, \mathcal{F}_{\mathcal{H}^{\mathcal{D}}}(f_n)\}$ is $O(|\mathcal{D}| \cdot C_K)$, where $C_K$ represents the cost of evaluating $\widehat{K}_{f_i}(\cdot)$.*

*Proof.* Consider Eq. (3). The cost of evaluating this Equation for a given $f$ involves the computation of a summation of $n$ terms, with $n$ equals to the number of different frequencies. Due Theorem 1 $n = O\left(\sqrt{|\mathcal{D}|}\right)$. Since we have to evaluate Eq. (3) for any distinct $f$ in the dataset, the Equation has to be computed for $n$ times. Then, since each evaluation costs $C_k$, the overall cost is $O\left(\sqrt{|\mathcal{D}|} \cdot \sqrt{|\mathcal{D}|} \cdot C_k\right) = O(|\mathcal{D}| \cdot C_k)$.

As for the kernel selection, interestingly we can take advantage of the peculiarity of the frequency domain to base our estimation on a very natural kernel definition. Indeed, as kernel $K_{f_i}(\cdot)$ we will exploit the binomial distribution $binopdf(f; n, p)$ with parameter $n$, denoting the number of independent trials, equal to $N(\mathcal{H})$, and parameter $p$, denoting the success probability, equal to $p = f_i/N(\mathcal{H})$.

Now we take into account the cost of computing the term $\widehat{K}_{f_i}(f)$ when $K_{f_i}(\cdot)$ is the binomial kernel.

**Theorem 3.** *Given a dataset $\mathcal{D}$ and two frequencies $f_i$ and $f_j$ in $\mathcal{H}^{\mathcal{D}}$, the cost of computing $\widehat{K}_{f_i}(f_j)$ is $O(1)$ with a pre-processing $O(|\mathcal{D}|)$.*

## 4   Categorical Outlierness

In this section we introduce the concept of *outlierness* and discuss about the measure we have designed to discover outlier properties in categorical datasets.

**Definition 4 (Cumulated frequency distribution).** *Given a frequency distribution $\mathcal{H} = \{f_1, \ldots, f_n\}$, the associated cumulated frequency distribution $H$ is*

$$H(f) = \sum_{f_j \le f} \mathcal{F}_{\mathcal{H}}(f_j).$$

*In the following, we refer to the value $H(f_i)$ also as to $H_i$.*

The idea behind the measure we will discuss in the following is that an object in a categorical dataset can be considered an outlier with respect to an attribute if the frequency of the value assumed by this object on such an attribute is rare if compared to the frequencies associated with the other values assumed on the same attribute by the other objects of the dataset.

We are interested in two relevant kinds of anomalies referring to two different scenarios.

**Lower Outlier.** An object $o$ is anomalous since for a given attribute $a$ the value that $o$ assumes in $a$ is rare (its frequency is low) while, typically, the dataset objects assume a few similar values, namely the frequencies of the other values are high.

**Upper Outlier.** An object $o$ is anomalous since for a given attribute $a$ the value that $o$ assumes in $a$ is usual (its frequency is high) while, typically, the dataset objects assume almost distinct values, namely the frequencies of the other values are low.

In order to discover outliers, we exploit the cumulated frequency distribution associated with the dataset. With this aim, we use the area above and below the curve of the cumulated frequency distribution to quantify the degree of anomaly associated with a certain frequency. Intuitively, the larger the area above the portion of the curve included from a certain frequency $f_i$ to the maximum frequency $f_{\max}$, and the more $f_i$ differs from frequencies that are greater than $f_i$. At the same time, the larger the area below the portion of the curve included from the minimum frequency $f_{\min}$ and a certain frequency $f_i$, and the more $f_i$ differs from frequencies that are smaller than $f_i$.

You can evaluate the contribution given by the area *above* the *cumulated frequency distribution curve* to the outlierness of a certain frequency $f_i$, using the following expression:

$$A^{\uparrow}(f_i) = \sum_{j > i} (f_j - f_{j-1}) \cdot (H_n - H_{j-1}). \tag{4}$$

The *lower outlier score* $out^{\downarrow}(f_i)$ is given by the the normalised area

$$out^{\downarrow}(f_i) = A^{\uparrow}(f_i) / A^{\uparrow}_{\max}(f_i), \tag{5}$$

obtained by dividing the area $A^{\uparrow}(f_i)$ by

$$A^{\uparrow}_{\max}(f_i) = (A^{\uparrow}(f_0) - A^{\uparrow}(f_i)) + (f_n - f_i) \cdot (H_n - H_{i-1}) \tag{6}$$

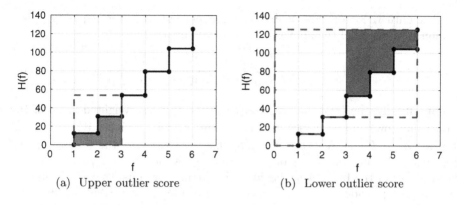

(a)  Upper outlier score                    (b)  Lower outlier score

**Fig. 1.** Outlierness computation example.

corresponding to the area above the cumulated frequency histogram up to the frequency $f_i$, represented by the term $(A^\uparrow(f_0) - A^\uparrow(f_i))$, plus an upper bound to the area above the cumulated frequency histogram starting from $f_i$, represented by the term $(f_n - f_i) \cdot (H_n - H_{i-1})$. Notice that the former term is minimised for $f_i \to 1$, while the latter term tends to $A^\uparrow(f_i)$ for $f_n \to \infty$ and, hence, in this case $out^\downarrow(f_i)$ tends to its maximum value 1.

The second scenario we are interested in aims to highlight the *upper outliers*, namely those objects that, for a given attribute, assume a value whose frequency is high, while typically, the dataset objects assume distinct values, that is the frequencies of the other values are low.

In order to discover such a kind of anomaly we take into account the area *below* the cumulated frequency distribution, starting from the lowest frequency up to the target frequency $f_i$. The bigger this area, the more this frequency can be highlighted as anomalous.

The contribution of the frequency $f_i$ is computed as

$$A^\downarrow(f_i) = \sum_{j \leq i} (f_j - f_{j-1}) \cdot H_{j-1}. \tag{7}$$

The *upper outlier score* $out^\uparrow(f_i)$ is given by the the normalised area

$$out^\uparrow(f_i) = A^\downarrow(f_i)/A^\downarrow_{\max}(f_i), \tag{8}$$

obtained by dividing the area $A^\downarrow(f_i)$ by the term

$$A^\downarrow_{\max}(f_i) = (f_i - 1) \cdot H_i \tag{9}$$

representing an upper bound to the area below the cumulated frequency histogram up to the frequency $f_i$. Notice that $A^\downarrow(f_i)$ tends to $A^\downarrow_{\max}(f_i)$ for $f_{i-1} \to 1$ and $H_i \to H_{i-1}$, or equivalently $\mathcal{F}(f_i) \ll \mathcal{F}(f_{i-1})$, so in this case $out^\uparrow(f_i)$ tends to its maximum value 1.

The outlierness, or abnormality score, associated with the frequency $f_i$ is a combined measure of the above two normalised areas:

$$out(f_i) = \frac{W_i^\uparrow \cdot out^\uparrow(f_i) + W_i^\downarrow \cdot out^\downarrow(f_i)}{W_i^\uparrow \cdot \phi(out^\uparrow(f_i)) + W_i^\downarrow \cdot \phi(out^\downarrow(f_i))} \tag{10}$$

Specifically, the (global) outlierness score of $f_i$ is the weighted mean of the upper and lower outliernesses associated with $f_i$, with weights $W_i^\uparrow = H_i$ and $W_i^\downarrow = (H_n - H_{i-1})$, respectively. Note that $H_i$ represents the fraction of the frequencies having value less or equal than $f_i$, while $(H_n - H_{i-1})$ represents the fraction of the frequencies having value greater or equal than $f_i$. Thus, the two weights provide the relative importance of the two contributions in terms of the fraction of the data population used to compute each of them. As for the function $\phi(x)$, it evaluates to 0 if $x = 0$, and to 1 otherwise. Thus, it serves the purpose of ignoring the weight associated with the lower or upper outlierness if it evaluates to 0 and, otherwise, of taking it into account in its entirety.

In order to clarify areas employed for outlierness computation, consider the following example. Consider a single attribute dataset whose associated set of distinct frequencies is $\{f_1 = 1, f_2 = 2, f_3 = 3, f_4 = 4, f_5 = 5, f_6 = 6\}$ and the set of weights is $\{w_1 = 3, w_2 = 2, w_3 = 1, w_4 = 2, w_5 = 1, , w_6 = 2\}$. Assume that we want to compute the outlierness associated with the frequency $f_3 = 3$. Figures 1a and b represent the areas exploited to compute such outlierness. On the left the area $A_i^\downarrow$ together with the area used for normalisation, $A_\downarrow^{max}$, is reported, while, on the right, the area $A_i^\uparrow$ together with the area used for normalisation, $A_\uparrow^{max}$, is reported.

If $W_i^\downarrow > W_i^\uparrow$ we say that the global score is of an *upper score*. Conversely if $W_i^\uparrow \leq W_i^\downarrow$ we say that the global score is a *lower score*.

We will use the notation $out_\mathcal{H}(\cdot)$ whenever it is needed to highlight the frequency distribution $\mathcal{H}$ used to compute the outlierness. The outlierness $out_a(v, D)$ of the value $v \in D[a]$ with respect to the attribute $a$ in the dataset $D$, is given by $out_{\mathcal{H}^{D[a]}}(f_v^{\mathcal{H}^{D[a]}})$.

Exceptional values $v$ for an attribute $a$, are those associated with large values of outlierness $out_a(v, D)$. Thus, we are interested in detecting such exceptional values. However, it must be pointed out that very often a value emerges as exceptional for a certain attribute only when we restrict our attention to a subset of the whole population. This intuition leads to the definition of the notion of explanation-property pair.

**Definition 5.** *An explanation-property pair $(E, p)$, or simply pair for the sake of conciseness, consists of condition $E$, also called explanation, and of an atomic condition $p = \{(p_a, p_v)\}$, also called property. By $p_a$ ($p_v$, resp.) we denote the attribute (value, resp.) involved in the atomic condition $p$.*

Given a pair $\pi = (E, p)$, $\mathcal{D}_\pi$ denotes the set of objects $\mathcal{D}_{E \cup \{p\}}$. The outlierness $out(\pi)$ of an explanation-property pair $\pi = (E, p)$ is the outlierness $out_{p_a}(p_v, D_E)$ of the value $p_v$ with respect the attribute $p_a$ in the dataset $D_E$.

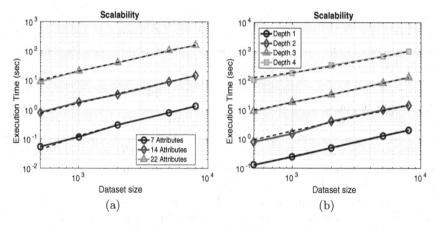

**Fig. 2.** Scalability analysis.

We implemented an algorithm that receives in input a dataset $\mathcal{D}$ and a depth parameter $\delta \geq 1$, and returns all the pairs $(E, p)$ among those composed of at most $\delta$ atomic conditions, that is such that $|E| \leq \delta$. The algorithm analyzes all possible explanations of length less or equal than $\delta$ according to a depth-first strategy. This strategy allows a very efficient selection of sub-populations exploiting an approach similar to the one described in [2].

## 5    Experimental Results

First of all, to study the applicability of our method to real datasets, we have tested its scalability by varying the number of objects, the number of attributes, and the depth parameter. Then, to clarify the different nature of our anomalies with those returned by other outlier detection methods, we compared our method with traditional distance-based and density-based outlier detection approaches and with a method tailored for categorical data. Finally, we discussed knowledge mined by means of our approach.

In the experiments we used the following dataset from the UCI ML Repository: *Zoo* ($n = 101$ objects and $m = 18$ attributes), *Mushrooms* ($n = 8{,}124$ objects and $m = 22$ attributes),

*Scalability.* Figure 2 shows the scalability analysis of our method on the *Mushrooms* dataset. In the experiment reported in Fig. 2a, we varied the number of objects $n$ in $\{500, 1000, 2000, 5000, 8000\}$ and the number of attributes $m$ in $\{7, 14, 22\}$, while the depth parameter has been held fixed to $\delta = 3$. The dashed lines represent the trend of the linear growth estimated exploiting regression. This estimation confirms that the algorithm scales linearly with respect to the dataset size. As for the number of attributes, as expected for a given number of objects, the execution time increases due to the growth of the associated search space. On the full dataset the execution time is very contained as it amounts to about 2 min. In the experiment reported in Fig. 2b, we varied both the number of objects $n$ and the depth parameter $\delta$ in $\{1, 2, 3, 4\}$, while considering the full feature space. Also in this case the linear growth is represented by the dashed lines, and similar considerations can be drawn.

*Comparison with outlier detection methods.* We compare our method with two of the main categories of outliers:

($i$) *distance-based* approaches, that are used to discover *global* outliers, i.e. objects showing abnormal behaviour when compared with the whole dataset population; ($ii$) *density-based* approaches, which are able to single out *local* outliers, i.e. objects showing abnormal behaviour when compared with a certain subset of the data with their neighbourhood.

As distance-based definition, we use the average KNN score, representing the average distance from the $k$-nearest neighbours of the object. As density-based, we use Local Outlier Factor or LOF [5]. Both methods employ the Hamming distance. Moreover, we compare our method with the ROAD algorithm [15] that exploits both distances and densities, namely it establishes two independent rankings: ($i$) each data object is assigned to a frequency score and objects with low scores are considered as outliers (Type-1 outliers); ($ii$) the $k$-mode clustering is performed in order to isolate those objects that are far from *big clusters* (i.e. clusters containing at least $\alpha\%$ of the whole dataset) according to Hamming distance (Type-2 outliers). The goal of these experiments is to highlight that we are able to detected anomalies of different nature and to provide evidence that our method is knowledge-centric, since it concentrates on anomalous values, as opposed to classical methods which instead are object-centric.

To compare the approaches, we ranked the dataset objects $o$ by assigning to each of them the largest outlierness of a pair $\pi$ such that $o \in \mathcal{D}_\pi$. We determined our top–10 outliers by selecting the objects associated with the largest outliernesses. Then we selected these objects, containing values deemed to be exceptional by our method, with the purpose of verifying how they are ranked by popular object-centric techniques. Hence, we computed their outlier scores according to the KNN, LOF and ROAD definitions.

All the chosen competitors require an input parameter $k$, representing the number of $k$ nearest-neighbors or the number of clusters to be taken into account. Since selecting the right value of $k$ is a challenging task we computed the KNN, LOF and ROAD outlier scores for all the possible values of $k$ and determined the ranking positions associated with our top–10 outliers. Particularly, all the integers from 1 to the number of objects $n$ have been considered for KNN while 30 log-space values between 1 and $n$ have been considered for LOF due to its higher temporal cost. For ROAD algorithm, we stopped at the value of $k$ such that at least a *big cluster* is obtained and use the frequency score to rank those objects having the same distance from their nearest big cluster.

Figures 3 and 4, report the box-plots for $k$ varying in $[1, n]$ of the KNN, LOF and ROAD Type-2 outliers rankings associated with our top–10 outliers. Plots on the top concern lower outliers, while plots on the bottom concern upper outliers. From these plots it can be seen that the median ranking associated with

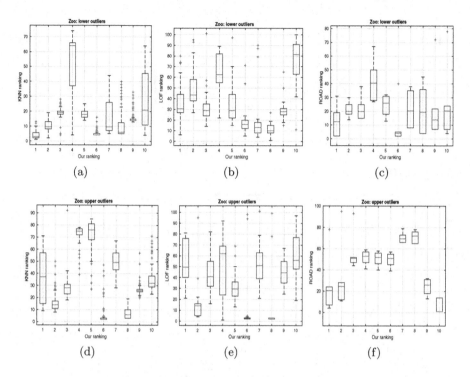

**Fig. 3.** Comparison with KNN, LOF and ROAD on *Zoo*.

our outliers can be far away from the top and also that, within the whole ranking distribution, the same outlier can be ranked in very different positions. In general, it seems that lower outliers are likely to be ranked better than upper outliers by our competitors, and this witnesses for the peculiar nature of upper outliers. On the *Zoo* dataset there is no apparent correlation between our outliers and KNN, LOF and ROAD outliers. On the *Mushrooms* dataset some of our lower outliers are, on the average, ranked very high also by the other algorithms. Some of them are almost always top outliers for all methods (see the top 1st, 2nd, 5th, and 7th outliers) thus witnessing that these outliers have both global and local nature. However, most of our outliers are not detected by these techniques.

Before concluding this comparison, it must be pointed out that the best rankings associated with the selected objects are obtained for very different values of the parameter $k$. Since, the output of the KNN, LOF and ROAD methods are determined for a selected value of $k$, it is very unlike that, even in presence of some agreement between our top outliers and local and global outliers, they are simultaneously ranked in high positions for the same provided value of $k$.

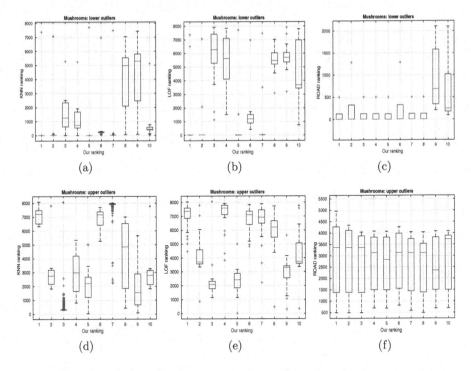

**Fig. 4.** Comparison with KNN, LOF and ROAD on *Mushrooms*.

*Knowledge mined.* In this section we discuss about the experiments conduced on Zoo dataset in order to highlight the kind of knowledge our technique is able to find out. Information provided by the top-10 lower top pairs is summarized below:

- Among non-acquatic animals, the **clam** is the only one that breathes;
- The **platypus** lays eggs although it provides milk;
- Among predators without feathers, the **ladybird** is the only airborne;
- The **stingray** is a catsize animal, but it is venomous;
- Among catsized animals, the **octopus** is the only invertebrate;
- Among airborne and providing milk animal, the **calf** is a domestic one;
- The **crab** is a four-legs animal and it is the only invertebrate;
- Among vertebrate breathing animals, the **pitviper** is among the very few venomous ones.

Instead, mining upper top pairs, we find out that the dataset contains the frog twice, the former is venomous and the latter is no-venomous. However, the animal names are like primary keys for the dataset, so having the same name twice can be pointed out as an anomalous behavior. Our technique is able to highlight such a situation. Finally, upper outliers highlight some curiosities about the animal world including the following:

- Among breathing not catsized predators the most frequent are **not flying birds**;
- Among no-feathers no-toothed animals most have **six legs**;
- Among no-flying breathing catsized animals, the most frequent are **mammals**;
- Most **gastropods** have no legs;
- Most **breathing**, venomous animals have six legs;
- Most **no-flying acquatic no-toothed** have four legs;
- Most **no-toothed** have two legs.

## 6   Conclusions

In this work we have provided a contribution to single out and explain anomalous values in categorical domains. We perceive frequencies of attribute values as samples of a distribution whose density has to be estimated. This lead to the notion of frequency occurrence we exploit to build our definition of outlier: an attribute value is suspected to be an outlier if its frequency occurrence is exceptionally typical or un-typical within the distribution of frequencies occurrences of any other attribute value. As a second contribution, our technique is able to provide interpretable explanations for the abnormal values discovered. Thus, the outliers we provide can be seen as a product of the knowledge mined, making the approach knowledge-centric rather than object-centric. The performances have been evaluated on some popular benchmark categorical datasets and a comparative view has been proposed.

## References

1. Angiulli, F., Fassetti, F., Manco, G., Palopoli, L.: Outlying property detection with numerical attributes. Data Min. Knowl. Discov. **31**(1), 134–163 (2017)
2. Angiulli, F., Fassetti, F., Palopoli, L.: Detecting outlying properties of exceptional objects. ACM Trans. Database Syst. (TODS) **34**(1), 7 (2009)
3. Angiulli, F., Fassetti, F., Palopoli, L.: Discovering characterizations of the behavior of anomalous subpopulations. IEEE TKDE **25**(6), 1280–1292 (2013)
4. Angiulli, F., Pizzuti, C.: Outlier mining in large high-dimensional data sets. IEEE Trans. Knowl. Data Eng. **17**(2), 203–215 (2005)
5. Breunig, M.M., Kriegel, H.P., Ng, R.T., Sander, J.: LOF: identifying density-based local outliers. In: ACM Sigmod Record, vol. 29, pp. 93–104. ACM (2000)
6. Chandola, V., Banerjee, A., Kumar, V.: Anomaly detection: a survey. ACM Comput. Surv. (CSUR) **41**(3), 15 (2009)
7. Chandola, V., Boriah, S., Kumar, V.: A framework for exploring categorical data. In: SIAM International Conference on Data Mining (SDM), pp. 187–198 (2009)
8. Dang, X.H., Assent, I., Ng, R.T., Zimek, A., Schubert, E.: Discriminative features for identifying and interpreting outliers. In: IEEE ICDE, pp. 88–99 (2014)
9. Das, K., Schneider, J.: Detecting anomalous records in categorical datasets. In: ACM International Conference on Knowledge Discovery and Data Mining (KDD), pp. 220–229 (2007)
10. He, Z., Deng, S., Xu, X.: An optimization model for outlier detection in categorical data. In: Huang, D.-S., Zhang, X.-P., Huang, G.-B. (eds.) ICIC 2005. LNCS, vol. 3644, pp. 400–409. Springer, Heidelberg (2005). https://doi.org/10.1007/11538059_42

11. Knorr, E.M., Ng, R.T.: Finding intensional knowledge of distance-based outliers. In: International Conference on Very Large Data Bases, VLDB, pp. 211–222 (1999)

12. Koufakou, A., Secretan, J., Georgiopoulos, M.: Non-derivable itemsets for fast outlier detection in large high-dimensional categorical data. Knowl. Inf. Syst. **29**(3), 697–725 (2011)

13. Micenková, B., Ng, R.T., Dang, X., Assent, I.: Explaining outliers by subspace separability. In: IEEE International Conference on Data Mining, pp. 518–527 (2013)

14. Pang, G., Cao, L., Chen, L.: Outlier detection in complex categorical data by modelling the feature value couplings. In: IJCAI, pp. 1902–1908 (2016)

15. Suri, N.R., Murty, M.N., Athithan, G.: An algorithm for mining outliers in categorical data through ranking. In: IEEE International Conference on Hybrid Intelligent Systems (HIS), pp. 247–252 (2012)

# A Framework for Human-Centered Exploration of Complex Event Log Graphs

Martin Atzmueller[1,2(✉)], Stefan Bloemheuvel[1,2], and Benjamin Kloepper[3]

[1] Department of Cognitive Science and Artificial Intelligence, Tilburg University,
Warandelaan 2, 5037 AB Tilburg, The Netherlands
{m.atzmuller,s.bloemheuvel}@uvt.nl
[2] Jheronimus Academy of Data Science (JADS),
Sint Janssingel 92, 5211 DA 's-Hertogenbosch, The Netherlands
[3] Corporate Research Center, ABB AG, Wallstadter Str. 59, 68526 Ladenburg, Germany
benjamin.kloepper@de.abb.com

**Abstract.** Graphs can conveniently model complex multi-relational characteristics. For making sense of such data, effective interpretable methods for their exploration are crucial, in order to provide insights that cover the relevant analytical questions and are understandable to humans. This paper presents a framework for human-centered exploration of attributed graphs on complex, i.e., large and heterogeneous event logs. The proposed approach is based on specific graph modeling, graph summarization and local pattern mining methods. We demonstrate promising results in the context of a real-world industrial dataset.

## 1 Introduction

The analysis of complex event logs, i.e., large amounts of log data collected from heterogeneous sources, is a challenging problem. For that, modeling the log data as attributed graphs is a promising direction, due to the powerful analysis and mining methods that are enabled by graphs capturing complex multi-relational data and information. Event logs, for example, provide time-stamped information of different events in a (complex) system, which can be enriched with further information, e.g., attributes describing properties of the respective events in a graph, leading to well-structured (feature-rich) graphs [1].

This paper provides a framework for human-centered exploration on graphs constructed from complex event log data, proposing methods for graph modeling, graph summarization and local pattern mining [2]. With that, we specifically tackle the problem of providing interpretable graph structures and patterns, allowing to reduce the complexity of the data, in order to obtain interesting patterns that are understandable to the user. In a mixed-initiative approach, the methods are applied by the user and provide support in order to model the graph and detect interesting patterns, which can be refined in an iterative approach – guided by the user. In particular, the user can also include background knowledge, constraints, and specific queries on the modeled complex event log graph, for exploration, filtering, refinement. The ultimate goal is the extraction of interesting insights and knowledge for computational sensemaking [3], and to aid in decision support for process diagnostics and optimization.

© Springer Nature Switzerland AG 2019
P. Kralj Novak et al. (Eds.): DS 2019, LNAI 11828, pp. 335–350, 2019.
https://doi.org/10.1007/978-3-030-33778-0_26

Our contributions are summarized as follows:

1. We present the proposed framework for human-centered exploration and its components, in particular graph summarization and local pattern mining methods.
2. Furthermore, we describe the modeling of a complex event log graph from respective event log data, using Markov Chain modeling and network science methods.
3. Finally, we provide a case study using a real-world industrial dataset demonstrating the application of the proposed framework, and present promising results in that context.

The rest of the paper is structured as follows: Sect. 2 discusses related work. After that, Sect. 3 presents the proposed framework for human-centered exploration of complex event log graphs. Next, Sect. 4 describes a case study using real-world industrial data and discusses promising results. Finally, Sect. 5 concludes with a summary and interesting directions for future work.

## 2   Related Work

Below, we first describe related work on the analysis and mining of event logs, before we outline graph summarization techniques.

### 2.1   Mining Event Logs

One prominent option for mining event logs is enabled by process mining approaches, which try to discover the process model of log data [4], i.e., aiming at the discovery of business process related events in a sequential event log. The assumption is that event logs contain fingerprints of business processes, which can be identified by sequence analysis. One task of process mining is conformance checking [5], which has been introduced to check the matching of an existing business process model with a segmentation of the log entries. Since event logs are inherently sequential, it is possible, e.g., to find the Petri-Net model that represents the structure of the events [6,7]. The results can then be used to improve the performance of processes, also considering declarative approaches [8,9] for including domain knowledge besides the given activity log. Based on the findings of standard process mining, also fault detection and anomaly detection can be implemented using log analysis [10]. Fault detection consists of creating a database of faulty message patterns. If an event (or a sequence of events) matches a pattern, the log system can take appropriate action. Anomaly detection then aims at building a model of the normal log behavior in order to detect unexpected behavior [10]. However, here the challenge is the specification and learning of normal behavior.

In contrast to the approaches discussed above, we do not focus on business process mining for obtaining structures of the business process. In contrast, we focus on flexible exploration of the data in a human-centered approach, which enables the inclusion of domain knowledge for the exploration, including constraints, queries, and expectations of the user. Furthermore, the standard approaches discussed above do not focus on different levels of analysis, which is enabled in our proposed framework through graph summarization and refinement. One further important focus is to provide an understandable and interpretable summary on the event log data. With that, and the automatic exploration methods, fault detection and anomaly detection can then be implemented.

## 2.2  Graph Summarization

Graph summarization speeds up the analysis of a graph by creating a lossy but concise representation of the graph [11]. Approaches include grouping-based methods, simplification-based methods, compression-based methods and influence based methods [12]. One of the most popular grouping-based methods is generating a supergraph, where nodes are recursively aggregated into supernodes and edges are aggregated into compressed or virtual nodes [13]. Simplification-based graph summarization consists of removing less important nodes or edges, which results in a sparsified graph. In comparison with supergraphs, a summary now consists of a subset of the original set of nodes and edges, e.g., [14]. However, there is no consensus so far, on what a graph summary should look like. It is always application dependent and can serve countless different goals: preserving the answers to graph queries, finding different graphs structures, merging nodes into supernodes and merging edges into superedges [12]. In [15] we have presented initial results on graph summarization using sequential pattern mining and sequence clustering focussing on simple graphs only. Here, we extend on that by considering graph clustering on attributed graphs.

Therefore, in contrast to typical approaches for graph summarization, the framework proposed in this paper enables a combination of approaches on attributed graphs. In a human-centered process, the respective graph modeling can be adapted and refined to the most suitable level of abstraction and information corresponding to the graph.

## 2.3  Local Pattern Mining on Attributed Graphs

In general, attributed (or labeled) graphs as rich graph representations enable approaches that specifically exploit the descriptive information of the labels assigned to nodes and/or edges of the graph. Exemplary approaches include density-based methods [16], distance-based methods [17], entropy-based methods [18], model-based methods [19], seed-centric methods [20] and finally (local) pattern mining approaches [21–24]. Local pattern mining, e.g., [25–27] has many flavors. Those typically are based in considering the support set of any pattern, i.e., the set of the covered objects, e.g., transactions, sequences, or nodes in the context of graph mining. The goal then is to enumerate the set of all patterns that satisfy some constraint, e.g., a minimal support in terms of the number of covered objects, or their (topological) connectivity.

In the field of complex networks, a popular approach consists of extracting a *core subgraph* from the network, i.e., some essential part of the graph whose nodes satisfy a local property. The $k$-core definition was first proposed in [28]. It requires all nodes in the core subgraph to have a degree of at least $k$. Furthermore, as an extension of the *closed pattern mining* methodology on attributed graphs, [29] proposed a generic method to enumerate the set of core closed patterns. The MinerLSD algorithm (which is included in our proposed framework) extends that, and adds efficient pruning using the local modularity of the pattern core subgraphs [2].

In contrast to the approaches and method mentioned above, our proposed framework does not primarily rely on an automatic approach. It includes the MinerLSD algorithm, as an efficient method for local pattern mining on attributed graphs, and extends it towards mining of complex event log graphs. However, pattern mining is then combined with graph summarization and refinement, in a human-centered approach.

## 3    Human-Centered Exploration Framework

In the following, we first provide an overview on the framework for human-centered exploration of complex event log graphs. After that, we provide some necessary background on graphs and patterns. Next, we discuss the central components in detail, i.e., modeling and summarizing a complex event log graph, as well as local pattern mining. In particular, we describe the adaptations of specific methods used for mining and analysis.

### 3.1    Overview

A schematic view on the framework is given in Fig. 1. The aim of the overall process is to explore the complex event data, and to obtain interesting patterns that indicate respective sequential/graph structures. As we will describe in the case study in Sect. 4 in detail, one example of an interesting (sequence) pattern is an unexpected sequence, i.e., one which is rare but not random, to be used for process diagnostics and optimization.

For the process, first the complex event log data is transformed into an attributed graph, where edges connect log events, and properties of the events given in the event log are used for the labeling of the nodes. In addition, for the modeling we can include further properties from network science, in order to provide additional features for labeling the nodes. This is included in an iterative process, where the user can refine and enrich the model. The result is an attributed *complex event log graph*. Complementary, graph summarization is performed such that the attributed graph can be presented and inspected at the appropriate level of detail, also facilitating the subsequent mining and analysis steps. As the next step, local pattern mining for detecting interesting graph structures is applied on the attributed graph. These graph structures, i.e., graph patterns inducing subgraphs on the attributed graph, are also assessed and refined by the user in an iterative process. Here, queries, constraints and domain knowledge of the user can be included. Finally, as a result of the framework, the detected subgraph structures as well as the corresponding sequences extracted from the respective subgraphs are utilized for applications such as fault/anomaly detection and process optimization.

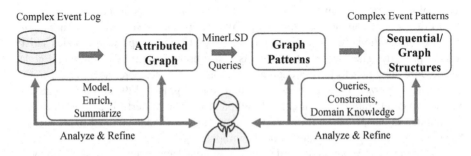

**Fig. 1.** Overview on the proposed framework: The complex event log is transformed into an attributed graph, which is modeled, enriched and summarized by the user as needed in an iterative approach. Here, already simple queries can be answered. After that, either query-driven or mixed-initiative pattern mining is performed for obtaining graph patterns, leading to sequential/graph structures which are then analyzed and refined.

## 3.2 Background: Graphs and Patterns

A *graph* $G = (V, E)$ is an ordered pair, consisting of a finite set $V$ containing the *vertices* (or *nodes*), and a set $E$ of *edges* (or *connections*) between the vertices, with $n := |V|$, $m := |E|$. A *weighted* graph is a graph $G = (V, E)$ together with a function $w : E \to \mathbb{R}^+$ that assigns a positive weight to each edge. For the *adjacency matrix* $A \in \mathbb{R}^{n \times n}$ with $n = |V|$ holds $A_{ij} = 1$ ($A_{ij} = w(i,j)$) iff $(i,j) \in E$ for $i, j \in V$, assuming a bijection from $1, \ldots, n$ to $V$. A *path* $v_0 \to_G v_n$ of *length* $n$ in a graph $G$ is a sequence $v_0, \ldots, v_n$ of nodes with $n \geq 1$ and $(v_i, v_{i+1}) \in E$ for $i = 0, \ldots, n-1$. A *shortest path* between nodes $u$ and $v$ is a path $u \to_G v$ of minimal length.

We distinguish different topological properties of a node, including local (degree, clustering coefficient) and global properties (pagerank and centrality measures), as indicators of how important a node is [30,31].

- The *degree* $\deg(i)$ of a node $i$ in a graph is the number of connections it has to other nodes, i.e., $\deg(i) := |\{j \mid A_{ij} = 1\}|$. For a directed graph, we further consider the indegree $d^{in}$ and outdegree $d^{out}$ distinguishing between incoming/outgoing edges.
- In a weighted graph, we complement the degree of a node $i$ by its strength $s(i) = \sum_j A_{ij}$, i.e., the sum of the weights of the attached edges.
- The *clustering coefficient* (or transitivity) [32] for a vertex $v \in V$ in a graph $G = (V, E)$ is defined as the fraction of possible links among $v$'s neighbors which are contained in $E$. It quantifies how densely the neighborhood of a node is connected.
- For ranking nodes, the PageRank [31] algorithm can be applied. For an $m \times m$ column stochastic adjacency matrix $A$ and damping factor $\alpha$, and uniform *preference vector* $p := (1/m, \ldots, 1/m)$, the *global* PageRank vector $w =: PR$ is given as the fixpoint of the following equation: $w = \alpha A w + (1 - \alpha) p$.
- The betweenness centrality *bet* measures the number of shortest paths of all node pairs that go through a specific node. $bet(v) = \sum_{s \neq v \neq t \in V} \frac{\sigma_{st}(v)}{\sigma_{st}}$. Hereby, $\sigma_{st}$ denotes the number of shortest paths between $s$ and $t$ and $\sigma_{st}(v)$ is the number of shortest paths between $s$ and $t$ passing through $v$. Thus, a vertex has a high *betweenness centrality* if it can be found on many shortest paths between other vertex pairs.
- The closeness centrality *clos* considers the length of these shortest paths. Then, the shorter its shortest path length to all other reachable nodes, the higher a vertex ranks: $clos(v) = \frac{1}{\sum_{t \in V \setminus v} d_G(v,t)}$. $d_G(v,t)$ denotes hereby the geodesic distance (shortest path) between the vertices $v$ and $t$.
- The eigenvector centrality *eig* of a node is an important measure of its influence, similar to the pagerank measure. Intuitively, a node is central, if it has many central neighbors. The eigenvector centrality $eig(v)$ of node $v$ is defined as $eig(v) = \lambda \sum_{\{u,v\} \in E} eig(u)$, where $\lambda \in \mathbb{R}$ is a constant.

An attributed graph $G = (V, E)$ includes a set of items $I$, and a dataset $D$ describing vertices as itemsets. Each vertex $v \in V$ is described by an itemset $D(v)$ taken from the set of items $I$. Then, for mining local patterns, we want to enumerate all (maximal) vertex subsets $W$ in $G$ such that there exists an itemset $q$ which is a subset of all itemsets $D(v), v \in W$. Furthermore, we can also require $W$ to satisfy some graph related constraints. So, in standard terminology, $q$ is a *pattern* that *occurs* in all element of $W$; this is also called the *support set* or *extension* $ext(q)$ of $q$.

### 3.3   Graph Modeling of Complex Event Log Data

An event log $L$ consists of a set of tuples $l = \{m, t, P\}$, where $m \in M$ is an event *message* or *type*, $t$ is a timestamp, enabling a temporal ordering, and $P$ is a set of attributes or properties of the event type $m$. For convenience, we assume that these properties can be modeled as an itemset, taken from a set of items $I$, such that $P(m) \subseteq I$, for event type $m$. The set $M$ contains all possible event types. Given such an event log $L$, we create an attributed (directed and weighted) graph as follows:

1. Since we are interested in the connections and sequence of the events, we model all event types $m \in M$ that occur in $L$ as nodes of the graph.
2. Furthermore, we create an edge $e = (u, v)$ connecting nodes $u \in M$ and $v \in M$ if the event type $v$ occurs directly after event type $u$ in the event log $L$. Therefore, we can simply traverse the event log with a window of size 2, considering the respective pairs of event types, and create according edges in the graph. The result is a directed multigraph, i.e., multiple edges can occur between the contained nodes. From that, we can directly obtain a weighted directed simple graph, by aggregating all (directed) edges between the nodes, calculating the weight by the number of the respective aggregated edges between two nodes.
3. For labeling the nodes, for each node $u$ we consider the union of all attribute (item) sets $A$, which occur in a log event $l = (u, t, A)$ for any $t$.

In addition to the node properties $P$ as given above, we can also utilize those using topological properties of the nodes as outlined above in Sect. 3.2, i.e., degree, strength, transitivity, pagerank, betweenness centrality, closeness centrality, and eigenvector centrality. Then, we add according (categorical) labels to $P$ based on the respective measures, applying appropriate binning on the numeric values. These properties are important to both consider the topological connectivity, as well as add information given the edges/transition. Weighted versions of these properties add information for the nodes, based on the weights of the incident edges.

In the case of a simple directed weighted graph, we add an additional modeling perspective. Since we model the graph $G$ according to the sequence of log events $L$, we can also take a sequential interpretation (according to the first order Markov property). We can then directly consider the resulting adjacency matrix $A$ of $G$ as a transition matrix, from a first order Markov Chain modeling perspective, if normalized accordingly. Thus, if we row normalize the transition counts as discussed above, then these correspond to event traversal probabilities from one event to another event, represented by the respective individual nodes.

It is important to note, that the graph modeling options introduced in this framework are quite flexible, since they already allow to view the event log graph from several perspectives: First of all, the attributes of the nodes denote their properties, where the given nodal properties can be extended and enriched using local and global topological features. In addition, regarding the edges we can consider the pure multigraph, analyzing the edges individually, or take an aggregated view on the simple graph, or a first-order Markov chain perspective, for which we calculate the transition probabilities according to the aggregated counts. From the latter, then also the probability of a sequence of nodes can be estimated by simple multiplication of the respective (transition) edges' weights.

## 3.4  Graph Summarization

For the graph summarization step, we have several options, to inspect the graph at different levels of detail. Below, we discuss graph contraction and community detection. In general, a summary $S$ of an input graph $G = (V, E)$ consists of the following:

- A partition of the nodes $V$ into parts $\{V_1, \ldots, V_k\}$, such that $V_i \subseteq V$ and $V_i \cap V_j = \emptyset$, for $i, j \in \{1, \ldots, k\}$ and $i \neq j$. A supernode is a group of nodes $V_i$ of summary $S$.
- For each of those supernodes $V_i \in V$, the summary $S$ describes the number of edges that are within the nodes in the supernode.
- For every pair of supernodes $V_i, V_j \in \mathbf{V}$, the summary $S$ gives the number of edges across the two supernodes.

**Graph Contraction.**  One popular method to summarize a graph is graph contraction. Graph contraction itself can work on nodes and/or edges, resulting in node contraction and edge contraction.

- The node contraction of a pair of vertices $V_i$ and $V_j$ of a graph produces a new graph in which the two nodes $V_i$ and $V_j$ are replaced by a new node $V$, such that $V$ is adjacent to the union of the original node. An example of this procedure can be seen in Fig. 2.
- Edge contraction is a procedure that removes an edge from a graph while also merging the two nodes that it originally joined.

In our framework different node and/or edge contraction mechanisms are possible. In the context of this paper, we applied HDBSCAN [33] – a hierarchical clustering algorithm that builds upon DBSCAN [34], allowing clusters of any arbitrary shape to be detected. HDBSCAN is applied on the respective event sequences (using a TF-IDF similarity measure, e.g., [35]), in order to group the event types. Utilizing the hierarchical clustering, the domain expert is integrated by inspecting, assessing, and labeling the respective clusters.

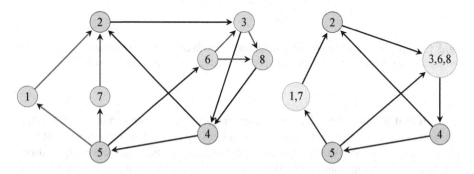

**Fig. 2.** Left: The path from node 5 to node 2 always has to cross node 1 and 7. A supernode-generating algorithm merges these two nodes into one supernode. In addition, there is a cluster of nodes in the top right corner (nodes 3, 6 and 8). Right: Nodes 3, 6 and 8 got merged into the supernode (3,6,8), nodes 1 and 7 got merged into the supernode (1,7).

For the exploration of complex log event graphs, according aggregation operators need to be defined when either nodes and or edges are contracted. Regarding the node properties, e.g., the union or intersection can be considered, while the topological properties can be recomputed in the contracted graph. Regarding the weights of the contracted edges, the aggregation is not necessarily straight-forward. While, in principle, the edge weights could also be summed up for count data, the resulting interpretation needs to take the semantics of the aggregation pattern into account. For the Markov Chain modeling, the transitions can in principle be recomputed considering transitions between sets of nodes. However, for the respective interpretation it is always important to link the contracted supernode to the detail nodal structure of its contained sub-nodes.

**Community Detection.** Another method to generate supernodes is community detection; in general, its aim is to find the optimal communities with more nodes and edges inside the community compared to nodes and edges that link to nodes in the rest of the graph. Figure 3 shows an example with three supernodes, one for each community. For community detection, there are very many potential algorithms, e.g., [36,37]. In the context of this paper, we focus on descriptive community detection, which allows to utilize attributive information of the nodes. A prominent algorithm, i.e., the COMODO algorithm [24] can be emulated by our local pattern mining method MinerLSD, when only considering the Modularity metric for assessing local communities. Furthermore, also partitioning approaches, like the GAMER algorithm [23] can be applied.

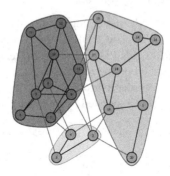

**Fig. 3.** Example of applying community detection for node contraction. The graph is compressed into a graph consisting of three nodes only, corresponding to the respective communities.

**Analysis and Refinement.** Using graph summarization, the user can inspect the graph at different levels of detail, in order to detect densely connected areas, important hotspots regarding the transitions and event log sequences. For interactive exploration, visualization is an important tool, such that e.g., also a drill down and roll up between different (hierarchical) aggregation levels is supported. In the proposed framework, this is a central step supporting the human-centered approach. Then, the domain specialist can also easily judge to merge clusters or keep them split. The complex event log graph can then be provided to the local pattern mining step, at the appropriate level of detail.

### 3.5   Local Pattern Mining on Attributed Graphs Using MinerLSD

Pattern mining on attributed graphs specifically aims at a description-oriented view, by including patterns on attributes, but also considering the topological structure. Here, we focus on such "nuggets in the data" [38] with an interpretable description.

The *MinerLSD* algorithm focuses both on local pattern mining using the local modularity metric [24], as well as graph abstraction that reduces graphs to k-core subgraphs [39]. In order to prevent the typical pattern explosion in (naive) pattern mining, MinerLSD employs closed patterns, cf. [2]. Below, we summarize the algorithm, and refer to [2] for a detailed discussion. As input parameters, MinerLSD requires a graph $G = (V, E)$, a set of items $I$, a dataset $D$ describing vertices as itemsets (as also discussed above) and a core operator $p$, e.g., focusing on k-cores. $p$ depends on $G$ and to any image $p(X) = W$ we associate the core subgraph $C$ whose vertex set is $V(C) = W$. As further parameters, MinerLSD considers the corresponding value $k$ as well as a frequency threshold $s$ (defaulting to 0) and a local modularity threshold $lm$. The algorithm outputs the frequent pairs $(c, W)$ where $c$ is a core closed pattern and $W = p \circ ext(c)$ its associated (k-)core. It is important to note, that in the enumeration step MinerLSD ensures that each pair $(c, W)$ is enumerated (at most) once. For pattern selection and ranking, MinerLSD applies the local modularity quality function $\text{MODL}(W)$. For a subgraph $W$,

$$\text{MODL}(W) = \frac{m_W}{m} - \sum_{u,v \in W} \frac{d(u)d(v)}{4m^2} = \frac{1}{m}\left(m_W - \sum_{u,v \in W} \frac{d(u)d(v)}{4m}\right),$$

where $m_W$ denotes the number of edges within the subgraph $W$, $m$ denotes the total number of edges, and $d(u)$ denotes the degree of node $u$.

For analyzing complex log event graphs using MinerLSD, we can then also utilize the directed version of the MODL measure for ranking the patterns (see [24] for details):

$$\text{MODL}^D(W) = \frac{1}{m} \sum_{u,v \in W} \left(A_{u,v} - \frac{d^{in}(u)d^{out}(v)}{m}\right) = \frac{m_W}{m} - \sum_{u,v \in C} \frac{d^{in}(u)d^{out}(v)}{m^2},$$

where $d^{in}$ and $d^{out}$ denote the indegree and outdegree of a node, respectively.

Intuitively, MODL and $\text{MODL}^D$ provide the prominent property of assigning a higher ranking to larger (core) subgraphs under consideration, if these are considerably more densely connected than expected by chance. This focuses pattern exploration on the statistically most unusual subgraphs. In addition, applying $k$-core constraints helps due to its focus on denser subgraphs, as also theoretically analyzed in [40] for $k$-cores. In addition, the local modularity neglects the importance of a minimal support threshold which is typically applied in pattern mining, since it directly includes the size of the pattern as a criterion. This enables a very efficient pattern mining approach, given either a suitable threshold for the local modularity, or by targeting the top-$k$ patterns. Furthermore, we can easily extract sequences from the (directed) subgraph regarding specific queries, and compute transition probabilities of those sequences. Moreover, using MinerLSD for explorative pattern mining, we can also utilize other core operators besides k-cores, e.g., triangles, or hubs and authorities. These core operators specify specific graph structures of interest, e.g., densely connected sets of events, as well as important events according to incoming/outgoing edges. We refer to [29,41] for a detailed discussion.

## 4  Case Study

The applied dataset was provided by ABB in anonymized version, providing a real-world event log of an industrial process. Information on the first few rows of the dataset is shown in Table 1. Besides the time the event happened, the set of attributes that describe an event are a message number, the message category and severity, the title of an event, the description that provides more information, and the location of the event (performed by a robot, which is in a cell or robots and that lies in a line of cells). The total number of events that were available in the dataset was about 4 million, capturing about 1.5 years of activity monitoring of about 200 robots. For preprocessing, we concatenated titles and message numbers into an *event type*. To generate nodes from the event log, these event types were used as nodes and the transitions between them were considered as edges, as discussed above in Sect. 3.3, also enriching the graph with the topological attributes.

**Table 1.** Example: Events of the real-world complex event log dataset used in this paper.

| Event_timestamp | Message | Title | Description | Line | Cell | Robot |
|---|---|---|---|---|---|---|
| 2014-6-3 12:29:37 | 10011 | Motors ON State | Motor on ... | 13 | 2 | 48 |
| 2014-6-3 12:29:33 | 10012 | Safety Guard Stop State | Entering guard state ... | 13 | 2 | 48 |
| 2014-6-3 12:29:36 | 10010 | Motors OFF State | Motor off ... | 13 | 2 | 7 |
| 2014-6-3 12:29:39 | 20205 | Auto Stop Open | Entering auto stop ... | 13 | 2 | 48 |

### 4.1  Graph Summarization and Visual Analysis

In our case study, the graph summarization process was applied on the modeled attributed graph. The original graph had 234 nodes (individual message values) and about 4 million edges (events). The nodes in the log data increased from 234 to 486 when the *event type* was created, concatenating title, see Table 2) of the event to the message number, in the data preprocessing phase, since that distinguished between different event subtypes with the same message number. Edge contraction, applying the HDBSCAN algorithm [15,33] resulted in a considerable reduction in the number of edges. Furthermore, when aggregating edges into a simple graph and also applying node contraction using core-pattern mining (using the MinerLSD algorithm with different $k$-core sizes) we further observe considerable reductions in complexity. Please note, that especially for the $k$-core contraction, the size of the graph can be flexibly tuned by utilizing adequate constraints (cf., Table 2).

**Table 2.** Results in each stage of the graph summarization process.

| Metric | Original graph | Preprocessed multigraph | Edge contracted multigraph | Edge aggregated simple graph | Pattern-driven node contracted graph |
|---|---|---|---|---|---|
| #Nodes | 234 | 486 | 499 | 499 | 1352, 121 |
| #Edges | 3,975,765 | 1,609,363 | 627,127 | 8,876 | $k = 10, 23$ (k-core) |

It is important to note that each operation on the graph retains the ability to perform queries on the event log data. For examining the last two days in the event log, for example, a total of 5067 events were executed, for which there was a collision event in the factory, and a factory restart, resulting in a significant downtime of the factory. So, what caused these downtimes in the factory? Using the graph model, it is possible to calculate all the paths to such events, in particular, to the event '20205:AutoStopOpen'. Details are visualized in Fig. 4. Furthermore, using the attributed graph, we can also make sense of the context, e.g., relating to the involved production cells/lines and robots.

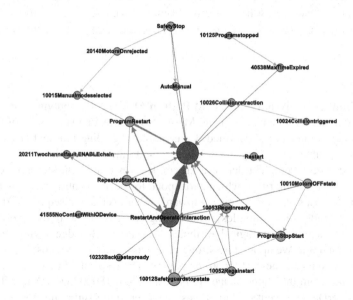

**Fig. 4.** Example graph focussing on the last two days of the event log, with all paths to the event '20205:AutoStopOpen' (center). A node/edge has a larger width if it occurs more often.

### 4.2 Query-Driven Analysis and Mixed-Initiative Pattern Mining

**Query-Driven Analysis.** The summarized graph provides the means for considerable reductions in complexity. Then, also specific queries of the user can be answered, e.g., regarding (expected) sequences in the graph. Calculating the Markov Chain probabilities from the transitions between the events, the probability of the expected sequences – according to domain knowledge – can then be checked. For example, the 10230:Backupstepready, 10231:Backupstepready and 10232:Backupstepready events form a sequence with three events that should always occur together (the order can vary). The sequence probabilities are shown in Table 3. The most common order in the event log was 10230 → 10231 → 10232. Another example is the (10271:CyclicBrakeCheckStarted → 10270:CyclicBrakeCheckDone) sequence, which should only occur in this order. However, both (10271 → 10270) and (10270 → 10271) are likely to appear. Such interesting observations then point to more detailed investigation of the event log.

**Table 3.** Examples of expected sequences according to domain knowledge.

| Sequence | Probability |
|---|---|
| 10230:Backupstepready → 10231:Backupstepready → 10232:Backupstepready | 0.0000377 |
| 10232:Backupstepready → 10230:Backupstepready → 10231:Backupstepready | 0.0000270 |
| 10231:Backupstepready → 10232:Backupstepready → 10230:Backupstepready | 0.0000216 |
| 10271:CyclidBrakeCheckStarted → 10270:CyclidBrakeCheckDone | 0.0056 |
| 10270:CyclicBrakeCheckDone → 10271:CyclicBrakeCheckStarted | 0.0042 |

**Mixed-Initiative Analysis Using Local Pattern Mining.** By applying the local pattern mining method implemented using MinerLSD, interesting subgraphs of the complex event log graph can be identified, for example, regarding minimal size (support) constraints, minimal (clustering) quality according to the local modularity measure, or focussing on specific graph structures, e.g., on k-cores (indicating densely connected events) or hubs and authorities, which correspond to specific central events that mostly serve as predecessor or sucessor events towards other events in a sequence. In the following, we provide some examples of interesting patterns that were discovered on the graph enriched with the topological attributes, focusing on the edge aggregated simple graph representation. We applied no strong core constraints, only focussing on 1-cores which require connectedness of the graph pattern. Then, specific interesting sequences were extracted from the pattern. One simple example is (80002:UserDefinedEvent13' →'80002:UserDefinedEvent3), which was, however, quite surprising for the domain specialist; this indicates that "waiting for subsystem tool A" causes an "error in tool B", which was unexpected and instigated further analysis.

For the graph pattern ("Warning", "cell6", "line20") which indicates warning conditions occurring in production lines 6 and 20, Table 4 shows some interesting sequences extracted from the pattern. Here, specifically sequences 1–3 indicate problems and their respective (root) causes, i.e., a misconfiguration of signal values of the robot or overload of the main computer, respectively. Sequences 4–5 are of a different type: Those occurred during maintenance and troubleshooting, and are interesting patterns for optimizing diagnostics. However, when looking at faults and anomalies during normal production they do not need to be considered further, and can be marked as such,

**Table 4.** Examples of sequences detected using local pattern mining, and their according probability: Sequences 1–3 indicate problems and causes which can be targeted for process optimization. Sequences 4–5 occurred during maintenance and troubleshooting.

| Sequence | Prob. |
|---|---|
| 71414:Concurrentchangesofsignalvalue → 80002:UserDefinedEvent3 | 6e-7 |
| 71414:Concurrentchangesofsignalvalue → 80002:UserDefinedEvent4 | 6e-7 |
| 20314:Enable2supervisionfault → 40538:MaxTimeExpired → 80002:UserDefinedEvent3 | 2e-9 |
| 20481:SCOVRactive → 40538:MaxTimeExpired → 80002:UserDefinedEvent3 | 2e-8 |
| 20481:SCOVRactive → 40538:MaxTimeExpired → 80002:UserDefinedEvent4 | 7e-9 |

to be added to the available domain knowledge. Overall, the domain specialist considered the detected patterns highly interesting, for which the iterative process on pattern mining and refinement was essential in order to arrive at interesting and relevant patterns and sequences. For example, a detailed analysis of the last day of the event log for '80002:UserDefinedEvent3' is shown in Fig. 5, visualizing the paths (focusing on length 2) leading to that event.

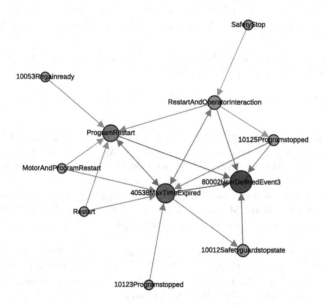

**Fig. 5.** Detail graph focussing on the event '80002:UserDefinedEvent3' (last day of the event log).

## 5  Conclusions

In this paper, we presented a framework for human-centered exploration of complex event logs, utilizing attributed graph modeling, graph summarization, and local pattern mining. In a real-world case study, we showed the implementation and application of the proposed framework, and presented promising results. Those indicate, that attributed graph modeling and graph summarization on complex event log data can be applied to effectively reduce the data complexity and has the potential to answer specific queries. Furthermore, using local pattern mining we can identify interesting graph patterns and sequential structures, yielding important and interesting insights, e.g., for process optimization and diagnostics. With the human-in-the-loop iterative feedback and refinement cycles are enabled, to include background knowledge and to adapt constraints. In the case study, context information provided by the domain expert was crucial in order to check the quality of the obtained patterns. Altogether, graph-based modeling allows explicit and detailed workflows to be visualized. Second, the adaptability of the graph based modeling is a strength of the method. Third, query-driven analysis and mixed-initiative mining of graph patterns and sequential structures complement each other well.

For future work, we aim to apply the framework on further datasets and domains. Furthermore, we aim to extend the framework towards model-based approaches, in order to capture richer constraints in the pattern mining and refinement cycles.

**Acknowledgements.** This work has been partially supported by Interreg NWE, project Di-Plast - Digital Circular Economy for the Plastics Industry.

## References

1. Interdonato, R., Atzmueller, M., Gaito, S., Kanawati, R., Largeron, C., Sala, A.: Feature-rich networks: going beyond complex network topologies. Appl. Netw. Sci. **4**, 4 (2019)
2. Atzmueller, M., Soldano, H., Santini, G., Bouthinon, D.: MinerLSD: efficient mining of local patterns on attributed networks. Appl. Netw. Sci. **4**, 43 (2019)
3. Atzmueller, M.: Declarative aspects in explicative data mining for computational sensemaking. In: Seipel, D., Hanus, M., Abreu, S. (eds.) WFLP/WLP/INAP -2017. LNCS (LNAI), vol. 10997, pp. 97–114. Springer, Cham (2018). https://doi.org/10.1007/978-3-030-00801-7_7
4. Van Der Aalst, W.: Process Mining: Discovery, Conformance and Enhancement of Business Processes, vol. 2. Springer, Heidelberg (2011). https://doi.org/10.1007/978-3-642-19345-3
5. Munoz-Gama, J., Carmona, J., van der Aalst, W.M.P.: Single-entry single-exit decomposed conformance checking. Inf. Syst. **46**, 102–122 (2014)
6. van Dongen, B.F., Van der Aalst, W.M.: Multi-phase process mining: aggregating instance graphs into EPCs and petri nets. In: PNCWB 2005 workshop, pp. 35–58. Citeseer (2005)
7. Wen, L., van der Aalst, W.M., Wang, J., Sun, J.: Mining process models with non-free-choice constructs. Data Min. Knowl. Discov. **15**(2), 145–180 (2007)
8. Chesani, F., Lamma, E., Mello, P., Montali, M., Riguzzi, F., Storari, S.: Exploiting inductive logic programming techniques for declarative process mining. In: Jensen, K., van der Aalst, W.M.P. (eds.) Transactions on Petri Nets and Other Models of Concurrency II. LNCS, vol. 5460, pp. 278–295. Springer, Heidelberg (2009). https://doi.org/10.1007/978-3-642-00899-3_16

9. Rovani, M., Maggi, F.M., de Leoni, M., van der Aalst, W.M.: Declarative process mining in healthcare. Expert Syst. Appl. **42**(23), 9236–9251 (2015)
10. Vaarandi, R.: A data clustering algorithm for mining patterns from event logs. In: Proceedings of IEEE Workshop on IP Operations & Management, pp. 119–126. IEEE (2003)
11. Riondato, M., García-Soriano, D., Bonchi, F.: Graph summarization with quality guarantees. Data Min. Knowl. Discov. **31**(2), 314–349 (2017)
12. Liu, Y., Safavi, T., Dighe, A., Koutra, D.: Graph summarization methods and applications: a survey. ACM Comput. Surv. (CSUR) **51**(3), 62 (2018)
13. LeFevre, K., Terzi, E.: Grass: graph structure summarization. In: Proceedings of SDM, SIAM, pp. 454–465 (2010)
14. Shen, Z., Ma, K.L., Eliassi-Rad, T.: Visual analysis of large heterogeneous social networks by semantic and structural abstraction. IEEE TVCG **12**(6), 1427–1439 (2006)
15. Bloemheuvel, S., Kloepper, B., Atzmueller, M.: Graph summarization for computational sensemaking on complex industrial event logs. In: Proceedings of Workshop on Methods for Interpretation of Industrial Event Logs, International Conference on Business Process Management (BPM 2019), Vienna, Austria (2019)
16. Zhou, Y., Cheng, H., Yu, J.X.: Graph clustering based on structural/attribute similarities. PVLDB **2**(1), 718–729 (2009)
17. Steinhaeuser, K., Chawla, N.V.: Community detection in a large real-world social network. In: Liu, H., Salerno, J.J., Young, M.J. (eds.) Social Computing, Behavioral Modeling, and Prediction, pp. 168–175. Springer, Boston (2008). https://doi.org/10.1007/978-0-387-77672-9_19
18. Zhu, L., Ng, W.K., Cheng, J.: Structure and attribute index for approximate graph matching in large graphs. Inf. Syst. **36**(6), 958–972 (2011)
19. Balasubramanyan, R., Cohen, W.W.: Block-LDA: jointly modeling entity-annotated text and entity-entity links. In: Proceedings of SDM, SIAM, pp. 450–461 (2011)
20. Kanawati, R.: Seed-centric approaches for community detection in complex networks. In: Meiselwitz, G. (ed.) SCSM 2014. LNCS, vol. 8531, pp. 197–208. Springer, Cham (2014). https://doi.org/10.1007/978-3-319-07632-4_19
21. Moser, F., Colak, R., Rafiey, A., Ester, M.: Mining cohesive patterns from graphs with feature vectors. In: Proceedings of SDM, SIAM, pp. 593–604 (2009)
22. Silva, A., Meira Jr., W., Zaki, M.J.: Mining attribute-structure correlated patterns in large attributed graphs. Proc. VLDB Endow. **5**(5), 466–477 (2012)
23. Günnemann, S., Färber, I., Boden, B., Seidl, T.: GAMer: a synthesis of subspace clustering and dense subgraph mining. KAIS **40**(2), 243–278 (2013)
24. Atzmueller, M., Doerfel, S., Mitzlaff, F.: Description-oriented community detection using exhaustive subgroup discovery. Inf. Sci. **329**, 965–984 (2016)
25. Morik, K.: Detecting interesting instances. In: Hand, D.J., Adams, N.M., Bolton, R.J. (eds.) Pattern Detection and Discovery. LNCS (LNAI), vol. 2447, pp. 13–23. Springer, Heidelberg (2002). https://doi.org/10.1007/3-540-45728-3_2
26. Knobbe, A.J., Cremilleux, B., Fürnkranz, J., Scholz, M.: From local patterns to global models: the lego approach to data mining. In: From Local Patterns to Global Models: Proceedings of the ECML/PKDD-08 Workshop (LeGo-08), pp. 1–16 (2008)
27. Atzmueller, M.: Subgroup discovery. WIREs DMKD **5**(1), 35–49 (2015)
28. Seidman, S.B.: Network structure and minimum degree. Soc. Netw. **5**, 269–287 (1983)
29. Soldano, H., Santini, G., Bouthinon, D., Lazega, E.: Hub-authority cores and attributed directed network mining. In: Proceedings of ICTAI, Boston, MA, USA, pp. 1120–1127. IEEE (2017)
30. Wasserman, S., Faust, K.: Social Network Analysis: Methods and Applications. Structural Analysis in the Social Sciences, 1 edn. vol. 8. Cambridge University Press, Cambridge (1994)

31. Brin, S., Page, L.: The anatomy of a large-scale hypertextual web search engine. Comput. Netw. ISDN Syst. **30**(1), 107–117 (1998)
32. Watts, D.J., Strogatz, S.H.: Collective dynamics of 'small-world' networks. Nature **393**(6684), 440–442 (1998)
33. McInnes, L., Healy, J., Astels, S.: hdbscan: Hierarchical density based clustering. J. Open Source Softw. **2**(11), 205 (2017)
34. Ester, M., Kriegel, H.P., Sander, J., Xu, X., et al.: A density-based algorithm for discovering clusters in large spatial databases with noise. In: Proceedings of KDD, pp. 226–231 (1996)
35. Aizawa, A.: An information-theoretic perspective of tf-idf measures. Inf. Process. Manag. **39**(1), 45–65 (2003)
36. Fortunato, S., Castellano, C.: Community Structure in Graphs. In: Encyclopedia of Complexity and System Science. Springer, Heidelberg (2007)
37. Bothorel, C., Cruz, J.D., Magnani, M., Micenkova, B.: Clustering attributed graphs: models measures and methods. Netw. Sci. **3**(03), 408–444 (2015)
38. Klösgen, W.: Explora: a multipattern and multistrategy discovery assistant. In: Advances in Knowledge Discovery and Data Mining, pp. 249–271. AAAI Press, Palo Alto (1996)
39. Soldano, H., Santini, G., Bouthinon, D.: Local knowledge discovery in attributed graphs. In: Proceedings of ICTAI, pp. 250–257. IEEE (2015)
40. Peng, C., Kolda, T.G., Pinar, A.: Accelerating Community Detection by Using k-core Subgraphs. arXiv preprint arXiv:1403.2226 (2014)
41. Soldano, H., Santini, G.: Graph abstraction for closed pattern mining in attributed networks. In: Proceedings of ECAI, FAIA, vol. 263, pp. 849–854. IOS Press (2014)

# Sparse Robust Regression
# for Explaining Classifiers

Anton Björklund[1(✉)], Andreas Henelius[1], Emilia Oikarinen[1],
Kimmo Kallonen[2], and Kai Puolamäki[1]

[1] Department of Computer Science, University of Helsinki, Helsinki, Finland
{anton.bjorklund,andreas.henelius,emilia.oikarinen,
kai.puolamaki}@helsinki.fi
[2] Helsinki Institute of Physics, University of Helsinki, Helsinki, Finland
kimmo.kallonen@helsinki.fi

**Abstract.** Real-world datasets are often characterised by outliers, points far from the majority of the points, which might negatively influence modelling of the data. In data analysis it is hence important to use methods that are robust to outliers. In this paper we develop a robust regression method for finding the largest subset in the data that can be approximated using a sparse linear model to a given precision. We show that the problem is NP-hard and hard to approximate. We present an efficient algorithm, termed SLISE, to find solutions to the problem. Our method extends current state-of-the-art robust regression methods, especially in terms of scalability on large datasets. Furthermore, we show that our method can be used to yield interpretable explanations for individual decisions by opaque, black box, classifiers. Our approach solves shortcomings in other recent explanation methods by not requiring sampling of new data points and by being usable without modifications across various data domains. We demonstrate our method using both synthetic and real-world regression and classification problems.

## 1 Introduction and Related Work

In analyses of real-world data we often encounter outliers, i.e., points which are far from the majority of the other data points. Such points are problematic as they may negatively influence modelling of the data. This is observed in, e.g., ordinary least-squares regression where already a single data point may lead to arbitrarily large errors [11]. It is hence important to use *robust methods* that effectively ignore the effect of outliers. A number of approaches have been proposed for robust regression, see, e.g., [27] for a review. Our proposed method is most closely related to *Least Trimmed Squares* (LTS) [2,26,28] that finds a subset of size $k$ minimising the sum of the squared residuals in this subset, in contrast to methods that de-emphasise [33] or penalise [20,30,34] outliers.

In this paper we present a sparse robust regression method that outperforms many of the existing state-of-the-art robust regression methods in terms of scalability on large datasets, termed SLISE (Sparse LInear Subset Explanations).

© The Author(s) 2019
P. Kralj Novak et al. (Eds.): DS 2019, LNAI 11828, pp. 351–366, 2019.
https://doi.org/10.1007/978-3-030-33778-0_27

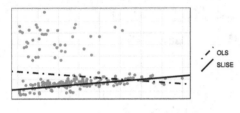

**Fig. 1.** Robust regression.

**Table 1.** Classifier probabilities for *high income*.

|  | Education | |
| --- | --- | --- |
| Age | *Low* | *High* |
| *Young* | 0.07 | 0.31 |
| *Old* | 0.22 | 0.61 |

Specifically, we consider *finding the largest subset of data items that can be represented by a linear model to a given accuracy*. Hence, there is an important difference between our method and LTS: with LTS the size of the subset is fixed and specified a priori. Furthermore, the linear models obtained from SLISE are sparse, meaning that the model coefficients are easier to interpret, especially for datasets with many attributes.

*Example 1: Robust Regression.* Figure 1 shows a dataset containing outliers in the top left corner. Here ordinary least-squares regression (OLS) finds the wrong model due to the influence of these outliers. In contrast, SLISE *finds the largest subset of points that can be approximated by a (sparse) linear model*, yielding high robustness by ignoring the outliers.

Interestingly, it turns out that our robust regression method can also be used to *explain individual decisions by opaque (black box) machine learning models*: e.g., why does a classifier predict that an image contains the digit 2? The need for interpretability stems from the fact that high accuracy is not always sufficient; we must understand *how* the model works. This is important in safety-critical real-world applications, e.g., in medicine [6], but also in science, such as in physics when classifying particle jets [18]. In terms of explanations we consider *post-hoc interpretation of opaque models*, i.e., understanding predictions from already existing models, in contrast to creating models directly aiming for interpretability (e.g., super-sparse linear integer models [32] or decision sets [19]). In general, model explanations can be divided into *global* explanations (for the entire model), e.g., [1,10,16,17], and *local* explanations (for a single classification instances), e.g., [5,13,21,25]. Here we are interested in the latter. For a survey of explanations see, e.g., [15].

To explain an instance, we need to find a (simple and interpretable) model that *matches the black box model locally* in the *neighbourhood* of the instance whose classification we want to explain. Defining this neighbourhood is important but non-trivial (for discussion, see, e.g., [14,24]). The two central questions are: (i) how do we find the local model and (ii) how do we define the neighbourhood? Our approach solves these two problems at the same time by finding the largest subset of data items such that the residuals of a linear model passing through the instance we want to explain are minimised.

*Example 2: Explanations.* Consider a simple toy dataset of persons with the attributes age $\in \{0,1\}$ and education $\in \{0,1\}$, where 0 denotes low age and education and 1 high age and education, respectively. Assume that the dataset consists mostly of people with high education, if we for example are studying factors affecting salaries within the faculty of a university department. Now, we are given a classifier that outputs the probability of high income (vs. low income), given these two attributes. Our task is to find the most important attribute used by the classifier when estimating the income level of an old professor in the dataset. Looking only at the class probabilities, shown in Table 1, it appears that education is the most significant attribute, and this is indeed what, e.g., the state-of-the art local explanation method LIME [25] finds. We, however, argue that this explanation is misleading: our toy data set contains very few instances of persons with low education, and therefore knowing the education level does not really give any information about the class. We argue that *in this dataset* age is a better determinant of high income, and this is found by SLISE.

The above example shows the importance of the interaction between the model and the data. The model in Table 1 is actually a simple logistic regression[1]. Hence, even if the model is simple, a complex structure in the data can make interpretation non-trivial. LIME found the simple logistic regression model, whereas we found the behaviour of the model in the dataset. This distinction is significant because it suggests that you cannot always cleanly separate the model from the data. An example of this is conservation laws in physical systems. Accurate data will never violate such laws, which is something the model can rely on. Without adhering to the data during the explanation you may therefore find explanations that violate the laws of physics. SLISE satisfies such constraints automatically by observing how the classifier performs in the dataset, instead of randomly sampling (possibly non-physical) points around the item of interest (as in, e.g., [5,13,21,25]). Another advantage is that we do not need to define a neighbourhood of a data item, which is especially important in cases where modelling the distance is difficult, such as with images.

*Contributions.* We develop a novel robust regression method with applications to local explanations of opaque machine learning models. We consider the problem of *finding the largest subset that can be approximated by a sparse linear model* which is **NP**-hard and hard to approximate (Theorem 1) and present an approximative algorithm for solving it (Algorithm 1). We demonstrate empirically using synthetic and real-world datasets that SLISE outperforms state-of-the-art robust regression methods and yields sensible explanations for classifiers.

*Organisation.* In Sect. 2 we formalise our problem for both robust regression and local explanations, and show its complexity. We then discuss practical numeric optimisation in Sect. 3. The algorithm is presented in Sect. 4, followed by the empirical evaluation in Sect. 5. We end with the conclusions in Sect. 6.

## 2    Problem Definition

Our goal is to develop a linear regression method with applications to both (i) robust *global linear regression model* and (ii) providing a *local linear regression*

---

[1] Probability of high income is given by $p = \sigma(-2.53 + 1.73 \cdot \text{education} + 1.26 \cdot \text{age})$.

*model* of the decision surface of an opaque model in the vicinity of a particular data item. In the second case the simple linear model thus provides an explanation for the (typically more) complex decision surface of the opaque model.

Let $(X, Y)$, where $X \in \mathbb{R}^{n \times d}$ and $Y \in \mathbb{R}^n$, be a dataset consisting of $n$ pairs $\{(x_i, y_i)\}_{i=1}^n$ where we denote the $i$th $d$-dimensional item (row) in $X$ by $x_i$ (the *predictor*) and similarly the $i$th element in $Y$ by $y_i$ (the *response*). Furthermore let $\varepsilon$ be the largest tolerable error and $\lambda$ be a regularisation coefficient. We now state the main problem in this paper:

*Problem 1.* Given $X \in \mathbb{R}^{n \times d}$, $Y \in \mathbb{R}^n$, and non-negative $\varepsilon, \lambda \in \mathbb{R}$, find the regression coefficients $\alpha \in \mathbb{R}^d$ minimising the *loss function*

$$\mathbf{Loss}(\varepsilon, \lambda, X, Y, \alpha) = \sum_{i=1}^n H(\varepsilon^2 - r_i^2)\left(r_i^2/n - \varepsilon^2\right) + \lambda\|\alpha\|_1, \tag{1}$$

where the residual errors are given by $r_i = y_i - \alpha^\mathsf{T} x_i$, $H(\cdot)$ is the Heaviside step function satisfying $H(u) = 1$ if $u \geq 0$ and $H(u) = 0$ otherwise, and $\|\alpha\|_1 = \sum_{i=1}^d |\alpha_i|$ denotes the L1-norm. If necessary, $X$ can be augmented with a column of all ones to accommodate the *intercept* term of the model.

Alternatively, the Lagrangian term $\lambda\|\alpha\|_1$ in Eq. (1) can be replaced by a constraint $\|\alpha_i\|_1 \leq t$ for some $t$. Note that Problem 1 is a combinatorial problem in disguise, where we try to find a maximal subset $S$, as can be seen by rewriting Eq. (1) as (using the shorthand $[n] = \{1, \ldots, n\}$)

$$\mathbf{Loss}(\varepsilon, \lambda, X, Y, \alpha) = \sum_{i \in S}\left(r_i^2/n - \varepsilon^2\right) + \lambda\|\alpha\|_1 \text{ where } S = \{i \in [n] \mid r_i^2 \leq \varepsilon^2\}. \tag{2}$$

The loss function of Eq. (1) (and Eq. (2)) thus consists of three parts; the maximisation of subset size $\sum_{i \in S} \varepsilon^2 = |S|\varepsilon^2$, the minimisation of the residuals $\sum_{i \in S} r_i^2/n \leq \varepsilon^2$, and the LASSO-regularisation $\lambda\|\alpha\|_1$. The main goal is to maximise the subset and this is reflected in the loss function, since any decrease of the subset size has an equal or greater impact on the loss than all the residuals combined. At the limit of $\varepsilon \to \infty$, it follows that $S = [n]$ and Problem 1 is equivalent to LASSO [31]. We now state the following theorem concerning the complexity of Problem 1.

**Theorem 1.** *Problem 1 is **NP**-hard and hard to approximate.*

*Proof.* We prove the theorem by a reduction to the MAXIMUM SATISFYING LINEAR SUBSYSTEM problem [4, Problem MP10], which is known to be **NP**-hard. In MAXIMUM SATISFYING LINEAR SUBSYSTEM we are given the system $X\alpha = y$, where $X \in \mathbb{Z}^{n \times m}$ and $y \in \mathbb{Z}^n$ and we want to find $\alpha \in \mathbb{Q}^m$ such that as many equations as possible are satisfied. This is equivalent to Problem 1 with $\varepsilon = 0$ and $\lambda = 0$. Also, the problem is not approximable within $n^\gamma$ for some $\gamma > 0$ [3]. $\qquad\square$

*Local Explanations.* To provide a local explanation for a data item $(x_k, y_k)$ where $k \in [n]$, we use an additional constraint requiring that the regression plane passes through this item, i.e., we add the constraint $r_k = 0$ to Problem 1. This constraint is easily met by centring the data on the item $(x_k, y_k)$ to be explained: $y_i \to y_i - y_k$ and $x_i \to x_i - x_k$ for all $i \in [n]$, in which case $r_k = 0$ and any potential intercept is zero. Hence, it suffices to consider Problem 1 both when finding the best global regression model and when providing a local explanation for a data item.

In practice, we employ the following procedure to generate local explanations for classifiers. If a classifier outputs class probabilities $P \in \mathbb{R}^n$ we transform them to linear values using the logit transformation $y_i = \log(p_i/(1 - p_i))$, yielding a vector $Y \in \mathbb{R}^n$. This new vector $Y - y_k$ is what we use for finding the explanation.

Now, the local linear model, $\alpha$ from Problem 1, and the subset, $S$ from Eq. (2), constitute the explanation for the data item of interest. Note that the linear model is comparable to the linear model obtained using standard logistic regression, i.e., we approximate the black box classifier by a logistic regression in the vicinity of the point of interest.

# 3   Numeric Approximation

We cannot effectively solve the optimisation problem in Problem 1 in the general case. Instead, we relax the problem by replacing the Heaviside function with a sigmoid function $\sigma$ and a continuous and differentiable rectifier function $\phi(u) \approx \min(0, u)$. This allows us to compute the gradient and find $\alpha$ by minimising

$$\beta\text{-Loss}(\varepsilon, \lambda, X, Y, \alpha) = \sum_{i=1}^{n} \sigma(\beta(\varepsilon^2 - r_i^2))\phi\left(r_i^2/n - \varepsilon^2\right) + \lambda\|\alpha\|_1, \quad (3)$$

where the parameter $\beta$ determines the steepness of the sigmoid and the rectifier function $\phi$ is parametrised by a small constant $\omega > 0$ such that $\phi(u) = u$ for $u < -\omega$, $\phi(u) = -(u^2/\omega + \omega)/2$ for $-\omega \leq u \leq 0$, and $\phi(u) = -\omega/2$ for $0 < u$. It is easy to see that Eq. (3) is a smoothed variant of Eq. (1) and that the two become equal when $\beta \to \infty$ and $\omega \to 0^+$.

We perform this minimisation using *graduated optimisation*, where the idea is to iteratively solve a difficult optimisation problem by progressively increasing the complexity [23]. A natural parametrisation for the complexity of our problem is via the $\beta$ parameter. We start from $\beta = 0$ which corresponds to a convex optimisation problem equivalent to LASSO, and gradually increase the value of $\beta$ towards $\infty$ which corresponds to the Heaviside solution of Eq. (1). At each step, we use the previous optimal value of $\alpha$ as a starting point for minimisation of Eq. (3). It is important that the optima of the consecutive solutions with increasing values of $\beta$ are close enough, which is why we derive an approximation ratio between the solutions with different values of $\beta$. We observe that our problem can be rewritten as a maximisation of $-\beta\text{-Loss}(\varepsilon, \lambda, X, Y, \alpha)$. The choice of $\beta$ does not affect the L1-norm and we omit it for simplicity ($\lambda = 0$).

**Theorem 2.** *Given $\varepsilon, \beta_1, \beta_2 > 0$, such that $\beta_1 \leq \beta_2$, and the functions $f_j(r) = -\sigma(\beta_j(\varepsilon^2 - r^2))\phi(r^2/n - \varepsilon^2)$, and $G_j(\alpha) = \sum_{i=1}^{n} f_j(r_i)$ where $r_i = y_i - \alpha^\mathsf{T} x_i$ and $j \in \{1, 2\}$. For $\alpha_1 = \arg\max_\alpha G_1(\alpha)$ and $\alpha_2 = \arg\max_\alpha G_2(\alpha)$ the inequality $G_2(\alpha_2) \leq K G_2(\alpha_1)$ always holds, where $K = G_1(\alpha_1)/(G_2(\alpha_1) \min_r f_1(r)/f_2(r))$ is the approximation ratio.*

**Parameters:** (1) Dataset $X \in \mathbb{R}^{n \times d}$, $Y \in \mathbb{R}^n$, (2) error tolerance $\varepsilon$,
                  (3) regularisation coefficient $\lambda$, (4) sigmoid steepness $\beta_{max}$,
                  (5) target approximation ratio $r_{max}$

1 **Function** SLISE($X$, $Y$, $\varepsilon$, $\lambda$, $\beta_{max}$, $r_{max}$)
2      $\alpha \leftarrow$ OrdinaryLeastSquares($X, Y$) and $\beta \leftarrow 0$
3      **while** $\beta < \beta_{max}$ **do**
4          $\alpha \leftarrow$ OWL-QN($\beta$-Loss, $\varepsilon$, $\lambda$, $X$, $Y$, $\alpha$)
5          $\beta \leftarrow \beta'$ such that AppoximationRatio($X$, $Y$, $\varepsilon$, $\beta$, $\beta'$, $\alpha$) $= r_{max}$
6      $\alpha \leftarrow$ OWL-QN($\beta$-Loss, $\varepsilon$, $\lambda$, $X$, $Y$, $\alpha$)
     **Result:** $\alpha$

**Algorithm 1:** The SLISE algorithm.

*Proof.* Let us first argue the non-negativity of $f_1$ and $f_2$. The inequalities $\sigma(z) > 0$ and $\phi(z) < 0$ hold for all $z \in \mathbb{R}$, thus $f_j(r) > 0$. Now, by definition, $G_1(\alpha_2) \leq G_1(\alpha_1)$. We denote $r_i^* = y_i - \alpha_2^\mathsf{T} x_i$ and $k = \min_r f_1(r)/f_2(r)$, which allows us the rewrite and approximate:

$$G_1(\alpha_2) = \sum_{i=1}^{n} f_1(r_i^*) = \sum_{i=1}^{n} f_2(r_i^*) f_1(r_i^*)/f_2(r_i^*) \geq k G_2(\alpha_2).$$

Then $G_2(\alpha_2) \leq G_1(\alpha_2)/k \leq G_1(\alpha_1)/k \leq G_2(\alpha_1) G_1(\alpha_1)/(k G_2(\alpha_1))$, and the inequality from the theorem holds. $\square$

We use Theorem 2 to choose the sequence of $\beta$ values ($\beta_1 = 0, \beta_1, \ldots, \beta_l = \beta_{max}$) so that at each step the approximation ratio as defined by $K$ stays within a bound specified by the parameter $r_{max}$ in Algorithm 1.

## 4 The SLISE Algorithm

In this section we describe an approximate numeric algorithm Algorithm 1 (SLISE) for solving Problem 1. As a starting point for the regression coefficients $\alpha$ we use the solution obtained from an ordinary least squares regression (OLS) on the full dataset (Algorithm 1, line 2). We now perform graduated optimisation (lines 3–5) in which we gradually increase the value of $\beta$ from 0 to $\beta_{max}$. At each iteration, we find the model $\alpha$ using the current value of $\beta$, such that $\beta$-Loss in Eq. (3) is minimised (line 4). To perform this optimisation we use OWL-QN [29], which is a quasi-Newton optimisation method with built-in L1-regularisation. We then increase $\beta$ gradually (line 5) such that the approximation ratio $K$ in Theorem 2 equals $r_{max}$.

The time complexity of SLISE is affected by the three main parts of the algorithm; the loss function, OWL-QN, and graduated optimisation. The evaluation of the loss function has a complexity of $\mathcal{O}(nd)$, due to the multiplication between the linear model $\alpha$ and the data-matrix $X$. OWL-QN has a complexity of $\mathcal{O}(dp_o)$, where $p_p$ is the number of iterations. Graduated optimisation is also an iterative method $\mathcal{O}(dp_g)$, but it only adds the approximation ratio calculation $\mathcal{O}(nd)$ (which is not dominant). Combining these complexities yields a complexity of $\mathcal{O}(nd^2p)$ for SLISE, where $p = p_o + p_g$ is the total number of iterations.

**Table 2.** The datasets. The synthetic dataset can be generated to the desired size.

|            | EMNIST | IMDB | PHYSICS | SYNTHETIC |
|------------|--------|------|---------|-----------|
| Items      | 40 000 | 25 000 | 260 000 | $n$ |
| Dimensions | 784    | 1000 | 5       | $d$ |
| Type       | Image  | Text | Tabular | - |
| Classifier | CNN    | LR, SVM | NN   | - |

## 5    Experiments

SLISE has applications in both robust regression and for explaining black box models, and the experiments are hence divided into two parts. In the first part (Sect. 5.1) we consider SLISE as a *robust regression* method and demonstrate that (i) SLISE scales better on high-dimensional datasets than competing methods, (ii) SLISE is very robust to noise, and (iii) the solution found using SLISE is optimal. In the second part (Sect. 5.2) we use SLISE to *explain predictions from opaque models*. The experiments were run using R (v. 3.5.1) on a high-performance cluster [12] (4 cores from an Intel Xeon E5-2680 2.4 GHz with 16 Gb RAM). SLISE and the code to run the experiments is released as open source and is available from http://www.github.com/edahelsinki/slise.

*Datasets.* We use real (EMNIST [9], IMDB [22], PHYSICS [8]) and synthetic datasets in our experiments (properties given in Table 2). Synthetic datasets are generated as follows. The data matrix $X \in \mathbb{R}^{n \times d}$ is created by sampling from a normal distribution with zero mean and unit variance. The response vector $Y \in \mathbb{R}^n$ is created by $y_i \leftarrow a^\mathsf{T} x_i$ (plus some normal noise with zero mean and 0.05 variance), where $a \in \mathbb{R}^d$ is one of nine linear models drawn from a uniform distribution between $-1$ and 1. Each model creates 10% of the $Y$-values, except one that creates 20% of the $Y$-values. This larger chunk should enable robust regression methods to find the corresponding model.

*Pre-processing.* It is important both for robust regression and for local explanations to ensure that the magnitude of the coefficients in $\alpha$ are comparable, since sparsity is enforced by L1-penalisation of the elements in $\alpha$. Hence, we normalize the PHYSICS datasets dimension-wise by subtracting the mean and dividing by the standard deviation. For EMNIST the data items are $28 \times 28$ images and we scale the pixel values to the range $[-1, 1]$. Some of the pixels have the same value for all images (i.e., the corners) so these pixels were removed and the images flattened to vectors of length 672. And for the text data in IMDB we form a bag-of-words model using the 1000 most common words after case normalisation, removal of stop words and punctuation, and stemming. The obtained word frequencies are divided by the most frequent word in each review to adjust for different review lengths, yielding real-valued vectors of length 1000. The $Y$-values for all datasets are scaled to approximately be within $[-0.5, 0.5]$ based on the 5th and 95th quantiles.

**Table 3.** Properties of regression methods. RR stands for robust regression.

| Algorithm | Robust | Sparse | R-Package | Description |
|---|---|---|---|---|
| SLISE | Yes | Yes | | RR with variable-size subsets |
| FAST-LTS [28] | Yes | No | robustbase | RR with fixed-size (50%) subsets |
| SPARSE-LTS [2] | Yes | Yes | robustHD | Sparse LTS solutions |
| MM-ESTIMATOR [34] | Yes | No | MASS | Maximum likelihood-based RR |
| MM-LASSO [30] | Yes | Yes | pense | Sparse MM-ESTIMATOR solutions |
| LAD-LASSO [33] | Maybe | Yes | MTE | Combines LAD (Least Absolute Deviation) and a LASSO penalty |
| LASSO [31] | Yes | No | glmnet | OLS with a L1-norm |

*Classifiers.* We use four high-performing classifiers; a convolutional neural network (CNN), a normal neural network (NN), a logistic regression (LR), and a support vector machine (SVM), see Table 2. The classifiers are used to obtain class probabilities $p_i$ of the given data instances. As described in Sect. 2 we transform $p_i$:s into linear values using the logit transformation $y_i = \log(p_i/(1 - p_i))$.

*Default Parameters.* The two most important parameters for SLISE are the error tolerance $\varepsilon$ and the sparsity $\lambda$. These, however, depend on the use-case and dataset and *must* be manually adjusted. The default is to use $\lambda = 0$ (no sparsity) and $\varepsilon = 0.1$ (10 % error tolerance due to the scaling mentioned above). The parameter $\beta_{\max}$ must only be large enough to make the sigmoid function essentially equivalent to a Heaviside function. As a default we use $\beta_{\max} = 30/\varepsilon^2$. The division by $\varepsilon^2$ is used to counteract the effects the choice of $\varepsilon$ has on the shape of the sigmoid. The maximum approximation ratio $r_{\max}$ is used to control the step size for the graduated optimisation. We used $r_{\max} = 1.2$, which for our datasets provided good speed without sacrificing accuracy.

## 5.1   Robust Regression Experiments

We compare SLISE to five state-of-the-art robust regression methods (Table 3, LASSO is included as a baseline). All algorithms have been used with default settings. Not all methods support sparsity, and when they do, finding an equivalent regularisation parameter $\lambda$ is difficult. Hence, unless otherwise noted, all sparse methods are used with almost no sparsity ($\lambda = 10^{-6}$).

*Scalability.* We first investigate the scalability of the methods. Most of the methods have similar theoretical complexities of $\mathcal{O}(nd^2)$ or $\mathcal{O}(nd^2p)$, but for the iterative methods the number of iterations $p$ might vary. We empirically determine the running time on synthetically generated datasets with (i) $n \in \{500, 1\,000, 5\,000, 10\,000, 50\,000, 100\,000\}$ items and $d = 100$ dimensions, and (ii) $d \in \{10, 50, 100, 500, 1\,000\}$ dimensions and $n = 10\,000$ items. The methods that support sparsity have been used with different levels of sparsity ($\lambda \in \{0, 0.01, 0.1, 0.5\}$) and the mean running times are presented. We use a cutoff-time of 10 min.

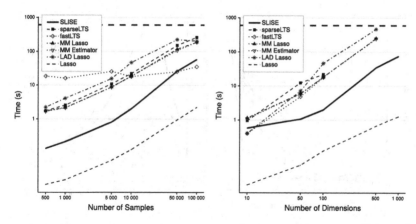

**Fig. 2.** Running times in seconds. Left: Varying the number of samples with fixed $d = 100$. Right: Varying the number of dimensions with fixed $n = 10\ 000$. The cutoff time of 600 s is shown using a dashed horizontal line at $t = 600$.

The results are shown in Fig. 2. We observe that SLISE scales very well in comparison to the other robust regression methods. In Fig. 2 (left) SLISE outperforms all methods except FAST-LTS, which uses subsampling to keep the running time fixed for varying sizes of $n$. In Fig. 2 (right) we see that SLISE consistently outperforms the other robust regression methods for all $d > 10$ and it is the only robust regression method that allows us to obtain results even for a massive $10\ 000 \times 1\ 000$ dataset in less than 100 s (the other robust regression algorithms did not yield results within the cutoff time).

*Robustness.* Next we compare the methods' robustness to noise. We start with a dataset $D$ in which a fraction $\delta$ of data items are corrupted by replacing the response variable with random noise (uniformly distributed between $\min(Y)$ and $\max(Y)$), yielding a corrupted dataset $D_\delta$. The regression functions are learned from $D_\delta$, after which the total sum of the residuals are determined in the clean data $D$. If a method is robust to noise the residuals in the clean data will be small, since the noise from the training data is ignored by the model. The results, using the PHYSICS data, are shown in Fig. 3 (left). Due to the varying subset size SLISE is able to reach higher noise fractions before breaking down than LTS. Note that at high noise fractions all methods are expected to break down.

*Optimality.* Finally, we demonstrate that the solution found using SLISE optimises the loss of Eq. (1). The SLISE algorithm is designed to find the largest subset such that the residuals are upper-bounded by $\varepsilon$. To investigate if the model found using SLISE is optimal, we determine a regression model (i.e., obtain a coefficient vector $\alpha$) using each algorithm. We then calculate the value of the loss-function in Eq. (1) for each model with varying values of $\varepsilon$. The results, using SYNTHETIC data with $n = 1\ 000$ and $d = 30$, are shown in Fig. 3 (right). All loss-values have been normalised with respect to the LASSO model at the corresponding value of

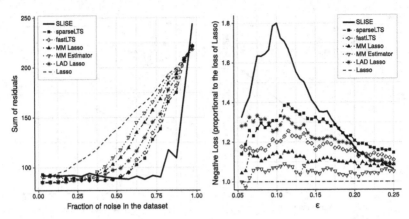

**Fig. 3.** Left: Robustness of SLISE to noise. The $x$-axis shows the fraction of noise and the $y$-axis the sum of the residuals. Small residuals indicate a robust method. Right: Optimality of SLISE. Negative loss-values are shown, normalised with respect to the corresponding loss for LASSO. Higher values are better.

$\varepsilon$ and the curve for LASSO hence appears constant. SLISE consistently has the smallest loss in the region around $\varepsilon = 0.1$, as expected.

### 5.2    Local Explanation Experiments

**Text Classification.** We first compare SLISE to LIME [25], which also provides explanations in terms of sparse linear models. We use the IMDB dataset and explain a logistic regression model. LIME was used with default parameters and the number of features was set to 8. SLISE was also used with default parameters, except using $\lambda = 0.75$ to yield a sparsity comparable to LIME. The results are shown Fig. 4. The LIME-explanation surprisingly shows that the word street is important. Street indeed has a positive coefficient in the global model, but the word is quite rare, only occurring in 2.6% of all reviews. SLISE, in contrast, takes this into account and focuses on the words great, fun, and enjoy. The results for both algorithms are practically unchanged when all reviews with the word street are removed from the test dataset, i.e., LIME emphasises this word *even though it is not a meaningful discriminator for this dataset.*

Figure 5 shows a second text example with an ambiguous phrase (not bad). The classification is incorrect (negative), since the SVM cannot take the interaction between the words not and bad into account. The explanation from SLISE reveals this by giving negative weights to the words wasn't and bad.

**Image Classification.** We now demonstrate how SLISE can be used to explain the classification of a digit from EMNIST, the 2 shown in Fig. 6a. We use SLISE with default parameters, except using a sparsity of $\lambda = 2$, and a dataset with 50% images of the digit 2 and 50% images of other digits (0, 1, 3–9).

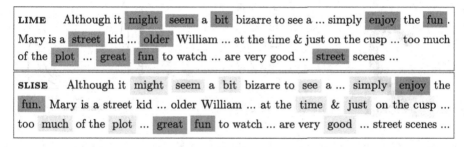

**Fig. 4.** Comparing LIME (top) and SLISE (bottom) with a logistic regression on the IMDB dataset. Parts without any weight from either model are left out for brevity.

> SLISE    ... But I do have to say that in reality, Shemp wasn't really that bad. ... ... At least he wasn't as bad as Joe Besser. ... ... The slapstick gags are hilarious, especially this one scene ...

**Fig. 5.** SLISE explaining how the SVM does not model **not bad** as a positive phrase.

*Approximation as Explanation.* The linear model $\alpha$ approximates the opaque function (here a CNN) in the region around the item being explained. The model weights allow us to deduce features that are important for the classification. Figure 6b shows a *saliency map* in terms of the weight vector $\alpha$. Each pixel corresponds to a coefficient in the $\alpha$-vector and the colour of the pixel indicates its importance in distinguishing a digit 2 from other digits. Purple denotes a pixel supporting positive classification of a 2, and orange a pixel not supporting a 2. More saturated colours correspond to more important weights. We see that the long horizontal line at the bottom is important in identifying 2s, as this feature is missing in other digits. Also, the empty space in the middle-left separates 2s from other digits (i.e., if there is data here the digit is unlikely a 2).

Figure 6c shows the class probability distributions for the test dataset and the found subset $S$. To deduce which features in $\alpha$ that distinguish one class (e.g., 2s from the other digits) we must ensure that the found subset $S$ contains items from both classes (as here in Fig. 6c), otherwise, the projection is to a linear subspace where the class probability is unchanged. During our empirical evaluation of the EMNIST dataset this did not happen.

*Subset as Explanation.* Unlike many other explanation methods the subset found by SLISE consists of real samples. This makes the subset interesting to examine. Figure 7a shows six digits from the subset and how the linear model interacts with them. We see why the 1 is less likely to be a 2 than the 8 (0.043 vs 0.188). Another interesting question is for which digits the approximation is not valid,

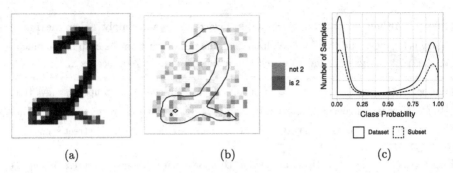

(a)                              (b)                              (c)

**Fig. 6.** (a) The digit being explained. (b) Salience map showing the regression weights of the linear model found using SLISE. The instance being explained is overlaid in the image. Purple colour indicates a weight supporting positive classification of a 2, and orange colour indicates a weight not in support of classifying the item as a 2. (c) Class probability distributions for the full dataset and for the found subset $S$.

in other words which digits are outside the subset. Figure 7b shows a scatterplot of the dataset used to find an explanation for the 2 (shown on a black background). The data items in the subset $S$ lie within the corridor marked with dashed green lines. The top right contains digits to which both SLISE and the classifier assign high likelihoods of being 2s. The bottom left contains digits unlike 2s. The data items in the top left and bottom right contain items for which the local SLISE model is not valid and they are not part of the subset. We see that Z-like 2s and L-like 6s are particularly ill-suited for this approximation.

*Modifying the Subset Size.* The subset size controls the locality of explanations. Large subsets lead to more general explanations, while small subsets may cause overfitting on features specific to the subset. Figure 7d shows a progression of explanations for a 2 (similar to Fig. 6b) in order of decreasing subset size (from $\varepsilon = 0.64$ to $\varepsilon = 0.02$). We observe that these explanations emphasise slightly different regions due to the change in locality (and hence in the model). Note that $\varepsilon \to \infty$ is equivalent to logistic regression through the item being explained.

*Modifying the Dataset.* The dataset used to find the explanation can be modified in order to answer specific questions. E.g., restricting the dataset to only 2s and 3s allows investigation of what separates a 2 from a 3. This is shown in Fig. 7c. We see that 3s are distinguished by their middle horizontal stroke and the 2s by the bottom horizontal strokes ("split" due to the bottom curve of 3s).

**Classification of Particle Jets.** Some datasets adhere to a strict generating model, this is the case for, e.g., the PHYSICS dataset, which contains particle jets extracted from simulated proton-proton collision events [8]. Here the laws of

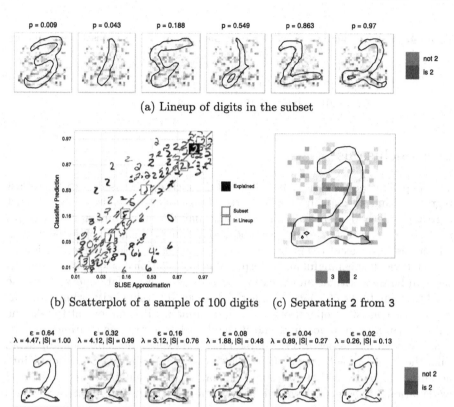

(a) Lineup of digits in the subset

(b) Scatterplot of a sample of 100 digits    (c) Separating 2 from 3

(d) Linear models with different parameters

**Fig. 7.** Exploring how SLISE's model interacts with other digits than the one being explained (a and b), how varying the parameters affects the explanation (d), and how modifying the dataset can answer specific questions (c).

physics must not be violated, and SLISE automatically adheres to this constraint by only using real data to construct the explanation. In Table 4 we use SLISE to explain a classification made by a neural network. The classification task in question is to decide whether the initiating particle of the jet was a *quark* or a *gluon*. The total energy of the jet is on average distributed differently among its constituents depending on the jet's origin [7]. Here, the SLISE explanation shows the importance of the energy distribution variable QG_ptD.

**Table 4.** SLISE explanation for why an example in the PHYSICS dataset is a quark jet.

|       | Pt   | Girth  | QG_ptD | QG_axis2 | QG_mult |
|-------|------|--------|--------|----------|---------|
| Jet   | 1196 | 0.020  | 0.935  | 0.002    | 16      |
| $\alpha$ | 0.01 | −0.05  | 0.18   | −0.02    | 0       |

## 6   Conclusions

This paper introduced the SLISE algorithm, which can be used both for robust regression and to explain classifier predictions. SLISE extends existing robust regression methods, especially in terms of scalability, important in modern data analysis. In contrast to other methods, SLISE finds a subset of variable size, adjustable in terms of the error tolerance $\varepsilon$. SLISE also yields sparse solutions.

SLISE yields meaningful and interpretable explanations for classifier decisions and can be used without modification for various types of data and without the need to evaluate the classifier outside the data set. This simplicity is important as it provides consistent operation across data domains. It is important to take the data distribution into account, and if the data has a strict generating model it is also crucial not to perturb the data. The local explanations provided by SLISE take the interaction between the model and the distribution of the data into account, which means that even simple global models might have non-trivial local explanations. Future work includes investigating various initialisation schemes for SLISE (currently an OLS solution is used).

**Acknowledgements.** Supported by the Academy of Finland (decisions 326280 and 326339). We acknowledge the computational resources provided by Finnish Grid and Cloud Infrastructure [12].

## References

1. Adler, P., et al.: Auditing black-box models for indirect influence. In: ICDM, pp. 1–10 (2016)
2. Alfons, A., Croux, C., Gelper, S.: Sparse least trimmed squares regression for analyzing high-dimensional large data sets. Ann. Appl. Stat. **7**(1), 226–248 (2013)
3. Amaldi, E., Kann, V.: The complexity and approximability of finding maximum feasible subsystems of linear relations. Theor. Comput. Sci. **147**(1), 181–210 (1995)
4. Ausiello, G., Crescenzi, P., Gambosi, G., Kann, V., Marchetti-Spaccamela, A., Protasi, M.: Complexity and Approximation: Combinatorial Optimization Problems and their Approximability Properties, 2nd edn. Springer, Heidelberg (1999). https://doi.org/10.1007/978-3-642-58412-1
5. Baehrens, D., Schroeter, T., Harmeling, S., Kawanabe, M., Hansen, K., Müller, K.: How to explain individual classification decisions. JMLR **11**, 1803–1831 (2010)
6. Caruana, R., Lou, Y., Gehrke, J., Koch, P., Sturm, M., Elhadad, N.: Intelligible models for healthcare: predicting pneumonia risk and hospital 30-day readmission. In: SIGKDD, pp. 1721–1730 (2015)

7. CMS Collaboration: Performance of quark/gluon discrimination in 8 TeV pp data. CMS-PAS-JME-13-002 (2013)
8. CMS Collaboration: Dataset QCD_Pt15to3000_TuneZ2star_Flat_8TeV_pythia6 in AODSIM format for 2012 collision data. CERN Open Data Portal (2017)
9. Cohen, G., Afshar, S., Tapson, J., van Schaik, A.: EMNIST: an extension of MNIST to handwritten letters. arXiv:1702.05373 (2017)
10. Datta, A., Sen, S., Zick, Y.: Algorithmic transparency via quantitative input influence: theory and experiments with learning systems. In: IEEE S&P, pp. 598–617 (2016)
11. Donoho, D.L., Huber, P.J.: The notion of breakdown point. In: A festschrift for Erich L. Lehmann, pp. 157–184 (1983)
12. Finnish Grid and Cloud Infrastructure, urn:nbn:fi:research-infras-2016072533
13. Fong, R.C., Vedaldi, A.: Interpretable explanations of black boxes by meaningful perturbation. arXiv:1704.03296 (2017)
14. Guidotti, R., Monreale, A., Ruggieri, S., Pedreschi, D., Turini, F., Giannotti, F.: Local rule-based explanations of black box decision systems. arXiv:1805.10820 (2018)
15. Guidotti, R., Monreale, A., Ruggieri, S., Turini, F., Giannotti, F., Pedreschi, D.: A survey of methods for explaining black box models. CSUR **51**(5), 93:1–93:42 (2018). https://doi.org/10.1145/3236009
16. Henelius, A., Puolamäki, K., Boström, H., Asker, L., Papapetrou, P.: A peek into the black box: exploring classifiers by randomization. DAMI **28**(5–6), 1503–1529 (2014)
17. Henelius, A., Puolamäki, K., Ukkonen, A.: Interpreting classifiers through attribute interactions in datasets. In: WHI, pp. 8–13 (2017)
18. Komiske, P.T., Metodiev, E.M., Schwartz, M.D.: Deep learning in color: towards automated quark/gluon jet discrimination. JHEP **01**, 110 (2017)
19. Lakkaraju, H., Bach, S.H., Leskovec, J.: Interpretable decision sets: a joint framework for description and prediction. In: SIGKDD, pp. 1675–1684 (2016)
20. Loh, P.L.: Scale calibration for high-dimensional robust regression. arXiv preprint arXiv:1811.02096 (2018)
21. Lundberg, S.M., Lee, S.I.: A unified approach to interpreting model predictions. In: NIPS, pp. 4765–4774 (2017)
22. Maas, A.L., Daly, R.E., Pham, P.T., Huang, D., Ng, A.Y., Potts, C.: Learning word vectors for sentiment analysis. In: ACL HLT, pp. 142–150 (2011)
23. Mobahi, H., Fisher, J.W.: On the link between gaussian homotopy continuation and convex envelopes. In: Tai, X.-C., Bae, E., Chan, T.F., Lysaker, M. (eds.) EMMCVPR 2015. LNCS, vol. 8932, pp. 43–56. Springer, Cham (2015). https://doi.org/10.1007/978-3-319-14612-6_4
24. Molnar, C.: Interpretable Machine Learning (2019). https://christophm.github.io/interpretable-ml-book
25. Ribeiro, M.T., Singh, S., Guestrin, C.: Why should I trust you? Explaining the predictions of any classifier. In: SIGKDD, pp. 1135–1144 (2016)
26. Rousseeuw, P.J.: Least median of squares regression. J. Am. Stat. Assoc. **79**(388), 871–880 (1984)
27. Rousseeuw, P.J., Hubert, M.: Robust statistics for outlier detection. WIRES Data Min. Knowl. Discov. **1**(1), 73–79 (2011)
28. Rousseeuw, P.J., Van Driessen, K.: An algorithm for positive-breakdown regression based on concentration steps. In: Gaul, W., Opitz, O., Schader, M. (eds.) Data Analysis. Studies in Classification, Data Analysis, and Knowledge Organization, pp. 335–346. Springer, Heidelberg (2000)

29. Schmidt, M., Berg, E., Friedlander, M., Murphy, K.: Optimizing costly functions with simple constraints: a limited-memory projected quasi-newton algorithm. In: AISTATS, pp. 456–463 (2009)
30. Smucler, E., Yohai, V.J.: Robust and sparse estimators for linear regression models. Comput. Stat. Data Anal. **111**, 116–130 (2017)
31. Tibshirani, R.: Regression shrinkage and selection via the Lasso. J. R. Stat. Soc. Series. B Stat. Methodol. **58**(1), 267–288 (1996)
32. Ustun, B., Traca, S., Rudin, C.: Supersparse linear integer models for interpretable classification. arXiv:1306.6677v6 (2014)
33. Wang, H., Li, G., Jiang, G.: Robust regression shrinkage and consistent variable selection through the LAD-Lasso. J. Bus. Econ. Stat. **25**(3), 347–355 (2007)
34. Yohai, V.J.: High breakdown-point and high efficiency robust estimates for regression. Ann. Stat. **15**(2), 642–656 (1987). https://doi.org/10.1214/aos/1176350366

**Open Access** This chapter is licensed under the terms of the Creative Commons Attribution 4.0 International License (http://creativecommons.org/licenses/by/4.0/), which permits use, sharing, adaptation, distribution and reproduction in any medium or format, as long as you give appropriate credit to the original author(s) and the source, provide a link to the Creative Commons license and indicate if changes were made.

The images or other third party material in this chapter are included in the chapter's Creative Commons license, unless indicated otherwise in a credit line to the material. If material is not included in the chapter's Creative Commons license and your intended use is not permitted by statutory regulation or exceeds the permitted use, you will need to obtain permission directly from the copyright holder.

# Efficient Discovery of Expressive Multi-label Rules Using Relaxed Pruning

Yannik Klein, Michael Rapp$^{(\boxtimes)}$, and Eneldo Loza Mencía

Knowledge Engineering Group, TU Darmstadt, Darmstadt, Germany
yannik.klein@hotmail.com, {mrapp,eneldo}@ke.tu-darmstadt.de

**Abstract.** Being able to model correlations between labels is considered crucial in multi-label classification. Rule-based models enable to expose such dependencies, e.g., implications, subsumptions, or exclusions, in an interpretable and human-comprehensible manner. Albeit the number of possible label combinations increases exponentially with the number of available labels, it has been shown that rules with multiple labels in their heads, which are a natural form to model local label dependencies, can be induced efficiently by exploiting certain properties of rule evaluation measures and pruning the label search space accordingly. However, experiments have revealed that multi-label heads are unlikely to be learned by existing methods due to their restrictiveness. To overcome this limitation, we propose a plug-in approach that relaxes the search space pruning used by existing methods in order to introduce a bias towards larger multi-label heads resulting in more expressive rules. We further demonstrate the effectiveness of our approach empirically and show that it does not come with drawbacks in terms of training time or predictive performance.

**Keywords:** Multi-label classification · Rule learning · Label dependencies

## 1 Introduction

As many real world problems require to assign a set of labels, rather than a single class, to instances, multi-label classification (MLC) has become an established topic in the recent machine learning literature. For example, text documents are often related to multiple subjects and media, such as images or music, can usually be associated with several tags at the same time (see [14] for an overview).

Rule-based methods are a well-researched approach to solve classification problems. Due to their interpretability, the use of rule learning algorithms in MLC has recently been proposed as an alternative to complex statistical methods such as support vector machines or artificial neural networks (see e.g. [6,8]). Rules provide a natural and simple representation of a learned model and can easily be understood, analyzed, and modified by human domain experts. Especially in safety-critical domains, such as medicine, power systems, or financial markets, the interpretability of machine learning models is an important requirement to be able to prevent malfunctions and unexpected behavior.

© Springer Nature Switzerland AG 2019
P. Kralj Novak et al. (Eds.): DS 2019, LNAI 11828, pp. 367–382, 2019.
https://doi.org/10.1007/978-3-030-33778-0_28

Rules do not only reveal patterns and regularities hidden in the data, but are also able to make *global* or *local* correlations between labels explicit [9]. Exploiting such correlations is considered crucial in MLC and it is commonly accepted that approaches that are able to take label dependencies into account can be expected to achieve better predictive results [5,9,14]. Existing multi-label rule learning approaches are able to exploit label correlations by inducing *label-dependent* rules, i.e., rules that may contain label conditions in their bodies [9]. In addition, rules with *multi-label heads* provide the ability to model local dependencies between labels by including multiple label assignments in their heads [12]. This enables to model co-occurrences—a frequent pattern in multi-label data—, as well as other types of interdependencies between labels, in a natural and compact form.

**Motivation and Goals.** The induction of multi-label heads is particularly challenging as the number of label combinations that can potentially be included in a head increases exponentially with the number of available labels. To mitigate the computational complexity that comes with a search for multi-label heads, certain properties of commonly used multi-label evaluation measures—namely *anti-monotonicity* and *decomposability*—have successfully been exploited for pruning the search space. Although this enables to efficiently induce multi-label heads in theory, experiments have revealed that such patterns are unlikely to be learned in practice [12]. This is due to the restrictiveness of existing methods that assess the quality of potential heads solely in terms of the used evaluation function. These functions tend to prefer single-label predictions to rules with multi-label heads, because the quality of the individual label assignments contained in such a head usually varies. For example, if two rules with the same body but different single-label heads reach a heuristic value of 0.89 and 0.88, respectively, predicting both labels usually results in a performance decline compared to the value 0.89—typically a value between 0.89 and 0.88. However, opting for the multi-label head would arguably be a good choice: First, the resulting rule would have greater coverage. Second, it evaluates to a heuristic value only slightly worse than that of the best single-label rule.

In this work, we present a relaxed pruning strategy to overcome the bias towards single-label predictions. We further argue that strict upper bounds in terms of computational complexity can still be guaranteed when relaxing the search for multi-label heads. As our empirical studies reveal, the training process even tends to terminate earlier due to the increased coverage of the induced rules. The experiments also show that the use of relaxed pruning results in more compact models that reach predictive results comparable to those of existing approaches. Moreover, we discuss whether our approach discovers more label dependencies, which is a major goal of our method.

**Structure of This Work.** We start with an introduction to the problem domain and a recapitulation of previous work in Sect. 2. As the main contribution of this work, in Sects. 3 and 4, we present a plug-in approach that relaxes the search space pruning used by existing methods in order to introduce a bias towards the induction of larger multi-label heads. To illustrate the effects of our extension, we present an empirical analysis focusing on the predictive performance, model characteristics and run-time efficiency of the proposed method in Sect. 5. Finally, we provide an overview of related work in Sect. 6 before we conclude by summarizing our results in Sect. 7.

## 2 Preliminaries

In contrast to binary or multi-class classification, in MLC an instance can be associated with several labels $\lambda_i$ out of a predefined label space $\mathbb{L} = \{\lambda_1, \ldots, \lambda_n\}$. The task is to learn a classifier function $g(.)$ that maps an instance $x$ to a predicted label vector $\hat{y} = (\hat{y}_1, \ldots, \hat{y}_n) = \{0, 1\}^n$, where each prediction $\hat{y}_i$ specifies the presence (1) or absence (0) of the corresponding label $\lambda_i$. An instance $x_j$ consists of attribute-value pairs given as a vector $x_j = (v_1, \ldots, v_l) \in \mathbb{D} = A_1 \times \cdots \times A_l$, where $A_i$ is a numeric or nominal attribute. We handle MLC as a supervised learning problem in which the classifier function $g(.)$ is induced from labeled training data $T = \{(x_1, y_1), \ldots, (x_m, y_m)\}$, containing tuples of training instances $x_j$ and true label vectors $y_j$.

### 2.1 Multi-label Classification Rules

We are concerned with the induction of conjunctive, propositional rules $r : \hat{Y} \leftarrow B$. On the one hand, the body $B$ of such a rule contains an arbitrary number of conditions that compare an attribute-value $v_i$ of an instance to a constant by using equality (nominal attributes) or inequality (numerical attributes) operators. If an instance satisfies all conditions in the body of a rule $r$, it is said to be *covered* by $r$. On the other hand, the head $\hat{Y}$ consists of one (*single-label head*) or several (*multi-label head*) label assignments $\hat{y}_i = \{0, 1\}$ that specify whether the label $\lambda_i$ should be predicted as present (1) or absent (0) for the covered instances. Multi-label heads enable to model local dependencies, such as co-occurrences or exclusions, that hold for the instance subspace covered by the rule's body.

In general, the head $\hat{Y}$ of a rule may have different semantics in a multi-label setting. We consider the predictions provided by an individual rule to be *partial*, because we believe that this particular strategy has several conceptual and practical advantages. When using partial predictions, each rule only predicts the presence or absence of a subset of the available labels and leaves the prediction of the remaining ones to other rules.

### 2.2 Bipartition Evaluation Functions

To evaluate the quality of multi-label predictions, we use bipartition evaluation functions $\delta : \mathbb{N}^{2 \times 2} \to \mathbb{R}$ that are based on comparing the difference between true label vectors (*ground truth*) and predicted labels (cf. [14]). Such a function maps a two-dimensional (label) confusion matrix $C$ to a heuristic value $h \in [0, 1]$. A confusion matrix consists of the number of *true positive* (*TP*), *false positive* (*FP*), *true negative* (*TN*), and *false negative* (*FN*) labels predicted by a rule or classifier.

Let the variables $y_i^j$ and $\hat{y}_i^j$ denote the absence (0) or presence (1) of the label $\lambda_i$ of instance $x_j$ according to the ground truth or a prediction, respectively. Given these variables, we calculate the atomic confusion matrix $C_i^j$ for the respective label $\lambda_i$ and instance $x_j$ as

$$C_i^j = \begin{pmatrix} TP_i^j & FP_i^j \\ FN_i^j & TN_i^j \end{pmatrix} = \begin{pmatrix} y_i^j \hat{y}_i^j & (1 - y_i^j)\hat{y}_i^j \\ (1 - y_i^j)(1 - \hat{y}_i^j) & y_i^j(1 - \hat{y}_i^j) \end{pmatrix} \tag{1}$$

Note that, in accordance with [12], we assess *TP*, *FP*, *TN*, and *FN* differently to evaluate candidate rules during training. To ensure that absent and present labels have the same impact on the performance of a rule, we always count correctly predicted labels as *TP* and incorrect predictions as *FP*. Labels for which no prediction is made are counted as *TN* if they are absent, or as *FN* if they are present.

When evaluating multi-label predictions which have been made for $m$ instances and $n$ labels it is necessary to aggregate the resulting $m \cdot n$ atomic confusion matrices. We restrict ourselves to *micro-* and *(label-based) macro-averaging*, which are defined as

$$\delta(C) = \delta\left(\sum_i \sum_j C_i^j\right) \quad \text{and} \quad \delta(C) = \text{avg}_i\left(\delta\left(\sum_j C_i^j\right)\right) \tag{2}$$

where the $\sum$ operator denotes the cell-wise addition of atomic confusion matrices $C_i^j$, corresponding to label $\lambda_i$ and instance $x_j$, and $\text{avg}_i$ calculates the mean of the heuristic values obtained for each label $\lambda_i$.

### 2.3 Multi-label Rule Learning Heuristics

In the following, we present the bipartition evaluation functions—also referred to as *heuristics*—that are used in this work to assess the quality of candidate rules in terms of a heuristic value $h$. According to these heuristics, rules with a greater heuristic value are preferred to those with smaller values.

Among the heuristics we use in this work is *Hamming accuracy* (HA). It measures the percentage of correctly predicted labels among all labels and can be computed using micro and macro-averaging with the same final result.

$$\delta_{hamm}(C) := \frac{TP + TN}{TP + FP + TN + FN} \tag{3}$$

Moreover, we use the *F-measure* (FM) to evaluate candidate rules. It calculates as the (weighted) harmonic mean of *precision* and *recall*. If $\beta < 1$, precision has a greater impact. If $\beta > 1$, the F-measure becomes more recall-oriented.

$$\delta_F(C) := \frac{(1 + \beta^2) \cdot TP}{(1 + \beta^2) \cdot TP + \beta^2 \cdot FN + FP} \quad \text{with } \beta \in [0, +\infty] \tag{4}$$

### 2.4 Pruning the Search for Multi-label Heads

We rely on the multi-label rule learning algorithm proposed by Rapp et al. [12] to learn rule-based models. It uses a *separate-and-conquer* strategy, where new rules are induced iteratively. Whenever a new rule is learned, the covered instances are removed from the training data set if enough of their labels are predicted by already induced rules ("separate" step). Afterwards, the next rule is induced from the remaining instances and labels ("conquer" step). The training process continues until only few training instances are left. To classify an instance, the rules contained in the model are applied in the order of their induction. If a rule covers the given instance, the labels in its head are applied unless they were already predicted by a previous rule.

To learn new rules, the algorithm performs a top-down greedy search, starting with the most general rule. By adding additional conditions to the rule's body it is successively specialized, resulting in fewer instances being covered. For each candidate rule, a corresponding single- or multi-label head, that models the labels of the covered instances as accurately as possible, must be found.

To find a suitable (multi-label) head for a given body, potential label combinations are evaluated with respect to a certain heuristic $\delta$ using a breadth-first search. Instead of performing an exhaustive search through the label space, which is unfeasible in practice due to its exponential complexity, the search is pruned by leaving out unpromising label combinations as illustrated in Fig. 1. To prune the search for multi-label heads, while still being able to find the best solution, Rapp et al. [12] exploit certain properties of multi-label evaluation measures—namely *anti-monotonicity* and *decomposability*. In this work, we focus on the latter for two reasons: First, decomposability is a stronger criterion compared to anti-monotonicity. It enables to prune the search space more extensively and comes with linear costs, i.e., the best multi-label head can be inferred from considering each label separately. Second, most common multi-label evaluation measures have been proved to be decomposable, including the ones used in this work (cf. Sect. 2.3). The definition of decomposability is given below.

**Definition 1 (Decomposability, cf. [12]).** *A multi-label evaluation function $\delta$ is* decomposable *if the following conditions are met:*

i) *If the multi-label head rule $\hat{Y} \leftarrow B$ contains a label attribute $\hat{y}_i \in \hat{Y}$ for which the corresponding single-label head rule $\hat{y}_i \leftarrow B$ does not reach $h_{max}$, the multi-label head rule cannot reach that performance either (and vice versa).*

$$\exists i \left( \hat{y}_i \in \hat{Y} \wedge h(\hat{y}_i \leftarrow B) < h_{max} \right) \iff h(\hat{Y} \leftarrow B) < h_{max}$$

ii) *If all single label head rules $\hat{y}_i \leftarrow B$ which correspond to the label attributes of the multi-label head $\hat{Y}$ reach $h_{max}$, the multi-label head rule $\hat{Y} \leftarrow B$ reaches that performance as well (and vice versa).*

$$h(\hat{y}_i \leftarrow B) = h_{max} , \forall \hat{y}_i \left( \hat{y}_i \in \hat{Y} \right) \iff h(\hat{Y} \leftarrow B) = h_{max}$$

According to Definition 1, we can safely prune the search space by restricting the evaluation to all possible single-label heads for a given body. To construct the best multi-label head, the highest heuristic value among all single-label heads is determined and those achieving the highest value are combined, while the others are discarded.

# 3   Dynamic Weighting of Rules Using Relaxation Lift Functions

The pruning strategy described in Sect. 2.4 completely neglects combinations of labels with similar, but not equal, heuristic values. As illustrated by the example given in Sect. 1, when pruning according to decomposability, single-label heads with marginally greater heuristic values are preferred to multi-label heads rated slightly worse. Relaxed pruning aims at tolerating minor declines in terms of a rule's heuristic value in favor of greater coverage. By relaxing the pruning constraints, and hence introducing a bias towards multi-label heads, more expressive rules are expected to be learned.

The main challenge of introducing such a bias revolves around two questions: First, the desired degree of the bias is unclear, i.e., how much of a decline in the heuristic value is tolerable. Second, the ideal number of labels in the head is unknown—especially if rules may also predict the absence of labels. As both factors highly depend on the data set at hand, providing any recommendations is difficult. Moreover, the training time potentially suffers from relaxed pruning, as more label combinations are taken into account.

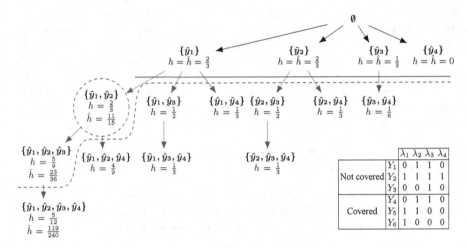

**Fig. 1.** Search for the best (relaxed) multi-label head given the labels $\lambda_1$, $\lambda_2$, $\lambda_3$, and $\lambda_4$. The instances corresponding to the label sets $Y_4$, $Y_5$, and $Y_6$ are assumed to be covered, whereas those of $Y_1$, $Y_2$, and $Y_3$ are not. The dashed line indicates label combinations that can be pruned with relaxed pruning, the solid line corresponds to standard decomposability (cf. [12], Fig. 1).

## 3.1   Lifting the Heuristic Values of Rules

We introduce a bias towards multi-label heads by multiplying the heuristic value $h$ of a rule with a dynamic weight $l \in \mathbb{R}$, which we refer to as a *relaxation lift*. To prefer larger multi-label heads $l$ must increase with the number of labels in the head. The relaxation lift, which we will simply refer to as *lift* in the remainder of this work, therefore specifies the decline in a rule's heuristic value that is acceptable in favor of predicting more labels.

To specify a relaxation lift for every number of labels $x \in [1, n]$ possibly contained in a head, we use *relaxation lift functions* $\rho : \mathbb{R}_+ \rightarrow \mathbb{R}$ mapping a given number of labels $x$ to a relaxation lift $l$. Although the function is only applied to natural numbers, defining $\rho$ in terms of real numbers facilitates the definition.

Given a rule $r : H \leftarrow B$ and a lift function $\rho$, the *lifted heuristic value* of the rule can be calculated as

$$\hat{h} = h \cdot \rho(x) \tag{5}$$

**Table 1.** Example of calculating the lifted heuristic value by multiplying the normal heuristic value and the relaxation lift.

| $|H|$ | $h$ | $\rho(|H|)$ | $\hat{h}$ |
|---|---|---|---|
| 1 | 0.70 | 1.00 | $0.70 \cdot 1.00 = 0.7000$ |
| 2 | 0.67 | 1.07 | $0.67 \cdot 1.07 = 0.7169$ |
| 3 | 0.63 | 1.12 | $0.63 \cdot 1.12 = 0.7056$ |

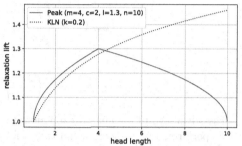

**Fig. 2.** The KLN and peak relaxation lift functions.

where $x = |H|$ corresponds to the number of labels in the rule's head and $h$ is the (normal) heuristic value of the rule as calculated using a certain evaluation function (cf. Sect. 2.3). An example of how to calculate lifted heuristic values $\hat{h}$ is given in Table 1. These values are meant to be used as a replacement of the (normal) heuristic values $h$ when searching for multi-label heads (cf. Sect. 2.4 and Fig. 1).

## 3.2 Relaxation Lift Functions

The proposed framework for relaxed pruning flexibly allows to utilize different relaxation lift functions with varying characteristics and effects on the rule induction process. In the following, we discuss two different types of functions used in this work. A visualization of these functions is given in Fig. 2.

**KLN Relaxation Lift Function.** This simple lift function calculates as the natural logarithm of the number of labels $x$, multiplied by a user-customizable parameter $k \geq 0$. Adding an offset of 1 to the calculated lift ensures that $l = 1$ in case of single-label heads.

$$\rho_{KLN}(x) = 1 + k \cdot \ln(x) \tag{6}$$

The extent of the lift increases with greater values for the parameter $k$. Due to the natural logarithm, the function becomes less steep as the number of labels increases. This is necessary to prevent the selection of heads with a very large number of labels.

**Peak Relaxation Lift Function.** This function also aims at preventing too many labels from being included in the head. With increasing number of labels $1, \ldots, m$, where $m$ is a configurable parameter referred to as the *peak*, the lift becomes greater, then decreases. This enables to introduce a bias towards heads with a specific number of labels, as they are lifted more than others. Given the peak $m$, the desired lift at the peak $l_{max}$, the total number of available labels $n$, and a parameter $c$ that determines the curvature, the peak lift function is defined as follows. Note that $c = 1$ corresponds to a linear gradient.

$$\rho_{peak}(x) = \begin{cases} f_{m,1}(x) & \text{if } x \leq m \\ f_{m,n}(x) & \text{otherwise} \end{cases} \tag{7}$$

$$f_{a,b}(x) = 1 + \left(\frac{x-b}{a-b}\right)^{\frac{1}{c}} \cdot (l_{max} - 1) \tag{8}$$

The advantage of the peak lift function is its efficiency, as more extensive pruning can be performed when using small values for the peak $m$. Compared to the KLN lift function, it is less susceptible to including too many labels in the heads. Moreover, the peak lift function can be adapted more flexibly via the parameters $m$, $l_{max}$, and $c$. However, as these parameters tend to have a significant impact on the learned model, this flexibility comes with a greater risk of misconfiguration.

## 4   Relaxed Pruning of the Label Search Space

As we assess the quality of potential heads in terms of their lifted heuristic value $\hat{h}$, rather than $h$, it is necessary to adjust the search through the label space. In the following, we show that strictly pruning according to decomposability, as suggested in [12], does not yield the best head in terms of $\hat{h}$. Hence, we propose *relaxed pruning* as an alternative and discuss the necessary changes in detail. We also provide an example that illustrates our approach.

**Suboptimal Pruning.** When pruning according to decomposability, the best (multi-label) head is obtained by combining all single-label heads that reach the best heuristic value (cf. Fig. 1). By giving a simple counter-example, we show that this is not possible when searching for the head with the highest lifted heuristic value. Consider two heads $\{\hat{y}_1\}$ and $\{\hat{y}_2\}$ with (macro-averaged) heuristic values $0.8$ and $0.75$, respectively. As we do not lift single-label heads, the lifted and normal heuristic values are identical. Exclusively employing decomposability for finding the best performing lifted head results in the head $\{\hat{y}_1\}$, because combining both heads yields a lower value $0.775$. However, assuming the lift for two labels is $1.1$, the lifted heuristic value evaluates to $\hat{h} = 0.775 \cdot 1.1 = 0.8525$. Consequently, combining both heads results in a higher lifted heuristic value in such case. As a result, we conclude that the search space pruning suggested by Rapp et al. [12] is not suited to find the best head in terms of its lifted heuristic value $\hat{h}$.

**Relaxed Pruning for Macro-averaged Measures.** We adjust the algorithm described in Sect. 2.4 based on two observations: First, the best lifted head of length $k$ results from applying the lift to the head with the highest normal heuristic value of length $k$. As all heads of length $k$ are multiplied with the same lift, a head of length $k$ with a worse heuristic value cannot possibly achieve a higher lifted heuristic value. Thus, we obtain the best lifted head of a certain length by finding the best unlifted head. Second, in case of decomposable evaluation measures that are computed via label-based macro-averaging, such as Hamming accuracy and macro-averaged F-measure (cf. Sects. 2.2 and 2.3), we can guarantee that the best unlifted head of length $k$ results from combining the $k$ best single-label heads.

The basic structure of our procedure is illustrated in Algorithm 1. Similar to pruning according to decomposability, we need to evaluate all single-label heads on the training set for a given rule body and evaluation function (cf. solid line in Fig. 1). In accordance with our observations, we start with an empty head and successively add the best remaining single-label head (cf. REFINECANDIDATE in Algorithm 1). Using this strategy, we obtain the best unlifted multi-label head for each head length. We can then apply the lift in order to get the lifted heuristic value. During this process, we keep track of the head with the best lifted heuristic value $\hat{h}_{best}$. When using a decomposable evaluation measure, including Hamming accuracy, we do not need to re-evaluate any multi-label heads on the training data but can calculate their normal heuristic value as the average of the single-label heads' heuristic values.

---

**Algorithm 1.** Search for the multi-label head with the greatest lifted heuristic value.

```
 1: procedure FINDBESTHEAD(∅ ← B)
 2:     S := sort({ŷ ← B : ŷ ∈ Ŷ})              ▷ sorted single label heads
 3:     c := ∅ ← B; c_best := c          ▷ current candidate and best lifted candidate
 4:     for i = 1, . . . , n do                   ▷ for all head lengths
 5:         c = REFINECANDIDATE(S, c)         ▷ add best remaining label to head
 6:         c.ĥ ← c.h · ρ(i)                     ▷ lift heuristic value
 7:         if c.ĥ ≥ c_best.ĥ then               ▷ update best lifted head
 8:             c_best = c
 9:         if PRUNABLE(c_best, c) then          ▷ check boundary
10:             return c_best                    ▷ return rule if TP ≥ FP
11:     end for
12:     return c_best                            ▷ return rule if TP ≥ FP
13: end procedure
```

---

Instead of generating each possible multi-label head, we calculate an upper bound $\hat{h}_{upper}$ of the lifted heuristic value that could still be achieved by larger multi-label heads. For this, we multiply the normal heuristic value $h_k$ of the current head with length $k$ by the highest remaining lift, i.e., $\hat{h}_{upper} = h_k \cdot \max_{k < i \leq n} \rho(i)$. If $\hat{h}_{upper} < \hat{h}_{best}$, we can prune as the highest performance cannot be achieved by longer heads (cf. PRUNABLE in Algorithm 1). This results from the fact that the normal heuristic value cannot increase by adding more single-label heads as we start with the best. Thus, upper bounds $\max_{k < i \leq n} \rho(i)$ of the lift and the heuristic value $h_k$ are multiplied in order to obtain an upper bound $\hat{h}_{upper}$ for the lifted heuristic value. This approach still guarantees finding the best performing lifted head for macro-averaged heuristics, i.e., also for the measures we are particularly interested in this work, namely macro F-measure and Hamming accuracy.

**(Approximate) Relaxed Pruning for Decomposable Measures.** Even though micro-averaged evaluation measures, such as the micro-averaged F-measure, are often decomposable, combining the best $k$ single-label heads does not necessarily result in the best unlifted head of length $k$ in such case. This is, because the labels are not weighted equally as it is the case for macro-averaged measures. As a consequence, we cannot guarantee to find the best lifted head. Instead, we consider the introduced strategy for finding the best head of length $k$ as an approximation. According to our experiments, this approximation seems to work well in practice—most likely because we relax the search for optimal heuristic values anyway.

**Complexity.** Compared to the original algorithm as described in Sect. 2.4, the use of relaxed pruning does not require any additional evaluations of rules on the training instances. The number of evaluations on the training instances is proportional to the number of labels $n$ (multiplied by the number of training instances $m$)—the same as for the original approach. However, in the worst case, it additionally requires to construct and evaluate $n - 1$ multi-label heads (cf. outer left path in Fig. 1). However, as these heads can be evaluated based on the confusion matrices of the corresponding single-label heads, these additional steps are computationally cheap and do not require any additional evaluation on the training instances.

**Example.** In this example, we illustrate the pruning procedure. Suppose we use the KLN lift function with $k = 0.14$ for the example depicted in Fig. 1. Then $\rho(2) = 1.1$, $\rho(3) = 1.15$ and $\rho(4) = 1.19$. As mentioned, we follow the outer left path. For the head $\{\hat{y}_1\}$ the lifted heuristic value and the maximum lifted value evaluate to $\hat{h} = \frac{2}{3} = \hat{h}_{best}$ and $\hat{h}_{upper} = \frac{2}{3} \cdot 1.19 = 0.793$.[1] Because $\hat{h}_{upper} \geq \hat{h}_{best}$, we cannot prune at this point. For the head $\{\hat{y}_1, \hat{y}_2\}$, $\hat{h}_{upper}$ stays the same, but the lifted heuristic value evaluates to $\hat{h} = \frac{2}{3} \cdot 1.1 = 0.733 = \hat{h}_{best}$. As $\hat{h}_{upper} \geq \hat{h}_{best}$, we still need to check the head $\{\hat{y}_1, \hat{y}_2, \hat{y}_3\}$, for which we calculate $\hat{h} = \frac{5}{9} \cdot 1.15 = 0.639$ and $\hat{h}_{upper} = \frac{5}{9} \cdot 1.19 = 0.661$. As the pruning criterion $\hat{h}_{upper} < \hat{h}_{best}$ holds, we terminate the search and return the best head. The dashed line in Fig. 1 indicates which heads need to be examined when using relaxed pruning. Note that the best lifted and unlifted head are the same in this example.

**Fixing the Head.** During the rule refinement process rules are specialized by adding additional conditions to the body. When searching for a new (multi-label) head each time a rule has been modified, as suggested in [12], previously found heads are often discarded in favor of single-label heads with lower coverage but a higher (lifted) heuristic

---

[1] We round to three decimal places.

value. However, keeping the original head and modifying the body accordingly might result in a better rule. To address this problem, we fix the head during the rule refinement process, i.e., we keep the original head instead of searching for a new head each time the body is modified. As a positive side effect of this modification the time required for building a model usually decreases as it is not necessary to frequently search for new heads.

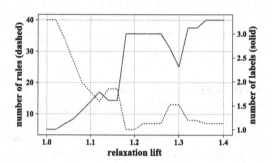

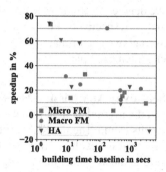

**Fig. 3.** Sensitivity analysis for the KLN lift function on FLAGS.

**Fig. 4.** Time comparison.

**Constraints on Rules.** In addition to fixing the head, as discussed in the previous section, we require each rule to predict at least as many *TP* as *FP*, which effectively imposes a lower bound on the quality of the rules. In preliminary experiments, we found this constraint to be helpful to prevent suboptimal label predictions from being included in the heads for the sake of increasing its lift.

Moreover, we require each label assignment in a head to result in at least one *TP*. This prevents label assignments that do not affect the (normal) heuristic value of a rule, but would result in a higher lift, from being added to a head. For example, such situation might occur if a label is already predicted for all training instances by previously induced rules.

## 5   Evaluation

In this section, we demonstrate the effectiveness of our approach empirically. For our analysis, we consider the predictive performance, the model characteristics, as well as the training time, and give examples of multi-label rules learned by our method.

**Experimental Setup.** We tested our method using relaxed pruning on seven multi-label data sets and compared it to the approach by Rapp et al. [12] using the same configuration.[2] To isolate the influence relaxed pruning has on the learned models, we

---

[2] We used the data sets BIRDS, FLAGS, CAL500, EMOTIONS, MEDICAL, SCENE and YEAST from http://mulan.sf.net/datasets-mlc.html. The source code and data sets are publicly available at https://github.com/keelm/SeCo-MLC/tree/relaxed-pruning.

transferred our additions to fix the heads and impose constraints on the learned rules as discussed earlier. For every data set and every target performance measure (HA, macro- and micro-averaged FM with $\beta = 0.5$, as suggested by [12]), we determined the best lift setting from a number of candidates using five-fold cross-validation on the training set. If two lift settings achieved the same performance, we chose the one with a higher lift as it will typically result in a more compact model. After evaluating all candidates, we trained a model using the best setting and validated it on the test set. We denote the ability to learn rules that predict the presence and absence of labels as $+$ and $-$, respectively. Further, we abbreviate micro-averaging (mic or micro) and macro-averaging (mac or macro). *Subset accuracy* (SA) measures the percentage of perfectly classified examples.

**Sensitivity Analysis and Model Characteristics.** Figure 3 depicts the number of rules and the average number of labels in their heads, depending on the extent of the lift that results from using the KLN lift function and macro-averaged FM. As expected, greater lifts tend to result in larger heads and fewer rules. This is, because rules that predict several labels, rather than a single one, have greater coverage. As a consequence, fewer rules are required to cover the entire training data. However, if the lift is too high, very generic rules, which predict the majority class label and are unable to model the training data accurately, are learned.

**Table 2.** Model characteristics for BIRDS. For each heuristic the left and right column shows the values for the normal and relaxed pruning approach, respectively.

| Number of | Mic FM$^+$ | | Mic F$^-$ | | Mac F$^+$ | | Mac F$^-$ | | HA$^+$ | | HA$^-$ | |
|---|---|---|---|---|---|---|---|---|---|---|---|---|
| Rules | 140 | 140 | 132 | 92 | 140 | 129 | 132 | 113 | 162 | 136 | 58 | 23 |
| Conditions | 219 | 213 | 184 | 146 | 220 | 199 | 184 | 175 | 254 | 204 | 58 | 29 |
| Label conditions | 7 | 4 | 1 | 2 | 7 | 3 | 1 | 1 | 3 | 0 | 1 | 0 |
| Multi-label heads | 1 | 5 | 0 | 22 | 1 | 8 | 0 | 14 | 1 | 30 | 0 | 12 |
| Labels per multi-label head | 2.0 | 2.0 | - | 2.59 | 2.0 | 2.0 | - | 2.57 | 2.0 | 2.1 | - | 17.0 |

**Table 3.** Number of wins/losses/ties of relaxed pruning.

| Heuristic | HA | Mic FM | Mac FM | SA |
|---|---|---|---|---|
| HA$^+$ | 2/4/1 | 1/5/1 | 2/4/1 | 0/4/3 |
| HA$^-$ | 5/1/1 | 4/2/1 | 2/4/1 | 5/0/2 |
| Mic FM$^+$ | 3/3/1 | 3/3/1 | 3/3/1 | 3/2/2 |
| Mic FM$^-$ | 3/1/3 | 1/3/3 | 1/3/3 | 2/2/3 |
| Mac FM$^+$ | 1/5/1 | 1/5/1 | 3/3/1 | 2/3/2 |
| Mac FM$^-$ | 2/1/4 | 2/1/4 | 2/1/4 | 2/1/4 |

Moreover, we observe that the average number of conditions in the bodies decreases with higher lifts. As we induce rules with larger multi-label heads, we would have expected label conditions to be used more frequently. Surprisingly, the percentage of label conditions approaches zero even for a moderate lift. For the peak lift function, the overall trend is identical. However, the maximum number of labels in the head is typically limited.

In addition to the sensitivity analysis, we list some characteristics of models learned during the evaluation in Table 2. We can observe similar phenomena as in the sensitivity analysis. The model characteristics, however, show that our observations also seem to hold for the best lift setting.

**Computational Costs.** As shown in Fig. 4, we compare the training times of our method to the baseline by Rapp et al. [12] using the same configuration. The horizontal axis corresponds to the times required by the baseline to build the models. The vertical axis denotes the relative speedup (or slowdown) that results from using relaxed pruning. Although it potentially evaluates more heads, our method is faster in most of the cases. Typically, a speedup between 10% and 25% can be observed. As we isolate relaxed pruning from the other changes, the speedup most likely results from fewer rules being learned due to their increased coverage as discussed above. Furthermore, we observe that the average number of conditions per rule decreases when using relaxed pruning, i.e., fewer refinement candidates must be taken into account.

**Predictive Performance.** Table 3 lists the number of wins and losses of the compared approaches. We conclude that relaxed pruning results in a predictive performance that is comparable to that of the baseline, despite learning more compact models. More precisely, we observe a decline in performance when using $HA^+$ or macro $FM^+$ as the objective for inducing the heads, but using macro $FM_-^+$ or, in particular, $HA_-^+$ results in an improvement. For micro $FM^+$, the performance is quite similar, despite missing the guarantees discussed in Sect. 4. Regarding an overall comparison between all approaches and heuristics, we can observe that learning the absence and presence of labels and seeking for relaxed Hamming accuracy ($HA\ R_-^+$) ranks highest in average among the 12 approaches w.r.t. Hamming accuracy, but also for subset accuracy, which no approach dedicatedly addresses. In contrast, for micro and macro F-measure, the best performing models are obtained by using relaxed pruning together with the micro F-measure and only predicting the presence of labels (Mic FM $R^+$). This reflects the focus of the F-measure on positive labels compared to HA. In conclusion, relaxing the pruning constraints and deliberately preferring rules with a worse heuristic value in favor of coverage and expressiveness does not seem to have a negative effect on the predictive performance of the models and even results in improvements in some cases.

As mentioned earlier, we determined the best lift settings on the training data. In the majority of the cases, the peak lift function is preferred to the KLN lift function. Due to the variety of possible parameter settings, the peak lift function is more difficult to tune. We observe a trend towards lifts clearly greater than 1. We assume that the parameter for specifying the lift mainly depends on the average number of labels per instance. Moreover, it may also be relevant whether the absence of labels is predicted by the rules in addition to the relevance of labels, as we expect that a greater peak might be beneficial in such case.

**Exemplary Rules.** In Fig. 5 we show exemplary rules as induced with and without the use of relaxed pruning. It can be seen that multi-label heads and label conditions are both suited to model label dependencies. Depending on the model, these different representations may even be equivalent in meaning (cf. first row). Whereas the use of relaxed pruning seems to result in fewer label conditions being learned, it often results in significantly more multi-label heads. This makes a quantitative analysis of the number of label dependencies discovered by the respective approaches more difficult. Nevertheless, our results suggest that relaxed pruning helps to model label dependencies in the form of multi-label heads. Such heads often provide a more compact representation of the discovered correlations. In contrast to label conditions, rules with multi-label heads provide useful information on their own. They do not require to take the order of the rules into account and must not be interpreted in the context of other rules. Due to these advantages, we argue that multi-label heads are easier to understand in many cases.

## 6  Related Work

Most approaches to multi-label rule learning found in the literature are based on association rule (AR) discovery. Alternatively, a few approaches use evolutionary algorithms or classifier systems for evolving multi-label classification rules [1–3]. Inducing rules with several labels in the head is usually implemented as a post-processing step. For example, [13] and similarly [7] induce single-label association rules that are merged to create multi-label rules. By using a separate-and-conquer strategy the step of inducing descriptive but often redundant models of the data is omitted. Instead, classification rules that aimed at providing accurate predictions are learned directly [9].

Most of the approaches mentioned above are restricted to expressing a certain type of relationship since labels are only allowed in the heads of the rules. Approaches that also use labels as antecedents are often restricted to global label dependencies, such as the approaches by [4, 10, 11] that use the relationships discovered by AR mining on the label matrix for refining the predictions of multi-label classifiers.

| | | | |
|---|---|---|---|
| $Class5 \leftarrow Att61$ | (112, 50) | $Class4, Class5 \leftarrow Att61$ | (230, 94) |
| $Class4 \leftarrow Class5$ | (118, 44) | | |
| $Class3 \leftarrow Att50$ | (84, 50) | $Class2, Class3 \leftarrow Att50$ | (174, 111) |
| $Class2 \leftarrow Class3$ | (146, 141) | | |
| $red \leftarrow colours_1, area_1, bars, crescent$ | (85, 9) | $red, white \leftarrow colours_1$ (166, 38) | |
| $red \leftarrow bars, crescent, colours_2, area_1, stripes$ (13, 6) | | $\overline{green}, red \leftarrow colours_2$ (71, 35) | |
| $\overline{red} \leftarrow area_2, circles$ | (9, 1) | $green, \overline{red}$ (2, 0) | |
| $\overline{red} \leftarrow sunstars_1$ | (4, 0) | | |
| $\overline{red} \leftarrow sunstars_2$ | (1, 0) | | |
| $red$ | (1, 0) | | |
| $\overline{RBN} \leftarrow ssd59$ (288, 0) | | $BHG, Warbling\ Vireo, MGW, Stellar's\ Jay, RBN \leftarrow ssd63$ (1012, 3) | |
| $\overline{RBN} \leftarrow ssd89$ (21, 0) | | $\overline{MGW}, \overline{RBN} \leftarrow ssd56$ (141, 0) | |
| $RBN \leftarrow ssd153$ (4, 0) | | $Common\ Nighthawk, \overline{RBN} \leftarrow ssd7, ssd145$ (190, 0) | |
| $\overline{RBN}$ (9, 0) | | $RBN \leftarrow ssd8$ (16, 0) | |
| | | $RBN \leftarrow ssd45$ (4, 0) | |
| | | $\overline{RBN}$ (1, 0) | |

**Fig. 5.** Selected learned normal (left) and relaxed (right) pruning rules regarding a specific label. We show $(TP, FP)$ and absence of label $x$ as $\bar{x}$. Top down: YEAST (macro $FM^+$) twice, FLAGS (micro $FM^+_-$) and BIRDS (macro $FM^+_-$). We abstract specific conditions and only represent attribute names, but indicate different values. For BIRDS we abbreviate Red-breasted Nuthatch (RBN), Black-headed Grosbeak (BHG), MacGillivray's Warbler (MGW) and audio-ssd (ssd).

# 7   Conclusions

In this work, we demonstrated the effectiveness of introducing a bias towards rules with larger multi-label heads. By deliberately preferring rules with a worse heuristic value, we are capable of learning more compact models with more expressive rules that are explicitly tailored to exploit label dependencies. In addition, we argued that strict upper bounds in terms of computational complexity still hold when using relaxed pruning and our experiments revealed that training time even tends to decrease due to the increased coverage of the induced rules. In general, we are able to achieve comparable predictive performance—observing gains in performance for 3 out of 6 tested objectives.

**Acknowledgments.** This research was supported by the German Research Foundation (DFG) (grant number FU 580/11). Calculations for this research were conducted on the Lichtenberg high performance computer of the TU Darmstadt.

# References

1. Allamanis, M., Tzima, F.A., Mitkas, P.A.: Effective rule-based multi-label classification with learning classifier systems. In: Tomassini, M., Antonioni, A., Daolio, F., Buesser, P. (eds.) ICANNGA 2013. LNCS, vol. 7824, pp. 466–476. Springer, Heidelberg (2013). https://doi.org/10.1007/978-3-642-37213-1_48
2. Arunadevi, J., Rajamani, V.: An evolutionary multi label classification using associative rule mining for spatial preferences. In: IJCA Special Issue on Artificial Intelligence Techniques - Novel Approaches and Practical Applications (2011)
3. Ávila-Jiménez, J.L., Gibaja, E., Ventura, S.: Evolving multi-label classification rules with gene expression programming: a preliminary study. In: Corchado, E., Graña Romay, M., Manhaes Savio, A. (eds.) HAIS 2010. LNCS (LNAI), vol. 6077, pp. 9–16. Springer, Heidelberg (2010). https://doi.org/10.1007/978-3-642-13803-4_2
4. Charte, F., Rivera, A.J., del Jesús, M.J., Herrera, F.: LI-MLC: a label inference methodology for addressing high dimensionality in the label space for multilabel classification. IEEE Trans. Neural Netw. Learn. Syst. 25(10), 1842–1854 (2014)
5. Dembczyński, K., Waegeman, W., Cheng, W., Hüllermeier, E.: On label dependence and loss minimization in multi-label classification. Mach. Learn. 88(1–2), 5–45 (2012)
6. Lakkaraju, H.M, Bach, S.H., Leskovec, J.: Interpretable decision sets: a joint framework for description and prediction. In: International Conference on Knowledge Discovery and Data Mining (2016)
7. Li, B., Li, H., Wu, M., Li, P.: Multi-label classification based on association rules with application to scene classification. In: The 9th International Conference for Young Computer Scientists (2008)
8. Mencía, E.L., Fürnkranz, J., Hüllermeier, E., Rapp, M.: Learning interpretable rules for multi-label classification. In: Escalante, H.J., et al. (eds.) Explainable and Interpretable Models in Computer Vision and Machine Learning. TSSCML, pp. 81–113. Springer, Cham (2018). https://doi.org/10.1007/978-3-319-98131-4_4
9. Mencía, E.L., Janssen, F.: Learning rules for multi-label classification: a stacking and a separate-and-conquer approach. Mach. Learn. 105(1), 77–126 (2016)
10. Papagiannopoulou, C., Tsoumakas, G., Tsamardinos, I.: Discovering and exploiting deterministic label relationships in multi-label learning. In: ACM SIGKDD International Conference on Knowledge Discovery and Data Mining (2015)

11. Park, S.-H., Fürnkranz, J.: Multi-label classification with label constraints. In: ECML PKDD 2008 Workshop on Preference Learning (2008)
12. Rapp, M., Loza Mencía, E., Fürnkranz, J.: Exploiting anti-monotonicity of multi-label evaluation measures for inducing multi-label rules. In: Phung, D., Tseng, V.S., Webb, G.I., Ho, B., Ganji, M., Rashidi, L. (eds.) PAKDD 2018. LNCS (LNAI), vol. 10937, pp. 29–42. Springer, Cham (2018). https://doi.org/10.1007/978-3-319-93034-3_3
13. Thabtah, F.A., Cowling, P.I., Peng, Y.: Multiple labels associative classification. Knowl. Inf. Syst. **9**(1), 109–129 (2006)
14. Tsoumakas, G., Katakis, I., Vlahavas, I.: Mining Multi-label Data. In: Maimon, O., Rokach, L. (eds.) Data Mining and Knowledge Discovery Handbook. Springer, Boston (2009)

# Networks

# Evolving Social Networks Analysis via Tensor Decompositions: From Global Event Detection Towards Local Pattern Discovery and Specification

Sofia Fernandes[1]([⊠]), Hadi Fanaee-T[2], and João Gama[1]

[1] LIAAD - INESC TEC, University of Porto, Porto, Portugal
sdsf@inesctec.pt
[2] Department of Biostatistics, University of Oslo, Oslo, Norway

**Abstract.** Existing approaches for detecting anomalous events in time-evolving networks usually focus on detecting events involving the majority of the nodes, which affect the overall structure of the network. Since events involving just a small subset of nodes usually do not affect the overall structure of the network, they are more difficult to spot. In this context, tensor decomposition based methods usually beat other techniques in detecting global events, but fail at spotting localized event patterns. We tackle this problem by replacing the batch decomposition with a sliding window decomposition, which is further mined in an unsupervised way using statistical tools. Via experimental results in one synthetic and four real-world networks, we show the potential of the proposed method in the detection and specification of local events.

**Keywords:** Time-evolving social networks · Tensor decomposition · Event detection

## 1 Introduction

In time-evolving social networks such as computer communications or phone calls, the detection of anomalous events is of great interest. For example, in an IP traffic network, the identification of an unexpected densification of the network is usually associated with malicious activities. Moreover, in phone call networks, the densification of a subnetwork may represent the organization of an event/meeting.

Given these networks, how to spot these anomalous/unexpected behaviors defines the scope of event detection in time-evolving networks. The events may occur at a global level, involving the majority of the nodes in the network (as it is the case of an email sent to all employees making an announcement); or at a local level involving just a subset of nodes. Currently, the existing methods for tackling this problem fail in (at least) one of the following features: (*i*) detecting

© Springer Nature Switzerland AG 2019
P. Kralj Novak et al. (Eds.): DS 2019, LNAI 11828, pp. 385–395, 2019.
https://doi.org/10.1007/978-3-030-33778-0_29

the nodes associated with the anomaly and (ii) capturing local anomalies. Therefore, our aim in this work is to develop a method encompassing these features. Given the multi-dimensional nature of time-evolving networks and the success in mining such networks using tensor decomposition (TD) [13], we exploit TD in a sliding window mode and then apply statistical tools to spot which of the communication patterns found correspond to events.

Our contributions are as follows:

- We propose a framework to spot irregular behaviors in time-evolving networks. Our method combines tensor decomposition with statistical tools to achieve an unsupervised method that (i) spots anomalies at local and global levels and (ii) identifies the sources of anomaly.
- We analyze the detection results provided by our method in one synthetic and four real-world networks and show its usefulness in spotting both local and global anomalies.

Our work is organized in the standard way.

## 2    Related Work

A wide range of strategies has been considered for the problem of event detection on time-evolving networks [1,15]. The most common approach to this problem consists of measuring the (dis)similarity between consecutive network states. In these approaches, when a significant similarity decrease is observed, such instant is flagged as anomalous. The metric value is used to obtain an anomaly score according to which the timestamp is ranked (top ranks are expected to be associated with events). The main difference between these methodologies is the (dis)similarity measure considered (ranging from the graph edit distance and its variations [9] to statistics on node and egonet features [2]). In [17], the authors combine some of the existing approaches to form an ensemble of anomaly detectors, thus achieving more accurate results. This type of approach allows us to detect global anomalies by spotting the time in which a global event was observed, however, generally, they do not spot local events nor provide information on the nodes involved in the event, on the contrary to our approach.

Given the multi-dimensional nature of time-evolving networks, the problem of event detection has also been addressed by matrix and tensor decomposition. The idea generally is to measure the anomalous level of a given timestamp based on the reconstruction error [11]. In other words, given the original network representation and its approximation, timestamps in which the reconstruction error is high suggest that an unexpected event occurred. The reconstruction error approach is applied in a batch mode, in which all the timestamps are available and consequently it is not suitable for real-time detection (while our method is). Additionally, in some works [13], TD is used to extract the patterns and the events are detected by analyzing the latent space. The major limitation of this strategy is that it is not automatic/unsupervised as it requires further analysis of the decomposition results. Moreover, Fanaee-T and Gama [7] proposed a hybrid TD model along with multivariate statistics to detect events from evolving traffic networks. Although their approach is automatic, it detects events only in the global scale and it is not developed for detection and specification of local patterns.

Contrary to the existing approaches, we process the networks using a sliding window, thus capturing more local dynamics of the network and events that occurred during a short time period and involved just a subset of the nodes in the network. We also mine the latent space provided by TD in each window, so that we are able to find the sources of the anomalous events, a feature which is not usually provided in other detectors.

## 3   Background

### 3.1   Tensors

Multi-dimensional numerical arrays are known as tensors. A $M$-order tensor refers to an array of M dimensions: $\mathcal{X} \in \mathbb{R}^{N_1 \times N_2 \times \dots \times N_M}$. The size of the tensor is given by $N_1 \times N_2 \times \dots \times N_M$. The density of a tensor refers to the rate of non-zero entries in it. $\mathcal{X}(i_1, i_2, \dots, i_M)$ denotes the value of the entry $(i_1, i_2, \dots, i_M)$ in the tensor. Similarly to matrices, there are decomposition methods for tensors. In this work we consider the well-known CANDECOMP/PARAFAC (CP) [10]. Given a 3-order tensor, $\mathcal{X} \in \mathbb{R}^{N_1 \times N_2 \times N_3}$, the goal of CP is to find $R$ (factor) vectors $\mathbf{a}_r \in \mathbb{R}^{N_1}$, $\mathbf{b}_r \in \mathbb{R}^{N_2}$, $\mathbf{c}_r \in \mathbb{R}^{N_3}$, where $1 \leq r \leq R$, such that:

$$\mathcal{X} \approx \sum_{r=1}^{R} \mathbf{a}_r \circ \mathbf{b}_r \circ \mathbf{c}_r \tag{1}$$

The vectors associated with each mode may be arranged into factor matrices, which are obtained as follows: $\mathbf{A} = [\mathbf{a}_1 | \mathbf{a}_2 | \dots | \mathbf{a}_R] \in \mathbb{R}^{N_1 \times R}, \mathbf{B} = [\mathbf{b}_1 | \mathbf{b}_2 | \dots | \mathbf{b}_R] \in \mathbb{R}^{N_2 \times R}$ and $\mathbf{C} = [\mathbf{c}_1 | \mathbf{c}_2 | \dots | \mathbf{c}_R] \in \mathbb{R}^{N_3 \times R}$.

There are multiple algorithms to fit the CP model. Nonetheless, since networks are usually sparse, we consider Alternating Poisson Regression (CP-APR) [3], a non-negative decomposition algorithm designed to model sparse count data.

## 4   Problem Addressed

Let $G$ denote a time-evolving network, described by the sequence of its adjacency matrices $\{A_G^t\}_1^L$, over time period 1 to $L$. Moreover, let $G'$ denote a subgraph formed by a subset of interacting entities in the network; $s_{G'}^t$ denotes the density of subgraph $G'$ (that is, the portion of existing edges among all possible edges) at time $t$ and

$$\Delta_{G'}^{t_1, t_2} = s_{G'}^{t_2} - s_{G'}^{t_1} \ ,$$

denotes the difference in the density of subgraph $G'$ between time instants $t_2$ and $t_1$. Then, we are interested in detecting interaction peaks in $G'$, for varying subgraphs $G'$. Formally, a timestamp $\tau$ is dubbed as an interaction peak of nodes in subgraph $G'$ if, for $\delta > 0$, [15]:

$$\Delta_{G'}^{t,\tau} > \delta, \forall t \neq \tau \; , \qquad (2)$$

that is, instant $\tau$ is flagged as event if it satisfies (2). Moreover, it is dubbed as global if subgraph $G'$ includes almost all nodes in the network or as local if it includes only a small subset of nodes.

## 5    Proposed Method

In this work, a network is modeled as a tensor of type *entities* $\times$ *entities* $\times$ *time* corresponding to the sequence of adjacency matrices over time. The network is processed using a sliding window $\mathcal{W}$ of length $L$, which corresponds to the number of timestamps in the window. Briefly, given a network time window $\mathcal{W}$, our method consists of three stages defined as follows.

1. Decomposition of the tensor window

We decompose $\mathcal{W}$ using $R$ components (using CP-APR): $\mathcal{W} \approx \sum_{r=1}^{R} \mathbf{a}_r \circ \mathbf{b}_r \circ \mathbf{t}_r$, with $\mathbf{a}_r, \mathbf{b}_r, \mathbf{t}_r \geq 0$. $\mathbf{a}_r$ and $\mathbf{b}_r$ are associated with entities dimensions and $\mathbf{t}_r$ is associated with the time dimension. Each concept $r$, defined by $\{\mathbf{a}_r, \mathbf{b}_r, \mathbf{t}_r\}$, induces a subgraph $G'_r$ formed by the nodes so that

$$\begin{cases} nodes_{a_r} = \{i : \mathbf{a}_r(i) > 0\} \\ nodes_{b_r} = \{i : \mathbf{b}_r(i) > 0\} \end{cases} ; \qquad (3)$$

(we refer to these nodes as the active nodes in concept $r$). Their time activity is described by $\mathbf{t}_r$, as shown in Fig. 1.

2. Identification of the anomaly candidates

Since each $\mathbf{t}_r$ may be interpreted as a communication/activity vector then an abrupt peak may be interpreted as an outlier in this vector. In this work, given a numerical vector $\mathbf{x} \in \mathbb{R}^N$, we consider the standard classification of the $i^{th}$ element of $\mathbf{x}$, $\mathbf{x}(i)$, as outlier [5]. In more detail, $\mathbf{x}(i)$ is dubbed as (high) outlier value if $\mathbf{x}(i) > Q3 + 1.5IQR$. Additionally, if

$$\mathbf{x}(i) > Q3 + 3IQR \qquad (4)$$

$\mathbf{x}(i)$ is dubbed as extreme outlier, for $Q1$ denoting the first quartile, $Q3$ denoting the third quartile and $IQR = Q3 - Q1$. Thus, we search for isolated values that satisfy (4) in $\mathbf{t}_r$. If an isolated extreme outlier is found then it satisfies (2) for some $\delta$ and component $r$ is flagged as being associated with an extreme peak. We illustrate the process in Fig. 1.

3. Event Characterization and Verification

The goal of this stage is to discard the anomaly candidates that do not represent an irregular/unexpected pattern. To achieve our aim, we construct the graph induced from the anomaly candidate and quantify its presence in the remaining timestamps. In more detail, the anomaly candidate $r$ induces a graph which is

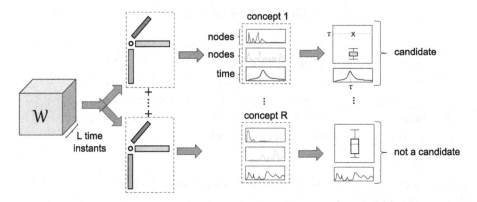

**Fig. 1.** Illustrative example of the pattern finding via TD and anomaly candidate identification in a given time window $\mathcal{W}$: candidates are the TD concepts for which the time factor vector has an isolated extreme outlier.

obtained as follows: from (3) we obtain the active nodes in mode 1 ($nodes_{a_r}$) and the active nodes in mode 2 ($nodes_{b_r}$), then the anomaly candidate is a subgraph $\hat{G}'_r = (V'_r, E'_r)$, where $V'_r = nodes_{a_r} \cup nodes_{b_r}$ and the edges are obtained from the TD approximation, that is, as $A_{\hat{G}}(nodes_{a_r}, nodes_{b_r})$ where ($i$) the indexes ($nodes_{a_r}, nodes_{b_r}$) correspond to the rows and columns associated with nodes in $nodes_{a_r}$ and $nodes_{b_r}$, respectively, and ($ii$) $A_{\hat{G}} = \mathbf{a}_r \circ \mathbf{b}_r \circ \mathbf{t}_r(\tau)$, for the anomalous instant $\tau$. Additionally, since we are looking for dense subgraphs, we discard the nodes in the anomaly candidate induced subgraph that are associated with few links in the original subnetwork. The goal is to remove nodes from the anomaly candidate pattern that do not provide much information. Finally, to assess the abnormality of the pattern, we consider three statistics on the graph induced by the set of nodes $V'_r$ in the original network window: ($i$) the density (per timestamp) of the induced subgraph in the original network; ($ii$) the average weighted node degree (per timestamp); ($iii$) the level of presence of the anomaly candidate in the original network (per timestamp). We quantify this as the rate of edges in the anomaly candidate subgraph $\hat{G}'_r$ which are also present in the original network at each timestamp. By measuring these statistics, we obtain three vectors of length $L$. The anomaly candidate is flagged as an event if for all those three vectors the anomalous instant $\tau$ exhibits a high value, being an isolated outlier (that is, satisfying (4)). If this criteria is not met then it means that the level of abnormality of the candidate is low and therefore the pattern does not correspond to an event.

***Ensemble.*** The proposed methodology has two main parameters: the number of components to be used in TD and the window length. Since selecting these parameters *a priori* may not be straightforward, we propose the usage of an ensemble of these models by varying these parameters. In other words, for a given set of number of components $\{R_i\}_{i=1}^{N}$ and a set of window lengths $\{L_i\}_{i=1}^{M}$, we generate a model $M_{ij}$ for each combination of the number components $R_i$

**Table 1.** Datasets Summary.

| Dataset | Type | Size |
|---|---|---|
| synth | Synthetic network | $500 \times 500 \times 60$ |
| stockmarket | Stock market network | $30 \times 30 \times 42$ |
| challengenet | Computer communication network | $125 \times 125 \times 1304$ |
| enron | E-mail exchange network | $184 \times 184 \times 44$ |
| manufacturing | E-mail exchange network | $167 \times 167 \times 272$ |

and window length $L_j$. Given the set of models, we define a schema to combine their results. Our strategy is based on two assumptions: ($i$) if the same event is detected by a large number of models than it is expected to model a relevant anomaly; ($ii$) given two events that were detected by the same number of models, then the events involving a larger number of nodes are less expected (that is, more abnormal). Thus, we assign scores based on the number of models that detected each anomaly: events that were detected by a larger number of models are associated with the top ranks. Given two events that were detected by the same number of models, we rank them based on the number of nodes participating in the anomaly: events involving a higher number of nodes are considered more anomalous. We note that these models may be run in parallel.

## 6   Experiments

### 6.1   Datasets

In this work we considered one synthetic network (synth) and four real-world time-evolving networks: stockmarket [4], challengenet [16], enron [14] and manufactoring [12], summarized in Table 1. In the weighted networks (all but stockmarket), we applied a logarithmic scale to the edge weights.

synth, stockmarket and challengenet datasets were used to validate our work while the remaining two datasets were used as case studies.

synth was generated using (1) (with $R = 3$) by combining the node factor matrices extracted from a sampled subnetwork of the real-world co-authorship network DBLP [6] with an artificial temporal factor matrix (modeling a periodical cosine, a linear trend and a white noise vector with 60 elements each). Then three local anomalies were injected into the network by replacing three subgraphs with less than 10 nodes with a dense subgraph extracted from a different network (InfectiousPatterns [8]).

In stockmarket, according to the analysis provided in [4], which is also supported by the known economic situation, there are two major events (at timestamps 24 and 30).

The challengenet was characterized by three events: abrupt node increase at instants 376, 377 and 1126. Nonetheless, in order to enrich this network, we injected anomalies by increasing at least $10\times$ the degree of one node at timestamps 500 and 612 and by injecting a clique subgraph at timestamp 1053.

## 6.2   Design of Experiments

We process each dataset using a sliding window with no overlap and apply the detection procedure to each time window. In our ensemble, we considered window lengths according to the time granularity of the datasets, we used: 5, 10 and 15 timestamps for synth; 8, 10 and 12 timestamps for stockmarket (where each timestamp represents a period of 6 months); 9, 18 and 36 timestamps for challengenet (since each timestamp represents a 10 min period); 8, 12 and 16 weeks in enron; and, finally, 7, 14 and 21 timestamps for manufactoring. Regarding the number of components, we used 15, 25, 35, 50 and 75 in all datasets.

## 6.3   Baselines

We considered two baselines: the TD reconstruction error (TDRE) [11] and the recent work of Rayana *et al.* [17], SELECTV. Regarding TDRE, in order to make a fair comparison, we also considered an ensemble of models with different number of components. The timestamps were ranked based on the reconstruction error and the ranking results of the multiple models were averaged. In SELECTV, for each ensemble we considered the time-series of one of three node features: weighted degree ($w$), unweighted degree ($uw$) and number of triangles in the node egonet ($t$). Thus, the method $SELECTV_w$ refers to the application of a SELECTV ensemble to the time-series of weighted degree. Likewise for the remaining features.

## 6.4   Evaluation Metrics

Since our method does not provide scores to all the timestamps, we considered the top-$k$ precision as evaluation metric. This metric consists of the rate of true events within the top-$k$ anomaly scores, where $k$ is the number of true events.

## 6.5   Results

**Synth.** According to the precision results (Table 2), our approach was able to spot the three injected local anomalies, while TDRE was able to spot one event and SELECTV failed at detecting all the known events. We analyzed the top-3 ranked events flagged by the baselines and verified that 2 of such instants, 9 and 12, corresponded to the interaction peaks of the network (and therefore, may be regarded as global events, which resulted from the usage of a white noise factor). Our method also flagged instants 9 and 12 as anomalous, however, a lower anomaly score was assigned to them and therefore they did not appear in the top-3 scores.

**Stock Market.** By analyzing Table 2, we observed that the two known anomalies were successfully detected by SELECTV and our method. It is noteworthy that the events in the network involved more than 75% of the nodes and therefore may be regarded as global. Nonetheless, TDRE was not able to spot the events. Such performance may be justified by the volatility of the network (the network is constantly changing and TD - in batch mode - failed to capture such changes, thus resulting in high reconstruction error for several timestamps, not necessarily associated with irregular behaviors).

**Challenge Network.** Based on Table 2, we observed that our method detected all events in `challengenet`. TDRE was able to spot the event at time 376 but failed to spot the remaining events. SELECTV failed mainly at detecting the event at time 500. We believe the methods had failed to detect anomalous behavior at time 500 due to its local character - at time 500 we increased the degree of node 6 substantially but, as it was not a very active node in the network, this increase may have remained unnoticed to the baseline detectors.

Regarding the anomaly characterization, we verified that the nodes which were the anomaly sources always exhibited a high activity level in one of the nodes modes, thus allowing its identification.

**Table 2.** Top-$k$ precision in the validation networks.

|              | TDRE | SELECTV$_w$ | SELECTV$_{uw}$ | SELECTV$_t$ | Our approach |
|--------------|------|-------------|----------------|-------------|--------------|
| synth        | 0,33 | 0,00        | 0,00           | 0,00        | **1,00**     |
| stockmarket  | 0,00 | **1,00**    | **1,00**       | **1,00**    | **1,00**     |
| challengenet | 0,17 | 0,67        | 0,83           | 0,83        | **1,00**     |

**Case Study 1: Enron.** Our method flagged as anomalous the instants 84, 90, 104, 107, 120, 125, 126, 127, 129, 144 and 145. We analyzed each of the detected anomalies and corresponding flagged individuals and verified that the subgraphs associated with the anomalous nodes had at least 20× more edges at the time of the anomaly than the average number of edges across all timestamps. Moreover, we verified that all the events detected by our method represented topological irregular behaviors either at a local level (such as weeks 90 and 125, whose events corresponded to the interaction peak of a subgraph containing less than 5 employees) or at a global level (such as weeks 104 and 120, whose events involved the majority of the employees). When applying TDRE, we also spotted instants 107 and 145 (which are global events, involving all employees). SELECTV flagged 7 instants in common with our approach (mainly the events involving all employees). We analyzed the other events flagged by these methods but we were not able to spot considerable irregularities in terms of node activity peaks.

**Case Study 2: Manufacturing.** On the contrary to the previous datasets, the number of events detected considerably increased as we increased the number of components in the `manufactoring`. This is expected since, as we increase the number of components, we are able to spot more local patterns (with a smaller number of participants). Thus, these results suggest that an anomalous event in this network should be a pattern with a large number of participants. Therefore, we restricted our analysis to the most relevant anomalies by considering only the top 12 ranked events. We verified that each of the events corresponded to the interaction peak between a given node and a subset of nodes. We also measured the number of edges in the subgraphs associated with the anomalies detected and verified that at the time of the anomaly ($\tau$), the number of edges was at least 20× larger than the average number of edges across all timestamps. It is noteworthy that none of the events detected by our method were also flagged by TDRE. Once more, we believe that this is due to the different network dynamics over time which are not modeled by TD when considering a batch mode. With respect to the SELECTV, there were 5 instants (among the top-12 ranked) which were flagged by both SELECTV and our approach. Regarding the other events detected by SELECTV, we verified that they revealed global anomalies (which were also detectable when tracking the density of the network over time).

**General Observations.** Our analysis showed that the events detected by our method in the datasets under study corresponded to structural irregularities - usually, node degree substantial increase or subgraph densification. In some cases the event was global, involving the majority of the nodes in the network (as it occurred in `stockmarket`, in which our method performed as good or better than the baselines). In other cases, the event involved just a small subset of nodes in the network, a scenario in which our method advantages were especially notorious: our method detected anomalies not easily detected when considering a global view of the network as in TDRE. We recall that in `synth` network the events involved just a small number of nodes and our method was the one exhibiting the best performance. Finally, our method was able to identify the anomaly participants in the networks under study.

# 7  Conclusions

In this work we present a new formulation of TD models for the detection and specification of both global and local events in time-evolving social networks. Our approach ($i$) is automatic, ($ii$) benefits from ensemble learning, thus providing a better coverage; and ($iii$) spots the nodes associated with the irregular behavior. We applied our method to one synthetic and four real-world datasets and showed that the events flagged by our approach were always associated with irregularities in the interaction patterns. In particular, our method was able to spot local irregularities, on the contrary to its competitors. A future work direction we are interested in resides in extending this framework to perform also change detection, thus empowering the applications and analysis power of our method.

**Acknowledgments.** This work is financed by National Funds through the Portuguese funding agency, FCT - Fundação para a Ciência e a Tecnologia within project: UID/EEA/50014/2019. Sofia Fernandes also acknowledges the support of FCT via the PhD grant PD/BD/114189/2016. The authors also acknowledge the SocioPatterns collaboration for making the dataset (in [8]) available.

# References

1. Akoglu, L., Tong, H., Koutra, D.: Graph based anomaly detection and description: a survey. Data Min. Knowl. Disc. **29**(3), 626–688 (2015)
2. Berlingerio, M., Koutra, D., Eliassi-Rad, T., Faloutsos, C.: Network similarity via multiple social theories. In: 2013 IEEE/ACM International Conference on Advances in Social Networks Analysis and Mining (ASONAM), pp. 1439–1440. IEEE (2013)
3. Chi, E.C., Kolda, T.G.: On tensors, sparsity, and nonnegative factorizations. SIAM J. Matrix Anal. Appl. **33**(4), 1272–1299 (2012)
4. Costa, P.: Online Network Analysis of Stock Markets. Master's thesis, University of Porto (2018)
5. Dawson, R.: How significant is a boxplot outlier? J. Stat. Educ. **19**(2) (2011)
6. Desmier, E., Plantevit, M., Robardet, C., Boulicaut, J.-F.: Cohesive co-evolution patterns in dynamic attributed graphs. In: Ganascia, J.-G., Lenca, P., Petit, J.-M. (eds.) DS 2012. LNCS (LNAI), vol. 7569, pp. 110–124. Springer, Heidelberg (2012). https://doi.org/10.1007/978-3-642-33492-4_11
7. Fanaee-T, H., Gama, J.: Event detection from traffic tensors: a hybrid model. Neurocomputing **203**, 22–33 (2016)
8. Isella, L., Stehlé, J., Barrat, A., Cattuto, C., Pinton, J.F., Van den Broeck, W.: What's in a crowd? Analysis of face-to-face behavioral networks. J. Theor. Biol. **271**(1), 166–180 (2011). http://www.sociopatterns.org/datasets/infectious-sociopatterns-dynamic-contact-networks/
9. Kapsabelis, K.M., Dickinson, P.J., Dogancay, K.: Investigation of graph edit distance cost functions for detection of network anomalies. ANZIAM J. **48**, 436–449 (2007)
10. Kolda, T.G., Bader, B.W.: Tensor decompositions and applications. SIAM Rev. **51**(3), 455–500 (2009)
11. Koutra, D., Papalexakis, E.E., Faloutsos, C.: TensorSplat: spotting latent anomalies in time. In: Proceedings of the 2012 16th Panhellenic Conference on Informatics, PCI 2012, pp. 144–149. IEEE Computer Society (2012)
12. Michalski, R., Palus, S., Kazienko, P.: Matching organizational structure and social network extracted from email communication. In: Abramowicz, W. (ed.) BIS 2011. LNBIP, vol. 87, pp. 197–206. Springer, Heidelberg (2011). https://doi.org/10.1007/978-3-642-21863-7_17
13. Papalexakis, E., Pelechrinis, K., Faloutsos, C.: Spotting misbehaviors in location-based social networks using tensors. In: Proceedings of the Companion Publication of the 23rd International Conference On World Wide Web Companion, pp. 551–552. International World Wide Web Conferences Steering Committee (2014)
14. Priebe, C.E., Conroy, J.M., Marchette, D.J., Park, Y.: Scan statistics on Enron graphs. Comput. Math. Organ. Theory **11**(3), 229–247 (2005)
15. Ranshous, S., Shen, S., Koutra, D., Harenberg, S., Faloutsos, C., Samatova, N.F.: Anomaly detection in dynamic networks: a survey. Wiley Interdisc. Rev.: Comput. Stat. **7**(3), 223–247 (2015)

16. Rayana, S., Akoglu, L.: An ensemble approach for event detection and characterization in dynamic graphs. In: ACM SIGKDD ODD Workshop (2014)
17. Rayana, S., Akoglu, L.: Less is more: building selective anomaly ensembles. ACM Trans. Knowl. Discov. Data (TKDD) **10**(4), 42 (2016)

# Efficient and Accurate Non-exhaustive Pattern-Based Change Detection in Dynamic Networks

Angelo Impedovo[1(✉)], Michelangelo Ceci[1], and Toon Calders[2]

[1] Department of Computer Science, University of Bari "Aldo Moro", Bari, Italy
{angelo.impedovo,michelangelo.ceci}@uniba.it
[2] Department of Computer Science, University of Antwerp, Antwerp, Belgium
toon.calders@uantwerpen.be

**Abstract.** Pattern-based change detectors (PBCDs) are non-parametric unsupervised change detection methods that are based on observed changes in sets of frequent patterns over time. In this paper we study PBCDs for dynamic networks; that is, graphs that change over time, represented as a stream of snapshots. Accurate PBCDs rely on exhaustively mining sets of patterns on which a change detection step is performed. Exhaustive mining, however, has worst case exponential time complexity, rendering this class of algorithms inefficient in practice. Therefore, in this paper we propose non-exhaustive PBCDs for dynamic networks. The algorithm we propose prunes the search space following a beam-search approach. The results obtained on real-world and synthetic dynamic networks, show that this approach is surprisingly effective in both increasing the efficiency of the mining step as in achieving higher detection accuracy, compared with state-of-the-art approaches.

**Keywords:** Change detection · Pattern mining

## 1 Introduction

Change detection in dynamic networks is the task of finding time points in which the behavior of the observed network begins to change from the ordinary situation. Once identified, the points provide temporal indications on the obsolescence of previously trained models, which should be adapted to new data. The problem, also known as concept drift detection [3], affects both supervised and unsupervised techniques and therefore is one of the most important problems in the analysis of data that are characterized by a temporal component. For example, in the supervised setting the detection is performed by controlling a quality measure of a previously learned model on new data (e.g. using the misclassification rate [3]). Whereas, in the unsupervised setting, it takes into account how new data deviate from the ordinary data distribution [6]. In both cases, if a significant peak on some appropriate measures is observed, then a change is detected and some actions are performed to update the model by considering new data.

© Springer Nature Switzerland AG 2019
P. Kralj Novak et al. (Eds.): DS 2019, LNAI 11828, pp. 396–411, 2019.
https://doi.org/10.1007/978-3-030-33778-0_30

One of the main challenges in change detection is that of detecting changes in an efficient and accurate manner. In the specific case of dynamic networks, existing approaches may fail to address both challenges because of the inherent complexity of the data [1]. In fact, (i) many approaches are designed for time series and categorical data and not for network-based data, and (ii) many network-based approaches are time consuming, and hence not scalable in both network size and number of graph snapshots. Moreover, the lack of a ground truth able to establish which data represents a change favors the adoption of unsupervised change detection methods.

Recently, non-parametric unsupervised change detection methods relying on frequent patterns have been proposed for transactional data [6,8] and dynamic networks [9]. Such methods are *pattern-based change detectors* (PBCDs hereafter) in which the change is sought on a descriptive model of the data, rather than on the data itself. More precisely, the quality measure tracked by PBCDs for detecting changes is the dissimilarity between sets of patterns (e.g. binary Jaccard measure [9], Levenshtein measure [6]) discovered before and after the arrival of new transactions. In PBCDs, the complete set of frequent patterns is mined upon the arrival of new transactions (*mining step*), before measuring the quality measure (*detection step*).

Typically, the mining step relies on exhaustive algorithms leading to *complete* pattern mining. The main intuition for this solution is that *completeness* is desirable to accurately model the data. However, in practical cases, only a small portion of patterns is likely to be relevant for the detection. Furthermore, exhaustive pattern mining may represent the major obstacle for any PBCD that needs to timely react to incoming data. Our main claim is, therefore, that it is possible to relax the completeness property and it is possible to adopt non-exhaustive mining methods in change detection without loosing in accuracy. Simultaneously, this relaxation can improve the efficiency of the mining step and allow the PBCD system to quickly react. To the best of our knowledge, the StreamKRIMP algorithm [8] is the only PBCD adopting a non-exhaustive mining method, which is based on the MDL principle. However, StreamKRIMP is designed for transactional data streams and not for dynamic networks. Moreover, few other attempts can be retraced in pattern-based anomaly detection methods [5] and subgroup discovery [10], where the search space is pruned according to heuristic evaluations. An alternative line of research for reducing the patterns in PBCDs is that of exhaustively mining condensed sets of patterns which are representative of all the possible patterns. In particular, both non-derivable patterns [7] and $\Delta$-closed patterns [11] have been proposed for anomaly detection and change detection, respectively. However, in such approaches, condensed sets of patterns are discovered by exhaustive mining procedures, and hence they do not provide solutions to the computational issues of PBCDs.

By taking into account the aforementioned reasons, this paper extends the PBCD methodology originally proposed in KARMA [9], based on exhaustive frequent connected subgraph mining, with non-exhaustive mining algorithms. In particular, we customize the general architecture inherited by the KARMA algorithm with improved mining step and detection step, then we perform an extensive study to select the most accurate and the most efficient PBCDs. Experiments show that the proposed approaches improve the efficiency and the detection accuracy with respect to their exhaustive counterpart, on both real world and synthetic dynamic networks.

## 2   Background

Let $N$ be the set of nodes, $L$ be the set of edge labels, and $I = N \times N \times L$ the alphabet of all the possible labeled edges, on which a lexicographic order $\geq$ is defined. A dynamic network is represented as the time-ordered stream of graph snapshots $D = \langle G_1, G_2, \ldots, G_n \rangle$. Each snapshot $G_i \subseteq I$ is a set of edges denoting a directed graph observed in $t_i$, which allows self-loops and multiple edges with different labels. $G_i$ is uniquely identified by id $i$. Let $G$ be a directed graph, a *connected subgraph* $S \subseteq G$ is a directed graph such that for any pair of nodes (in $S$) there exists a path connecting them. Then, a *subtree* $S \subseteq G$ is a connected subgraph in which every node (in $S$) is connected to a unique parent node, except for the root node.

The data representation fits with the one adopted in transactional data mining, allowing the mining of frequent patterns by adapting traditional frequent itemset mining algorithms. In this perspective a snapshot $G_{tid} \in D$ is a transaction uniquely identified by $tid$, whose items are labeled edges from $I$. While a pattern $P \subseteq I$, with *length* $|P|$, can be seen as a word $P = \langle i_1 \ldots i_n \rangle$ of $n$ lexicographic sorted items, with prefix $S = \langle i_1 \ldots i_{n-1} \rangle$ and suffix $i_n$. The *tidset* of $P$ in the network $D$ is defined as $tidset(P, D) = \{tid \mid \exists G_{tid} \in D \land P \subseteq G_{tid}\}$, while the *support* of $P$ in $D$ is $sup(P, D) = \frac{|tidset(P,D)|}{|D|}$. $P$ is *frequent* in $D$ if $sup(P, D) > \alpha$, where $\alpha \in [0, 1]$.

In PBCDs designed for network data, we deem as interesting two types of patterns: (i) *frequent connected subgraphs* (FCSs) and, (ii) *frequent subtrees* (FSs). Both FCSs and FSs are mined from snapshots belonging to time windows. A window $W = [t_i, t_j]$, with $t_i < t_j$, is the sequence of snapshots $\{G_i, \ldots, G_j\} \subseteq D$. Consequently, the width $|W| = j - i + 1$ is equal to the number of snapshots collected in $W$. For our convenience we term $F_W$ the set of all the FCSs (FSs) in the window $W$.

### 2.1   Problem Statement

Let $D = \langle G_1, G_2, \ldots, G_n \rangle$ a dynamic network, $\alpha \in [0, 1]$ be the minimum support threshold, $\beta \in [0, 1]$ the minimum change threshold. Then, pattern-based change detection *finds* pairs of windows $W = [t_b, t_e]$ and $W' = [t'_b, t'_e]$, where $t_b \leq t'_b \leq t_{e+1}$ and $t_e < t'_e$, satisfying $d(F_W, F_{W'}) > \beta$, where (i) $F_W$ and $F_{W'}$ are the sets of patterns discovered on $W$ and $W'$ according to $\alpha$, and (ii) $d(F_W, F_{W'}) \in [0, 1]$ is a dissimilarity measure between sets of patterns. In this perspective, changes correspond to significant variations in the set of patterns discovered on two windows, which denote *stable* features exhibited by the graph snapshots.

# 3   Architecture of a PBCD

The aforementioned change detection problem can be solved by various computational solutions. In this section we provide the general architecture of a PBCD for network data, by generalizing the algorithm KARMA proposed in [9].

In general, a PBCD forms a two-step approach in which: (i) a pattern mining algorithm extracts the set of patterns observed from the incoming data, and (ii) the amount of change is quantified by adopting a dissimilarity measure defined between sets of patterns. Practically speaking, a PBCD is an iterative algorithm that consumes data coming from a data source, in our case a dynamic network, and produces quantitative measures of changes. In particular, the KARMA algorithm is a PBCDs based on exhaustive mining of FCSs, whose general workflow can be seen in Fig. 1. The algorithm iteratively consumes blocks $\Pi$ of graph snapshots coming from $D$ (Step 2) by using two successive landmark windows $W$ and $W'$ (Step 3). This way, it mines the complete sets of FCSs, $F_W$ and $F_{W'}$, necessary to the detection step (Steps 4–5). The window grows ($W = W'$, Step 8) with new graph snapshots, and the associated set of FCSs is kept updated (Step 9) until the Tanimoto coefficient $d(F_W, F_{W'})$ exceeds $\beta$ and a change is detected. In that case, the algorithm drops the content of the window by retaining only the last block of transactions ($W = \Pi$, Steps 6–7). Then, the analysis restarts. The KARMA algorithm offers a general architecture for building custom PBCD, which is made of 4 components: (i) the *window model* (Fig. 1, Steps 3, 8 and 6), (ii) the *feature space* (FCSs or FSs), (iii) the *mining step* (Fig. 1, Steps 4, 9 and 7), and (iv) the *detection step* (Fig. 1, Step 5). In the following sections we will focus on both the mining step and the detection step, also by commenting their contribution to the efficiency of the PBCD strategy.

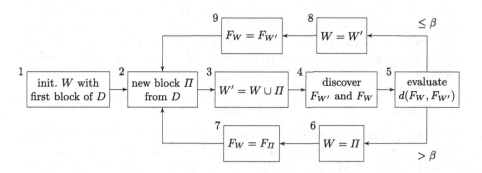

**Fig. 1.** The KARMA algorithm workflow

Here, we briefly discuss the choice of an appropriate time window model of a PBCD. We deem as interesting 3 models: the *landmark model*, the *sliding model* and the *mixed model*. They differ in the way they consume the incoming block $\Pi$ of graph snapshot. In its original version, KARMA uses the *landmark model*. Here, $\Pi$ is added to the window $W$, forming the successive window $W' = W \cup \Pi$

(Fig. 1, Step 3). In this model, the window grows until a change is detected and when a change is detected, old data are discarded. In the *sliding model*, the detection is performed on two successive windows $W$ and $\Pi$ of fixed size. The windows are non overlapping and they always slides forward, both in case of change detected and not detected ($W = \Pi$). Therefore, old data are always discarded. In the *mixed model*, the detection is performed on $W$ and $\Pi$, as in the sliding model. However, as in the landmark model, $\Pi$ is added to $W$, forming $W' = W \cup \Pi$ until a change is detected. In that case, old data are discarded.

## 4    Exhaustive and Non-exhaustive Mining in PBCDs

The main difference between exhaustive and non-exhaustive PBCDs lies in the *exhaustiveness* of the mining step used to discover the patterns, which is the major bottleneck of any exhaustive PBCD approach. In fact, the discovery of an exponentially large number of patterns affects the efficiency of both the mining and detection step, hence rendering this class of algorithms not efficient in practice. The main objective of this paper is to reduce the computational complexity of exhaustive approaches by adopting non-exhaustive ones. In particular, we propose a mining algorithm able to prune the search space of patterns following a beam-search approach.

Being based on beam-search, the proposed approach relies on a parameter $k$ which controls the *beam size* of the mining step when traversing the search space of patterns, that is a lattice $L = (2^I, \subseteq)$ ordered by the generality relation $\subseteq$, conveniently represented in a SE-Tree data structure. Since an exhaustive search can be achieved with non-exhaustive procedures by setting $k = |I|$, we refer to Algorithm 1 in both cases. In particular, the algorithm implements a pattern-growth approach for mining patterns $F_W$ in a time window $W$, and it is initially called with empty prefix $\emptyset$. An important remark is that exhaustive PBCDs rely on complete pattern sets, discovered by the exhaustive mining procedure, as the feature sets for the detection problem. On the contrary, non-exhaustive PBCDs rely only on limited pattern sets discovered by the non-exhaustive mining procedure.

### 4.1    Exhaustive FCSs and FSs Mining

The mining procedure (Algorithm 1) takes 4 input parameters, that is the content of the window $W$, the minimum support threshold $\alpha$, the beam-size $k$, and the pattern prefix (initially equals to $\emptyset$). The algorithm exhaustively traverses the search space of FCSs and FSs by setting $k = |I|$, following a recursive DFS approach. In particular, it is able to *(i)* build patterns with a pattern-growth approach in which items are appended as suffix to a pattern prefix, and *(ii)* evaluate the supports through tidset intersection. The result is the complete set of the frequent patterns $F_W$ in $W$ according to $\alpha$.

The procedure considers the window $W$ as an $i$-conditional database of transactions in which every item $j \leq i$ has been removed, as done in [4]. At the beginning of each recursive call, Line 2 initializes the set $F[P]$ of frequent patterns on

---

**Algorithm 1:** Mining procedure based on beam search

---

**Output:** F[P] the set of frequent patterns having prefix P
1   **minePatterns** $(W, \alpha, k, P)$
2     $F[P] = \emptyset, Items = \emptyset, Beam = \emptyset$
3     **for** $(i, tidset)$ occurring in $W$ **do**
4       **if** $\frac{|tidset|}{|W|} \geq \alpha$ **then**
5         **if** $isValid(P \cup \{i\})$ **then**
6           $F[P] = F[P] \cup (P \cup \{i\})$
7         **end**
8         $Items = Items \cup \{i\}$
9       **end**
10    **end**
11    $Beam = topKSortedBySupport(Items, k)$
12    **for** all $i$ occurring in $Beam$ **do**
13      $W^i = \emptyset$
14      **for** $j$ occurring in $Beam \mid j > i$ **do**
15        $C = tidset(i, W) \cap tidset(j, W)$
16        $W^i = W^i \cup \{(j, C)\}$
17      **end**
18      **if** $isValid(P \cup \{i\})$ **then**
19        $F[P \cup \{i\}] = minePatterns(W^i, \alpha, k, P \cup \{i\})$
20        $F[P] = F[P] \cup F[P \cup \{i\}]$
21      **end**
22    **end**
23    **return** $F[P]$

---

prefix $P$ as empty. Then, Lines 3–10 exploit the vertical layout of $W$, and test the supports against the threshold $\alpha$. The FCS (FS) $P \cup \{i\}$ is built by appending the item $i$ to the prefix $P$ only when allowed by the predicate $isValid$ (Line 4), which checks whether $P \cup \{i\}$ is a connected subgraph (when mining FCSs) or a subtree (when mining FSs), respectively. Lastly, they are added to the set $F[P]$. The algorithm adds the suffix $i$ of any pattern discovered to $Items$.

Line 11 selects only the most promising subset of $k$ patterns, according to their support. In practice, this line is irrelevant in exhaustive mining as it will always select all the patterns, since $k = |I|$. Then, lines 12–22 build the $i$-conditional databases on which to perform recursive calls. In particular, the algorithm iterates over each item $i$ in $Beam$, and Line 13 initialize the associated $i$-conditional database $W^i$ as empty. Lines 14–17, iterate on items $j$ from $Beam$ such that $j > i$. This way, Line 15 computes the tidset $C$ as the set intersection between the tidsets of $i$ and $j$ in the database $W$, respectively. Then, $C$ is the tidset of $j$ in the newly created $i$-conditional database $W^i$. The mining procedure is recursively called at Line 19 for mining the set $F[P \cup \{i\}]$ of FCSs (or FSs) with valid prefix $P \cup \{i\}$, according to the pattern language. This way, subgraphs which are not connected (or do not represent trees) are pruned at Line 18. Finally, the patterns in $F[P \cup \{i\}]$ are added to $F[P]$, which is returned as the final result.

The exhaustive mining of FCSs and FSs requires time proportional to $O(2^{|I|})$ in the worst case scenario, in line with that of traditional frequent itemset mining. Moreover, due to the constraints imposed by the pattern language, the number of FCSs and FSs is in practice much lower than the number of itemsets,

FSs $<$ FCSs $< 2^{|I|}$. However, the mining time is still exponential in the number of edges $|I|$, thus resulting inefficient.

## 4.2   Non-exhaustive FCSs and FSs Mining

The non-exhaustive mining is achieved by pruning the search space of FCSs and FSs according to some heuristic criteria. In particular, when calling Procedure 1 with $k < |I|$, the intermediate selection step (Line 11) selects only the most promising subset of $k$ patterns to further advance the process (Lines 12–22). As in traditional beam search-based algorithms, frequent patterns in a recursive call are evaluated by means of a heuristic evaluation function and sorted in increasing order. Then, only the first top-$k$ of them are further considered.

Many heuristic evaluation functions can be used to select the most promising subset of patterns, among them we adopt the support of patterns. In particular, Line 11 sorts the patterns in $F[P]$ according to their support, then only $k$ of them with the greatest support are kept to further advance the process, while the remaining ones are ignored. However, since the sorting is performed between patterns having *same length*, the evaluation based on the support is consistent with the evaluation based on the *area*.

*Proof.* Let $S$ be a pattern, and $W$ be a window. Then, the support $sup(S, W) = \frac{|tidset(S,W)|}{|W|}$ and the area $area(S, W) = |S| \cdot |tidset(S, W)|$ are linearly proportional to $|tidset(S, W)|$ with two constant factors $\frac{1}{|W|}$ and $|S|$.

The area of the FCS (or FS) $S$ in the window $W$, $area(S, W)$, is an *interestingness* measure adopted in *tile mining* [4]. In our case, it is used to restrict the search space by considering only the most *interesting* patterns at each recursion step. This simple, yet effective approach allows us to significantly prune the search space when mining the limited sets $F_W$ and $F_{W'}$. In particular, the mining is more focused towards patterns covering large portions of the window, *tiles* from a transactional database point of view [4], and hence more interesting.

The non-exhaustive mining procedure is more efficient than the exhaustive one, requiring time proportional to $O(2^k)$ in the worst-case scenario. In fact, the algorithm restricts the attention on only $k$ items from $I$ in the base recursion step, with $k \ll |I|$.

## 5   Detecting Changes on Pattern Sets

Once the complete or limited pattern sets, $F_W$ and $F_{W'}$, have been discovered by either the exhaustive or non-exhaustive procedure, respectively, the detection step can be executed and the dissimilarity score $\beta = d(F_W, F_{W'})$ computed. We recall that $d(F_W, F_{W'})$ is a binary dissimilarity measure defined on sets of patterns. For our convenience we define it as operating on the vector encoding $\mathbf{w}$ and $\mathbf{w}'$ of $F_W$ and $F_{W'}$, respectively.

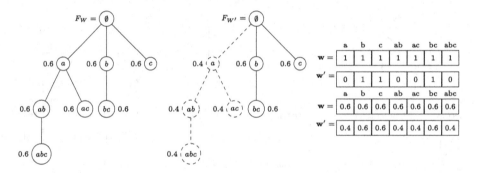

**Fig. 2.** Example of binary (top) and real-valued (bottom) vector encoding of $F_W$ and $F_{W'}$. Dashed circles denote infrequent patterns with $\alpha = 0.5$.

## 5.1  Detecting Changes on Complete Pattern Sets

When detecting changes on complete pattern sets, the encoding is built by enumerating the patterns in $F_W \cup F_{W'}$. More specifically, $\mathbf{w}$ ($\mathbf{w'}$) is a vector of size $n = |F_W \cup F_{W'}|$, where the $i$-th element is a weight associated to the $i$-th pattern in the enumeration of $F_W \cup F_{W'}$ with respect to $W$ (or $W'$, respectively). Then, a change is detected if the dissimilarity score exceeds the minimum threshold $\beta$, that is when $d(F_W, F_{W'}) > \beta$.

In the case of KARMA, as shown in Fig. 2, $\mathbf{w}$ and $\mathbf{w'}$ are binary vectors indicating whether each FCS from the enumeration is frequent or not in $W$ and $W'$, respectively. Then the algorithm computes the *Tanimoto coefficient* $d(F_W, F_{W'}) = 1 - \frac{\mathbf{w} \cdot \mathbf{w'}}{\|\mathbf{w}\|^2 + \|\mathbf{w'}\|^2 - \mathbf{w} \cdot \mathbf{w'}}$. By doing so, KARMA quantifies the fraction of FCSs which have crossed the minimum support threshold, thus indicating a relevant change in the underlying graph data distribution. However, this solution does not take into account the FCSs not crossing the minimum support threshold, although exhibiting a potentially significant support spread.

To overcome this limitation, an alternative approach also shown in Fig. 2 is to build the vector encoding as real-valued vectors of supports in $W$ and $W'$, respectively. Then, it is possible to compute the *weighted Jaccard dissimilarity* $d(F_W, F_{W'}) = 1 - \frac{\sum_i min(\mathbf{w}_i, \mathbf{w'}_i)}{\sum_i max(\mathbf{w}_i, \mathbf{w'}_i)}$. We deem this measure as relevant because relates the dissimilarity to the *absolute growth-rate* [2] of each pattern $S$, defined as $GR(S, W, W') = |sup(S, W) - sup(S, W')|$. The *absolute growth-rate* is a contrast measure adopted in *emerging pattern mining* to discover contrast patterns between two datasets [2].

*Proof.* Given the analytic formulations for $max(a, b) = \frac{1}{2}(a + b + |a - b|)$ and $min(a, b) = \frac{1}{2}(a + b - |a - b|)$, and the vector encoding $\mathbf{w}$ and $\mathbf{w'}$ of $F_W$ and $F_{W'}$. The weighted Jaccard dissimilarity can be rewritten as $d(F_W, F_{W'}) = 1 - \frac{\sum_i min(\mathbf{w}_i, \mathbf{w'}_i)}{\sum_i max(\mathbf{w}_i, \mathbf{w'}_i)} = 1 - \frac{\sum_i \mathbf{w}_i + \mathbf{w'}_i - GR(S_i, W, W')}{\sum_i \mathbf{w}_i + \mathbf{w'}_i + GR(S_i, W, W')}$.

In exhaustive PBCDs the number of patterns grows exponentially in the number of items. Therefore, regardless from the measure $d(F_W, F_{W'})$ adopted, the

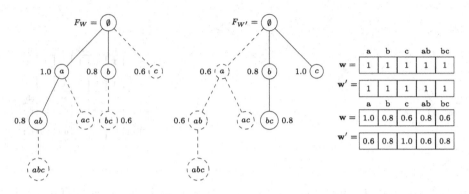

**Fig. 3.** Example of limited sets of frequent patterns $F_W$ (left) and $F_{W'}$ (right) discovered with $k = 2$ and $\alpha = 0.1$. Dashed circles denote pruned non-interesting patterns.

detection step on complete pattern sets requires an amount of time proportional to the number of patterns in the enumeration.

### 5.2 Detecting Changes on Limited Pattern Sets

Non-exhaustive PBCDs detect changes in the same way exhaustive PBCDs do, that is by computing the score $d(F_W, F_{W'})$ in terms of the *Tanimoto coefficient* or the *unweighted Jaccard dissimilarity*, and testing it against the minimum change threshold $\beta$. Although the detection approach remains the same, a subtle difference in the meaning of the detection is present. In fact, while the dissimilarity measures adopted on complete pattern sets quantify how much the supports of FCSs (FSs) change between $W$ and $W'$, they do not consider the *interestingness* of patterns on $W$ and $W'$, respectively, as intended by the non-exhaustive mining algorithm.

Non-exhaustive mining based on the interestingness prunes the search space of patterns in two different ways for $W$ and $W'$, respectively. Thus restricting the search only to the most interesting FCSs and FSs, while discarding the less interesting ones (Fig. 3). By doing this, the detection relies on a considerably low number of patterns, hence resulting more efficient, while losing information associated to patterns which have been pruned. This affects the construction of the vector encoding $\mathbf{w}$ and $\mathbf{w'}$, which is built according to the enumeration $F_W \cup F_{W'}$ consisting of a reduced number of patterns. The example reported in Fig. 3 depicts a scenario in which every pattern is frequent in both $W$ and $W'$, although with different supports, thus determining different interestingness. Any information related to the patterns "*ac*" and "*abc*" is lost, as they are not present in the enumeration of $F_W \cup F_{W'}$, and therefore they do not contribute to the change.

Therefore, the detection step becomes non-exhaustive itself, by focusing the detection only on the most interesting frequent patterns. In particular, as the number of patterns discovered by the non-exhaustive mining procedure grows exponentially with the parameter $k \ll |I|$, the detection step requires in practice much smaller time than that required on complete pattern sets.

# 6    Computational Complexity

In this section we study the computational complexity of PBCDs in the worst-case scenario. In particular, the analysis takes into account the influence of the *feature space*, the *mining step*, the *detection step* and *the window model*. Given the dynamic network $D = \langle G_1, G_2, \ldots, G_n \rangle$ where $I$ denotes the possible labeled edges observed over the time, and $|\Pi|$ the size of blocks, then every PBCD built according to the architecture in Fig. 1 consumes exactly $e = \frac{n}{|\Pi|}$ blocks of transactions, thus requiring $O(e)$ iterations. The time complexity $O(a + b)$ required during every iteration depends on the cost $a$ of the mining step, and the cost $b$ of the detection step.

The mining step requires time complexity $a = O(2^c) \cdot d$ in the worst case scenario. $O(2^c)$ denotes the number of patterns discovered according to the *feature space* and to the *exhaustiveness* of the mining step. In the exhaustive setting, all the edges $(c = |I|)$ are considered to discover $O(2^{|I|})$ patterns. Since $FSs < FCSs < 2^{|I|}$, we refer to $O(2^{|I|})$ as the maximum number of patterns discovered in the worst-case scenario. However, it reduces to $O(2^k)$ in the non-exhaustive setting $(c = k)$, where $k \ll |I|$. The term $d$ denotes a multiplicative factor describing the amount of work spent by the algorithm in tidset intersections, which depends on the *time window model* adopted. In the case of landmark and mixed model it is $O(|W| + |\Pi|)$, while in the case of the sliding model it is $O(|\Pi|)$. As for the detection step, the computation of the $d(F_W, F_{W'})$ requires time complexity $O(b)$ proportional to the enumeration of patterns $|F_W \cup F_{W'}|$, which is $O(2^{|I|})$ in exhaustive setting, and $O(2^k)$ in the non-exhaustive one.

Then, the computational complexity in the worst case scenario of exhaustive PBCDs is $O(e \cdot (d2^{|I|} + 2^{|I|}))$, while for non-exhaustive PBCDs is $O(e \cdot (d2^k + 2^k))$. Therefore, it is exponential in $|I|$ and $k$, with $k \ll |I|$, respectively.

# 7    Experimental Results

The experiments are organized alongside different perspectives concerning both synthetic and real-world dynamic networks. In particular, we answer the following research question: **(Q1)** What is the best PBCD in terms of *efficiency* and *accuracy* when tuning the minimum change threshold $\beta$ on synthetic networks? **(Q2)** How much the parameter $k$ affects the *efficiency* and the *accuracy* of non-exhaustive PBCD on synthetic networks? **(Q3)** How much the parameter $k$ affects the *efficiency* of non-exhaustive PBCD on real-world networks?

For experiments on synthetic networks, we generated 40 networks, 20 with frequent drifts and 20 with rare drifts. Every network consists of 200 hourly blocks made of 120 graph snapshots, one observed every 30 s, for a total amount of 24000 snapshots. Each hourly block is built by randomly choosing with

**Table 1.** Most accurate (top) and most efficient (bottom) PBCD when tuning $\beta$.

| PBCD component | Most accurate PBCD @ $\beta$ | | | | | | | | | | | |
|---|---|---|---|---|---|---|---|---|---|---|---|---|
| | 0.10 | | 0.20 | | 0.30 | | 0.40 | | 0.50 | | 0.60 | |
| | choice | p-val. | choice | p-val. | choice | p-val. | choice | p-val. | choice | p-val. | choice | p-val. |
| Features (fcs/fs) | fs | 0.0025 | fs | 0.0001 | fs | 0.0001 | fs | 0.0317 | fs | 0.2357 | fs | 0.6816 |
| Mining (ex/nex) | nex | 0.0001 | nex | 0.0001 | nex | 0.0001 | nex | 0.0828 | nex | 0.4657 | nex | 0.7154 |
| Detection (tan/wj) | tan | 0.3618 | wj | 0.0001 | wj | 0.0001 | wj | 0.0001 | wj | 0.0001 | wj | 0.1004 |
| Windows (lan/sli/mix) | lan | 0.0001 | mix | 0.0001 | mix | 0.0001 | mix | 0.0001 | mix | 0.0001 | mix | 0.0001 |

| PBCD component | Most efficient PBCD @ $\beta$ | | | | | | | | | | | |
|---|---|---|---|---|---|---|---|---|---|---|---|---|
| | 0.10 | | 0.20 | | 0.30 | | 0.40 | | 0.50 | | 0.60 | |
| | choice | p-val. | choice | p-val. | choice | p-val. | choice | p-val. | choice | p-val. | choice | p-val. |
| Features (fcs/fs) | fs | 0.0001 | fs | 0.0001 | fs | 0.0001 | fs | 0.0001 | fs | 0.0001 | fs | 0.0001 |
| Mining (ex/nex) | nex | 0.0001 | nex | 0.0001 | nex | 0.0001 | nex | 0.0001 | nex | 0.0001 | nex | 0.0001 |
| Detection (tan/wj) | tan | 0.0001 | tan | 0.0001 | tan | 0.0001 | tan | 0.0001 | tan | 0.0079 | tan | 0.3618 |
| Windows (lan/sli/mix) | sli | 0.0001 | sli | 0.0001 | sli | 0.0001 | sli | 0.0001 | sli | 0.0001 | sli | 0.0001 |

replacement (i) one out of 10 different generative models in the case of frequent drifts, and (ii) one out of 2 different generative models in the case of rare drifts. As a consequence, it is more likely that two consecutive hourly blocks are built according to different generative models, thus denoting a change, in the dataset with frequent drifts than in the one with rare drifts. Every generative model builds a first snapshot made of 50 nodes by adopting a random scale-free network generator, which is then replicated for the remaining snapshots of the block. Every graph snapshot of a block is then perturbed by adding new edges and removing existing ones with a probability equals to 2%. A random perturbation is required to test the false alarm rate of the two approaches.

### 7.1  Q1: The Most Accurate and Most Efficient PBCD When Tuning $\beta$

In this paper, we discussed various components of PBCDs, that is (i) 2 mining steps (exhaustive and non-exhaustive), (ii) 2 feature spaces (FCSs and FSs), (iii) 2 detection steps (with the Tanimoto dissimilarity score and the weighted Jaccard score), and (iv) 3 time window models (landmark, sliding and mixed). These components can be combined to form 24 possible PBCDs, and hence determining variants of the original KARMA algorithm. Here, we evaluate which one performs statistically better, by measuring the efficiency (*running times*) and the change detection accuracy (*Accuracy*). We executed the 24 variants on 40 randomly generated synthetic networks, by tuning $\beta$ 6 times (resulting in 5760

**Table 2.** Running times of PBCD-1 and PBCD-2, when tuning $k$, against KARMA and StreamKRIMP on synthetic data ($\alpha = 0.5$, $\beta = 0.20$, $|\Pi| = 15$).

| Dataset | Running times (s) @ $k$ | | | | | | | | | | | | KARMA | KRIMP |
|---|---|---|---|---|---|---|---|---|---|---|---|---|---|---|
| | PBCD-1 | | | | | | PBCD-2 | | | | | | | |
| | 5 | 10 | 15 | 20 | 25 | 30 | 5 | 10 | 15 | 20 | 25 | 30 | | |
| freq-drifts-1 | 6.016 | 8.536 | 12.913 | 16.156 | 19.234 | 23.745 | 3.881 | 5.612 | 6.194 | 7.870 | 9.091 | 10.372 | 60.763 | 86.130 |
| freq-drifts-2 | 6.022 | 9.518 | 12.284 | 15.758 | 19,270 | 23.084 | 4.147 | 5.083 | 6.522 | 7.826 | 9.141 | 10.182 | 55.982 | 77.138 |
| freq-drifts-3 | 6.689 | 8.882 | 12.603 | 15.710 | 19.653 | 22.988 | 3.195 | 4.497 | 6.463 | 7.667 | 9.818 | 10.599 | 58.137 | 76.750 |
| freq-drifts-4 | 6.107 | 9.739 | 12.792 | 17.029 | 20.976 | 23.980 | 3.778 | 4.529 | 6.712 | 8.129 | 9.800 | 10.419 | 78.240 | 23.213 |
| rare-drifts-1 | 7.438 | 12.578 | 18.358 | 24.582 | 29.271 | 35.250 | 3.341 | 4.709 | 5.854 | 7.556 | 9.005 | 10.034 | 1775.234 | 109.816 |
| rare-drifts-2 | 7.302 | 12.326 | 17.549 | 23.911 | 29.702 | 33.339 | 4.182 | 5.156 | 6.435 | 7.563 | 8.928 | 10.365 | 1971.625 | 112.303 |
| rare-drifts-3 | 7.181 | 12.696 | 18.259 | 25.245 | 29.642 | 35.035 | 3.933 | 5.132 | 6.228 | 7.236 | 8.608 | 10.102 | 1971.367 | 109.395 |
| rare-drifts-4 | 7.273 | 12.159 | 17.901 | 23.821 | 28.416 | 31.260 | 4.059 | 4.306 | 6.452 | 7.488 | 9.072 | 10.559 | 2026.794 | 116.141 |

executions) and fixing the value of $k$ to 20 items. Then, we selected the most accurate and the most efficient PBCD (Table 1) by using (i) a Wilcoxon post-hoc test when deciding about the feature space, the mining strategy and the detection step, and (ii) a Nemenyi-Friedman post-hoc test when deciding about the best time-windows model, both at significance level $\alpha = 0.05$.

As for the accuracy, the results show that FSs are more appropriate features than FCSs. Furthermore, the PBCDs equipped with non-exhaustive mining step always outperforms exhaustive ones. The Tanimoto measure, as originally used by the KARMA algorithm, outperforms the weighted Jaccard measure for low values of $\beta$. From this set of experiments it is clear that the factors that impact the most on the PBCDs accuracy are the change detection measure and the time windows model. In particular, it is strongly evident that a mixed model outperforms the landmark model, which is preferred only when $\beta = 0.10$.

As for the efficiency, the very low p-values indicate a strong evidence that non-exhaustive PBCD based on FSs and the Tanimoto distance in the sliding model, outperforms every other PBCD approach. In particular, this is an expected result since: (i) the mining of FSs is less time consuming than FCSs, (ii) a non-exhaustive mining strategy is more efficient than exhaustive one.

An aspect worth to be considered is that the original KARMA algorithm (*FCSs + EX + Tanimoto + Landmark*) is never selected as the best PBCD. In this perspective the adoption of a new feature set, the FSs, jointly with a non-exhaustive mining step generally improves the detection accuracy and the efficiency. However, the test suggests that the landmark model adopted by the KARMA algorithm is a bad choice leading to poor accuracy and efficiency. While the Tanimoto coefficient leads to poor detection accuracy.

## 7.2 Q2: Efficiency and Accuracy of Non-exhaustive PBCDs on Synthetic Networks

We report the results of a comparative evaluation in which we compare the running time (Table 2), the accuracy and the false alarm rate (Table 3) of two non-exhaustive PBCDs against the KARMA [9] and the StreamKRIMP [8] state-of-the-art PBCDs. In particular, StreamKRIMP treats the network as a data stream of labeled edges, and adopts a compression-based mining step. We select two non-exhaustive PBCDs emerged in the last section and test their performances on 8 synthetic networks, when tuning the parameter $k$. In particular, we chose the most efficient PBCD and the most accurate PBCD when $\beta = 0.2$. We denote them as PBCD-1 (*FSs + NEX + Weighted Jaccard + Mixed*), and PBCD-2 (*FSs + NEX + Tanimoto + Sliding*), respectively. The results in Tables 2 and 3 shows increasing efficiency and accuracy for decreasing values of $k$.

Results in Table 2 show an improved efficiency for both PBCD-1 and PBCD-2 with respect to KARMA and StreamKRIMP. This is an expected result, also confirmed by the statistical significance test in Sect. 7.1, because non-exhaustive mining of FSs is more efficient than (i) the exhaustive mining of FCSs performed by the KARMA algorithm, and (ii) the non-exhaustive mining of itemsets performed by StreamKRIMP. In particular, the running times of both PBCD-1 and PBCD-2 increases with $k$, as high values of $k$ lead to the discovery of an increasing number of patterns. Moreover, the results show that PBCD-2 is more

**Table 3.** Accuracy and false alarm rate of PBCD-1 and PBCD-2, when tuning $k$, against KARMA and StreamKRIMP on synthetic data ($\alpha = 0.5$, $\beta = 0.20$, $|\Pi| = 15$).

| Dataset | Accuracy @ $k$ | | | | | | | | | | | | KARMA | KRIMP |
|---|---|---|---|---|---|---|---|---|---|---|---|---|---|---|
| | PBCD-1 | | | | | | PBCD-2 | | | | | | | |
| | 5 | 10 | 15 | 20 | 25 | 30 | 5 | 10 | 15 | 20 | 25 | 30 | | |
| freq-drifts-1 | 1.0 | 1.0 | 0.987 | 0.987 | 0.967 | 0.957 | 1.0 | 1.0 | 0.918 | 0.891 | 0.824 | 0.751 | 0.8041 | 0.9299 |
| freq-drifts-2 | 1.0 | 1.0 | 0.991 | 0.991 | 0.965 | 0.952 | 1.0 | 1.0 | 0.916 | 0.889 | 0.819 | 0.735 | 0.7985 | 0.9105 |
| freq-drifts-3 | 1.0 | 1.0 | 0.988 | 0.988 | 0.958 | 0.948 | 1.0 | 1.0 | 0.918 | 0.888 | 0.819 | 0.755 | 0.7960 | 0.9155 |
| freq-drifts-4 | 1.0 | 1.0 | 0.987 | 0.987 | 0.963 | 0.952 | 1.0 | 1.0 | 0.936 | 0.907 | 0.831 | 0.757 | 0.7860 | 0.923 |
| rare-drifts-1 | 1.0 | 1.0 | 1.0 | 1.0 | 1.0 | 0.971 | 1.0 | 1.0 | 0.816 | 0.816 | 0.691 | 0.598 | 0.9362 | 1.0 |
| rare-drifts-2 | 1.0 | 1.0 | 1.0 | 1.0 | 1.0 | 0.967 | 1.0 | 1.0 | 0.799 | 0.799 | 0.674 | 0.571 | 0.9399 | 1.0 |
| rare-drifts-3 | 1.0 | 1.0 | 1.0 | 1.0 | 1.0 | 0.969 | 1.0 | 1.0 | 0.816 | 0.816 | 0.691 | 0.599 | 0.9368 | 1.0 |
| rare-drifts-4 | 1.0 | 1.0 | 1.0 | 1.0 | 1.0 | 0.965 | 1.0 | 1.0 | 0.799 | 0.799 | 0.674 | 0.576 | 0.9293 | 1.0 |

| Dataset | False alarm rate @ $k$ | | | | | | | | | | | | KARMA | KRIMP |
|---|---|---|---|---|---|---|---|---|---|---|---|---|---|---|
| | PBCD-1 | | | | | | PBCD-2 | | | | | | | |
| | 5 | 10 | 15 | 20 | 25 | 30 | 5 | 10 | 15 | 20 | 25 | 30 | | |
| freq-drifts-1 | 0 | 0 | 0.014 | 0.014 | 0.036 | 0.048 | 0 | 0 | 0.092 | 0.123 | 0.197 | 0.279 | 0.1099 | 0.0399 |
| freq-drifts-2 | 0 | 0 | 0.011 | 0.0105 | 0.039 | 0.053 | 0 | 0 | 0.095 | 0.125 | 0.204 | 0.297 | 0.1131 | 0.0513 |
| freq-drifts-3 | 0 | 0 | 0.013 | 0.0134 | 0.047 | 0.058 | 0 | 0 | 0.092 | 0.126 | 0.203 | 0.275 | 0.1146 | 0.0478 |
| freq-drifts-4 | 0 | 0 | 0.015 | 0.0148 | 0.042 | 0.054 | 0 | 0 | 0.072 | 0.105 | 0.190 | 0.273 | 0.1205 | 0.0437 |
| rare-drifts-1 | 0 | 0 | 0 | 0 | 0 | 0.0312 | 0 | 0 | 0.195 | 0.195 | 0.328 | 0.427 | 0.0073 | 0 |
| rare-drifts-2 | 0 | 0 | 0 | 0 | 0 | 0.032 | 0 | 0 | 0.214 | 0.214 | 0.347 | 0.457 | 0 | 0 |
| rare-drifts-3 | 0 | 0 | 0 | 0 | 0 | 0.0326 | 0 | 0 | 0.196 | 0.196 | 0.329 | 0.427 | 0.0027 | 0 |
| rare-drifts-4 | 0 | 0 | 0 | 0 | 0 | 0.0377 | 0 | 0 | 0.216 | 0.216 | 0.351 | 0.456 | 0.0013 | 0 |

efficient than PBCD-1, as the sliding window model leads to increasing efficiency with respect to the mixed model. This is explained by the forgetful nature of the sliding window, in which old graph snapshots are immediately discarded with the arrival of a new block of snapshots. In this way, the mining step requires reduced computational efforts, as patterns are mined from reduced sets of transactions. Thus intersecting small tidsets when computing the support of each pattern.

As for the accuracy (Table 3), both PBCD-1 and PBCD-2 are optimal change detection solutions, for the considered synthetic networks, for low values of $k$ ($k = 5$ and $k = 10$, respectively). Moreover, the results show a decreasing tendency in the accuracy of both PBCD-1 and PBCD-2. In particular, PBCD-1 always outperforms KARMA and StreamKRIMP (except for $k = 30$), while this is not the case of PBCD-2. These are expected results, again confirmed by the significance test in Sect. 7.1, as the non-exhaustive mining of FSs with a detection step based on the weighted Jaccard dissimilarity takes into account the absolute growth-rate and the interestingness of patterns. This is not the case of PBCD-2, in which the Tanimoto coefficient computed in the sliding window setting, on large sets of patterns, exhibits higher false positive rates. For high values of $k$, the mining step discover patterns representing behavior local to the snapshots collected in two successive sliding windows of equal size. Thus, injecting noisy features in the detection step. We note that (i) PBCD-1 exhibits moderately lower false alarm rates than PBCD-2, also outperforming KARMA for high values of $k$, and StreamKRIMP for low values of $k$, and (ii) PBCD-2 outperforms KARMA and StreamKRIMP for low values of $k$ only.

**Table 4.** Running times of PBCD-1 and PBCD-2, when tuning $k$, against KARMA on real-world networks ($\beta = 0.20$, $|\Pi| = 10\%$ of each dataset).

| Dataset | Running times (s) @ k | | | | | | | | | | | | KARMA |
|---|---|---|---|---|---|---|---|---|---|---|---|---|---|
| | PBCD-1 | | | | | | PBCD-2 | | | | | | |
| | 5 | 10 | 15 | 20 | 25 | 30 | 5 | 10 | 15 | 20 | 25 | 30 | |
| mawi | 6.769 | 9.68 | 1.919 | 14.886 | 17.554 | 18.847 | 5.169 | 6.17 | 6.82 | 7.977 | 8.547 | 8.736 | 86.493 |
| noaa | 14.794 | 16.513 | 18.068 | 19.935 | 19.603 | 23.217 | 15.164 | 16.35 | 17.18 | 19.061 | 20.533 | 20.263 | 65.697 |
| nodobo | 1.72 | 1.525 | 2.11 | 2.915 | 3.728 | 4.371 | 1.825 | 1.582 | 1.822 | 2.587 | 3.276 | 3.988 | 35.253 |
| keds | 0.955 | 0.924 | 0.873 | 0.946 | 1.109 | 1.034 | 0.979 | 0.892 | 0.795 | 0.982 | 1.075 | 1.014 | 1.108 |
| wikitalks | 83.846 | 78.454 | 80.182 | 77.904 | 81.605 | 79.131 | 77.128 | 76.77 | 76.021 | 80.45 | 77.299 | 75.876 | 94.914 |

We conclude that both the accuracy and the efficiency of non-exhaustive PBCDs benefits from the limited pattern sets which have been discovered. In particular, the combination of a mixed window model with the weighted Jaccard dissimilarity leads to accurate detection, while the combination of sliding windows and the Tanimoto coefficient leads to efficient detection, while improving the detection accuracy for very low values of $k$. From this perspective, the two approaches offers two efficient alternatives to the KARMA algorithm, in which the running times can be greatly reduced (up to two orders of magnitude in this set of experiments).

### 7.3   Q3: Efficiency of Non-exhaustive PBCDs on Real-World Networks

We also provide a practical idea of the efficiency on real-world networks by reporting the results of a comparative evaluation (Table 4) between PBCD-1, PBCD-2 and KARMA on 5 real-world networks, when tuning $k$. More specifically, we used the same networks adopted in [9]: the *keds*, *mawi*, *noaa*, *nodobo* and the *wikitalks* dataset. To guarantee a fair comparison, we fixed the value $\beta = 0.20$ in each experiment. However, since the networks span different periods, we independently fixed the size of block $|\Pi|$ to the 10% of each dataset, this guaranteed 100 iterations of each PBCD on every dataset. The minimum support $\alpha$ has been fixed to 0.05 for keds, nodobo and mawi, 0.20 for noaa, and 0.40 for wikitalks. From the obtained results it is evident that both PBCD-1 and PBCD-2 are always more efficient than KARMA for all the values of $k$. Furthermore, as observed on synthetic networks in Sect. 7.2 and as confirmed in Sect. 7.1, PBCD-2 continues to be more efficient than PBCD-1. The increasing tendency of the running times with $k$ is verified in the *mawi*, *noaa* and *nodobo* datasets.

## 8   Conclusions

In this paper, we have collected several improvements contributing to the efficiency and the accuracy of traditional PBCDs. This have been possible by inheriting the general PBCD schema from the KARMA algorithm, and extending it. Specifically, we have relaxed the exhaustiveness of the PBCDs mining step with a non-exhaustive mining strategy, inspired by beam search algorithms. The effect is that the mining algorithm discovers now limited sets of patterns by pruning the search space according to the interestingness of patterns. Moreover, we proposed an extended detection step which takes into account the growth-rate of the discovered patterns. Ultimately, we have conducted an extensive exploratory evaluation on both real and synthetic networks.

The experiments have shed some lights on the most accurate and on the most efficient PBCDs among the possible approaches. Furthermore, they have shown that non-exhaustive PBCDs are more efficient than exhaustive PBCDs, while achieving comparable levels of accuracy. Future directions of research involve the evaluation of the performances when adopting more sophisticated feature spaces, for example by considering graph embedding.

## References

1. Akoglu, L., Tong, H., Koutra, D.: Graph based anomaly detection and description: a survey. Data Min. Knowl. Discov. **29**(3), 626–688 (2015)
2. Bailey, J.: Statistical measures for contrast patterns. In: Contrast Data Mining: Concepts, Algorithms, and Applications, pp. 13–20. CRC Press (2013)

3. Gama, J., Medas, P., Castillo, G., Rodrigues, P.: Learning with drift detection. In: Bazzan, A.L.C., Labidi, S. (eds.) SBIA 2004. LNCS (LNAI), vol. 3171, pp. 286–295. Springer, Heidelberg (2004). https://doi.org/10.1007/978-3-540-28645-5_29

4. Geerts, F., Goethals, B., Mielikäinen, T.: Tiling databases. In: Proceedings of 7th International Conference Discovery Science, DS 2004, 2–5 October 2004, Padova, Italy, pp. 278–289 (2004)

5. He, Z., Xu, X., Huang, J.Z., Deng, S.: FP-outlier: frequent pattern based outlier detection. Comput. Sci. Inf. Syst. $2(1)$, 103–118 (2005)

6. Koh, Y.S.: CD-TDS: change detection in transactional data streams for frequent pattern mining. In: 2016 International Joint Conference on Neural Networks, IJCNN 2016, 24–29 July 2016, Vancouver, BC, Canada, pp. 1554–1561 (2016)

7. Koufakou, A., Secretan, J., Georgiopoulos, M.: Non-derivable itemsets for fast outlier detection in large high-dimensional categorical data. Knowl. Inf. Syst. $29(3)$, 697–725 (2011)

8. van Leeuwen, M., Siebes, A.: Streamkrimp: detecting change in data streams. In: Proceedings of European Conference on Machine Learning and Knowledge Discovery in Databases (Part I), ECML/PKDD 2008, 15–19 September 2008, Antwerp, Belgium, pp. 672–687 (2008)

9. Loglisci, C., Ceci, M., Impedovo, A., Malerba, D.: Mining microscopic and macroscopic changes in network data streams. Knowl. Based Syst. $161$, 294–312 (2018)

10. Padillo, F., Luna, J.M., Ventura, S.: Subgroup discovery on big data: pruning the search space on exhaustive search algorithms. In: 2016 IEEE International Conference on Big Data, BigData 2016, 5–8 December 2016, Washington DC, USA, pp. 1814–1823 (2016)

11. Trabold, D., Horváth, T.: Mining strongly closed itemsets from data streams. In: Yamamoto, A., Kida, T., Uno, T., Kuboyama, T. (eds.) DS 2017. LNCS (LNAI), vol. 10558, pp. 251–266. Springer, Cham (2017). https://doi.org/10.1007/978-3-319-67786-6_18

# A Combinatorial Multi-Armed Bandit Based Method for Dynamic Consensus Community Detection in Temporal Networks

Domenico Mandaglio and Andrea Tagarelli[✉]

Department of Computer Engineering, Modeling, Electronics, and Systems Engineering (DIMES), University of Calabria, Rende, Italy
d.mandaglio@dimes.unical.it, andrea.tagarelli@unical.it

**Abstract.** Community detection in temporal networks is an active field of research, which can be leveraged for several strategic decisions, including enhanced group-recommendation, user behavior prediction, and evolution of user interaction patterns in relation to real-world events. Recent research has shown that combinatorial multi-armed bandit (CMAB) is a suitable methodology to address the problem of dynamic consensus community detection (DCCD), i.e., to compute a single community structure that is conceived to be representative of the knowledge available from community structures observed at the different time steps.

In this paper, we propose a CMAB-based method, called CreDENCE, to solve the DCCD problem. Unlike existing approaches, our algorithm is designed to provide a solution, i.e., dynamic consensus community structure, that embeds both long-term changes in the community formation and newly observed community structures. Experimental evaluation based on publicly available real-world and ground-truth-oriented synthetic networks, with different structure and evolution rate, has confirmed the meaningfulness and key benefits of the proposed method, also against competitors based on evolutionary or consensus approaches.

## 1 Introduction

Community detection and evolution in temporal networks has been largely studied in the last few years, mainly focusing on graph-based unsupervised learning paradigms (e.g., [3,8,10,24]). Nonetheless, detecting and tracking the evolution of the change events that occur in the communities remains challenging [5], which is partly due to the uncertainty and dynamicity underlying the different types (e.g., birth/death, growth/decay, merge/split) and evolution rates of structural changes in time-evolving network systems.

In this regard, we have recently explored the opportunity of adopting the multi-armed bandit (MAB) paradigm, which is conceived to learn how to perform actions in an uncertain environment [17]. Indeed, in the problem under consideration, the uncertainty is inherent to the temporal network system and

ⓒ Springer Nature Switzerland AG 2019
P. Kralj Novak et al. (Eds.): DS 2019, LNAI 11828, pp. 412–427, 2019.
https://doi.org/10.1007/978-3-030-33778-0_31

the structural changes of its communities, while actions correspond to node assignments to communities. Moreover, each action is associated with a notion of "reward" that determines how much benefit is gained by (a set of) node assignments to communities. Within this view, MAB methods are well-suited to model the *exploitation-exploration* trade-off [13,21], i.e., balancing between making decisions that yield high current rewards or making decisions that sacrifice current gains with the prospect of better future rewards. Moreover, to deal with choosing *a set of* actions, i.e., a set of community assignments that constitute a whole community-structure, a particular extension of MAB problems is needed, which is called *combinatorial multi-armed bandit* (CMAB) [4,7].

In this work, we focus on the *dynamic consensus community detection* (DCCD) problem, that is, given a sequence of temporal snapshots of a time-evolving network, we want to compute a single community structure to be representative of the knowledge available from community structures detected in the different snapshot networks. Remarkably, unlike in *consensus community detection* [14,22], the knowledge on the community structures from which a consensus must to be inferred is not available at a given initial time, but it evolves over time along with the associated temporal network. In this respect, here we follow the directions outlined in [17] for the CMAB-based DCCD problem, and propose a fully defined instantiation of the algorithmic scheme.

Note that existing approaches to related problems involving a notion of community representative in temporal networks [6,12] may suffer from restrictions on the network model, such as fixed set of nodes and number of communities for each snapshot of the temporal network [12], or on selected types of community dynamics [6]. By contrast, our proposed approach does not incur such issues.

Our contributions can be summarized as follows:

- We develop CreDENCE – CMAB-based **D**ynamic Cons**EN**sus **C**ommunity **DE**tection method. To achieve the exploration-exploitation trade-off, our algorithm is designed to balance over time between the need for embedding long-term changes observed in the community formation and the need for capturing short-term effects and newly observed community structures. Moreover, CreDENCE is conceived to be versatile in terms of the static community detection approach used to identify the communities at each snapshot, and robust in terms of a number of parameters that control the CMAB-learning rate, temporal smoothness factors, and the node-relocation bias.
- We provide insights into technical as well as computational complexity aspects of CreDENCE; upon this, we propose an enhancement of CreDENCE to ensure its linear complexity in the size of the temporal network.
- Our experimental evaluation was conducted using 5 real-world networks and ground-truth-oriented synthetically generated networks, including comparison with 3 competing methods. Results have provided useful indications about the quality of the consensus solutions obtained by CreDENCE, which is able to cope with temporal networks having different evolution rates.

## 2    Problem Statement

We are given a set $\mathcal{V}$ of *entities* (e.g., users in a social environment) and a *temporal network* $\mathcal{G}$ as a series of graphs over discrete time steps $(G_1, G_2, \ldots, G_t, \ldots)$, where $G_t = \langle V_t, E_t \rangle$ is the graph at time $t$, with set of nodes $V_t$ and set of undirected edges $E_t$. We denote with $\mathcal{G}_{\leq t}$ a *series of graphs observed until time* $t$. Each node in $V_t$ corresponds to a specific instance from the set $\mathcal{V}_t \subseteq \mathcal{V}$ of entities that occur at time $t$. The snapshot graphs can share different subsets of entities.

Given any $G_t$, we denote with $\mathcal{C}^{(t)}$ a *community structure* for $G_t$, which is a set of non-overlapping communities, and is assumed to be unrelated to any other $\mathcal{C}^{(t')}$ ($t' \neq t$), both in terms of number of communities and set of entities involved. Let $\mathcal{E}_{\leq t} = \{\mathcal{C}^{(1)}, \ldots, \mathcal{C}^{(t)}\}$ be a *dynamic ensemble* at time $t$, i.e., a set of community structures associated to the snapshot graphs. We consider the following problem:

*Problem 1* (**Dynamic Consensus Community Detection (DCCD)**). *Given a temporal graph sequence $\mathcal{G}_{\leq t}$ and associated dynamic ensemble $\mathcal{E}_{\leq t}$, for any time $t \geq 1$ compute a community structure, called dynamic consensus community structure and denoted as $\mathcal{C}^*_{\leq t}$, which is designed to encompass the information from $\mathcal{G}_{\leq t}$ to be representative of the knowledge available in $\mathcal{E}_{\leq t}$.*

Given $\mathcal{G}_{\leq t}$ and $\mathcal{E}_{\leq t}$, the dynamic consensus being discovered over time can be represented as a matrix $\mathbf{M}$ we call *dynamic co-association* (or *consensus*) *matrix* (DCM). Its size is initially $\mathcal{V}_t \times \mathcal{V}_t$ with $t = 1$, and at a generic time $t$ is $|\mathcal{V}| \times |\mathcal{V}|$. The $(i, j)$-th entry of $\mathbf{M}$, denoted as $m_{ij}$, stores the probability of co-association for entities $v_i, v_j \in \mathcal{V}$, i.e., the probability that $v_i$ and $v_j$ are assigned to the same community, in the observed timespan.

Given the incremental nature of Problem 1, unlike in conventional consensus community detection [14,22], we want to avoid (re)computation of the consensus from scratch at any time $t$. We also do not want to depend on any mechanism of tracking of the evolution of communities [5]. More importantly, the dynamic consensus community structure should be able to embed long-term changes in the community formation as well as to capture short-term effects and newly observed community structures. To address Problem 1, in [17] we proposed a CMAB-based methodology, whose principles are recalled in the next section.

### 2.1    Dynamic Consensus Community Detection as a CMAB Problem

**Review of CMAB.** We are given $m$ *base arms*, where each arm $i$ is associated with a set of random variables $\{X_{i,t} \mid 1 \leq i \leq m, t \geq 1\}$, where $X_{i,t} \in [0,1]$ indicates the random outcome of *triggering*, or playing, the $i$-th arm in the $t$-th round. The random variables $\{X_{i,t} \mid t \geq 1\}$ of the $i$-th arm are independent and identically distributed. Moreover, in a *non-stationary context*, those variables may change [9]. Variables of different arms may not be independent.

At each round $t$, a *superarm* (a set of base arms) $A$ is chosen and the outcomes of the random variables $X_{i,t}$, for all $a_i \in A$, are revealed. Moreover, the base arms belonging to $A$ may probabilistically trigger other base arms not in $A$ [4,7], thus revealing their associated outcomes. Playing a superarm $A$ at round $t$ gives a reward $R_t(A)$ modeled as a random variable, which is a function of the outcomes of the triggered base arms. The objective of a CMAB method is to select at each round $t$ the superarm $A$ that maximizes the *expected reward* $\mathbb{E}[R_t(A)]$, in order to maximize the *cumulative expected reward* over all rounds. At each round, the bandit may decide to choose the superarm with the highest expected reward (exploitation) or to select a superarm discarding information from earlier rounds (exploration) with the aim of discovering the benefit from adopting some previously unexplored arm(s) [4,7].

**Adaptation to DCCD.** In our context, each pair of entities $\langle v_i, v_j \rangle$ in $\mathcal{G}_{\leq t}$ is a base arm and it is hypothetically associated with an unknown distribution (with unknown mean $\mu_{ij}$) for the probabilities of co-association over time, whose mean estimate is the entry $m_{ij}$ in DCM. Each observation of a community structure of a snapshot network can be considered as a sample from such distributions. Moreover, since the network and community structures can vary, the co-association distributions may also change their mean over time, thus the DCCD setting is non-stationary. Including a base arm $\langle v_i, v_j \rangle$ in a superarm corresponds to "*assign $v_i$ and $v_j$ to the same community at a given time*". If we denote with $c_i^{(t)}$ the community of $v_i$ at round $t$, a superarm $A$ at round $t$ is a set of pairs $\langle v_i, v_j \rangle$ such that $c_i^{(t)} = c_j^{(t)}$.

According to the framework in [17], playing a superarm $A$ at each round $t$ consists of stochastic optimization that considers node relocations to neighbor communities. The stochastic nature of the process depends on both the random order with which we consider the node relocations and on the fact that, according to the optimization of a quality criterion, an improvement due to relocation is accepted with a certain probability. Intuitively, this allows us to account for uncertainty in the long-term overall quality improvement of the consensus due to local relocations at a given time; for instance, it is unknown if the relation that explains two users share the same community at a given time could become meaningless in subsequent times. After playing a superarm $A$, the rewards associated to the entity pairs (base arms) corresponding to the status of communities after the relocation phase, are revealed; these pairs include both the nodes that did not move from their community and the arms $\langle v_i, v_j \rangle$ triggered with the accepted relocations, i.e., such that node $v_i$ was moved to the community of $v_j$. For the base arms that were neither selected nor triggered (i.e., pairs of nodes that were not in the same community before and after the relocation phase), we assume an implicit reward of zero that corresponds to the observation of the "no-coassociation" event. (This is in line with the possibility in CMAB of enabling the probabilistic triggering of *all* base arms.)

The reward of a superarm corresponds to the *quality* of the community structure at the end of the relocation phase, which is a non-linear function of the base arms' rewards. More specifically, we resort to *modularity* as quality criterion for a community structure. Let $X_{ij}^t$ be the reward associated to the base arm corresponding to node pair $\langle v_i, v_j \rangle$ at time step $t$. The reward of the played superarm

$A$ (leading to a consensus structure $\mathcal{C}^*_{\leq t}$ after the stochastic relocation of nodes) can be defined in terms of the base arms' rewards as follows:

$$R_t(A) = \frac{1}{d(\mathcal{V}_{[1..t]})} \sum_{i,j} \sum_{\ell=1}^{t} \beta^{t-\ell} \Big(A_{ij}^l - \frac{k_i^\ell k_j^\ell}{d(\mathcal{V}_{[1..t]})}\Big) \delta(X_{ij}^t) \qquad (1)$$

where $k_i^\ell$ is the degree of $v_i$ in the $\ell$-th snapshot, $A_{ij}^l$ is the $(i,j)$-th entry of the adjacency matrix of the $\ell$-th snapshot graph, $d(\mathcal{V}_{[1..t]})$ is the total degree of the multiplex graph including snapshots from the first one to the $t$-th (i.e., $d(\mathcal{V}_{[1..t]}) = \sum_{\ell=1}^{t} \sum_{v \in V_\ell} d(v))$, $\beta \in (0,1)$, and $\delta(X_{ij}^t) = 1$ if $X_{ij}^t > 0$, 0 otherwise. The stochastic nature of the above defined reward is determined by the random variables $X_{ij}^t$. In Sect. 3.2, we will define a generalization of the above reward equation that allows us to focus on selected snapshots of the networks.

## 3   The *CreDENCE* Method

To solve the dynamic consensus community detection problem, we develop a CMAB-based method called CreDENCE – CMAB-based **D**ynamic Cons**EN**sus Community D**E**tection, which is sketched in Algorithm 1.

Initially, the dynamic consensus matrix **M** is set as an identity matrix (Line 1), which reflects that no information has been processed yet, and hence each entity-node has co-association with itself only. At each round $t$, the algorithm chooses to perform either exploration or exploitation, according to a given bandit strategy ($\mathcal{B}$). Intuitively, in the exploitation phase, we seed an oracle (i.e., a conventional method for community detection) with the mean estimates of co-association of the current DCM to infer the communities in the new snapshot graph observed at time $t$; by contrast, in the exploration phase, the new communities are identified using the $t$-th graph only. In either phase, the community structure generated at time $t$ is finally used to produce a superarm that will correspond to the dynamic consensus community structure up to $t$ ($\mathcal{C}^*_{\leq t}$).

Besides the involvement of a conventional community detection method $\mathcal{A}$ and a bandit strategy $\mathcal{B}$ to control the exploration-exploitation trade-off, we introduce a few parameters to ensure robustness in the algorithmic scheme of CreDENCE: (i) the learning rate $\alpha$ for the update of the mean estimates (i.e., $m_{ij}$ entries), (ii) the relocation bias $\lambda$, and (iii) the temporal smoothness factor $\beta$ and window size $\omega$ to control the amount of past knowledge for the step of node-relocations. Nonetheless, some of these parameters are interrelated, or reasonable values can be chosen as default.

Another remark on CreDENCE concerns its *incremental* nature: whenever a new step of evolution is observed, say at $T+1$, the last-update status of the DCM matrix along with $\mathcal{G}_{\leq T+1}$ will become the input for a further CMAB round.

### 3.1   Finding Communities

At each round $t$, CreDENCE invokes a community detection method $\mathcal{A}$. This is just required to deal with (*static*) *simple graphs*. While in the exploration phase

**Algorithm 1.** CMAB-based **D**ynamic **C**ons**EN**sus **C**ommunity **DE**tection (CreDENCE)

---

**Input:** Temporal graph sequence $\mathcal{G}_{\leq T}$ ($T \geq 1$), (static) community detection method $\mathcal{A}$, bandit strategy $\mathcal{B}$, learning rate $\alpha \in (0,1)$, relocation bias $\lambda \in [0,1]$, temporal smoothness $\beta \in (0,1)$, temporal window width $\omega \geq 1$.
**Output:** Dynamic consensus community structure $\mathcal{C}^*_{\leq T}$.

1: $\mathbf{M} \leftarrow I_{|\mathcal{V}_1| \times |\mathcal{V}_1|}$
2: **for** $t = 1$ **to** $T$ **do**
3:     **if** $\mathcal{B}$ decides for EXPLORATION **then**
4:         $\mathcal{C}^{(t)} \leftarrow$ findCommunities$(G_t, \mathcal{A})$
5:     **else** {EXPLOITATION}
6:         $G_{\mathbf{M}} \leftarrow$ buildDCMGraph$(\mathbf{M})$
7:         $\mathcal{C}_{\mathbf{M}} \leftarrow$ partitionDCMGraph$(G_{\mathbf{M}}, \mathcal{A})$
8:         $\mathcal{C}^{(t)} \leftarrow$ inferCommunities$(G_t, \mathcal{C}_{\mathbf{M}})$
9:     **end if**
10:     $\mathcal{C}^*_{\leq t} \leftarrow$ project$(\mathcal{C}^{(t)}, \mathcal{G}_{\leq t})$
11:     $\mathcal{C}^*_{\leq t} \leftarrow$ evalRelocations$(\mathcal{G}_{\leq t}, \mathcal{C}^*_{\leq t}, \lambda, \beta, \omega)$        {*Using Eq. (3)*}
12:     $\mathbf{M} \leftarrow$ updateDCM$(\mathbf{M}, \mathcal{C}^*_{\leq t}, \alpha)$        {*Using Eq. (4)*}
13: **end for**
14: **return** $\mathcal{C}^*_{\leq T}$

---

it directly applies to the snapshot graph $G_t$ (Line 4), to handle the exploitation phase, the method should also be able to deal with *weighted* graphs: in this case, $\mathcal{A}$ is executed on the graph $G_{\mathbf{M}}$ (Line 7), which is built from the current DCM matrix in such a way that the edge weights in $G_{\mathbf{M}}$ correspond to the entries of $\mathbf{M}$ (Line 6). Next, from the obtained partitioning $\mathcal{C}_{\mathbf{M}}$ of $G_{\mathbf{M}}$ (Line 7), the knowledge about the community memberships of entity nodes in $\mathcal{C}_{\mathbf{M}}$ is used to infer a community structure $\mathcal{C}^{(t)}$ on $G_t$ (Line 8). Each community in $\mathcal{C}^{(t)}$ will have node set corresponding to exactly one community in $\mathcal{C}_{\mathbf{M}}$, and edge set consistent with the topology of $G_t$. Any entity $v$ that newly appears in $G_t$ (i.e., $v \in \mathcal{V}_t \wedge v \notin \mathcal{V}_{t'}, \forall t' < t$) and is disconnected will form a community in its own.

It should be noted that, although *any* method can in principle be used as $\mathcal{A}$, our preferred choice is towards efficient, modularity-optimization-based methods, such as [1]. This is motivated for consistency with our choice of using (multiplex) modularity as quality criterion in the (consensus) community structure refinement, as discussed next in Sect. 3.2.

## 3.2 Generating the Dynamic Consensus Community Structure

The dynamic consensus community structure $\mathcal{C}^*_{\leq t}$, for each $t$, is generated in two steps. The first step (Line 10) corresponds to a simple projection of the community memberships from $\mathcal{C}^{(t)}$ onto $\mathcal{G}_{\leq t}$. The second step (Line 11) corresponds to *stochastic refinement* of the candidate $\mathcal{C}^*_{\leq t}$ obtained at the previous step. This stochastic refinement is performed through local search optimization, which is designed to relocate some nodes from their assigned community in $\mathcal{C}^*_{\leq t}$ to a neighboring one by acting greedily w.r.t. a quality criterion.

As previously anticipated, one appropriate choice refers to *modularity*. However, to account for the multiplexity of $\mathcal{G}_{\leq t}$ as well as the dynamic aspects, we modify the definition of modularity to include the temporal window by which the modularity context is set. The reason behind this choice is twofold: (i) to focus on a limited number of latest snapshots of the network, and (ii) to reduce the computational burden in the local search optimization.

Given $\mathcal{C}^*_{\leq t}$, temporal-window width $\omega$ and temporal smoothness factor $\beta$, we denote with $d(\mathcal{V}_{[t-\omega+1..t]})$ the total degree of the multiplex graph including snapshots from the $(t-\omega+1)$-th to the $t$-th and, for any community $c$, $d_\ell(c)$ and $d_\ell^{int}(c)$ are the total degree and the internal degree of $c$, respectively, measured w.r.t. edges of the $\ell$-th snapshot network only. We define the $(\omega, \beta)$-*multiplex modularity* of $\mathcal{C}^*_{\leq t}$ as follows:

$$Q(\mathcal{C}^*_{\leq t}, \omega, \beta) = \frac{1}{d(\mathcal{V}_{[t-\omega+1..t]})} \sum_{c \in \mathcal{C}^*_{\leq t}} \sum_{\ell=t}^{t-\omega+1} \beta^{t-\ell} \left( d_\ell^{int}(c) - \frac{(d_\ell(c))^2}{d(\mathcal{V}_{[t-\omega+1..t]})} \right) \quad (2)$$

As previously mentioned, the meaning of $\beta$ is to smooth the contribution of earlier snapshots in the computation of the quality of the dynamic consensus, i.e., lower values of $\beta$ will penalize older snapshots. It is worth noting that $\beta$ may take a role that is opposite to that of the learning rate $\alpha$ in Algorithm 1. Therefore, by default, we set $\beta = 1 - \alpha$.

The local search optimization, at any time $t$, evaluates the possible improvement in terms of modularity due to the relocation of nodes $v_i$ that lay on the boundary of their assigned communities towards one of the communities that at time $t$ contain nodes linked to $v_i$. By denoting with $c_i$ the initial community of a boundary node $v_i$, and simplifying the modularity notation with function symbol $Q$, the *modularity variation*, denoted as $\Delta Q_i$, corresponding to moving $v_i$ to a neighbor community is as follows:

$$\Delta Q_i = Q(c_i \setminus \{v_i\}) - Q(c_i) + \max_{c_j \in NC_i^{(t)}} (Q(c_j \cup \{v_i\}) - Q(c_j)) \quad (3)$$

where $NC_i^{(t)}$ is the set of neighbor communities for node $v_i$ at time $t$. If $\Delta Q_i > 0$, then there is a single chance to accept the relocation of $v_i$ to $c_j$ with probability $1 - \lambda e^{-\lambda \Delta Q_i}$, with $\lambda \in [0,1]$ to control the bias towards relocations.

### 3.3   Updating the Dynamic Consensus

The DCM-update scheme in Algorithm 1 (Lines 12) follows a standard principle in reinforcement learning, whereby as the agent explores further, it is capable of updating its current estimate according to a general scheme of the form $newEstimate \leftarrow oldEstimate + \alpha(target - oldEstimate)$, which intuitively consists in moving the current estimate in the direction of a "target" value, with slope $\alpha$. In our setting, we want to control the update of co-associations by subtracting a quantity $\alpha$ of resource from the co-associations of each node, at time $t$, and redistributing this quantity among the nodes in $c_i^{(t)}$, for each $v_i$. This redistribution corresponds to the *reward* of a single co-association, i.e., given $v_i$, the reward of assigning any $v_j$ to the same community of $v_i$. Upon this, given $\alpha \in [0, 1]$ and any $(i, j)$-th entry of $\mathbf{M}$, we define the *update* equation as:

$$m_{ij}^{(t+1)} = m_{ij}^{(t)} + \alpha \left( \frac{1}{|c_i^{(t)}|} [v_j \in c_i^{(t)}] - m_{ij}^{(t)} \right) = \frac{\alpha}{|c_i^{(t)}|} [v_j \in c_i^{(t)}] + (1 - \alpha) m_{ij}^{(t)} \quad (4)$$

where $[x \in X]$ denotes the Iverson-bracket notation for the indicator function.

**Properties of the Update Equation.** It should be noted that the reward $1/|c_i^{(t)}|$ produces the effect of making it stronger the co-association between nodes belonging to smaller communities. This is consistent with a major finding in a recent study proposed in [11] whereby co-memberships of nodes in larger communities are statistically less significant (than in smaller ones), because members in such communities have limited influence upon each other in the network. A further reason to favor co-associations in smaller communities is to compensate for a typical bias relating to a tendency of producing large communities (e.g., resolution limit in modularity-optimization based methods).

Another property of the update rule in Eq. (4) is the *exponential smoothing of earlier actions*, with constant $\alpha$ [21], i.e., the update scheme leads to weight recently obtained rewards more heavily than earlier ones, and the reward of a past co-association between two nodes decreases exponentially in time.

**Proposition 1.** *Equation (4) ensures that the rewards of past co-association between any two nodes $v_i, v_j$ decreases by a factor $(1 - \alpha)^{t-s}$, with $s \leq t$.*

PROOF. Let us assume that nodes $v_i, v_j$ are assigned to the same community and remain therein over time. By repeated substitutions, we derive that:

$$
\begin{aligned}
m_{ij}^{(t+1)} &= \frac{\alpha}{|c_i^{(t)}|} + (1 - \alpha) m_{ij}^{(t)} = \frac{\alpha}{|c_i^{(t)}|} + \left(1 - \alpha\right) \left[ \frac{\alpha}{|c_i^{(t-1)}|} + \left(1 - \alpha\right) m_{ij}^{(t-1)} \right] \\
&= \frac{\alpha}{|c_i^{(t)}|} + \frac{(1 - \alpha)\alpha}{|c_i^{(t-1)}|} + (1 - \alpha)^2 m_{ij}^{(t-1)} \\
&= \frac{\alpha}{|c_i^{(t)}|} + \frac{(1 - \alpha)\alpha}{|c_i^{(t-1)}|} + ... + \frac{(1 - \alpha)^{t-1}\alpha}{|c_i^{(1)}|} + (1 - \alpha)^t m_{ij}^{(1)} \\
&= (1 - \alpha)^t m_{ij}^{(1)} + \sum_{s=1}^{t} (1 - \alpha)^{t-s} \frac{\alpha}{|c_i^{(s)}|}.
\end{aligned}
$$

Also, it can easily be shown that Eq. (4) ensures that $\mathbf{M}$ is a *stochastic matrix*.

## 3.4   Speeding up *CreDENCE*

The time complexity of the basic version of CreDENCE is determined by the update operations on $\mathbf{M}$ given by Eq. (4) and by the community detection step.

As concerns the update step, we first observe that the relocation of nodes can be executed in $O(|V_t| + \omega|E_t|) = O(|\mathcal{V}| + \omega|E_t|)$, since for each node we look at its neighbor communities, which are bounded by the degree of the node. Evaluating the modularity improvement (Eq. (3)) is $O(\omega)$, provided that $\omega$ indexes are maintained to store the degree of communities for each of the last $\omega$ time steps, and to store the number of links of $v$ with nodes in community $c$ at time $t$, for each node $v$, time $t$ and community $c$. Therefore, since we constrain the number of relocation trials to be of the order of the number of nodes, the overall time cost of relocation of nodes is $O(|\mathcal{V}| + \omega|E_t|)$. However, the update of $\mathbf{M}$ involves a number of entries that is at least equal to $\sum_{c_i \in \mathcal{C}^{(t)}} |c_i|^2$. This could lead to a cost that becomes quadratic in the number of entities as soon as some of the communities have size of the order of $\mathcal{V}$. Moreover, the spatial complexity of CreDENCE is determined by the number of non-zero entries of $\mathbf{M}$, which again could be quadratic in the number of entities.

As discussed above, maintaining and updating the DCM matrix represents a computational bottleneck of CreDENCE. By definition, $\mathbf{M}$ can easily become dense, yet noisy, since many co-associations may be weak (e.g., outdated co-associations), thus corresponding to poorly significant consensus memberships. One way to alleviate this issue is to prune the matrix by zeroing those entries that are below a predefined threshold; in practice, this will unlikely be enough to solve the issue. Rather, we notice that it is more appropriate to introduce a *constraint of linkage between nodes* when evaluating Eq. (4): this is not only consistent with the requirement of having as high density as possible within a (consensus) community (as studied in [22]), but it will also impact on making $\mathbf{M}$ sparser. However, one drawback would be the loss of symmetry in $\mathbf{M}$.

We hence propose a modification to the update equation that both integrates the linkage constraint and preserves the stochasticity property of the matrix:

$$m_{ij}^{(t+1)} = \frac{\alpha}{|c_i^{(t)} \cap N_i^{(t)}|} [v_j \in c_i^{(t)} \cap N_i^{(t)}] + (1 - \alpha) m_{ij}^{(t)}, \tag{5}$$

where $N_i^{(t)}$ denotes the set of neighbors of $v_i$ in $G_t$. The entry $m_{ij}$ now is meant to store the strength of co-association of $v_i$ conditionally to the topological link with $v_j$. Moreover, the graph representation of $\mathbf{M}$ becomes directed: to keep the scheme presented in Algorithm 1, we simply modify the definition of the consensus graph $G_{\mathbf{M}}$ so that the weight of an edge $(v_i, v_j)$ is set to $\max\{m_{ij}, m_{ji}\}$. This allows us to preserve the importance of a co-association between any two entities when finding a community structure in $G_{\mathbf{M}}$.

We incorporate the above modifications into Algorithm 1 to obtain an enhanced, efficient version of CreDENCE. It can be noticed that the time complexity of CreDENCE now becomes $O(T \times (|\mathcal{V}| + |E_{\leq T}|))$, while the spatial cost is determined by the size of $\mathbf{M}$, i.e., $O(|E_{\leq T}|)$, with $E_{\leq T} = \bigcup_{t=1}^{T} E_t$.

## 4    Evaluation Methodology

**Data.** We used 5 real-world, publicly available temporal networks: *Epinions* [18], *Facebook* [23], *Wiki-Conflict* [2], *Wiki-Elections* [15], *YouTube* [19]. Table 1

**Table 1.** Main characteristics of our evaluation data. Mean ± standard deviation values refer to all snapshots in a network.

| | #entities ($\|\mathcal{V}\|$) | #edges | #time steps | node set coverage | edge semantics | % static (nodes, edges) | % hapax (nodes, edges) | % dynamic (nodes, edges) |
|---|---|---|---|---|---|---|---|---|
| Epinions | 131 828 | 727 344 | 32 | 0.05 | trust/distrust | (0.1, 0) | (80.8, 95.6) | (19, 2.2) |
| Facebook | 63 731 | 17 676 817 | 30 | 0.87 | friendship birth | (82.9, 2.7) | (0.2, 0) | (16.9, 1.9) |
| Wiki-Conflict | 118 100 | 2 272 276 | 82 | 0.05 | wikipage editing | (0, 0) | (60.1, 83.4) | (38.9, 5.8) |
| Wiki-Election | 7 118 | 102 906 | 44 | 0.08 | vote assignment | (0, 0) | (49.7, 95.7) | (50.3, 2.2) |
| YouTube | 3 223 589 | 41 955 741 | 8 | 0.62 | friendship birth | (33.4, 6.7) | (12.4, 4) | (54.2, 11.6) |

| | network evolution rate | | | |
|---|---|---|---|---|
| | $e_t^+ = \frac{\|E_t \setminus E_{t-1}\|}{\|E_t\|}$ | $e_t^- = \frac{\|E_{t-1} \setminus E_t\|}{\|E_{t-1}\|}$ | $v_t^+ = \frac{\|V_t \setminus V_{t-1}\|}{\|V_t\|}$ | $v_t^- = \frac{\|V_{t-1} \setminus V_t\|}{\|V_{t-1}\|}$ |
| Epinions | 0.97 ± 0.007 | 0.98 ± 0.008 | 0.65 ± 0.08 | 0.69 ± 0.06 |
| Facebook | 0.02 ± 0.01 | 0 | 0.006 ± 0.006 | 0 |
| Wiki-Conflict | 0.95 ± 0.02 | 0.95 ± 0.02 | 0.52 ± 0.1 | 0.51 ± 0.12 |
| Wiki-Election | 0.99 ± 0.004 | 0.99 ± 0.005 | 0.5 ± 0.07 | 0.49 ± 0.08 |
| YouTube | 0.16 ± 0.06 | 0 | 0.14 ± 0.06 | 0 |

reports statistics for each evaluation network. Note that with terms 'static', 'hapax', and 'dynamic' we mean nodes/edges that are present in all snapshots, present in only one snapshot, and present in multiple, not necessarily contiguous snapshots, respectively. Also, symbols $e_t^+$ and $e_t^-$ refer to the fraction of new edges and disappeared edges, respectively, when transitioning from the $t-1$-th to the $t$-th snapshot; analogously for nodes corresponding to symbols $v_t^+$, $v_t^-$. Note also that, while the friendship-based networks (i.e., Facebook and YouTube) evolve very smoothly, the other selected networks undergo to drastic changes in terms of disappearing/appearing edges and nodes. Preprocessing of the networks and statistics about the temporal width resolution are available at http:// people.dimes.unical.it/andreatagarelli/cmab-dccd.

We also used synthetic networks generated through *RDyn* [20], which is designed to handle community dynamics and change events (merge/split). RDyn adopts the notion of stable iteration to mimic **ground-truth** communities; in particular, when a community structure reaches a minimum quality (i.e., conductance), then it is recognized as ground-truth. We believe that the latter property of RDyn is important since it fills a lack in the literature about the unavailability of ground-truth data for (large) time-evolving multilayer networks.

**Competing Methods.** We conducted a comparative evaluation of CreDENCE with the following three methods, which are also based on modularity optimization and do not require an input number of communities:

- DynLouvain [10]: it applies Louvain method [1] to a condensed network based on the topology of the snapshot at current time $t$ and community structure at time $t - 1$.
- EvoAutoLeaders [8]: this is an evolutionary method based on a notion of community as a set of follower nodes congregating close to a potential leader (i.e., the most central node in the community).
- M-EMCD* [16]: this is a parameter-free enhanced version of the consensus-based method in [22], which filters noisy co-associations via marginal

likelihood filter and optimize the multilayer modularity of the consensus w.r.t. a static ensemble of community structures.

**Evaluation Settings.** We varied the learning rate $\alpha$ in $\{0.15, 0.5, 0.85\} \cup \{\alpha^*\}$, where $\alpha^*$ is an *adaptive* learning rate set to the fraction of times a base arms is used, and the temporal-window width $\omega$ from 2 to 10; however, unless otherwise specified, we used the setting $\omega = 2$, $\beta = 1 - \alpha$ to emphasize the importance of few, more recent snapshots. We set the relocation bias $\lambda$ to 0, i.e., a relocation is accepted if it leads to an improvement in modularity (Eq. (3)). To reduce sensitivity issues due to the randomness in the exploration-exploitation interleaving, we averaged the CreDENCE performance scores over 100 runs.

To detect communities from each snapshot (i.e., $\mathcal{A}$ in Algorithm 1), we used the classic Louvain method [1]. This choice is not only consistent with our modularity-optimization-based relocation phase, but also with the choice of static algorithm in most approaches for dynamic community detection [5].

As for the bandit strategy $\mathcal{B}$, we resorted to $\epsilon$-*greedy*, i.e., with a small probability $\epsilon$ we take an exploration step, otherwise (i.e., with probability $1 - \epsilon$) an exploitation step. We set $\epsilon = 0.1$, which revealed to lead to a performance stability trade-off for networks having different evolution rates.

## 5    Results

**Impact of Learning Rate.** As shown in Fig. 1, the number of detected consensus communities generally increases for higher values of $\alpha$, because this more quickly leads to lose memory of past co-associations, thus causing proliferation of communities in the consensus solution. Moreover, on the networks having high rate of structural change, the trends for the various settings of $\alpha$ tend to deviate in correspondence of the time steps associated with most change events; by contrast, in the networks characterized by a smooth evolution (i.e., Facebook and YouTube), the consensus sizes are very similar while varying $\alpha$.

Figure 1 also shows *multilayer modularity* [22] results by varying $\alpha$. Lower values of $\alpha$ generally lead to higher modularity except for Facebook and YouTube networks. This is explained since, in networks having high rate of structural change, a lower learning rate helps remember past co-associations, thus information about older snapshots. Moreover, in such networks we observe a decreasing trend in modularity since the consensus must embed an increasing number of snapshots, each very different from the others (cf. Table 1). By contrast, for Facebook and YouTube, a high learning rate reveals to be beneficial to discovering consensus communities with higher modularity.

We also measured the Strehl and Ghosh's *NMI* between the dynamic consensus and the community structure of snapshot, for each time step (Fig. 1). As expected, the two structures are more similar (i.e., higher NMI values) as $\alpha$ increases, which implies weighting more the current snapshot in the consensus generation. Analogous remarks were drawn for the *average cumulative NMI*, which is computed at each $t$ by averaging the NMI between the dynamic consensus at $t$ and the community structures over all snapshots at any time $t' \leq t$.

**Impact of Temporal-Window Width.** Higher values of $\omega$ will lead to better modularity performance, which is explained since the criterion function optimized in the relocation phase becomes closer to the measured modularity as $\omega$ increases. We indeed observed modularity improvements up to 0.04 (at any time

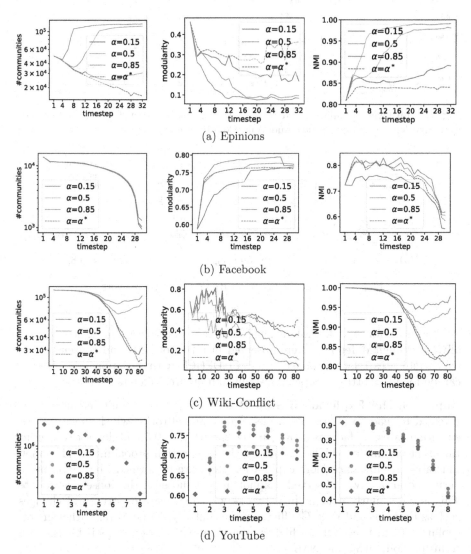

**Fig. 1.** Size of the dynamic consensus by CreDENCE (left), multilayer modularity of the CreDENCE solutions (mid), and NMI between the CreDENCE consensus community structure and the snapshot's community structure, at each $t$ (right).

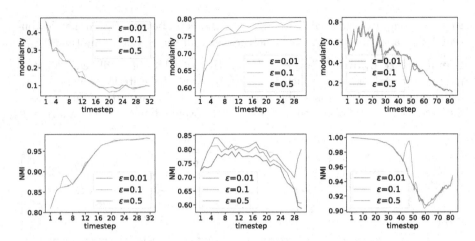

**Fig. 2.** Multilayer modularity and NMI by varying exploration-step probability $\epsilon$, on Epinions (left), Facebook (mid), and Wiki-Conflict (right).

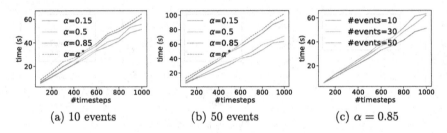

(a) 10 events          (b) 50 events          (c) $\alpha = 0.85$

**Fig. 3.** Time performance on RDyn synthetic networks.

step) already for $\omega = 4$, while negligible increments occurred as $\omega > 4$. Generally, no evident differences were observed in terms of NMI and consensus size, which indicates relative robustness of CreDENCE with variation in $\omega$. Results can be found at http://people.dimes.unical.it/andreatagarelli/cmab-dccd.

**Impact of the Exploration Step Probability.** The default setting $\epsilon = 0.1$ revealed to lead to a suitable trade-off for our networks, which have different evolution rates. In fact, as shown in Fig. 2, with $\alpha = 0.5$ and default setting for the other parameters, higher values for the exploration probability lead to more "unstable" results since more exploration steps are performed, thus information derived from a newly observed snaphot impact more on the consensus update. This is particularly evident in networks with high rate of structural changes, such as Epinions and Wiki-Conflict. On the contrary, for networks with a smooth evolution (e.g., Facebook), a higher number of exploration steps is beneficial in terms of modularity and NMI.

**Efficiency Evaluation.** To assess the scalability of CreDENCE, we used RDyn [20] to generate different synthetic networks by varying the number of

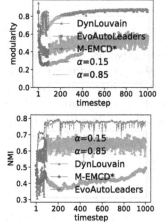

**Table 2.** Increment percentages of CreDENCE w.r.t. DynLouvain and M-EMCD*. Values correspond to the increment percentages averaged over all snapshots in a network, using the average best-performing $\alpha$.

| | DynLouvain | | M-EMCD* | |
|---|---|---|---|---|
| | Modularity | NMI | Modularity | NMI |
| *Epinions* | 1789.0% | −2.2% | 13.9% | 37.6% |
| *Facebook* | 3.5% | 9.4% | 60.0% | 37.5% |
| *Wiki-Conflict* | >1.0 E+05% | −1.8% | −6.8% | 37.6% |
| *Wiki-Election* | 660.5% | −2.1% | 32.0% | 58.5% |
| *YouTube* | −0.1% | 8.4% | 21.1% | 11.6% |
| *RDyn* | 2.0% | 24.97% | 103.22% | 81.1% |

**Fig. 4.** Competitors vs. CreDENCE on RDyn: modularity (top), NMI (bottom).

snapshots and community events.[1] Figure 3 reports the execution times for different settings of $\alpha$, over a temporal network with 1K entities and 1K time steps.

We observe that, for different change rate of community events (Fig. 3(a)–(b)), CreDENCE always scales linearly with the number of considered timesteps, which is consistent with our complexity analysis (cf. Sect. 3.4). Also, the execution time is generally higher for the adaptive learning rate $\alpha^*$ as well as for lower values of $\alpha$ (i.e., as the past co-associations are preserved longer), thus making the DCM matrix denser and more costly to process. Figure 3(c) shows the execution times of our method with $\alpha = 0.85$, on three synthetic networks with 10, 30, 50 community events, respectively. As expected, the higher the evolution rate, the higher the execution time; nonetheless, CreDENCE again shows to scale linearly with the size of the network.

**Comparison with Competing Methods.** Table 2 and Fig. 4 compare CreDENCE with the other methods. Concerning modularity results, our method outperforms both DynLouvain and M-EMCD*, where performance gains vs. the former (resp. latter) are outstanding for networks with high (resp. low) rate of structural change. NMI by CreDENCE is always significantly higher than the competitors' ones, especially against M-EMCD*; one exception is represented by a gap of just 2% w.r.t. DynLouvain for three networks with high evolution rate. Moreover, we emphasize that CreDENCE also outperforms the evolutionary EvoAutoLeaders, as long as the competitor results were available—indeed, it incurred in processing-time issues (tens hours) in all networks but the smallest ones, i.e., Wiki-Election and RDyn.

---

[1] Experiments were carried out on a Linux (Mint 18) machine with 2.6 GHz Intel Core i7-4720HQ processor and 16 GB ram.

# 6    Conclusion

In this paper, we proposed CreDENCE, a CMAB-based method for the problem of dynamic consensus community detection in temporal networks. Experimental evidence on real and synthetic networks has shown the meaningfulness of the consensus solutions produced by CreDENCE, also revealing its unique ability of dealing with temporal networks that can have different evolution rate.

We plan to further investigate on the impact of different bandit strategies (e.g., UCB, Thompson sampling), and on learning our model parameters to best fit the community structure and evolution in a given temporal network.

# References

1. Blondel, V.D., Guillaume, J.L., Lambiotte, R., Lefebvre, E.: Fast unfolding of communities in large networks. J. Stat. Mech. **10**, P10008 (2008)
2. Brandes, U., Kenis, P., Lerner, J., van Raaij, D.: Network analysis of collaboration structure in Wikipedia. In: Proceedings of the ACM WWW, pp. 731–740 (2009)
3. Brodka, P., Saganowski, S., Kazienko, P.: GED: the method for group evolution discovery in social networks. Social Netw. Analys. Mining **3**(1), 1–14 (2013)
4. Chen, W., Wang, Y., Yuan, Y.: Combinatorial multi-armed bandit: general framework and applications. In: Proceedings of the ICML, pp. 151–159 (2013)
5. Dakiche, N., Tayeb, F.B.-S., Slimani, Y., Benatchba, K.: Tracking community evolution in social networks: a survey. Inf. Process. Manag. **56**(3), 1084–1102 (2019)
6. La Fond, T., Sanders, G., Klymko, C., Van Emden, H.: An ensemble framework for detecting community changes in dynamic networks. In: Proceedings of the IEEE HPEC, pp. 1–6 (2017)
7. Gai, Y., Krishnamachari, B., Jain, R.: Combinatorial network optimization with unknown variables: multi-armed bandits with linear rewards and individual observations. IEEE/ACM Trans. Netw. **20**(5), 1466–1478 (2012)
8. Gao, W., Luo, W., Bu, C.: Adapting the TopLeaders algorithm for dynamic social networks. J. Supercomput. 23 (2017)
9. Gur, Y., Zeevi, A.J., Besbes, O.: Stochastic multi-armed-bandit problem with non-stationary rewards. In: Proceedings of the NIPS, pp. 199–207 (2014)
10. He, J., Chen, D.: A fast algorithm for community detection in temporal network. Physica A Stat. Mech. Appl. **429**, 87–94 (2015)
11. Wagenseller, P., Wang, F., Wu, W.: Size matters: a comparative analysis of community detection algorithms. IEEE Trans. Comput. Soc. Syst. **5**, 951–960 (2018)
12. Jiao, P., Wang, W., Jin, D.: Constrained common cluster based model for community detection in temporal and multiplex networks. Neurocomputing **275**, 768–780 (2018)
13. Katehakis, M.N., Veinott Jr., A.F.: Multi-armed bandit problem: decomposition and computation. Math. Oper. Res. **12**, 262–268 (1987)
14. Lancichinetti, A., Fortunato, S.: Consensus clustering in complex networks. Sci. Rep. **2**, 336 (2012)
15. Leskovec, J., Huttenlocher, D.P., Kleinberg, J.M.: Governance in social media: a case study of the Wikipedia promotion process. In: Proceedings of the ICWSM (2010)

16. Mandaglio, D., Amelio, A., Tagarelli, A.: Consensus community detection in multilayer networks using parameter-free graph pruning. In: Phung, D., Tseng, V.S., Webb, G.I., Ho, B., Ganji, M., Rashidi, L. (eds.) PAKDD 2018. LNCS (LNAI), vol. 10939, pp. 193–205. Springer, Cham (2018). https://doi.org/10.1007/978-3-319-93040-4_16

17. Mandaglio, D., Tagarelli, A.: Dynamic consensus community detection and combinatorial multi-armed bandit. In: Proceedings of the IEEE/ACM ASONAM (2019). https://doi.org/10.1145/3341161.3342910

18. Massa, P., Avesani, P.: Controversial users demand local trust metrics: an experimental study on epinions.com community. In: Proceedings of the AAAI, pp. 121–126 (2005)

19. Mislove, A.E.: Online social networks: measurement, analysis, and applications to distributed information systems. Ph.D. thesis, Rice University (2009)

20. Rossetti, G.: RDyn: graph benchmark handling community dynamics. J. Complex Netw. 5(6), 893–912 (2017)

21. Sutton, R.S., Barto, A.G.: Introduction to Reinforcement Learning, vol. 135. MIT press, Cambridge (1998)

22. Tagarelli, A., Amelio, A., Gullo, F.: Ensemble-based community detection in multilayer networks. Data Min. Knowl. Discov. 31(5), 1506–1543 (2017)

23. Viswanath, B., Mislove, A., Cha, M., Krishna Gummadi, P.: On the evolution of user interaction in Facebook. In: Proceedings of the ACM WOSN, pp. 37–42 (2009)

24. Wang, Z., Li, Z., Yuan, G., Sun, Y., Rui, X., Xiang, X.: Tracking the evolution of overlapping communities in dynamic social networks. Knowl. Based Syst. 157, 81–97 (2018)

# Resampling-Based Framework for Unbiased Estimator of Node Centrality over Large Complex Network

Kazumi Saito[1,2], Kouzou Ohara[3(✉)], Masahiro Kimura[4], and Hiroshi Motoda[5]

[1] Faculty of Science, Kanagawa University, Hiratsuka, Japan
k-saito@kanagawa-u.ac.jp
[2] Center for Advanced Intelligence Project, RIKEN, Tokyo, Japan
kazumi.saito@riken.jp
[3] College of Science and Engineering, Aoyama Gakuin University, Sagamihara, Japan
ohara@it.aoyama.ac.jp
[4] Department of Electronics and Informatics, Ryukoku University, Kyoto, Japan
kimura@rins.ryukoku.ac.jp
[5] Institute of Scientific and Industrial Research, Osaka University, Suita, Japan
motoda@ar.sanken.osaka-u.ac.jp

**Abstract.** We address a problem of efficiently estimating value of a centrality measure for a node in a large network, and propose a sampling-based framework in which only a small number of nodes that are randomly selected are used to estimate the measure. The error estimator we derived is an unbiased estimator of the approximation error defined as the expectation of the difference between the true and the estimated values of the centrality. We experimentally evaluate the fundamental performance of the proposed framework using the closeness and betweenness centralities on six real world networks from different domains, and show that it allows us to estimate the approximation error more tightly and more precisely than the standard error estimator traditionally used based on i.i.d. sampling, *i.e.*, with the confidence level of 95% for a small number of sampling, say 20% of the total number of nodes.

**Keywords:** Error estimation · Resampling · Node centrality · Complex network

## 1 Introduction

One common approach to analyze a large complex network is investigating its characteristics through a measure called centrality [2,5,8,11,25]. Various kinds of centralities are used according to what we want to know. For example, if our goal is to know the topological characteristics of a network, degree, closeness, and betweenness centralities [8] can be used. If it is to know the importance of nodes that constitute a network, HITS [6] and PageRank [5] centralities are often used. Influence degree centrality [12] is another one to measure the importance of nodes.

The size of network keeps increasing and, thus, it is becoming pressingly important that we are able to efficiently compute these centrality values. If a centrality measure

© Springer Nature Switzerland AG 2019
P. Kralj Novak et al. (Eds.): DS 2019, LNAI 11828, pp. 428–442, 2019.
https://doi.org/10.1007/978-3-030-33778-0_32

is based not only on local structure around a target node, *e.g.*, its neighboring nodes, but also on global structure of a network, *e.g.*, paths between arbitrary node pairs, its computation becomes harder as the size of the network increases. Thus, it is crucial to reduce the computational cost of such centralities for large networks. Typical examples are the closeness and the betweenness centralities which we consider in this paper.

In order to efficiently compute such centralities for large networks, Ohara et al. [17, 18] have proposed a resampling-based framework for estimating node centrality, and performed gap analysis for detecting nodes with high centrality in this framework, especially focusing on social networks. However, in order to estimate resampling errors by their methods, the true standard deviations of node centralities are assumed to be available although, in fact, we are not able to know them in advance. In this paper, we substantially improve this resampling-based framework by deriving an unbiased estimator for these resampling errors without assuming such true statistics. In our experiments which employ six complex networks: the same three social networks used by Ohara et al. [17, 18], and other three types of networks, *i.e.*, information, spatial, and cognitive networks, we demonstrate that our proposed estimator has desirable properties and better performances in comparison to a standard estimator based on the i.i.d. (identically independent distribution) assumption, which is widely employed in standard machine learning problems.

The paper is organized as follows. Section 2 describes related work. Section 3 gives the formal definitions of both the resampling-based framework that we propose and the traditional bound of approximation error. Section 4 explains the closeness and the betweenness centralities we used to evaluate our framework and presents how to estimate their approximation error. Section 5 reports experimental results for these centralities on six real world networks. Section 6 concludes this paper and addresses the future work.

## 2 Related Work

Sampling is a practical approach and often used when analyzing a large network. Many kinds of sampling methods have been investigated and proposed so far [9, 14, 15]. In addition, much efforts have been devoted to evaluate sampling methods (*e.g.* [24]) by using a variety of benchmark networks (*e.g.* [1]). Non-uniform sampling techniques give different probabilities of sampling to different nodes, *e.g.*, higher probabilities to specific nodes with high-degree. Similarly, results by traversal/walk-based sampling are biased towards high-degree nodes. In our problem setting the goal is to accurately estimate centralities of an original network and thus uniform sampling that selects nodes of a given network uniformly at random is essential because biased samples might skew centrality values that are derived from a resulting network. This motivates us to propose the framework that ensures the accuracy of the approximations of centrality values under uniform sampling. Although we use a simple random sampling here, our framework can adopt a more sophisticated technique such as MH-sampling [9] in so far as it falls under uniform sampling. In this sense, our framework can be regarded as a generic method that is applicable to any uniform sampling technique.

For these sampling techniques, we want to know the necessary number of sampling times, as a bound, in order to guarantee the performance of estimation results. For instance, in case of betweenness centrality, Brandes and Pich [4] advocated a bound based on Chernoff-Hoeffding inequality [7,10], and Riondato and Kornaropoulos [22] derived a bound based on the VC (Vapnik-Chervonenkis) dimension [23]. However, these criteria might have an intrinsic limitation for precisely evaluating the estimation error because their framework assumes that infinite samples are obtainable under the i.i.d. setting. In fact, the number of samples used for estimating centralities are typically limited to some finite number. Here we should note that another resampling framework proposed by Ohara et al. [20] in which the problem domain is information diffusion is substantially different from this work because the number of samples can be infinitely large and i.i.d. assumption has to be made.

## 3    Resampling-Based Estimation Framework

For a given set of objects $S$ whose number of elements is $L = |S|$, and a function $f$ which calculates some associated value of each object, we first consider a general problem of estimating the mean $\mu$ of the set of all values $\{f(s) \,|\, s \in S\}$ from only its arbitrary subset of partial values $\{f(t) \,|\, t \in T \subset S\}$. For a subset $T$ whose number of elements is $N = |T|$, we denote its partial mean by $\mu(T) = (1/N)\sum_{t \in T} f(t)$. Below, we formally derive an expected estimation error $RE(N)$ of the squared difference between $\mu$ and $\mu(T)$, with respect to the number of elements $N$. Hereafter, $RE(N)$ is simply referred to as resampling error.

Now, let $\mathcal{T} \subset 2^S$ be a family of subsets of $S$ whose number of elements is $N$, that is, $|T| = N$ for $T \in \mathcal{T}$. Then, we can compute the resampling error $RE(N)$ as follows:

$$
\begin{aligned}
RE(N) &= \left\langle (\mu - \mu(T))^2 \right\rangle = \left\langle \left( \mu - \frac{1}{N}\sum_{t \in T} f(t) \right)^2 \right\rangle = \binom{L}{N}^{-1} \frac{1}{N^2} \sum_{T \in \mathcal{T}} \left( \sum_{t \in T} (f(t) - \mu) \right)^2 \\
&= \binom{L}{N}^{-1} \frac{1}{N^2} \left( \binom{L-1}{N-1} \sum_{s \in S} (f(s) - \mu)^2 + \binom{L-2}{N-2} \sum_{s \in S} \sum_{t \in S, t \neq s} (f(s) - \mu)(f(t) - \mu) \right) \\
&= \binom{L}{N}^{-1} \frac{1}{N^2} \left( \left( \binom{L-1}{N-1} - \binom{L-2}{N-2} \right) \sum_{s \in S} (f(s) - \mu)^2 + \binom{L-2}{N-2} \left( \sum_{s \in S} (f(s) - \mu) \right)^2 \right) \\
&= \frac{L-N}{(L-1)N} \sigma^2.
\end{aligned}
\tag{1}
$$

Here the variance $\sigma^2$ is given by $\sigma^2 = (1/L)\sum_{s \in S}(f(s) - \mu)^2$. Evidently, we cannot directly compute $RE(N)$ from only a given subset $T$ because we are assuming that $|S| = L$ is too large to compute $\sigma^2$. Otherwise, sampling is not needed.

For a given subset $T \in \mathcal{T}$ with size $N$, we denote its partial variance by $\sigma^2(T) = (1/N)\sum_{t \in T}(f(t) - \mu(T))^2$. Then, in order to estimate $RE(N)$ from $T$, we propose an estimator defined by

$$
REE(T) = \frac{L-N}{L(N-1)} \sigma^2(T) = \frac{L-N}{LN(N-1)} \sum_{t \in T} (f(t) - \mu(T))^2,
\tag{2}
$$

For this estimator $REE(T)$, we obtain the following Theorem 1.

**Theorem 1.** *REE(T) is an unbiased estimator of RE(N).*

*Proof.* We first note the following equality holds:

$$\left\langle \frac{1}{N} \sum_{t \in T} (f(t) - \mu)^2 \right\rangle = \left( \frac{L}{N} \right)^{-1} \frac{1}{N} \left( \frac{L-1}{N-1} \right) \sum_{s \in S} (f(s) - \mu)^2 = \sigma^2.$$

Then, by using the following expected value of $\sigma^2(T)$,

$$\left\langle \sigma^2(T) \right\rangle = \left\langle \frac{1}{N} \sum_{t \in T} (f(t) - \mu)^2 - (\mu - \mu(T))^2 \right\rangle = \sigma^2 - RE(N) = \frac{L(N-1)}{(L-1)N} \sigma^2,$$

we obtain

$$\left\langle REE(T) \right\rangle = \frac{L-N}{L(N-1)} \left\langle \sigma^2(T) \right\rangle = \frac{L-N}{(L-1)N} \sigma^2 = RE(N).$$

$\square$

Hereafter, $REE(T)$ is referred to as resampling error estimator.

In this paper we consider a huge network consisting of millions of nodes as a collection of a large number of objects, and propose a framework in which we use the partial mean as an approximate solution with an adequate confidence level using the above estimation formula, Eq. (2). More specifically, we can expect that for a given subset $T$ with size $N$, and its partial mean $\mu(T)$, the probability that $|\mu(T) - \mu|$ is larger than $2 \times \sqrt{REE(T)}$, is less than 5%. This is because the estimated error by Eq. (2) is regarded as the variance with respect to the number of elements $N$. Hereafter this framework is referred to as the resampling estimation framework.

In order to confirm the effectiveness of the proposed resampling estimation framework, we also consider a standard approach based on the i.i.d. (independently identical distribution) assumption for comparison purpose. More specifically, for a given subset $T$ with size $N$, we assume that each element $t \in T$ is independently selected according to some distribution $p(T) = \prod_{t \in T} p(t)$. Here, we assume the simplest empirical distribution $p(t) = 1/L$. Then, by expressing elements of $T$ as $T = \{t_1, \cdots, t_N\}$, we obtain the following estimation formula for the expected error:

$$SE(N) = \left\langle (\mu - \mu(T))^2 \right\rangle = \frac{1}{L^N} \sum_{t_1 \in S} \cdots \sum_{t_N \in S} \frac{1}{N^2} \left( \sum_{n=1}^{N} (f(t_n) - \mu) \right)^2$$

$$= \frac{1}{L^N} \sum_{t_1 \in S} \cdots \sum_{t_N \in S} \frac{1}{N^2} \left( \sum_{n=1}^{N} (f(t_n) - \mu)^2 + \sum_{n=1}^{N} \sum_{m=1, m \neq n}^{N} (f(t_n) - \mu)(f(t_m) - \mu) \right)$$

$$= \frac{1}{N} \sigma^2, \tag{3}$$

Hereafter, $SE(N)$ is referred to as standard error. The difference between Eqs. (1) and (3) is only their coefficients. We note that $RE(N) = ((L - N)/(L - 1))SE(N)$, $RE(L) = 0$ and $SE(L) \neq 0$. Moreover, by using the following estimator,

$$SEE(T) = \frac{1}{N-1} \sigma^2(T) = \frac{1}{N(N-1)} \sum_{t \in T} (f(t) - \mu(T))^2, \tag{4}$$

we can easily prove that $SEE(T)$ is an unbiased estimator of $SE(N)$. Here note that $REE(T) = (1 - (N/L))SEE(T)$, $REE(S) = 0$ and $SEE(S) \neq 0$. For more details, we empirically compare these resampling error estimator $REE(T)$ and standard error estimator $SEE(T)$ through experiments on node centrality calculation of different networks as described below.

## 4    Application to Node Centrality Estimation

We investigate our proposed resampling framework on node centrality estimation of a network represented by a directed graph $G = (V, E)$, where $V$ and $E$ ($\subset V \times V$) are the sets of all the nodes and the links in the network, respectively. When there is a link $(u, v)$ from node $u$ to node $v$, $u$ is called a *parent node* of $v$ and $v$ is called a *child node* of $u$. For any node $v \in V$, let $A(u)$ and $B(v)$ respectively denote the set of all child nodes of $u$ and the set of all parent nodes of $v$ in $G$, i.e., $A(u) = \{v \in V; (u, v) \in E\}$ and $B(v) = \{u \in V; (u, v) \in E\}$. For readers' convenience, by following the paper by Ohara et al. [17], we describe the methods for computing centrality values in our proposed framework.

### 4.1    Closeness Centrality Estimation

The closeness $cls_G(u)$ of a node $u$ on a graph $G$ is defined as

$$cls_G(u) = \frac{1}{(|V| - 1)} \sum_{v \in V, v \neq u} \frac{1}{spl_G(u, v)}, \tag{5}$$

where $spl_G(u, v)$ stands for the shortest path length from $u$ to $v$ in $G$. Namely, the closeness of a node $u$ becomes high when a large number of nodes are reachable from $u$ within relatively short path lengths. Here note that we set $spl_G(u, v) = \infty$ when node $v$ is not reachable from node $u$ on $G$. Thus, in order to naturally cope with this infinite path length, we employ the inverse of the harmonic average as shown in Eq. (5).

The burning algorithm [16] is a standard technique for computing $cls_G(u)$ of each node $u \in V$. More specifically, after initializing a node subset $X_0$ to $X_0 \leftarrow \{u\}$, and path length $d$ to $d \leftarrow 0$, this algorithm repeatedly calculates a set $X_{d+1}$ of newly reachable nodes from $X_d$ and set $d \leftarrow d + 1$ unless $X_d$ is empty. Here, newly reachable nodes from $X_{d-1}$ is defined by $X_d = (\bigcup_{v \in X_{d-1}} A(v)) \setminus (\bigcup_{c < d} X_c)$. Then the shortest path length of node $v \in X_d$ from $u$ is obtained as $spl_G(u, v) = d$. Here recall that $spl_G(u, v) = \infty$ if $v$ is not reachable from $u$. Since the computational complexity of computing $cls_G(u)$ for each node $u \in V$ become $O(|E|)$, it takes a large amount of computation time for a huge networks consisting of millions of nodes.

Now, we present a method for computing $cls_G(u)$ of each node $u \in V$ under our resampling estimation framework. The method first constructs the reverse network of $G = (V, E)$ by reversing the direction of each link from $(u, v)$ to $(v, u)$. Namely, the reverse network is defined by $H = (V, F)$ and $F = \{(v, u) | (u, v) \in E\}$. Then, by using the burning algorithm starting from node $v$ over the reverse network, we can calculate each shortest path length from $v$ to $u$ as $spl_H(v, u)$. Clearly, $spl_H(v, u)$ is the shortest path length from node $u$ to $v$, i.e., $spl_G(u, v)$. Namely, for each node $u \in V$, by setting

$S_u = V \setminus \{u\}$ and $f_u(v) = spl_H(v, u)$, we can calculate partial mean from an arbitrary subset $T \subset S_u$. Here note that, due to the nature of the burning algorithm from $v$ to the other arbitrary node $u \in V$, we can obtain each partial mean $\mu_u(T) = (1/|T|) \sum_{v \in T} f_u(v)$ simultaneously for every node $u \in V$.

## 4.2 Betweenness Centrality Estimation

The betweenness $btw_G(u)$ of a node $u$ on a graph $G$ is defined as

$$btw_G(u) = \frac{1}{(|V| - 1)(|V| - 2)} \sum_{v \in V, v \neq u} \left( \sum_{w \in V, w \neq u, w \neq v} \frac{nsp_G(v, w; u)}{nsp_G(v, w)} \right), \qquad (6)$$

where $nsp_G(v, w)$ is the number of the shortest paths from $v$ to $w$ in $G$ and $nsp_G(v, w; u)$ is the number of the shortest paths from $v$ to $w$ in $G$ that pass through node $u$. Namely, the betweenness of a node $u$ becomes high when a large number of shortest paths between two nodes pass through node $u$. Here note that although $cls_G(u)$ and $cls_H(u)$ are not generally equal, since any node pair $(v, w)$ is examined in Eq. (6) we can easily see that $btw_G(u) = btw_H(u)$.

The Brandes algorithm [3] is a standard technique for computing $btw_G(u)$ of each node $u \in V$. The algorithm utilizes a series of node subsets $(X_0, \cdots, X_D)$ produced by the burning algorithm described in Sect. 4.1 starting from node $v \in V$, where $D$ stands for the maximum burning step. Then, after setting $nsp_G(v, w) \leftarrow 1$ for $w \in X_1$, the algorithm in turn computes $nsp_G(v, w) \leftarrow \sum_{x \in B(w) \cap X_{d-1}} nsp_G(v, x)$ for $w \in X_d$ from $d = 2$ to $D$. Next, we define the following betweenness $btw_G(u; v)$ of node $u$, which restricts its starting node to $v$,

$$btw_G(u; v) = \sum_{w \in V, w \neq u, w \neq v} \frac{nsp_G(v, w; u)}{nsp_G(v, w)}. \qquad (7)$$

Then, after setting $btw_G(u; v) \leftarrow 0$ for $u \in X_D$, the algorithm in turn computes $btw_G(u; v) \leftarrow \sum_{x \in A(u) \cap X_{d+1}} (nsp_G(v, u)/nsp_G(v, x))(1 + btw_G(x; v))$ for $u \in X_d$ from $d = D - 1$ to 1. Finally, by computing and summing $btw_G(u; v)$ by changing the starting node $v$, we can obtain the betweenness $btw_G(u)$ of each node $u \in V$. Again, the computational complexity of computing $btw_G(u)$ for each node $u \in V$ becomes $O(|E|)$.

Now, we present a method based on the Brandes algorithm for computing $btw_G(u)$ of each node $u \in V$ under our resampling estimation framework. Namely, for each node $u \in V$, by setting $S_u = V \setminus \{u\}$ and $f_u(v) = btw_G(u; v)/(|V| - 2)$, we can calculate partial mean from an arbitrary subset $T \subset S_u$. Again note that, due to the nature of the Brandes algorithm, we can obtain such partial mean simultaneously for all nodes $u \in V$.

## 5 Experiments

To experimentally evaluate our framework and methods in the previous sections, we employed six datasets of real networks, which were referred to as "Ameblo", "Cosme", "Enron", "Citation", "Road", and "Word", where the first three social networks were

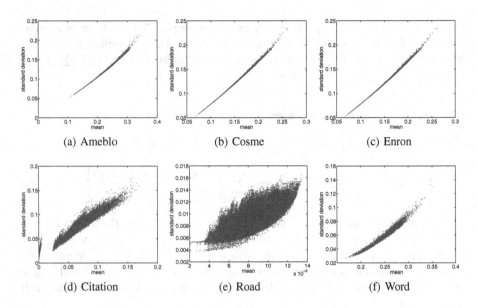

**Fig. 1.** Results for "closeness centrality value vs. standard deviation"

also employed in the previous study [17], while the rest were newly added as the other types of complex networks, *i.e.*, information, spatial, and cognitive networks. Here the first four networks are directed, and the rest are undirected. Below we summarize some details of these networks.

The Ameblo network is a reader network extracted from a Japanese blog service site "Ameba"[1], which has 56, 604 nodes (persons) and 734, 737 directed links. The Cosme network is a fan-link network extracted from a Japanese word-of-mouth communication site for cosmetics, "@cosme"[2], which has 45, 024 nodes (persons) and 351, 299 directed links. The Enron network is a communication network derived from the Enron Email Dataset [13], which has 19, 603 nodes (persons) and 210, 950 directed links. The Citation network is a high-energy physics citation network from the e-print arXiv obtained from SNAP (Stanford Network Analysis Project)[3], which has 34, 546 nodes (papers) and 421, 578 directed links. The Road network [19] is an urban street network of Washington D.C. extracted from OSM (OpenStreetMap) data "Metro Extracts"[4], which has 114, 758 nodes (junctions) and 128, 746 undirected links. The Word network [21] is a word association network that was retrieved and converted from the University of South Florida word association norms[5], which has 7, 207 nodes (words) and 31, 784 undirected links.

---

[1] http://www.ameba.jp/.

[2] http://www.cosme.net/.

[3] https://snap.stanford.edu/.

[4] https://mapzen.com/data/metro-extracts/.

[5] http://w3.usf.edu/FreeAssociation/.

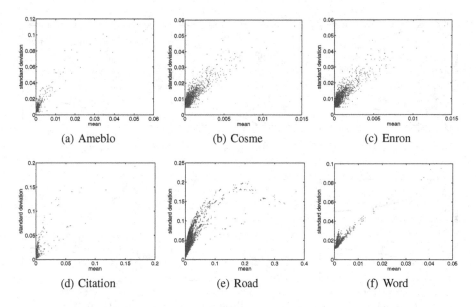

**Fig. 2.** Results for "betweenness centrality value vs. standard deviation"

## 5.1 Statistical Analysis

For each of the six real networks, $G = (V, E)$, we first computed the value of the closeness centrality $cls_G(u)$ and betweenness centrality $btw_G(u)$ of each node $u \in V$ by means of the algorithms presented in Sects. 4.1 and 4.2, respectively. In addition, we investigated their standard deviations given by

$$\sigma_{cls}(u) = \sqrt{\frac{1}{|V| - 1} \sum_{v \in V, v \neq u} \left( \frac{1}{spl_G(u, v)} - cls_G(u) \right)^2}$$

for the closeness centrality, and

$$\sigma_{btw}(u) = \sqrt{\frac{1}{|V| - 1} \sum_{v \in V, v \neq u} \left( \frac{btw_G(u; v)}{|V| - 2} - btw_G(u) \right)^2}$$

for the betweenness centrality. Figure 1 shows the results for closeness centrality, where Figs. 1(a) to (f) plot the pair $(cls_G(u), \sigma_{clc}(u))$ for the Ameblo, Cosme, Enron, Citation, Road and Word networks. Similarly, Fig. 2 shows the results for betweenness centrality, where Figs. 2(a) to (f) plot the pair $(btw_G(u), \sigma_{btw}(u))$ for the same six networks. In each figure, the horizontal and vertical axes indicate the values of corresponding centrality, $cls_G(u)$ or $btw_G(u)$, computed by all samples, and its standard deviation, $\sigma_{cls}(u)$ or $\sigma_{btw}(u)$, respectively.

We can observe that there exists positive correlation between the centrality value of each node and its standard deviation. This tendency can be found more clearly in the results for the closeness centrality, and we can observe almost line-like shapes except

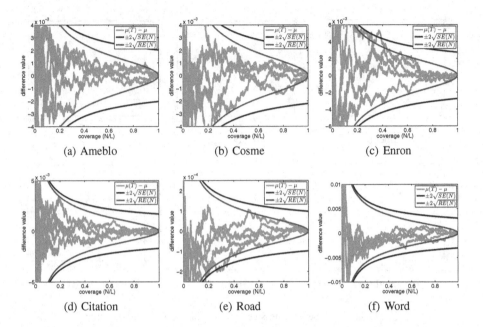

**Fig. 3.** Fluctuation of the estimated value of the closeness centrality as a function of the coverage for the node with the largest true centrality value by showing the results of five independent trials. (Color figure online)

for the Road network, which has relatively large variances around middle range of the closeness centrality values. In contrast, in case of the betweenness centrality, we can observe relatively similar results for all the networks.

Next, we evaluated the fundamental performance of the resampling error $RE(N)$, *i.e.*, how tightly and accurately it estimates the approximation error, using the closeness and betweenness centralities on the six networks. To this end, we considered a problem of estimating $\mu$, the true value of a centrality measure for node $u$ in network $G(V, E)$ using a set of partial values $\{f_u(v) \mid v \in T \subset V\}$ generated by sampling $N$ nodes from $V$, where $\mu$ stands for either $cls_G(u)$ or $btw_G(u)$ and $f_u(v)$ stands for either $spl_H(v, u)$ or $btw_G(u; v)$. More specifically, we empirically investigated whether or not the estimation $\mu(T) - \mu$, the difference between the mean derived from $T$ and true ones, falls within the range of $\pm 2 \times \sqrt{RE(N)}$. In addition, we considered the range of $\pm 2 \times \sqrt{SE(N)}$ for comparison.

Figures 3 and 4 show the results for the closeness and betweenness centralities, respectively. In this experiment, we considered the top nodes in each network that respectively have the largest true values of the corresponding centrality in Figs. 1 and 2. Here note that we can obtained almost the same results for the other rank nodes. In each figure, the horizontal axis "coverage" means the ratio of the number of sampled nodes $N$ to the total number of nodes $L$, *i.e.*, $N/L$, in each network, while the vertical axis means the difference from the true centrality value, and how the estimated value fluctuates as a function of the coverage is depicted. We conducted five independent trials

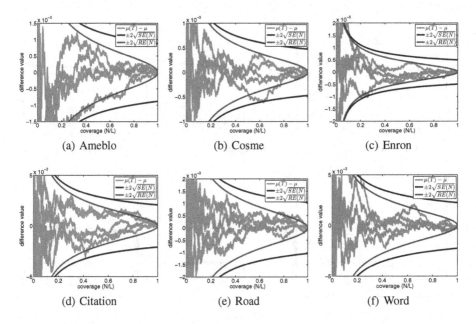

(a) Ameblo          (b) Cosme          (c) Enron

(d) Citation          (e) Road          (f) Word

**Fig. 4.** Fluctuation of the estimated value of the betweenness centrality as a function of the coverage for the node with the largest true centrality value by showing the results of five independent trials. (Color figure online)

for each of the top node in each network, and plotted estimated values $\mu(T) - \mu$ for a given coverage $N/L$ with green jagged lines. The red and blue lines show the ranges of $\pm 2 \times \sqrt{RE(N)}$ and $\pm 2 \times \sqrt{SE(N)}$, respectively.

From these results, we can confirm that the boundary determined by $RE(N)$ estimates the approximation error more tightly and converges to 0.0 as the coverage approaches 1.0, while the boundary by $SE(N)$ is looser and does not converge to 0.0 even if the coverage becomes 1.0. Furthermore, in most cases, the estimated value falls within the range of $2 \times \sqrt{RE(N)}$ for every network regardless of the centrality used. From these results, we can say that the resampling error $RE(N)$ provides us with a better error bound with the confidence level of 95% compared to the standard error $SE(N)$. Namely, it was shown that we can obtain quite similar results not only for social networks, but also for the other types of complex networks.

## 5.2 Results

For each set of partial values generated from $T$ consisting of $N$ sampled nodes, we quantitatively evaluated the difference between the resampling error $RE(N)$ and our proposed unbiased estimator $REE(T)$. Figures 5 and 6 show the results for the closeness and betweenness centralities, respectively, for our six networks, where each result is shown in Figs. 5(a) to (f) and Figs. 6(a) to (f). In this experiment, we used five independent trials for each of the top node in each network depicted in Figs. 3 and 4. In each figure, the horizontal axis also means the sample coverage, while the vertical axis means

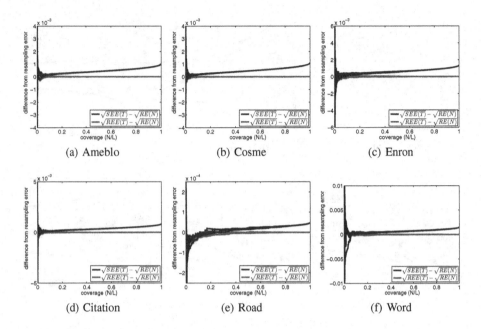

**Fig. 5.** Estimation precision of $RE(N)$ by $REE(T)$ for the closeness centrality by showing the results of five independent trials. (Color figure online)

the difference measure by $\sqrt{REE(T)} - \sqrt{RE(N)}$. Here we added $\sqrt{SEE(T)} - \sqrt{RE(N)}$ for comparison purpose, where the red and blue lines show $\sqrt{REE(T)} - \sqrt{RE(N)}$ and $\sqrt{SEE(T)} - \sqrt{RE(N)}$, respectively.

From these results, we can observe that the difference fluctuates when the value of coverage is less than 0.2 in both cases of $REE(T)$ and $SEE(T)$, but for a larger coverage it becomes remarkably stable and almost equal to 0.0 in the case of $REE(T)$, while it increases as the value of coverage becomes larger in the case of $SEE(T)$. This tendency is common to every network regardless of the centrality used. These results show that the proposed resampling error estimator $REE(T)$ can precisely estimate the resampling error $RE(N)$ if the coverage is larger than a certain threshold, say 0.2, while the standard error estimator $SEE(T)$ tends to overestimate the resampling error $RE(N)$.

Next, we quantitatively evaluated the empirical performance computed by

$$\langle | \sqrt{REE(T)} - |\mu(T) - \mu|| \rangle = \frac{1}{R} \sum_{r=1}^{R} | \sqrt{REE(T_r)} - |\mu(T_r) - \mu||,$$

where $T_r$ means a randomly generated set of nodes with sample size $N$ to calculate partial values, and we set $R$ to $R = 1,000$. Figures 7 and 8 show the results for the closeness and betweenness centralities, respectively, for our six networks. In each figure, the horizontal axis also means the sample coverage $N/L$, while the vertical axis means the differences computed by $\langle | \sqrt{REE(T)} - |\mu(T) - \mu|| \rangle$. We also added the following for comparison purpose,

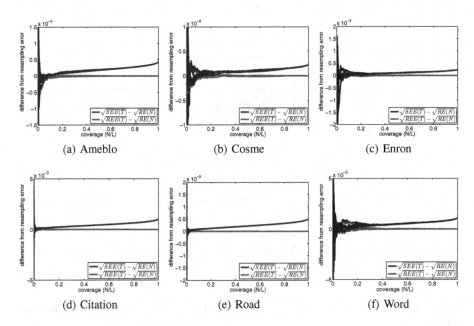

**Fig. 6.** Estimation precision of $RE(N)$ by $REE(T)$ for the betweenness centrality by showing the results of five independent trials. (Color figure online)

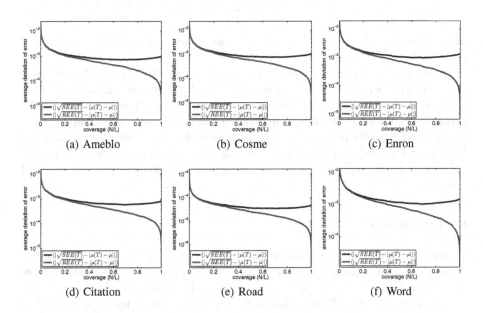

**Fig. 7.** Empirical convergence property of $REE(T)$ for the closeness centrality. (Color figure online)

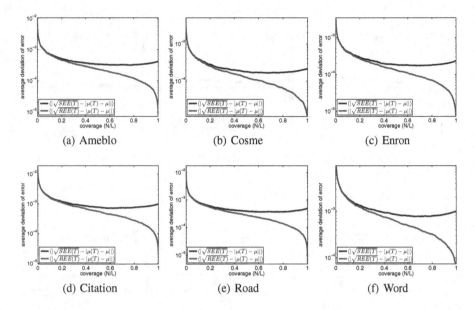

**Fig. 8.** Empirical convergence property of $REE(T)$ for the betweenness centrality. (Color figure online)

$$\langle | \sqrt{SEE(T)} - |\mu(T) - \mu|| \rangle = \frac{1}{R} \sum_{r=1}^{R} | \sqrt{SEE(T_r)} - |\mu(T_r) - \mu||,$$

where the red and blue lines show $\langle | \sqrt{REE(T)} - |\mu(T) - \mu|| \rangle$ and $\langle | \sqrt{SEE(T)} - |\mu(T) - \mu|| \rangle$, respectively.

As expected, for a larger coverage, we can observe that $REE(T)$ rapidly decreases to 0.0, but $SEE(T)$ even slightly increases. Here we should note that these convergence properties are surprisingly similar for all the six network regardless of the types of the networks and the centrality used. Consequently, we can say that the resampling error estimator $REE(T)$ we proposed is more promising than the standard error estimator $SEE(T)$ in this kind of estimation problem, and can give a tighter and more precise estimate of the resampling error $RE(N)$ with high confidence level than the standard error does.

## 6    Conclusion

We addressed a problem of estimating the value of a centrality measure for a node in a network. Centrality measure plays an important role in network analysis since it characterizes nodes in a network and its values indicate the importance of nodes in certain aspects. Thus, it is crucially important to be able to efficiently calculate the value of a centrality measure for each node for a network of large size. Its computation could be intractable for such centrality measures that require use of a global network structure for their computation. We have to rely on a sampling-based approach to deal with the scalability problem, in which we approximate the true value of a centrality from only a small number of nodes that are randomly selected from the whole network. Sampling must be uniform for our problem to avoid distorted results. What is important is that we ensure the accuracy of the approximations without knowing the truth. The proposed resampling-based framework can evaluate the approximation error of the estimated values of a centrality measure for each node. We proved that the proposed error estimator is an unbiased estimator of the approximation error which is defined as the expectation of the difference between the unknown true and the estimated values of the centrality. We have conducted extensive experiments on six real world networks varying the coverage ratio of nodes to be sampled, and compared the performance with the standard error based on i.i.d. sampling known in statistics using two representative centrality measures, the closeness and betweenness centralities, both of which need a global structure for their computation. The six networks come from different domains, *i.e.*, social, information, spatial and cognitive, each with different topological characteristics. We empirically confirmed that the proposed framework enables us to estimate the approximation error more tightly and more precisely with the confidence level of 95% even for a set of sampled nodes whose coverage is small, say 0.2, than using the standard error estimate. It is noted that the framework we proposed is not specific to computation of centrality measures. Indeed, it is very generic and applicable to any other estimation problems that require aggregation of many (but a finite number of) primitive computations. We believe that the conclusion obtained in this paper can generalize but we have yet to test out the proposed framework in a broader setting and also in many more different domains, too.

**Acknowledgments.** This material is based upon work supported by JSPS Grant-in-Aid for Scientific Research (C) (No. 17K00314).

# References

1. AlGhamdi, Z., Jamour, F., Skiadopoulos, S., Kalnis, P.: A benchmark for betweenness centrality approximation algorithms on large graphs. In: Proceedings of the 29th International Conference on Scientific and Statistical Database Management (2017)
2. Bonacichi, P.: Power and centrality: a family of measures. Am. J. Sociol. **92**, 1170–1182 (1987)
3. Brandes, U.: A faster algorithm for betweenness centrality. J. Math. Sociol. **25**, 163–177 (2001)
4. Brandes, U., Pich, C.: Centrality estimation in large networks. Int. J. Bifurcat. Chaos **17**(7), 303–318 (2007)
5. Brin, S., Page, L.: The anatomy of a large-scale hypertextual web search engine. Comput. Netw. ISDN Syst. **30**, 107–117 (1998)
6. Chakrabarti, S., et al.: Mining the web's link structure. IEEE Comput. **32**, 60–67 (1999)
7. Chernoff, H.: A measure of asymptotic efficiency for tests of a hypothesis based on the sum of observations. Ann. Math. Stat. **23**(4), 493–507 (1952)
8. Freeman, L.: Centrality in social networks: conceptual clarification. Soc. Netw. **1**, 215–239 (1979)
9. Henzinger, M.R., Heydon, A., Mitzenmacher, M., Najork, M.: On near-uniform URL sampling. Int. J. Comput. Telecommun. Network. **33**(1–6), 295–308 (2000)

10. Hoeffding, W.: Probability inequalities for sums of bounded random variables. J. Am. Stat. Assoc. **58**(301), 13–30 (1963)
11. Katz, L.: A new status index derived from sociometric analysis. Sociometry **18**, 39–43 (1953)
12. Kimura, M., Saito, K., Ohara, K., Motoda, H.: Speeding-up node influence computation for huge social networks. Int. J. Data Sci. Anal. **1**, 1–14 (2016)
13. Klimt, B., Yang, Y.: The Enron corpus: a new dataset for email classification research. In: Boulicaut, J.-F., Esposito, F., Giannotti, F., Pedreschi, D. (eds.) ECML 2004. LNCS (LNAI), vol. 3201, pp. 217–226. Springer, Heidelberg (2004). https://doi.org/10.1007/978-3-540-30115-8_22
14. Kurant, M., Markopoulou, A., Thiran, P.: Towards unbiased BFS sampling. IEEE J. Sel. Areas Commun. **29**(9), 1799–1809 (2011)
15. Leskovec, J., Faloutsos, C.: Sampling from large graphs. In: Proceedings of the 12th ACM SIGKDD International Conference on Knowledge Discovery and Data Mining (KDD 2006), pp. 631–636 (2006)
16. Newman, M.E.J.: Scientific collaboration networks. II. Shortest paths, weighted networks, and centrality. Phys. Rev. E **64**, 016132 (2001)
17. Ohara, K., Saito, K., Kimura, M., Motoda, H.: Resampling-based framework for estimating node centrality of large social network. In: Džeroski, S., Panov, P., Kocev, D., Todorovski, L. (eds.) DS 2014. LNCS (LNAI), vol. 8777, pp. 228–239. Springer, Cham (2014). https://doi.org/10.1007/978-3-319-11812-3_20
18. Ohara, K., Saito, K., Kimura, M., Motoda, H.: Resampling-based gap analysis for detecting nodes with high centrality on large social network. In: Cao, T., Lim, E.-P., Zhou, Z.-H., Ho, T.-B., Cheung, D., Motoda, H. (eds.) PAKDD 2015. LNCS (LNAI), vol. 9077, pp. 135–147. Springer, Cham (2015). https://doi.org/10.1007/978-3-319-18038-0_11
19. Ohara, K., Saito, K., Kimura, M., Motoda, H.: Accelerating computation of distance based centrality measures for spatial networks. In: Calders, T., Ceci, M., Malerba, D. (eds.) DS 2016. LNCS (LNAI), vol. 9956, pp. 376–391. Springer, Cham (2016). https://doi.org/10.1007/978-3-319-46307-0_24
20. Ohara, K., Saito, K., Kimura, M., Motoda, H.: Resampling-based predictive simulation framework of stochastic diffusion model for identifying top-k influential nodes. Int. J. Data Sci. Anal (2019, online first)
21. Palla, G., Derényi, I., Farkas, I., Vicsek, T.: Uncovering the overlapping community structure of complex networks in nature and society. Nature **435**, 814–818 (2005)
22. Riondato, M., Kornaropoulos, E.M.: Fast approximation of betweenness centrality through sampling. Data Min. Knowl. Disc. **30**(2), 438–475 (2016)
23. Vapnik, V.N., Chervonenkis, A.Y.: On the uniform convergence of relative frequencies of events to their probabilities. Theory Probab. Appl. **16**(2), 264–280 (1971)
24. Wandelt, S., Shi, X., Sun, X.: Scalability of betweenness approximation algorithms: an experimental review. IEEE Access **7**, 104057–104071 (2019)
25. Zhuge, H., Zhang, J.: Topological centrality and its e-Science applications. J. Am. Soc. Inf. Sci. Technol. **61**, 1824–1841 (2010)

# Pattern Discovery

# Layered Learning for Early Anomaly Detection: Predicting Critical Health Episodes

Vitor Cerqueira[1,3]([✉]), Luis Torgo[1,2,3], and Carlos Soares[1,3]

[1] University of Porto, Porto, Portugal
vitor.cerqueira@fe.up.pt
[2] Dalhousie University, Halifax, Canada
[3] LIAAD-INESCTEC, Porto, Portugal

**Abstract.** Critical health events represent a relevant cause of mortality in intensive care units of hospitals, and their timely prediction has been gaining increasing attention. This problem is an instance of the more general predictive task of early anomaly detection in time series data. One of the most common approaches to solve this problem is to use standard classification methods. In this paper we propose a novel method that uses a layered learning architecture to solve early anomaly detection problems. One key contribution of our work is the idea of pre-conditional events, which denote arbitrary but computable relaxed versions of the event of interest. We leverage this idea to break the original problem into two layers, which we hypothesize are easier to solve. Focusing on critical health episodes, the results suggest that the proposed approach is advantageous relative to state of the art approaches for early anomaly detection. Although we focus on a particular case study, the proposed method is generalizable to other domains.

**Keywords:** Time series · Early anomaly detection · Healthcare · Layered learning

## 1 Introduction

Healthcare is one of the domains which has witnessed a significant growth in the application of machine learning approaches [1]. For instance, intensive care units (ICUs) evolved considerably in recent years due to technological advances such as the widespread adoption of bio-sensors [16]. This lead to new opportunities for predictive modelling in clinical medicine. One of these opportunities is the early detection of critical health episodes (CHE), such as acute hypotensive episode [8] (AHE) or tachycardia episode [7] (TE) prediction problems. CHEs such as these remain a significant mortality risk factors in ICUs [8], and their timely anticipation is fundamental for improving healthcare.

AHE or TE prediction can be regarded as a particular instance of early anomaly detection in time series data. Fawcett and Provost designated this kind

© Springer Nature Switzerland AG 2019
P. Kralj Novak et al. (Eds.): DS 2019, LNAI 11828, pp. 445–459, 2019.
https://doi.org/10.1007/978-3-030-33778-0_33

of prediction tasks as *activity monitoring* [5]. Essentially, the goal behind these problems is to issue accurate and timely alarms about interesting future events requiring action.

One of the most common ways to address early anomaly detection problems is to view them as conditional probability estimation problems [5,19]. Standard supervised learning classification methods can be used for that purpose. The idea is to approximate a function $f$ that maps a set of input observations $X = (x_1, x_2, \ldots, x_n)$ to a binary variable $y$, which represents whether an anomaly occurs or not. In the case of CHE prediction, the predictor variables $(X)$ summarize the recent physiological signals of a patient assigned to the ICU, while the target $(y)$ represents whether or not there is an impending event in the near future.

In many domains of application, the anomaly or event of interest is defined according to some rule derived from the data by professionals. For example in healthcare, CHEs are often defined as events where the value of some physiological signal exceeds a pre-defined threshold. Similar approaches for formalizing anomalies can be found in predictive maintenance [14], or wind power prediction [6]. In these scenarios we can also define pre-conditional events, which are arbitrary but computable relaxed versions of the event of interest. These pre-conditional events occur simultaneously with the anomaly one is trying to model, but are more frequent and, in principle, a good indication for these. To be more precise, a pre-conditional event (i) represents a less extreme version of the anomalies we are trying to detect (main events); and (ii) occurs simultaneously with anomalies (i.e. there can not be an anomaly without a pre-conditional event). This concept is illustrated in the right side of Fig. 1 as a Venn diagram for classes.

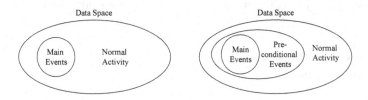

**Fig. 1.** Venn diagram for the classes in an early anomaly detection problem. The main event represents a small part of the data space; pre-conditional events are more frequent and include the occurrence of the main events.

Our working hypothesis in this paper is that modelling these pre-conditional events can be advantageous to capture the actual events of interest. To achieve this we adopt a layered learning methodology [17]. Layered learning denotes a learning approach in which a predictive task is split into two or more layers (simpler predictive tasks) where the learning process within a layer affects the learning process of the next layer.

We propose a layered learning method to address early anomaly detection problems by splitting the predictive task in two layers (c.f. right side of Fig. 1).

We first model pre-conditional events relative to normal activity. A subsequent model is applied to distinguish pre-conditional events from the actual anomalies. Effectively, the first layer affects the learning process of the second layer by decreasing the scope of its data space. Since the model in the second layer is created to distinguish the events of interest from pre-conditional events, it does not train on observations of what is designated as normal activity.

We apply the proposed approach to tackle the problem of CHE prediction. In the experiments, the layered learning model shows a better predictive performance relative to state of the art approaches, including a direct classification approach (without layered learning, see the left side of Fig. 1). In short, the contributions of this paper are:

- the idea of pre-conditional events in time series;
- a general layered learning approach to the early detection of events in time series data;
- the application of the proposed approach to AHE and TE prediction.

All work and results presented in the paper are reproducible. The data is publicly available [16], and the code for the methods can be found at https://github.com/vcerqueira/layered_learning_time_series.

## 2 Early Anomaly Detection in Time Series

Let $\mathcal{E} = \{E_1, \ldots, E_{|\mathcal{E}|}\}$ denote a set of time series. For example, $\mathcal{E}$ may represent a set of patients being monitored at the ICU of an hospital. Each time series $E \in \mathcal{E}$ can be represented as a set of subsequences $E = \{e_1, e_2, \ldots, e_i, \ldots, e_{t-1}, e_t\}$, where $e_i$ represents the $i$-th subsequence. A subsequence denotes a tuple $e_i = (t_i, X_i, y_i)$, where $t_i$ denotes the time stamp that marks the beginning of the subsequence, $X_i \in \mathbb{X}$ represents the input (predictor) variables, which summarize the recent past dynamics of the time series; and $y_i \in \mathbb{Y}$ denotes the target variable, which is a binary value ($y_i \in \{0, 1\}, \forall\ i \in \{1, \ldots, t\}$) that represents whether or not there is an impending anomaly or event of interest in the near future of the respective time series. How near in the future is typically a domain-dependent parameter.

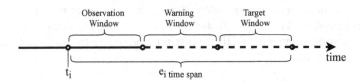

**Fig. 2.** Splitting a subsequence $e_i$ into observation window, warning window, and target window. The features $X_i$ are computed during the observation window, while the outcome $y_i$ is determined in the target window.

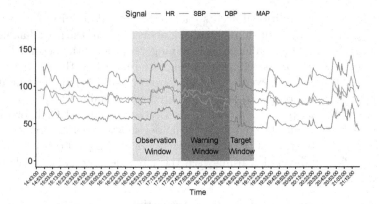

**Fig. 3.** The physiological signal of patients are monitored over time. Each subsequence, denoted by the shaded areas, is split in an OW, a WW, and a TW.

For each subsequence $e_i$ we construct the feature-target pair $(X_i, y_i)$ as follows. As illustrated in Fig. 2, each subsequence has three associated windows: (i) the target window (TW), which is used to determine the value of $y_i$; (ii) an observation window (OW), which is the period available for computing the values of $X_i$; and (iii) a warning window (WW), which is the lead time necessary for a prediction to be useful. For instance, in clinical medicine physicians need some time after an alarm is launched, for example to decide the most appropriate treatment. The sizes of these windows are domain-dependent. In principle, the problem will be easier as the OW is closer to the TW, that is, a smaller WW is required [12,20].

### 2.1   Event Prediction in ICUs

In this paper we focus on a particular instance of early anomaly detection problems: CHE prediction in ICUs, namely AHE and TE. Ghosh et al. [8] state that prolonged hypotension leads to a critical health damage, from cellular dysfunction to severe injuries in multiple organs. In turn, sustained tachycardia significantly increases the risk of stroke or cardiac arrest.

Patients assigned to the ICU are typically monitored constantly, with biosensors capturing several physiological signals, such as heart rate, or mean arterial blood pressure. This is illustrated in Fig. 3, where the data of a patient is depicted. A subsequence for CHE prediction is given as example in the shaded area of the graphic.

**Acute Hypotensive Episodes.** Hypotension episodes denote a prolonged drop in the blood pressure. More formally, AHE is an event defined as "a 30-min window having at least 90% of its mean arterial blood pressure (MAP) values below 60 mmHg [millimeters of mercury]" [12,19]. In this context, the target variable value is computed as follows:

$$y_i = \begin{cases} 1, & \text{if an AHE happens in } TW_i, \\ 0, & \text{otherwise.} \end{cases}$$

In other words, we consider that the $i$-th subsequence represents an anomaly if its TW represents an AHE (c.f. Fig. 3). Since AHEs are rare, the target vector $y$ is dominated by the negative class (i.e. $y = 0$), where a patient shows a normotensive status. For the target window of 30 min, we consider an OW and a WW of 60 min each. These values are typically used in the literature of AHE prediction models [8].

**Tachycardia Episodes.** Tachycardia denotes a high heart rate (HR). Generally, an HR over 100 beats per minute (bpm) under a resting state is considered as tachycardia. In order to consider a more robust definition for the purpose of discovering tachycardia episodes we follow a similar intuition to AHEs. We define TE as "a 30-min window having at least 90% of its HR values above 100 bpm". The respective target variable is computed as follows:

$$y_i = \begin{cases} 1, & \text{if an TE happens in } TW_i, \\ 0, & \text{otherwise.} \end{cases}$$

TEs are defined similarly to AHEs. Moreover, TEs also denote rare events since ICU patients usually show a HR below 100 bpm. We consider identical window sizes (OW, WW, TW) for both problems.

## 2.2 Discriminating Approaches to Early Anomaly Detection

Naturally, one of the most common approaches to solve the problem defined previously is to view it as a conditional probability estimation problem and use standard supervised learning classification methods [5, 19]. The idea is to build a model $f : \mathbb{X} \to \mathbb{Y}$, which can be used to predict the target values associated with unseen feature attributes. In other words, $f$ is a discriminating model that explicitly distinguishes normal activity from anomalous activity (c.f. left side Fig. 1).

Notwithstanding the widespread of this approach, early anomaly detection problems often comprise complex target variables whose definition is derived from the data. In such cases, it is possible to decompose the target variable into partial and less complex concepts, which may be easier to model. In this context, our working hypothesis is that we can leverage a layered learning approach to model these partial concepts, and obtain an overall better model for capturing the actual events of interest.

## 3 Layered Learning for Early Anomaly Detection

### 3.1 Layered Learning

Layered learning is designed for predictive tasks whose mapping from inputs to outputs is complex. In essence, layered learning consists in breaking a predictive task into several layers. The approach assumes that the problem addressed in each layer is simpler than the original one. As Stone and Veloso explain, "the key defining characteristic of layered learning is that each layer directly affects the learning of the next" [17]. This effect can occur in several ways. For example, by affecting the set of training examples, or by providing features used for learning the original concept.

## 3.2   Pre-conditional Events

The definition of an anomalous event in time series data is in many cases determined according to some rule derived from the data. As an example from the healthcare domain presented in the previous section, an AHE is defined as a percentage of numeric values within a time interval which are below some threshold (c.f. Sect. 2.1). TEs are defined in a similar manner. This type of approach for defining anomalous events is also common in other domains. For example in predictive maintenance [14], where numerical information from sensor readings is transformed into a class label which denotes whether or not an observation is anomalous. Or wind ramp detection, where a ramp event is a rare occurrence that denotes a large percentage change in wind power in a short time interval [6].

Since these anomalous events are defined according to the value of an underlying variable we can also define pre-conditional events: relaxed versions of the actual events of interest, but which are more frequent. A more precise definition can be given as follows. A pre-conditional event is an arbitrary but computable event that is expected to simultaneously occur with the main event taking place. If the main event occurs, the pre-conditional event must occur, but the latter can occur without the main event.

An example can be provided using the case study of AHE prediction. In Sect. 2.1, we defined the main event (AHE) as "a 30-min window having at least 90% of its mean arterial blood pressure (MAP) values below 60 mmHg". A possible pre-conditional event for this scenario could be "a 30-min window having at least **45**% of its mean arterial blood pressure (MAP) values below 60 mmHg". In summary, pre-conditional events should have the following two characteristics: (i) pre-conditional events should have a higher relative frequency than the main events; and (ii) pre-conditional events always happen when the main events happen. The inverse is not a necessary condition.

## 3.3   Our Approach

We can leverage the idea of pre-conditional events and use a layered learning strategy to tackle early anomaly detection problems in time series data. Our idea is to decompose the main predictive task into two layers, each denoting a predictive subtask. Pre-conditional events are modelled in the first layer, while the main events are modelled in the subsequent one.

The intuition behind this idea is given in Fig. 1. The figure presents two Venn diagrams for classes. Focusing on the left-hand side, the anomalies or main events (e.g. AHE) represent a small part of the data space. This is one of the issues that makes them difficult to model. In the typical classification approach main events are directly modelled with respect to normal activity.

Our idea is represented on the right-hand side. An initial pre-conditional concept is considered, which is more common than the main target concept, while also including it. The higher relative frequency of the pre-conditional events with respect to the main events helps to mitigate the problem of having an imbalanced distribution, which is the case in early anomaly detection tasks. This phenomenon can compromise the performance of learning algorithms [9]. In effect, we first model the pre-conditional events with respect to the normal activity. These pre-conditional events are, in principle, easier to learn relative to the main concept as they are more frequent and thus the classification algorithms will not suffer so much from an imbalanced distribution. Afterwards, the main target events are modelled with respect to the pre-conditional events, which is also a less imbalanced distribution than the original on the left diagram.

**Pre-conditional Events Sub-task.** Let $S$ denote a pre-conditional event. The target variable when modelling these events is defined as:

$$y_i^S = \begin{cases} 1 & if \ S \text{ happens,} \\ 0 & \text{otherwise.} \end{cases} \tag{1}$$

For this task a subsequence $e_i^S$ is a tuple $e_i^S = (t_i, X_i, y_i^S)$. The difference to the original set of subsequences $E$ is the target variable, which replaces $y$ with $y^S$. Finally, the goal of this first predictive task is to build a function $f^S$ that maps the input variables $X$ to the output $y^S$.

**Main Events Sub-task.** Provided that we solve the pre-conditional events sub-task, in order to predict impending main events the remaining problem is to find out whether or not, when $S$ happens, the main event also happens.

Let $\mathcal{F}$ be defined as the occurrence: "given $S$, there is an impending main event in the target window of the current subsequence". Effectively, the target variable for this task is defined as follows:

$$\text{Given } y^S = 1, \ y_i^{\mathcal{F}} = \begin{cases} 1 & \text{if a main event happens in } TW_i, \\ 0 & \text{otherwise.} \end{cases} \tag{2}$$

The target variable for this subtask ($y^{\mathcal{F}}$) is formalized in Eq. 2. Given that the class of $y^S$ is positive (which means that there is an impending pre-conditional event), the class of $y^{\mathcal{F}}$ is positive if a main event also happens in that same target window, or negative otherwise.

The goal of this second predictive task is to build a function $f^{\mathcal{F}}$, which maps $X$ to $y^{\mathcal{F}}$. Formally, a subsequence $e_i^{\mathcal{F}}$ is represented by $e_i^{\mathcal{F}} = (t_i, X_i, y_i^{\mathcal{F}})$. In this scenario however, the set of available subsequences $E$ is considerably less than in the pre-conditional sub-task because only the subsequences for which $y^S$ equals 1 are accounted for. Effectively, this aspect represents how the learning in the pre-conditional events sub-task affects the learning on the main events sub-task, i.e., by influencing the data examples used for training. In the main events sub-task, a predictive model is concerned with the distinction between pre-conditional events and main events. Essentially, it assumes that the distinction between normal activity and pre-conditional events is carried out by the previous layer. Given this independence, the training of the two layers can occur in parallel.

**Forecasting Impending Anomalies.** To make predictions about impending events of interest we combine the models $f^S$ with $f^F$ with a function $g : \mathbb{X} \times \mathbb{X} \to \mathbb{Y}$.

$$g(X_i) = f^S(X_i) \cdot f^F(X_i) \tag{3}$$

Essentially, according to Eq. 3 the function $g$ predicts that there is an impending main event in a given subsequence $e_i$ according to the multiplication of the outcome predicted by both $f^S$ and $f^F$.

### 3.4 Application of Layered Learning to CHE Prediction

As mentioned before (c.f. Sect. 2.1) an AHE is defined as a 30-min time period where 90% of the blood pressure values are below 60 mmHg. We propose to relax this threshold and define the pre-conditional event $S$ as follows. We define $S^{AHE}$ to represent "a 30-min window having at least **45%** of its mean arterial blood pressure values below 60 mmHg".

The event $S$ is consistent with the two above-mentioned characteristics: the frequency of $S$ across the database is considerably higher than an AHE – note that the blood pressure level can drop below 60 mmHg for some time period without being considered as an hypotensive episode. Consequently, the occurrence $S$ is simultaneous to the occurrence of an AHE (if 90% of the values are below 60 mmHg, so are 45%).

We apply the same reasoning to the TE prediction task. In Sect. 2.1, we defined a TE as "a 30-min window having at least 90% of its HR values above 100 bpm". In order to define $S^{TE}$ we again relaxate the percentage threshold as follows. $S^{TE}$ is defined as "a 30-min window having at least 45% of its HR values above 100 bpm". In both situations, the value of 45% was chosen arbitrarily. We attempted to make the pre-conditional events much more frequent relative to the main events. Nevertheless, this parameter can be optimized.

## 4     Empirical Evaluation

### 4.1     Case Study: MIMIC II

In the experiments we used the database Multi-parameter Intelligent Monitoring for Intensive Care (MIMIC) II [16], which is a benchmark for a number of predictive tasks in healthcare, including CHE prediction.

As inclusion criteria of patients and general database pre-processing steps, we follow Lee and Mark closely [12]. For example, the sampling frequency of the physiological data of each patient in the database is one minute. Moreover, the following physiological signals are collected: heart rate (HR), systolic blood pressure (SBP), diastolic blood pressure (DBP), and mean arterial blood pressure

(MAP). As described in Sect. 2.1, the TW size is 30 min. For each TW, there is a 60-min OW and a 60-min WW. For a comprehensive read regarding the data compilation we refer to the work by Lee and Mark [12]. Considering this setup, the number of patients is 1,072, leading to a data size of 1,975,936 subsequences. 71,035 of those events represent an AHE (about 3.5%). In turn, 13.6% of the subsequences represent a TE.

Regarding feature engineering, we follow previous work in the literature [11,19]. Using the observation window of each subsequence of each physiological signal, the feature engineering process was carried out using statistical, cross-correlation, and wavelet functions. The statistical metrics include skewness, kurtosis, slope, median, minimum, maximum, variance, mean, standard deviation, and inter-quartile range. For each observation window we also compute the cross-correlation of each pair of signals at lag 0. We also carry a wavelet transformation to capture the relative energies in different spectral bands.

## 4.2   Experimental Design

The experiments were designed to compare the proposed layered learning approach to state of the art methods for early anomaly detection. To estimate the predictive performance of each method we used a $5 \times 10$-fold cross-validation, in which folds are split by patient. To be more precise, in each iteration of the cross-validation procedure, one fold of the set of patients $\mathcal{E}$ is used for validation, another fold of different patients is used for testing, and the remaining patients are used for training the predictive model. The set of time series $\mathcal{E}$ only comprises a temporal dependency within each patient, and we assume the data across patients to be independent. In this context, the application of cross-validation in this setting is valid. Finally, the subsequences of the patients chosen for training are concatenated together to fit the predictive model.

The goal behind early anomaly detection problems is not to classify each subsequence as positive or negative [5]. Instead, the main goal is to detect, in a timely manner, when there is an impending anomalous event. In this context, we follow Weiss and Hirsh [20] regarding the evaluation metrics. Specifically, two measures are computed: Event Recall (ER), and Reduced Precision (RP). These two metrics follow the same intuition of the widely used Recall and Precision metrics, but are tailored for time-dependent data.

Let $T$ denote the total number of events of interest in a test data set, and let $\hat{T}_m$ represent the total number of those events correctly predicted by a model $m$. The ER for model $m$ is given by the following equation:

$$\mathrm{ER}_m = \frac{\hat{T}_m}{T} \tag{4}$$

ER differs from the classical recall metric because a single correct prediction within an observation window leading to an event is enough to consider that event correctly anticipated. As Fawcett and Provost put it, "alarming earlier may be more beneficial, but after the first alarm, a second alarm on the same event may add no value" [5].

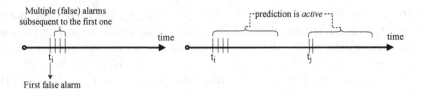

**Fig. 4.** Left: A sequence of consecutive false alarms – the first alarm is useful, but the subsequent ones may add no information; right: false alarms (denoted as vertical bars) over a time interval – there are 6 false positives, but only two discounted false positives.

The classical precision metric measures the percentage of positive predictions that are correct. Similarly to recall, in a time-dependent domain the classical precision may be misleading because multiple predictions on the same event are counted multiple times. This idea is intuited in Fig. 4 (left). This graphic shows a subsequence in which predictions are being produced over time. Starting from time $t_i$, four false alarms are triggered. Performance evaluation should take the first wrong prediction into account as a false positive. However, the subsequent false alarms (as shown in the left side of Fig. 4) are not meaningful since they add no information – assuming some action is taken after the first alarm.

RP overcomes this problem by considering a prediction to be *active* for some time period. Specifically, in this work we consider a time interval with the same size as the observation window. Notwithstanding, this is usually a domain dependent parameter. Effectively, the RP metric replaces the number of false positives with the number of discounted false positives – the number of non-overlapping observation periods associated with a false prediction. This idea is illustrated in Fig. 4 (right), where each vertical bar in the time line denotes an issued false alarm. There are a total of 6 false positives, but, if taking into account the time interval a prediction is active, there are only two discounted false positives (DFP). Finally, RP also considers the number of target events correctly identified ($\hat{T}_m$), instead of the number of correct predictions (true positives). In effect, RP for model $m$ is given by the following equation:

$$\mathrm{RP}_m = \frac{\hat{T}_m}{\hat{T}_m + \mathrm{DFP}_m} \tag{5}$$

### 4.3  Learning Algorithm and State of the Art Methods

In the experiments we tested different predictive models, namely a random forest, a support vector machine, a deep feed-forward neural network, and an extreme gradient boosting model [3]. We only show the results of the latter in these experiments, since it provides a better performance than the remaining methods for both AHE prediction and TE prediction.

The classifiers used in the experiments output a probability. The decision threshold is optimized following previous work in the literature of AHE prediction [12,19], which recommends selecting the threshold that maximizes the average of *classical* recall and specificity (true negative rate).

We compare the proposed layered learning approach (henceforth denoted as LL) with the following four methods:

CL a standard classification method that does not apply a layered learning approach and directly models the events of interest with respect to normal activity (c.f. Fig. 1) – in order to cope with the class imbalance problem this approach includes a random under-sampling of the majority class;

IF the Isolation Forest [13] method, which is a state of the art model-based approach to anomaly detection;

RG We include a regression-based alternative both for AHE prediction and TE prediction [12,15]. We apply a multi-step forecasting model to predict the future values of MAP (for AHEs) and HR (for TEs). Regarding the former, and following up on the definition of an AHE (Sect. 2.1), an alarm for an AHE is triggered if 90% of the forecasted values for the MAP variable are below 60 mmHg [15]. Likewise, an alarm for an TE is triggered if 90% of the forecasted values for the HR variable are above 100 bpm. The multi-step forecasting model follows a *direct* approach [18];

AH While there is an increasing number of machine learning applications in healthcare, many of the currently deployed systems still rely on simple *ad-hoc* rules to support the decision making process of professionals. Taking AHE prediction as an example, a simple rule is to trigger an alarm if the MAP of a patient drops below 60 mmHg in a given time step. A similar approach can be used for TE prediction, where an alarm is launched if the HR variable exceeds 100 bpm.

### 4.4 Results

Table 1 presents the average results, and respective standard deviation, for each method across the 50 folds (5 × 10-fold cross-validation). We analyse the significance of the results according to the Bayesian correlated t-test [2] (Figs. 5 and 6). In the Bayesian correlated t-test we consider the region of practical equivalence to be the interval $[-0.01, 0.01]$. In other words, two methods are practically equivalent if their difference in performance is below 0.01.

**Table 1.** Average of results for the CHE prediction problem across the 50 folds

| Method | AHE | | TE | |
|--------|-----|-----|-----|-----|
| | ER | RP | ER | RP |
| AH | $0.625 \pm 0.05$ | $0.129 \pm 0.02$ | $0.749 \pm 0.05$ | $\mathbf{0.204} \pm 0.02$ |
| CL | $0.794 \pm 0.05$ | $0.095 \pm 0.01$ | $0.909 \pm 0.03$ | $0.135 \pm 0.02$ |
| IF | $0.700 \pm 0.18$ | $0.035 \pm 0.01$ | $0.756 \pm 0.31$ | $0.051 \pm 0.01$ |
| LL | $\mathbf{0.830} \pm 0.05$ | $0.090 \pm 0.01$ | $\mathbf{0.925} \pm 0.02$ | $0.140 \pm 0.01$ |
| RG | $0.250 \pm 0.06$ | $\mathbf{0.205} \pm 0.04$ | $0.646 \pm 0.04$ | $0.195 \pm 0.02$ |

In terms of ER, on average the proposed method LL captures 83% of AHEs and 92.5% of TEs. These values are significantly better relative to the remaining

methods, including CL which is the typical approach to solve these predictive
tasks. Maximizing ER in this particular domain of application is important,
because the events of interest are disruptive. Regarding RP, overall the value
of this metric is generally low for all methods, which suggests an high number
of DFP. In relative terms, AH and RG show the best results on this metric. The
proposed approach (LL) shows a comparable RP with CL. Expectedly, there is
a trade-off between ER and RP: greater ER leads to lower RP, and vice-versa.
Notwithstanding, relative to CL, LL is able to significantly improve ER while
keeping a comparable (i.e., within the region of practical equivalence) RP. While
LL shows a significantly worse RP relative to AH and RG, it compensates with a
considerably better ER. In comparison with IF, LL is significantly better in both
metrics.

## 4.5   Discussion

In the experiments above we showed the competitiveness of the proposed method
for early anomaly detection in a case study from the healthcare domain. The
main challenge behind layered learning is the assumption that the task decomposition is a domain-dependent function. This can be regarded as an opportunity for domain experts to embed their domain expertise in predictive models. Notwithstanding, nowadays there is an increasing interest for end-to-end
automated machine learning technologies, and a manual decomposition can be
regarded as a bottleneck. In this context, future work includes the study of an
automated methodology for identifying or learning the pre-conditional events
from the data.

Although we focus on CHE prediction problems, our ideas for layered learning
can be generally applied to other early anomaly detection problems, for example
problems with complex targets, which can be decomposed into partial, simpler
targets. While the task decomposition is dependent on the domain, we describe
some guidelines which can facilitate its implementation.

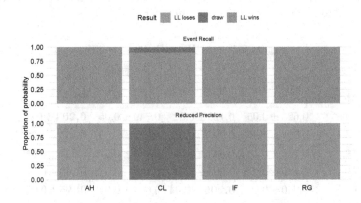

**Fig. 5.** Comparing CL with LL with a Bayesian correlated t-test for ER and RP metrics
(AHE prediction)

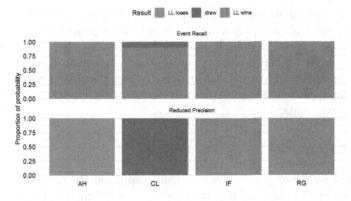

**Fig. 6.** Comparing CL with LL with a Bayesian correlated t-test for ER and RP metrics (TE prediction)

# 5    Related Work

## 5.1    Early Anomaly Detection and CHE Prediction

According to Fawcett and Provost [5], there are two classes of methods for activity monitoring: profiling methods, and discriminating methods. In a profiling strategy a model is constructed using only the normal activity of the data, without reference to abnormal cases. Consequently, an alarm is triggered if the current activity deviates significantly from the normal activity. On the other hand, a discriminating method constructs a model about anomalies with respect to the normal activity, handling the problem as a classification one. A system then uses a model to examine the time series and look for anomalies. We focus on the latter strategy, which is the one followed by the proposed layered learning method for early anomaly detection. Notwithstanding, we compare our approach to IF, which is a method that follows the profiling strategy.

Like other early anomaly detection problems, the typical approach to tackle CHE prediction problems is to use standard classification methods. This is the case of Lee and Mark, which use a feed-forward neural network as predictive algorithm [12]. Tsur et al. follow a similar approach, and also propose an enhanced feature extraction approach before applying an extreme gradient boosting algorithm [19]. In turn, Rocha et al. propose a regression approach (RG) by forecasting future values of blood pressure [15]. In their approach, alarms for impending AHE are launched according to a deterministic function which receives as input the numeric predictions. TE prediction also is a relevant task. For example, Forkan et al. [7] propose a predictive model for detecting several health conditions, including tachycardia and hypotension.

## 5.2    Layered Learning

Layered learning was proposed by Stone and Veloso, and was specifically designed for scenarios with a complex mapping from inputs to outputs [17].

In particular, they applied this approach to improve several processes in robotic soccer. Decroos et al. [4] apply a similar approach for predicting goal events in soccer matches. Instead of directly modelling such events, they first model goal attempts as what we call in this paper as a surrogate task. Layered learning is part of the family of hierarchical models. To our knowledge, this is the first time such an approach is applied to early anomaly detection using the idea of pre-conditional events.

## 6    Final Remarks

In this paper we developed a layered learning approach for the early detection of anomalies in time series data. We create an initial model that is designed to distinguish normal activity from a relaxed version of anomalous behavior (pre-conditional events). A subsequent model is created to distinguish such pre-conditional events from the actual events of interest.

We have focused on predicting critical health conditions in ICUs. Compared to standard classification, which is a common solution to this type of predictive tasks, the proposed model is able to capture significantly more anomalous events with a comparable number of false alarms.

Future work includes: (i) a better understanding of how layered learning works, how to tune its parameters; (ii) its application to tackle other early anomaly detection problems; (iii) automatic identification of pre-conditional events– we are studying the usage of subgroup discovery [10] to this effect.

**Acknowledgements.** Vitor Cerqueira is supported by a FCT PhD research grant (SFRH/BD/135705/2018).

## References

1. Bellazzi, R., Zupan, B.: Predictive data mining in clinical medicine: current issues and guidelines. Int. J. Med. Informatics **77**(2), 81–97 (2008)
2. Benavoli, A., Corani, G., Demšar, J., Zaffalon, M.: Time for a change: a tutorial for comparing multiple classifiers through bayesian analysis. J. Mach. Learn. Res. **18**(1), 2653–2688 (2017)
3. Chen, T., He, T., Benesty, M., Khotilovich, V., Tang, Y.: xgboost: eXtreme gradient boosting, 2017. R package version 0.6-4 (2015)
4. Decroos, T., Dzyuba, V., Van Haaren, J., Davis, J.: Predicting soccer highlights from spatio-temporal match event streams. In: AAAI, pp. 1302–1308 (2017)
5. Fawcett, T., Provost, F.: Activity monitoring: noticing interesting changes in behavior. In: Proceedings of the Fifth ACM SIGKDD International Conference on Knowledge Discovery and Data Mining, pp. 53–62. ACM (1999)
6. Ferreira, C., Gama, J., Matias, L., Botterud, A., Wang, J.: A survey on wind power ramp forecasting. Technical report, Argonne National Lab. (ANL), Argonne, IL (United States) (2011)
7. Forkan, A.R.M., Khalil, I., Atiquzzaman, M.: ViSiBiD: a learning model for early discovery and real-time prediction of severe clinical events using vital signs as big data. Comput. Netw. **113**, 244–257 (2017)

8. Ghosh, S., Feng, M., Nguyen, H., Li, J.: Hypotension risk prediction via sequential contrast patterns of ICU blood pressure. IEEE J. Biomed. Health Inform. **20**(5), 1416–1426 (2016)
9. He, H., Ma, Y.: Imbalanced Learning: Foundations, Algorithms, and Applications. Wiley, Hoboken (2013)
10. Lavrač, N., Kavšek, B., Flach, P., Todorovski, L.: Subgroup discovery with CN2-SD. J. Mach. Learn. Res. **5**, 153–188 (2004)
11. Lee, J., Mark, R.: A hypotensive episode predictor for intensive care based on heart rate and blood pressure time series. In: Computing in Cardiology, pp. 81–84. IEEE (2010)
12. Lee, J., Mark, R.G.: An investigation of patterns in hemodynamic data indicative of impending hypotension in intensive care. Biomed. Eng. Online **9**(1), 62 (2010)
13. Liu, F.T., Ting, K.M., Zhou, Z.H.: Isolation-based anomaly detection. ACM Trans. Knowl. Discov. Data (TKDD) **6**(1), 3 (2012)
14. Ribeiro, R.P., Pereira, P., Gama, J.: Sequential anomalies: a study in the railway industry. Mach. Learn. **105**(1), 127–153 (2016)
15. Rocha, T., Paredes, S., De Carvalho, P., Henriques, J.: Prediction of acute hypotensive episodes by means of neural network multi-models. Comput. Biol. Med. **41**(10), 881–890 (2011)
16. Saeed, M., Lieu, C., Raber, G., Mark, R.G.: MIMIC II: a massive temporal ICU patient database to support research in intelligent patient monitoring. In: Computers in Cardiology, pp. 641–644. IEEE (2002)
17. Stone, P., Veloso, M.: Layered learning. In: López de Mántaras, R., Plaza, E. (eds.) ECML 2000. LNCS (LNAI), vol. 1810, pp. 369–381. Springer, Heidelberg (2000). https://doi.org/10.1007/3-540-45164-1_38
18. Taieb, S.B., Bontempi, G., Atiya, A.F., Sorjamaa, A.: A review and comparison of strategies for multi-step ahead time series forecasting based on the NN5 forecasting competition. Expert Syst. Appl. **39**(8), 7067–7083 (2012)
19. Tsur, E., Last, M., Garcia, V.F., Udassin, R., Klein, M., Brotfain, E.: Hypotensive episode prediction in ICUs via observation window splitting. In: Brefeld, U., et al. (eds.) ECML PKDD 2018. LNCS (LNAI), vol. 11053, pp. 472–487. Springer, Cham (2019). https://doi.org/10.1007/978-3-030-10997-4_29
20. Weiss, G.M., Hirsh, H.: Learning to predict rare events in event sequences. In: KDD, pp. 359–363 (1998)

# Ensemble Clustering for Novelty Detection in Data Streams

Kemilly Dearo Garcia[1,2(✉)], Elaine Ribeiro de Faria[3], Cláudio Rebelo de Sá[1],
João Mendes-Moreira[4], Charu C. Aggarwal[5], André C. P. L. F. de Carvalho[2],
and Joost N. Kok[1]

[1] University of Twente, Enschede, The Netherlands
k.dearogarcia@utwente.nl
[2] University of São Paulo, São Paulo, Brazil
[3] Fed. University of Uberlandia, Uberlandia, Brazil
[4] LIAAD-INESC TEC, Faculty of Engineering, University of Porto, Porto, Portugal
[5] IBM T.J. Watson Research Center, Yorktown, USA

**Abstract.** In data streams new classes can appear over time due to
changes in the data statistical distribution. Consequently, models can
become outdated, which requires the use of incremental learning algorithms capable of detecting and learning the changes over time. However,
when a single classification model is used for novelty detection, there is
a risk that its bias may not be suitable for new data distributions. A
solution could be the combination of several models into an ensemble.
Besides, because models can only be updated when labeled data arrives,
we propose two unsupervised ensemble approaches: one combining clustering partitions using the same clustering technique; and other using
different clustering techniques. We compare the performance of the proposed methods with well known novelty detection algorithms. The methods were tested on datasets commonly used in the novelty detection literature. The experimental results show that proposed ensembles have
competitive performance for novelty detection in data streams.

**Keywords:** Novelty detection · Ensembles · Clustering · Data streams

## 1 Introduction

In many real world scenarios, data continuously arrives at a high rate in a non
stationary way, named *data streams*. As new data arrives, models previously
induced can become outdated [6], causing predictive loss. In addition, due to the
great amount of data generated, it is impossible to store it in the main memory,
requiring the elimination of previous outdated data and online processing of
incoming data [4]. In data streams, three types of changes can be found in the
literature: concept drift [6], recurring concepts [2] and novel concepts [4]. Concept
drift refers to changes in the statistical properties of the concept, such, i.e., a
change in the stochastic process generating the data [6]. Recurring concepts are

© Springer Nature Switzerland AG 2019
P. Kralj Novak et al. (Eds.): DS 2019, LNAI 11828, pp. 460–470, 2019.
https://doi.org/10.1007/978-3-030-33778-0_34

a special type of concept drift in which concepts that appeared in the past may recur in the future [2]. Novelty concepts are patterns that were not present during the training of a classification model [4], which appear in the data stream.

Novelty detection is a machine learning task based on the identification of new concepts [5]. Several state of the art approaches [12] consider novelty detection as a binary classification task, composed by *normal* and *abnormal* classes. However, more recent approaches address novelty detection as a multi-class classification task [4]. The *abnormal* classes can be also named as *not normal* [12], *anomaly* [10] or *novel/new* [4] classes. We follow the notation from [4]. In the latter, normal concepts are a set of classes used to train the classification model and novelty concepts are the *new classes* that emerge over time.

In this work, we propose an ensemble of clustering partitions for novelty detection in data streams. We consider one ensemble obtained by a combination of different hyperparameter setting of the CluStream algorithm [1], referred as Homogeneous ensemble Clustering for data Streams (HoCluS). We also consider another ensemble with different clustering techniques, referred as Heterogeneous ensemble Clustering for data Streams (HeCluS). Each clustering technique can independently create and update a pre-defined number of partitions as new data arrive. This approach allows the use of clustering techniques with different bias, in order to obtain more robust classification models. In order to compare the performance of the different approaches, we implemented the two proposed methods in MINAS (MultI-class learNing Algorithm for data Streams) [4], a single classifier novelty detection algorithm for data streams. We conducted a set of experiments using datasets commonly referred in the novelty detection literature.

This paper is organized as follows. In Sect. 2 we present related work on novelty detection in data streams. In Sect. 3 we describe the proposed approaches and how they are incorporated into the MINAS algorithm. Section 4 presents the experiments performed. Finally, we conclude and discuss future research in Sect. 5.

## 2   Related Work

Several machine learning approaches have addressed novelty detection in data streams. Following, we describe the principal approaches according to two aspects: (i) number of classification models and (ii) strategy to update the classification model.

Considering the first aspect, we can divide the existing approaches in: single classification model or ensemble of models. Most of the single classification approaches use a $k$NN classification model based on a clustering approach [12]. This type of model can forget old clusters, insert new clusters and update the existing ones. Even though it is computationally less costly to train and update a single classification model, it may not be the most suitable to all time periods of a stream.

In contrast, other works focus on ensemble models. Ensemble classification for novelty detection in data streams are usually formed by combining classification models induced by the same algorithm [11]. In most approaches, the update of the ensemble consists of replacing the worst accurate model by a new one, obtained from the last labeled data chunk. However a disadvantage is that the ensemble in some cases has to wait for a long period before the labels are known. During this waiting period the predictive performance of the ensemble could decrease drastically.

Considering the second, the update strategy, these strategies are: *supervised* or *unsupervised*; according to presence/absence of labeled instances. Supervised approaches assume that the true label of all instances will eventually be available to update the model. Some examples of models using supervised learning are decision trees [11] and $k$NN [8].

On the other hand, unsupervised learning approaches assume that the true label will not be available. Therefore, they need to update the classification model without external feedback [5]. In general, they use the $k$-means algorithm to extract clusters to represent the current classes. As a result, they have some limitations: find only hyper-spherical clusters, have a fixed number of clusters and are sensitive to outliers.

## 3   Ensemble Clustering for Data Streams

The idea of Ensemble Clustering for Data Streams is similar to the general concept of combining classification models to construct an ensemble [13]. For that reason, it can be used to find the most suitable partition for a dataset. In this work, because we are constructing an ensemble of clustering, we are dividing the process in: the generation of a set of partitions and their combination into an ensemble of clustering.

Formally a data stream $D_{tr}$ is a potentially infinite sequence of instances arriving in a time $tr$, $tr \in \{1, ..., \infty\}$. Where, each instance, $X$, contains $d$ dimensions denoted by $X_{tr} = (X^1, ..., X^d)$ and a target class $y_{tr}$. A data stream can be represented as [4,7]: $D_{tr} = \{(X_1, y_1), (X_2, y_2), ..., (X_{tr}, y_{tr})\}$.

Novelty detection in data streams can be divided in two phases: the *offline* and the *online phase*. Assuming that in the offline phase a dataset has $m$ classes. Then, $Y^{Nor} = \{y_1, y_2, ..., y_m\}$ represents the set of Normal Classes. These class labels and the corresponding data samples are used to build the initial classification model. When during the online phase a novel class with label $y_{m+1}$ emerges, a novelty detection approach needs to detect this new class (concept) as quickly as possible and update the classification model accordingly.

### 3.1   The MINAS Algorithm

To test the proposed approaches, we implemented HoClus and HeCluS into the algorithm MINAS. The algorithm MINAS, in the offline phase, has a single model built with labeled data from the *normal* classes. This phase happens only once at the initial stage. The dataset with the labeled instances is split into subsets of data, each one containing data from one class in $Y^{Nor}$. Then, a clustering algorithm is applied on each subset to create a partition for each class. A cluster $C_j$ is defined by a centroid $c_j$, a radius $r_j$ and a class label $y$. The radius of a cluster is the Euclidean distance between the centroid and the farthest data point in that cluster [11].

In the online phase the model calculates the Euclidean distance between each instance and the centroids of the clusters from the *normal* classes. If the smallest distance is less then the radius of the closest cluster, then the instance gets the label from that cluster. Otherwise, the instance is labeled as *Unknown* and stored in a buffer for future analysis. When the buffer is full of instances labeled as *Unknown*, a clustering algorithm is applied to obtain new clusters. A cluster is considered as *concept drift* if the distance between its centroid and the centroid from the nearest cluster from the *normal* class is bellow a given threshold. Otherwise, the cluster is considered a *novelty*. The instances that are not similar to any cluster, the outliers, are removed. Finally, the buffer is empty and this process is repeated every time the buffer is full.

## 3.2   Ensemble Clustering Applied to MINAS Algorithm

In this section we will explain how the two Ensemble Clustering for Data Streams were embedded in the MINAS algorithm. Both offline and online phase use two steps to build the ensemble of partitions: generation and combination. In the generation step, a user defined number of $P$ partitions from $N$ clustering techniques is generated, from the dataset $D_{tr}$. The output is an ensemble, $L_N$, containing $P \times N$ clustering partitions.

Given the ensemble $L_N$, we need to verify which clusters, from different partitions, are similar. For that the consensus function computes the Euclidean distance between the centroids of each cluster from the different partitions. If the distance between them is smaller than the sum of their radius, then those clusters will share the same label.

In the combination step, to use the partitions as an ensemble, it is necessary that similar clusters from different partitions have the same label. A cluster $C_j$ with centroid $c_j$ is similar to a cluster $C_k$ if the Euclidean distance between them is less than the sum of their radius: $EuclideanDistance(c_j, c_k) < (r_j + r_k)$.

In the offline phase, as in MINAS (See Sect. 3.1), the labeled instances are separated by labels in subsets. For each subset an Ensemble Clustering Generator is applied and $P$ partitions from $N$ clustering techniques are generated. In Fig. 1, the figures in gray represent MINAS algorithm. The ones with dashed lines indicate the parts that were adapted with the proposed method. Finally, the colored figures represent the steps of the Ensemble Clustering for Data Streams.

In the online phase, Fig. 2, the ensemble of clusters, blue diamond figure, using the majority vote of partitions, decides if a new instance is classified as *normal* or as *Unknown*. The instances classified as *Unknown* are stored in a buffer for future analysis. When the buffer reaches a given size $W$, new partitions are generated. This represented in Fig. 2 by circles/ellipses represent clusters and each color a label. After that, a consensus function, blue rectangle, is applied to combine all the clusters. The latter, will maintain only the clusters more similar

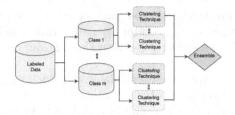

**Fig. 1.** Offline phase and ensemble clustering (adapted from [4])

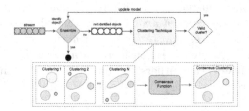

**Fig. 2.** Online phase and ensemble clustering (adapted from [4]) (Color figure online)

to other clusters from different partitions. Finally, these clusters are considered as novelties and then they are incorporated into the ensemble.

We define Homogeneous ensemble Clustering for data Streams (HoCluS) as an ensemble clustering obtained by the combination of $P$ partitions from the same algorithm. In this work we will use the algorithm for clustering in data streams CluStream [1]. CluStream is based on the $k$-means algorithm. Because of the random initialization phase of $k$-means, different partitions can be obtained. Because of that, an ensemble of CluStream partitions can be more robust for novelty detection than a single partition of CluStream.

We define a Heterogeneous Ensemble Clustering for data Streams (HeCluS) as the combination of $P$ partitions from $N$ different clustering techniques. In this work, our HeCluS has one model induced by each one of the following clustering algorithms for data streams: CluStream [1], DenStream [3] and ClusTree [9]. The main motivation to use DenStream, is because it is a stream clustering algorithm that is able to find clusters with arbitrary shape. Besides, it can also handle outliers [3]. On the other hand, we also use the ClusTree algorithm which also has a different bias from the other two. ClusTree builds clusters in a hierarchical data structure and can automatically set a number of clusters with arbitrary shape.

## 4   Experimental Setup and Results

We present in this section the experiments carried out for this study. We start by describing the datasets, then 2 we detail the experiment setup and finally we discuss the results.

The experiments were performed with synthetic and real datasets, both commonly used in novelty detection studies [1,3,5,11]. The synthetic datasets are: MOA, SynD, 1CDT, UG_2C_2D and Gear. The *MOA* dataset [4] has concept drift, appearance of new classes, recurrence and disappearance of existing classes. The clusters in this dataset are shaped as normally distributed hyper-spheres. The *SynD* [11] does not contain new classes, but does include concept drift. Finally, *1CDT, UG_2C_2D* and *Gear* are non stationary datasets[1]. In these datasets a single novelty occurs and concept drifts happen every 400 instances in 1CDT, 1000 instances in UG_2C_2D and 2000 instances in Gear. We also use two real datasets, *Forest Cover* and *KDD-CUP'99 NetWork Intrusion*[2]. The *KDD* dataset was used by [1] and [3].

We assume that the instances in the training data are the normal classes and new classes can appear during the online phase. In the offline phase, all methods are initialized with a batch of labeled data representing 10% of the data. We represent the results for each dataset with a confusion matrix (e.g. Table 1), which contains the percentage of: correctly classified classes (in gray), misclassified classes (in white), novelties detected (in gray) and *Unknown* instances (in white). The *Unknown* is the percentage of instances that the model was not able to classify. The sum of each column of the matrix should be 1. However, since we represent the average of 30 runs, the sum might not actually be 1. Whenever an instance is labelled as *Unknown*, it is considered as a classification error and counts as a *false negative*, which is a different approach adopted by [5]. For this reason, a high percentage of instances labelled as *Unknown*, will force the recall to be lower. This will make the comparison of the models more fair. The *F-Measure* is the weighted harmonic mean of precision and recall [5].

For datasets with more than two classes we used graphics to analyse the predictive performance of MINAS, HoCluS and HeCluS over time. We computed the F-Measure and *Unknown* every 10.000 instances. The algorithms CLAM, MINER and SAND, are not used in this analysis due difficulties to obtain the information necessary to calculate the F-Measure over time.

## 4.1 Results

We compare the predictive performance of HoCluS and HeCluS with the original MINAS and with three other supervised novelty detection methods: Miner [11], CLAM [2] and SAND [8]. For simplicity, we use the default hyper-parameters of the existing algorithms. In **Gear** dataset C1 and C2 are *normal* classes, both with concept drift. In Table 1 we can observe that the unsupervised method HeCluS has the highest *F-Measure*. Moreover, this was the only unsupervised method that does not misclassified C2 as a novelty. This can be due to the fact that HeCluS is able to build models with non-spherical clusters, which can better represent the classes of this dataset. We note that HoCluS and MINAS misclassify some instances as *novelties*, however this is less evident with HoCluS

---

[1] https://www.sites.google.com/site/nonstationaryarchive/.

[2] http://archive.ics.uci.edu/ml/index.php.

**Table 1.** Confusion matrix for gear dataset

| | Supervised | | | | | | Unsupervised | | | | | |
| | SAND | | CLAM | | MINER | | MINAS | | HoCluS | | HeCluS | |
| | C1 | C2 | C1 | C2 | C1 | C2 | C1 | C2 | C1 | C2 | C1 | C2 |
|---|---|---|---|---|---|---|---|---|---|---|---|---|
| C 1 | 1.00 | 0.00 | 1.00 | 0.25 | 0.97 | 0.04 | 0.82 | 0.03 | 0.88 | 0.06 | 0.94 | 0.06 |
| C 2 | 0.00 | 0.00 | 0.00 | 0.00 | 0.03 | 0.96 | 0.08 | 0.90 | 0.06 | 0.86 | 0.06 | 0.94 |
| Novelty | 0.00 | 0.57 | 0.00 | 0.00 | 0.00 | 0.00 | 0.10 | 0.07 | 0.02 | 0.03 | 0.00 | 0.00 |
| Unknown | 0.00 | 0.43 | 0.00 | 0.75 | 0.00 | 0.00 | 0.00 | 0.00 | 0.04 | 0.03 | 0.02 | 0.03 |
| F-Measure | 0.50 | | 0.80 | | 0.98 | | 0.92 | | 0.93 | | 0.97 | |

than with MINAS. In terms of the supervised methods, MINER has the best *F-Measure* with few misclassification. However, SAND and CLAM misclassify the majority data from C2 as *Unknown* or as C1. SAND classifies the drifts from C2 as *novelties* and CLAM does not learn from the *Unknown*.

In the **SynD** dataset C1 and C2 are the normal classes and both have concept drift. We observe, Table 2, that the methods show a similar behavior on the Gear dataset. In the group of unsupervised methods, HeCluS and HoCluS obtained a slightly better performance than MINAS. Considering the supervised methods, SAND, CLAM and MINER learn part of the concept drifts. SAND has *F-Measure* 0.77, however it has the highest percentage of *Unknown*. CLAM does not considered any instance as *Unknown*, but has more classification errors. MINER had the highest score and does not classify any instance as *Unknown*.

**Table 2.** Confusion matrix for Synd dataset

| | Supervised | | | | | | Unsupervised | | | | | |
| | SAND | | CLAM | | MINER | | MINAS | | HoCluS | | HeCluS | |
| | C1 | C2 | C1 | C2 | C1 | C2 | C1 | C2 | C1 | C2 | C1 | C2 |
|---|---|---|---|---|---|---|---|---|---|---|---|---|
| C 1 | 0.64 | 0.00 | 0.50 | 0.45 | 0.88 | 0.18 | 0.63 | 0.34 | 0.66 | 0.31 | 0.69 | 0.34 |
| C 2 | 0.00 | 0.60 | 0.50 | 0.55 | 0.12 | 0.82 | 0.37 | 0.66 | 0.30 | 0.70 | 0.32 | 0.65 |
| Novelty | 0.00 | 0.00 | 0.00 | 0.00 | 0.00 | 0.00 | 0.00 | 0.00 | 0.00 | 0.00 | 0.00 | 0.00 |
| Unknown | 0.36 | 0.40 | 0.00 | 0.00 | 0.00 | 0.00 | 0.01 | 0.01 | 0.01 | 0.01 | 0.00 | 0.00 |
| F-Measure | 0.77 | | 0.69 | | 0.92 | | 0.76 | | 0.78 | | 0.79 | |

The **1CDT** dataset, Table 3, has a normal class (C1) and a new class (C2) with concept drift. With this dataset we can evaluate the performance of the models with regard to novel concept detection in the online phase. We note that MINAS, HoCluS and HeCluS even during the online phase, they are not informed of the true class of the novelties detected. Because of that, even though they detect C2 as a novel class, their predictions are only represented as *novelty*. All models correctly classify most data from C1. On the other hand, in terms of the novel class, C2, they have different predictive performance. In terms of unsupervised approaches, HeCluS does not distinguish very well the normal concept, C1, from the novel concept, C2. On the other hand, HoCluS and MINAS, never misclassified C1 as C2. However, both have the highest percentage of *Unknown*. In terms of supervised approaches, SAND presents the lowest *F-Measure* because misclassifies most data from C2 as C1. SAND gives a confidence score for each model depending on their previous performance. Because of that the models representing the normal class have higher score than the novelty. This might explain the high percentage of misclassification of C2. The methods CLAM and MINER show high performance for C2, combining the percentage of correct classification and *novelty*.

**Table 3.** Confusion matrix for 1CDT dataset

| | Supervised | | | | | | Unsupervised | | | | | |
|---|---|---|---|---|---|---|---|---|---|---|---|---|
| | SAND | | CLAM | | MINER | | MINAS | | HoCluS | | HeCluS | |
| | C1 | C2 | C1 | C2 | C1 | C2 | C1 | C2 | C1 | C2 | C1 | C2 |
| C 1 | 1.00 | 0.84 | 1.00 | 0.00 | 1.00 | 0.00 | 0.94 | 0.00 | 0.94 | 0.00 | 1.00 | 0.20 |
| C 2 | 0.00 | 0.03 | 0.00 | 0.73 | 0.00 | 0.71 | 0.00 | 0.00 | 0.09 | 0.00 | 0.00 | 0.00 |
| Novelty | 0.00 | 0.11 | 0.00 | 0.27 | 0.00 | 0.27 | 0.04 | 0.87 | 0.00 | 0.87 | 0.00 | 0.75 |
| Unknown | 0.00 | 0.02 | 0.00 | 0.00 | 0.00 | 0.02 | 0.06 | 0.12 | 0.00 | 0.13 | 0.00 | 0.06 |
| F-Measure | 0.62 | | 0.92 | | 0.99 | | 0.95 | | 0.95 | | 0.93 | |

In the **Forest Cover** dataset, we have 3 normal classes and 5 novel classes. We can observe in Fig. 3(a) that HeCluS has the highest *F-Measure* over time and it is more stable. In terms of *Unknown* data, Fig. 3(b), HoCluS and MINAS have higher percentage of data classified as *Unknown* than HeCluS, specially from time 20 to 40. In terms of F-Measure, SAND has performance of 0.32 due confusion between the normal and novel classes. Possibly this happens because of the confidence factor, a hyperparamenter, that tends to privilege majority classes, which is the case for the classes from the *normal* class. CLAM had performance of 0.72 because it was able to detect most of the *novel* classes. Finally, MINER 0.59 detected a high number of class, but misclassify the novelties as the normal classes.

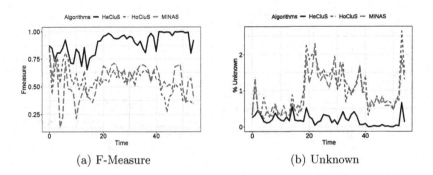

(a) F-Measure                    (b) Unknown

**Fig. 3.** F-measure and *Unknown* for forest cover dataset

In the **KDD** dataset we consider 18 *normal* classes and 5 *novel* classes. All methods start in time 0 with a low *F-Measure*, Fig. 4(a). During the time 15 to 30, HeCluS has better performance and is more stable, but after that period shows great instability. MINAS and HoCluS have less instability, but with low performance and periods with not correctly classify classes. HoCluS and MINAS have higher *Unknown* data, Fig. 4(b). However, the low *F-Measure* and *Unknown* show that MINAS and HoCluS misclassify the classes. In terms of the final *F-Measure*, SAND has performance of 0.27, CLAM 0.32 and MINER 0.30.

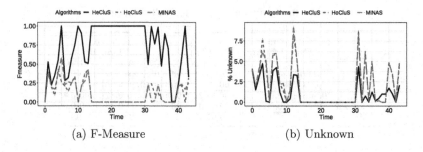

(a) F-Measure                    (b) Unknown

**Fig. 4.** F-measure and unknown for KDD99 dataset

The **MOA** dataset has 2 *normal* classes and 2 *novel* classes. The normal classes have concept drifts from time 0 to 90 and from time 30 to 55 they overlap. The first new class emerge at time 35 and second new class emerge after time 75. In Fig. 5(a), we can see that HeCluS is better and more stable than HoCluS and MINAS. In terms of performance, HoCluS is not different then MINAS. However, observing the peaks in Fig. 5(b), we can see that HeCluS updated less times that HoCluS and MINAS. In that case, MINAS needed to update more times than HoCluS. In terms of F-Measure, SAND had performance of 0.88, CLAM 0.44 and MINER 0.90.

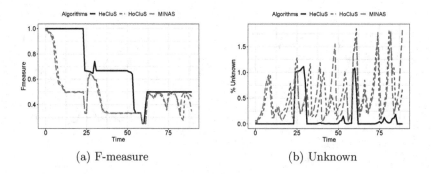

(a) F-measure                    (b) Unknown

**Fig. 5.** F-measure and Unknown for MOA dataset

The **UG_2C_2D** dataset has a *normal* class and a *new* class, both with concept drift and overlap. Analysing the unsupervised methods in Table 4, HeCluS classifies all data as C1 and does not learn the new class. HoCluS has slightly higher *F-Measure* than MINAS, both methods misclassify great percentage of C2 and C1, however they learn the new class. For the supervised methods, SAND and MINER also misclassify C2 as C1. MINER misclassify more C1 as C2 than the others methods. CLAM presents the best performance because does not misclassifies C2 as C1.

**Table 4.** Confusion matrix for UG_2C_2D dataset

| | Supervised | | | | | | Unsupervised | | | | | |
| | SAND | | CLAM | | MINER | | MINAS | | HoCluS | | HeCluS | |
| | C1 | C2 | C1 | C2 | C1 | C2 | C1 | C2 | C1 | C2 | C1 | C2 |
|---|---|---|---|---|---|---|---|---|---|---|---|---|
| C 1 | 0.71 | 0.50 | 1.00 | 0.02 | 0.50 | 0.58 | 0.42 | 0.45 | 0.45 | 0.50 | 1.00 | 1.00 |
| C 2 | 0.10 | 0.07 | 0.00 | 0.70 | 0.46 | 0.37 | 0.00 | 0.00 | 0.00 | 0.00 | 0.00 | 0.00 |
| Novelty | 0.28 | 0.44 | 0.00 | 0.00 | 0.01 | 0.01 | 0.51 | 0.46 | 0.50 | 0.43 | 0.00 | 0.00 |
| Unknown | 0.01 | 0.06 | 0.00 | 0.28 | 0.03 | 0.04 | 0.07 | 0.09 | 0.05 | 0.07 | 0.00 | 0.00 |
| F-Measure | 0.72 | | 0.91 | | 0.60 | | 0.43 | | 0.45 | | 0.50 | |

# 5    Conclusions

In this work we proposed the methods HeCluS and HoCluS for detection of novelties and concept drift in data streams. These ensembles combine several partitions from one or more clustering techniques. This allows the use of clustering techniques with different bias at the same time, in order to obtain more robust classification models.

In the experiment with the datasets with only concept drift, we demonstrated that HoCluS and HeCluS are competitive with state of the art supervised methods. Observing the performance of HoCluS and HeCluS over time, we conclude that HeCluS has lower percentage of *Unknown* instances. This shows that HeCluS takes more risks in the classification decision than HoCluS. Because of that HeCluS was better in most datasets. On the other hand, this behavior also gives the model less chances to be updated.

The use of ensembles with different clustering techniques is a promising strategy, because the inductive bias of each classification model can be more suitable for a given data stream or only for during certain periods in the same data stream. The experiments showed that during the online phase, the performance of all tested algorithms were affected by the changes in the data distribution.

# References

1. Aggarwal, C.C., Han, J., Wang, J., Yu, P.S.: A framework for clustering evolving data streams. In: VLDB 2003, Proceedings of 29th International Conference on Very Large Data Bases, 9–12 September 2003, Berlin, Germany, pp. 81–92 (2003)
2. Al-Khateeb, T., Masud, M.M., Khan, L., Aggarwal, C.C., Han, J., Thuraisingham, B.M.: Stream classification with recurring and novel class detection using class-based ensemble. In: 12th IEEE International Conference on Data Mining, ICDM 2012, Brussels, Belgium, 10–13 December 2012, pp. 31–40 (2012)

3. Cao, F., Ester, M., Qian, W., Zhou, A.: Density-based clustering over an evolving data stream with noise. In: Proceedings of the Sixth SIAM International Conference on Data Mining, 20–22 April 2006, Bethesda, MD, USA, pp. 328–339 (2006)
4. Faria, E.R., Gama, J., Carvalho, A.C.P.L.F.: Novelty detection algorithm for data streams multi-class problems. In: Proceedings of the 28th Annual ACM Symposium on Applied Computing, SAC 2013, Coimbra, Portugal, 18–22 March 2013, pp. 795–800 (2013)
5. Faria, E.R., Gonçalves, I.J.C.R., de Carvalho, A.C.P.L.F., Gama, J.: Novelty detection in data streams. Artif. Intell. Rev. 45(2), 235–269 (2016)
6. Gama, J., Zliobaite, I., Bifet, A., Pechenizkiy, M., Bouchachia, A.: A survey on concept drift adaptation. ACM Comput. Surv. 46(4), 44:1–44:37 (2014)
7. Garcia, K.D., de Carvalho, A.C.P.L.F., Mendes-Moreira, J.: A cluster-based prototype reduction for online classification. In: Yin, H., Camacho, D., Novais, P., Tallón-Ballesteros, A.J. (eds.) IDEAL 2018. LNCS, vol. 11314, pp. 603–610. Springer, Cham (2018). https://doi.org/10.1007/978-3-030-03493-1_63
8. Haque, A., Khan, L., Baron, M.: Semi supervised adaptive framework for classifying evolving data stream. In: Cao, T., Lim, E.-P., Zhou, Z.-H., Ho, T.-B., Cheung, D., Motoda, H. (eds.) PAKDD 2015. LNCS (LNAI), vol. 9078, pp. 383–394. Springer, Cham (2015). https://doi.org/10.1007/978-3-319-18032-8_30
9. Kranen, P., Assent, I., Baldauf, C., Seidl, T.: The clustree: indexing micro-clusters for anytime stream mining. Knowl. Inf. Syst. 29(2), 249–272 (2011)
10. Masud, M.M., et al.: Classification and adaptive novel class detection of feature-evolving data streams. IEEE Trans. Knowl. Data Eng. 25(7), 1484–1497 (2013)
11. Masud, M.M., Gao, J., Khan, L., Han, J., Thuraisingham, B.M.: Classification and novel class detection in concept-drifting data streams under time constraints. IEEE Trans. Knowl. Data Eng. 23(6), 859–874 (2011)
12. Spinosa, E.J., de Leon Ferreira de Carvalho, A.C.P., Gama, J.: Novelty detection with application to data streams. Intell. Data Anal. 13(3), 405–422 (2009)
13. Vega-Pons, S., Ruiz-Shulcloper, J.: A survey of clustering ensemble algorithms. IJPRAI 25(3), 337–372 (2011)

# Mining Patterns in Source Code Using Tree Mining Algorithms

Hoang Son Pham[1]([✉]), Siegfried Nijssen[1], Kim Mens[1], Dario Di Nucci[2], Tim Molderez[2], Coen De Roover[2], Johan Fabry[3], and Vadim Zaytsev[3]

[1] ICTEAM, UCLouvain, Louvain-la-Neuve, Belgium
hoang.s.pham@uclouvain.be
[2] Software Languages Lab, Vrije Universiteit Brussel, Brussels, Belgium
[3] Raincode Labs, Brussels, Belgium

**Abstract.** Discovering regularities in source code is of great interest to software engineers, both in academia and in industry, as regularities can provide useful information to help in a variety of tasks such as code comprehension, code refactoring, and fault localisation. However, traditional pattern mining algorithms often find too many patterns of little use and hence are not suitable for discovering useful regularities. In this paper we propose FREQTALS, a new algorithm for mining patterns in source code based on the FREQT tree mining algorithm. First, we introduce several constraints that effectively enable us to find more useful patterns; then, we show how to efficiently include them in FREQT. To illustrate the usefulness of the constraints we carried out a case study in collaboration with software engineers, where we identified a number of interesting patterns in a repository of Java code.

**Keywords:** Pattern mining · Frequent tree mining · Source code regularities

## 1 Introduction

During software development, many design and coding conventions get encoded in program source code, either explicitly or implicitly, through regularities such as API usage protocols, design patterns, coding idioms or conventions. Being able to discover such source code regularities in software systems is of great interest to software engineers, to help understanding, analysing, transforming, improving, maintaining or evolving a particular system, or to improve best practices for the development of new systems.

A data type of particular interest in the context of source code is the Abstract Syntax Tree (AST). ASTs capture not only the textual content, but also the structure of the code. However, to analyse these trees, algorithms capable of operating on tree structures are needed. *Frequent* tree mining algorithms [1,2] support this task. However, they typically find large numbers of patterns. This makes their output often useless in practice to software engineers. Several

© Springer Nature Switzerland AG 2019
P. Kralj Novak et al. (Eds.): DS 2019, LNAI 11828, pp. 471–480, 2019.
https://doi.org/10.1007/978-3-030-33778-0_35

solutions to this problem have been proposed, ranging from constraint-based pattern mining approaches and condensed representations, to statistically motivated pattern set mining approaches [3]. Among these, constraint-based data mining approaches are of particular interest, as they allow developers to specify easy to interpret constraints on the patterns to include in the output of the algorithm, and are guaranteed to find all patterns satisfying the constraints, contrary to sampling based approaches [4].

However, applying constraint-based data mining and condensed representations on the ASTs of software repositories is not straightforward. While frequent pattern mining has been studied extensively in the frequent tree mining literature, *constraint-based tree mining algorithms* did not receive a similar attention.

In this paper, we therefore propose a novel constraint-based tree mining algorithm, specifically designed for the analysis of software repositories. It combines two ideas: (i) *maximal frequent subtree mining* to ensure that a condensed representation of only large patterns is found, (ii) *constraint-based data mining*, in which additional constraints are imposed on the patterns to be found. Our approach is based on the addition of a number of novel constraints to the FREQT algorithm [5], combined with a new approach to find maximal subtrees.

In collaboration with software engineers we analysed in detail the quality of the patterns found. The results show (i) a significant reduction of the execution time and number of discovered patterns with respect to the original FREQT algorithm, (ii) that many of the discovered patterns highlight relevant code regularities, (iii) that some of the patterns found are significantly larger than the simpler coding idioms found in earlier studies [4].

The paper is organised as follows. Section 2 introduces frequent subtree mining and FREQT. Section 3 presents the key ideas of our solution, which is implemented by the FREQTALS algorithm described in Sect. 4. In Sect. 5 we conduct a case study to validate FREQTALS. Section 6 overviews related literature on pattern mining of semi-structured data and pattern mining applied to software. Section 7 concludes the paper.

## 2 Background

### 2.1 Frequent Subtree Mining

Abstract Syntax Trees are labelled, ordered, and rooted trees; they are produced by programming language parsers. We adopt a previously studied definition [1,5] for ordered trees $T = (V, E, \lambda, \Sigma)$; $V = \{1, 2, \ldots, n\}$ is the set of node identifiers; $E \subseteq V \times V$ is the set of edges; $\lambda : V \mapsto \Sigma$ is a function that associates labels to nodes of $V$; $\Sigma$ is the set of allowed labels. We assume the nodes are identified using contiguous integers listed in the order of a depth-first, left-to-right traversal of the tree; node $n$ is called the *rightmost node* of the tree. The shortest path from node 1 to node $n$ is its *rightmost path*.

Given two trees $T_1 = (V_1, E_1, \lambda_1, \Sigma)$ and $T_2 = (V_2, E_2, \lambda_2, \Sigma)$, $T_2$ is an induced subtree of $T_1$ ($T_1 \succeq T_2$) if there is an injective function $f : V_2 \mapsto V_1$ such that: (1) edges are preserved: for all $(v, v') \in E_2$: $(f(v), f(v')) \in E_1$; (2) labels are preserved: for all $v \in V_2$: $\lambda_2(v) = \lambda_1(f(v))$; (3) order is preserved: if $v_1 < v_2$ for a pair of nodes in $V_2$, then $f(v_1) < f(v_2)$.

The *support* of a pattern tree is the number of trees in a database in which the pattern occurs. *Frequent subtree mining* is the problem of finding all patterns of which the support is higher than a given *minimum support* threshold. These patterns satisfy the *minimum support constraint*, which we will refer to as constraint **C0** in this paper.

The number of frequent subtree patterns can become large, in particular for small minimum support thresholds. To deal with this issue, one solution is to mine *condensed representations* [1,2] and *maximal frequent subtrees*. Let $\mathcal{T}_c$ denote the set of all patterns that satisfy **C0**. We can define the problem of finding maximal frequent subtrees as the problem of finding all patterns not dominated by other patterns:

$$\max(\mathcal{T}_c) = \{T \in \mathcal{T}_c \mid \nexists T' \in \mathcal{T}_c : T' \succ T\}.$$

### 2.2  FREQT

FREQT was designed to mine frequent patterns from labelled ordered trees [5]. It searches for patterns using a depth-first search, where it grows patterns using *rightmost path extension*. The idea is to add new nodes only to the right of the rightmost path of a pattern. Hence, a pattern is created by adding its nodes in the order of a depth-first, left-to-right traversal.

| **Algorithm 1: FREQT** | **Algorithm 2: expand procedure** |
|---|---|
| 1 $\mathcal{FP} = \emptyset$ | 1 function expand($f$): |
| 2 $C \longleftarrow$ findLabels() | 2 $\quad$ $C \longleftarrow$ findCandidates($f$) |
| 3 prune($C$) | 3 $\quad$ prune ($C$) |
| 4 for *each* $c \in C$ do | 4 $\quad$ for *each* $c \in C$ do |
| 5 $\quad$ add ($\mathcal{FP}, c$) | 5 $\quad\quad$ add ($\mathcal{FP}, c$) |
| 6 $\quad$ expand($c$) | 6 $\quad\quad$ expand($c$) |
| 7 output($\mathcal{FP}$) | |

The structure of the depth-first search algorithm is described in Algorithms 1 and 2. By default FREQT only uses minimum support (**C0**) and *maximum size* (referred to as **C1**) constraints in the prune function. Anti-monotonic constraints are used to effectively reduce the size of the search space. A constraint is anti-monotonic iff for all pairs of patterns with $T_1 \succeq T_2$, if $T_2$ does not satisfy the constraint, then $T_1$ does not satisfy the constraint either. In addition, to avoid undesirable patterns from being added to the set $\mathcal{FP}$, a *minimum size* constraint (referred to as **C2**) is used in the add function.

Algorithms for finding only maximal frequent subtrees exist [1,2]; they usually reduce the search space by checking *all* extensions for *all* occurrences of a pattern. This is problematic for trees with a large fanout, such as ASTs.

# 3    Maximal Constraint-Based Frequent Subtree Mining

In this section, we will show how the AST representation allow us to impose additional meaningful constraints on patterns. There are two important ideas underlying these constraints.

First, given that programming languages are typically well-structured, parts of the data have a very predictable structure. Patterns that either reflect only this predictable structure, or that only include part of it, are not useful. For instance, by definition of the Java programming language, a node with label InfixExpression always has the same three children, leftOperand, operator and rightOperand. Clearly, a pattern composed of these four nodes is a frequent pattern but it is not interesting, as it is a consequence of the language and not the particular source code that is being mined. Patterns including the InfixExpression label, but not its three children, are not meaningful either, as by definition, these child nodes must be present.

Second, small fragments of ASTs are hard to interpret. In practice, we found that many software engineers find easier to interpret a code fragment if it is sufficiently large, allowing to put a pattern in its context. In terms of the patterns that we find, this means that our patterns need to satisfy minimum size criteria.

To find a small set of patterns which are sufficiently large and correctly reflect interesting program structures we propose the following constraints.

**Constraints on Labels.** To limit the number of patterns considered, the use of labels is a straightforward solution. The key benefit of label-based constraints is that they are easy to configure by software engineers. We consider the following constraints:

**C3.** Limit the set of labels allowed to occur in the root of patterns;
**C4.** Provide labels forbidden from occurring in the pattern;
**C5.** Limit the number of siblings in a pattern that can have the same label.

**Constraints on Leafs.** It is desirable that patterns not only represent the structure of the language, but also provide program-specific information. As such specific information can be found in the leaf nodes of ASTs in the database, we add this constraint:

**C6.** All leaf nodes in a pattern must have a label that is included in $\Sigma_{leaf}$, where $\Sigma_{leaf}$ is the set of labels that occur in the leafs of the trees in the database.

**Constraint on Obligatory Children.** Given a node, some of its children can be mandatory because they reflect a specific programming language construct (*e.g.*, the InfixExpression shown before). To avoid unnecessarily small patterns, we first need to characterise which labels are *structural*. We consider a label to be structural iff:

- in each of its occurrences, no two children have the same label;
- for all pairs of occurrences of the label, the order of the common child labels is the same.

For every label $a \in \Sigma$, we define its *obligatory child labels* $g(a)$ as the set of child labels common to all its occurrences. We added the follow constraint on obligatory child labels:

**C7.** Let $L$ be the structural labels in $\mathcal{D}$. For all nodes with a label $a \in L$, we require that its set of children includes nodes with all *obligatory* labels $g(a)$. Note that in combination with the leaf constraint, this constraint enforces that all structural nodes have leaf nodes as descendant.

**Maximal Subtree Mining.** We wish to ensure that the patterns found are as large as possible, while also being nonredundant. We propose to solve this using the following new idea: in a first phase, we find all patterns under the earlier mentioned constraints, combined with a maximum size constraint. This constraint limits the size of the search space. Subsequently, we grow the patterns found under these constrains as large as possible, and return the maximal patterns among these large patterns.

More formally, let $\mathcal{T}_{cm}$ be the set of subtrees identified using constraints C0–C7, including a maximum size constraint and let $\mathcal{T}_c$ be the set of trees that satisfies constraints C0–C7, without maximum size constraint. Let $occ(T)$ be the root occurrences of a particular tree. Let $C(T) = \{T' \in \mathcal{T}_c \mid occ(T) = occ(T')\}$. Then we wish to find: $\max(\cup_{T \in \mathcal{T}_{cm}} C(T))$.

# 4 The FREQTALS Tree Mining Algorithm

In this section, we present FREQTALS, an extension of the FREQT algorithm that takes into account the novel ideas described in Sect. 3.

**Constraints C3–C5** are all *anti-monotonic* in the following sense: if a tree does not satisfy the constraint, any supertree with the same root will not satisfy it either. To deal with such constraints we modified the **prune** function: extensions that do not satisfy the constraints are not added as candidates.

**Constraints C6 and C7** are harder to implement, as these constraints are not *anti-monotonic*. For instance, when we start the search process, the pattern will certainly not contain leaf labels; they will only be added later. However, FREQT grows patterns only by adding nodes to the right of the rightmost path. This enables us to deal with C6 and C7 as follows.

For **C6**, we know that the only leaf that we can still add a child to, is the rightmost node. Hence, if any leaf other than the rightmost node has a label not in the set of permitted leaf labels $\Sigma_{leaf}$, the search process will not be able to resolve this violation. Hence, in **prune** we add a condition that any tree in which a leaf other than the rightmost node has a label not in $\Sigma_{leaf}$ is pruned.

For **C7**, we exploit that obligatory child nodes of a structural node must occur in a specific order. Consider a structural label with three obligatory child labels $\sigma_1, \sigma_2, \sigma_3$. If a pattern already includes $\sigma_1$ and $\sigma_3$, the algorithm will not be able to add $\sigma_2$. In **prune** we add a condition so that any tree with such a situation is pruned.

---

**Algorithm 3:** FREQTALS algorithm

---

**input** : $\mathcal{D}$, constraints C0–C7.

**output:** $\mathcal{MP}$.

/* Step 1: mine subtrees under constraints C0-C7 using FREQT with modified Add and Prune functions                    */

1  $\mathcal{FP} = \text{FREQT}(\mathcal{D})$

/* Step 2: group the subtrees                                        */

2  $\mathcal{ROM} \longleftarrow \text{groupRootOccurrence}(\mathcal{FP})$

/* Step 3: find the maximal subtrees under constraints C2-C7    */

3  $\mathcal{MP} = \emptyset$

4  **for** *each* $r \in \mathcal{ROM}$ **do**

5      $\quad c \longleftarrow$ root label of $r$

6      $\quad \text{mineMaximalSubtrees}(c, r, \mathcal{MP})$

7  $\text{output}(\mathcal{MP})$

---

**Maximal Subtree Mining.** The most naïve algorithm to find maximal patterns would be one in which we grow a maximal pattern for each pattern satisfying the earlier constraints. While correct, this algorithm would also be time consuming. Algorithm 3 shows an outline of FREQTALS, which solves the problem more efficiently, while finding the same set of patterns. It has three phases:

1. search frequent subtrees under constraints C0–C7;
2. group frequent subtrees by root occurrences;
3. for each set of root occurrences identified, run a search process (without C0 and C1) to identify the maximal subtrees having these root occurrences.

Delving into more detail, in Line 1 we call the FREQT algorithm, using the modified **add** and **prune** functions. Furthermore, we add an optimisation so that any tree having a frequent extension, will not be put in $\mathcal{FP}$.

In Line 2 we group the root occurrences. Essentially, for all frequent patterns found, we first compute the set: $\mathcal{RO} = \{occ(T) \mid T \in \mathcal{FP}\}$. Note that multiple trees in $\mathcal{FP}$ may have the same $occ(T)$. Hence, this set is smaller than the original set of patterns. Subsequently, we only keep those sets of root identifiers that are minimal: $\mathcal{ROM} = \{r \in \mathcal{RO} \mid \nexists r' \in \mathcal{RO} : r' \subset r\}$. This optimisation does not affect our results. The key idea is that a pattern appearing in the larger set of occurrences, will also appear in the smaller set of occurrences.

Subsequently, in line 6 we start a search for maximal patterns for each set of root occurrences $r \in \mathcal{ROM}$. Here, for reasons of simplicity we made the choice to use a modified version of FREQT:

- we start the search with the root label appearing in the root occurrences $r$;
- the root occurrences considered during the search are only those in $r$, even if the root label has more occurrences in the original data;
- instead of using the minimum support constraint, we impose the constraint that the patterns searched for appear in all the given root occurrences;
- we do not apply a maximum size constraint;

**Table 1.** FREQTALS configuration for CheckStyle

| Constraint | Variable | Value |
|---|---|---|
| C0 | Minimum Support Threshold | 5 |
| C1 | Maximum # of Leaves | 4 |
| C2 | Minimum # of Leaves | 2 |
| C3 | Root Labels | TypeDeclaration, Block |
| C4 | Black List Labels | Javadoc, Modifiers, Annotations, ... |
| C5 | Maximum # of Similar Siblings | 10 |

– for each pattern that is generated, we check whether it should be included in $\mathcal{MP}$ and we update $\mathcal{MP}$ accordingly.

Note that patterns considered by `mineMaximalSubtrees` may have more occurrences in the original data. This is not harmful, as any such pattern will still be maximal and frequent. The key idea is that running FREQT on a smaller set of root occurrences, with a constraint that does not allow to lose any root occurrence, makes the search more efficient.

## 5    Empirical Evaluation

To evaluate FREQTALS, we carried out an empirical study on source code written in Java. We analysed the results from a qualitative (Sect. 5.1) and a quantitative (Sect. 5.2) point of view. More specifically, we analyse CHECKSTYLE, a well-documented static code analysis tool for Java that was selected from the the Qualitas Corpus [6].

Table 1 reports how we configured the algorithm for our evaluation. A minimum support threshold of 5 was chosen as for lower values the number of patterns exploded. We also focused only on AST sub-trees with root nodes of type `TypeDeclaration` (*i.e.,* a Java method definition) or `Block` (*i.e.,* a Java method body), because we were interested in the program logic.

### 5.1    Qualitative Analysis

The main purpose of our qualitative analysis is to determine whether the patterns identified by our algorithm are indeed useful.

With the given configuration, FREQTALS found 147 patterns that we manually analysed. To illustrate the characteristics of the patterns mined by FREQTALS, below we provide a detailed analysis of some of the patterns shown in Fig. 1. The CHECKSTYLE tool implements several design patterns such as the Visitor. Thus, some combinations of abstract methods are reused among many different classes (*e.g.,* `getDefaultTokens()` and `visitToken()`) and it is not surprising that our algorithm discovers many patterns with this pair of methods. Overall, 83 out of 147 mined patterns contained this pair. Pattern 34 shows

```
public final class ReturnCountCheck extends AbstractFormatCheck {
    ...
    @Overwrite
    public int[] getDefaultTokens(){
        return new int[] {
            TokenTypes.CTOR_DEF,
            TokenTypes.METHOD_DEF,
            TokenTypes.LITERAL_RETURN,
        };
    }
    ...
    @Overwrite
    public int[] getRequiredTokens() { … }
    ...
    @Overwrite
    public void visitToken(DetailAST aAST){
        switch (aAST.getType()) {
            case TokenTypes.CTOR_DEF:
            case TokenTypes.METHOD_DEF:
            ...
            break;
            ...
            default:
            ...
        }
    }
    ...
    @Overwrite
    public void leaveToken(DetailAST aAST) {
        switch (aAST.getType()) {
            case TokenTypes.CTOR_DEF:
            case TokenTypes.METHOD_DEF:
            ...
            default:
            ...
        }
    }
    ...
}
```
**Pattern 34:** An instance of Checkstyle's Visitor design pattern

```
private void visitMethod(final DetailAST aMethod)
    ...
    DetailAST child = objBlock.getFirstChild();
    while (child != null) {
        if (child.getType() == TokenTypes.METHOD_DEF {
            ... }
        child = child.getNextSibling();
    }
}
```
**Pattern 27:** AST traversal.

```
@Override
public void leaveToken(DetailAST aAST) {
    switch(aAST.getType()) {
        case TokenTypes.OBJBLOCK:
        case TokenTypes.SLIST:
        case TokenTypes.LITERAL_FOR:
        ...
    }
}
```
**Pattern 9:** Check for different blocks.

```
public int[] getDefaultTokens(){
    return new int[] {
        TokenTypes.ASSIGN,        // '='
        TokenTypes.DIV_ASSIGN,    // "/="
        TokenTypes.PLUS_ASSIGN,   // "+="
        ...
    };
}
```
**Pattern 18:** Method structure.

```
private boolean checkParams(DetailAST aMethod){
    ...
    if ((aAST.getType() == TokenTypes.VARIABLE_DEF) ||
        (aAST.getType() == TokenTypes.PARAMETER_DEF))
    {
        ...
    }
}
```
**Pattern 140:** IF Statement.

**Fig. 1.** Examples of patterns found in CHECKSTYLE

an example of such a design pattern instance that contains 4 reused methods: getDefaultTokens, getRequiredTokens, visitToken, and leaveToken. Pattern 27 shows a recurrent code snippet that checks every node of a given AST. Pattern 9 is an interesting example of a code structure that checks for different types of AST objects. This structure is quite frequent in CHECKSTYLE since it allows developers to customise the framework to write their own kinds of source code checks. Pattern 140 is also a typical code idiom recurring in CHECKSTYLE.

## 5.2  Quantitative Analysis

We limit our quantitative comparison between FREQT and FREQTALS to CHECKSTYLE, the dataset already considered in Sect. 5.1. We set C0 to 8, while keeping the same values, shown in Table 1, for the other settings. It is worth noting that for a more fair comparison, FREQT was modified to add a constraint

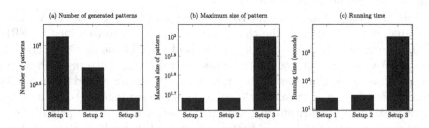

**Fig. 2.** Comparisons of three setups

on the minimum and maximum number of leaves. FREQTALS discovered 1,288 frequent patterns in 23 s, while the original FREQT did not finish within the search budget (*i.e.,* 60 min); it found 717,859 patterns within these 60 min.

To evaluate the different steps of FREQTALS, we executed it in three setups: the first applies constraints C0–C7; the second filters the results obtained by the first to keep only the maximal frequent patterns; the third applies all the constraints (except C0 and C1) combined with maximal subtree mining. We ran these three setups on entire CHECKSTYLE project. Similar to previous experiment, C0 was set to 8, and other settings were kept the same as shown in Table 1. Figure 2 shows the results. Note that to compare the results more easily we used a logarithmic scale for the plots in this Figure. In the first plot, we observe that the number of patterns discovered by the third setup is much smaller, as intended. Similarly, in the second plot, we see that the maximum size of the patterns mined by the third setup is larger, as desired. Nevertheless, there is no free lunch, as in the third plot we can observe that the third setup is more time consuming.

# 6    Related Work

This section discusses related work concerning pattern mining of (i) general semi-structured data and (ii) source code regularities and idioms.

**Pattern Mining of Semi-structured Data.** Extensive, but rather old literature exists on frequent tree mining algorithms [1,2]. Such algorithms can be categorised according to their input data, type of output patterns, and the approach taken for mapping patterns to data. Only algorithms designed for ordered, rooted trees, using an induced subtree relation are relevant to this work. The most well-known such algorithm is FREQT [5]. A benefit of FREQT is that it is a conceptually simple algorithm in which it is easy to add constraints. However, a major problem of frequent tree mining algorithms is that the number of output patterns is often very large. To tackle this problem, maximal frequent tree mining algorithms, i.e., CMTreeMiner [1], were developed. However, none of these algorithms operate on ordered trees. In recent years few new algorithms have been proposed, due to a lack of applications of such algorithms. Our work addresses this weakness by showing how pattern mining algorithms can indeed find useful patterns, as validated by software engineers. A notable exception is an algorithm that operates on attributed trees [7]. Our trees are not attributed, and hence we could not apply this algorithm.

**Mining Software Patterns.** There is an extensive literature on applying mining algorithms to software artefacts in general. Early examples include applications of formal concept analysis [8] and of association rule mining [9] for discovering design regularities. Narrowing down to the discovery of source code regularities, Allamanis *et al.* [4] describe an approach that mines for code idioms in a corpus of idiomatic code using non-parametric Bayesian methods. Similar approaches, like Bhatia *et al.* [10] mine for idioms using recurrent neural networks, aiming to correct incorrect uses of coding idioms. An advantage of FREQTALS is that the criteria used to include patterns in the output of the algorithm remain easy to understand, even for experts without background in statistics.

## 7    Conclusion and Future Work

In this paper, we proposed the FREQTALS algorithm, an extension of FREQT that combines *maximal frequent subtree mining* and *constraint-based data mining* to mine structural source code regularities. Experimental results show that (i) a significant reduction of the execution time and the number of discovered patterns with respect to the original frequent tree mining algorithm, (ii) that many of the discovered patterns highlight relevant code regularities, (iii) that some of the patterns found are significantly larger than expected. However, choosing appropriate configurations for a programming language is a difficult task. We envision to replicate our empirical evaluation on a larger set of systems and to define new guidelines to help software engineers in configuring our algorithm.

**Acknowledgments.** This work was conducted in the context of an industry-university research project between UCLouvain, Vrije Universiteit Brussel and Raincode Labs, funded by the Belgian Innoviris TeamUp project INTiMALS (2017-TEAM-UP-7).

## References

1. Chi, Y., Muntz, R.R., Nijssen, S., Kok, J.N.: Frequent subtree mining-an overview. Fundamenta Informaticae **66**(1–2), 161–198 (2005)
2. Jiménez, A., Berzal, F., Talavera, J.C.C.: Frequent tree pattern mining: a survey. Intell. Data Anal. **14**(6), 603–622 (2010)
3. Aggarwal, C.C., Han, J. (eds.): Frequent Pattern Mining. Springer, Cham (2014). https://doi.org/10.1007/978-3-319-07821-2
4. Allamanis, M., Sutton, C.: Mining idioms from source code. In: Proceedings of the 22nd ACM SIGSOFT International Symposium on Foundations of Software Engineering, pp. 472–483. ACM (2014)
5. Asai, T., Abe, K., Kawasoe, S., Sakamoto, H., Arimura, H., Arikawa, S.: Efficient substructure discovery from large semi-structured data. IEICE Trans. Inf. Syst. **87**(12), 2754–2763 (2004)
6. Tempero, E., et al.: The qualitas corpus: a curated collection of java code for empirical studies. In: 2010 17th AsiaPacific Software Engineering Conference, pp. 336–345. IEEE (2010)
7. Pasquier, C., Sanhes, J., Flouvat, F., Selmaoui-Folcher, N.: Frequent pattern mining in attributed trees: algorithms and applications. Knowl. Inf. Syst. **46**(3), 491–514 (2016)
8. Mens, K., Tourwé, T.: Delving source code with formal concept analysis. Comput. Lang. Syst. Struct. **31**(3–4), 183–197 (2005)
9. Lozano, A., Kellens, A., Mens, K., Arevalo, G.: Mining source code for structural regularities. In: Proceedings of the 2010 17th Working Conference on Reverse Engineering, pp. 22–31. IEEE Computer Society (2010)
10. Bhatia, S., Singh, R.: Automated correction for syntax errors in programming assignments using recurrent neural networks. arXiv preprint arXiv:1603.06129 (2016)

# KnowBots: Discovering Relevant Patterns in Chatbot Dialogues

Adriano Rivolli[1,2]([✉]), Catarina Amaral[2,4], Luís Guardão[2],
Cláudio Rebelo de Sá[3], and Carlos Soares[4]

[1] ICMC-USP/UTFPR, Cornélio Procópio, Brazil
rivolli@utfpr.edu.br
[2] INESC TEC, Porto, Portugal
[3] University of Twente, Enschede, The Netherlands
[4] Faculty of Engineering, University of Porto, Porto, Portugal

**Abstract.** Chatbots have been used in business contexts as a new way of communicating with customers. They use natural language to interact with the customers, whether while offering products and services, or in the support of a specific task. In this context, an important and challenging task is to assess the effectiveness of the machine-to-human interaction, according to business' goals. Although several analytic tools have been proposed to analyze the user interactions with chatbot systems, to the best of our knowledge they do not consider user-defined criteria, focusing on metrics of engagement and retention of the system as a whole. For this reason, we propose the KnowBots tool, which can be used to discover relevant patterns in the dialogues of chatbots, by considering specific business goals. Given the non-trivial structure of dialogues and the possibly large number of conversational records, we combined sequential pattern mining and subgroup discovery techniques to identify patterns of usage. Moreover, a friendly user-interface was developed to present the results and to allow their detailed analysis. Thus, it may serve as an alternative decision support tool for business or any entity that makes use of this type of interactions with their clients.

**Keywords:** Chatbot analytics · Chatbot analysis · Logs analysis · Sequence mining · Subgroup discovery

## 1 Introduction

Chatbots have been used in a variety of contexts by providing a natural language interface with an increasingly sophisticated design [14]. Their use in business contexts, as a way of communicating with customers, is becoming more common nowadays [15]. They have been used to address several tasks, like assistance in banking [1], customer service [3], educational tutoring [11,13], language learning [9] and online sales [10], to name a few.

Regarding the development of chatbot systems, the analytics dimension aims to monitor chatbot usage [13]. Developers can create their custom control-panel

© Springer Nature Switzerland AG 2019
P. Kralj Novak et al. (Eds.): DS 2019, LNAI 11828, pp. 481–492, 2019.
https://doi.org/10.1007/978-3-030-33778-0_36

or use a generic analytic tool that tracks the users' interactions and get metrics of them. Thus, analytics tools are valuable instruments for assessing the quality of the chatbot system and, ultimately, users' behavior.

Although many chatbot analytics tools have been developed, they focus on metrics of engagement and retention of the system as a whole. Their use can help chatbot maintainers to understand the behavior of users as well as to discover bottlenecks in the system. However, they cannot be used to explore behaviors in terms of goals, business criteria and unusual patterns, which are the aim of the KnowBots tool.

A category of chatbot systems uses rules to guide the conversation flow. Thus, business criteria can be defined in terms of reaching specific goals described by particular rule(s), whereas unusual patterns are characterized by usage patterns that deviate from the others regarding the attainment of business goals. It is performed by combining sequential pattern mining [6] and subgroup discovery [8] techniques. The former identifies the frequent subsequences and the latter filters the most relevant of them by a quality measure, which is defined according to business' interests. KnowBots also has a friendly user-interface to present the results and to allow their detailed analysis.

The KnowBots is the main contribution of this paper. To the best of our knowledge, the analytics dimension of the chatbots have not been explored scientifically, which is justified for the lack of references on this matter. On the other hand, many commercial tools are available to support the analysis of the chatbot interactions. They explore concepts from data mining, machine learning and information visualization domains.

The remaining sections of the paper are organized as follows: Sect. 2 formally defines the main concepts used in this work. Section 3 summarizes the main analytics tool available currently. Section 4 details the KnowBots tools, presenting their architecture and main features. Section 5 presents the exploratory analysis conducted, describing and discussing the obtained results. The paper ends with Sect. 6, that summarizes the relevant findings and future work directions.

## 2  Background

This section briefly presents the concepts that this study is based on. It covers chatbot systems, sequential pattern mining and subgroup discovery.

### 2.1  Chatbot System

Chatbot systems are computer programs designed to use natural language to interact with users, simulating a human conversation [15]. With the popularization of instant messages services and smartphones, chatbot systems started to be explored together with them [13]. Thereby, chatbot has received a lot of attention as a research topic, given the growing number of scientific publications about the subject. Also, it has been used as a business solution, whether in the prospection of new customers [2,10], in the service of these [1,3] or as internal services for employees [16,17].

A chatbot system can be designed using programed script and/or Natural Language Processing (NLP) approach [13]. The former follows a rule-based paradigm, thus it has a limited conversational scope. The latter uses artificial intelligence concepts to simulate a human-based behavior, which supposedly covers a broad conversational scope.

The KnowBots supports the analysis of the rule-based chatbot interaction, which can be defined as a set of state machines. For this work's purpose, a chatbot system is a triple $(S, r_1, \delta)$ where $S = \{r_1, \ldots, r_n\}$ is a set of $n$ rules $r_i$; $r_1$ is the starting point of the chat; and, $\delta = \{(r_1, r_2), \ldots, (r_i, r_j) \mid r_i, r_j \in S\}$ is a set of connected pairs of rules, which define the paths of conversation. Figure 1 illustrates the representation of a simple chatbot rules, they define the input, output and decision points. In this example, $S = \{r_1, \ldots, r_9\}$ and $\delta = \{(r_1, r_2), (r_2, r_3), (r_3, r_4), (r_4, r_5), (r_5, r_6), (r_5, r_8), (r_6, r_7), (r_8, r_9), (r_9, r_7)\}$. Business' goals may be defined by the nodes $r_3, r_4$ and $r_9$, for instance.

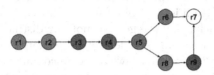

**Fig. 1.** Illustrative representation of chatbot rules.

## 2.2 Sequential Pattern Mining

Sequential pattern mining is a data mining field that aims to analyze frequent subsequences in a database of sequences [12]. A sequence $\alpha = \langle \alpha_1 \rightarrow \alpha_2 \ldots \rightarrow \alpha_q \rangle$ is an ordered set of events, where each event $\alpha_i$ is a non-empty and non-sorted collection of items $(i_1, \ldots, i_d)$. A subsequence of $\alpha$ is a sequence $\beta = \langle \beta_1 \rightarrow \beta_2 \ldots \rightarrow \beta_k \rangle$, such that there are integers $i_1 < i_2 < \ldots < i_k$ in which $\beta_1 \subseteq \alpha_{i_1}, \beta_2 \subseteq \alpha_{i_2}, \ldots, \beta_k \subseteq \alpha_{i_k}$. A subsequence observed repeatedly in the set of sequences with a minimum support threshold is a frequent subsequence, here named pattern.

The representation of a chatbot session, a conversation between a user and the bot, can be defined in terms of the rules triggered during the chat. Without loss of generality, an event is a single rule $\alpha_i \in S$ and a sequence is a chain of rules. For instance, $\alpha = \langle r_1 \rightarrow r_2 \rightarrow r_3 \rightarrow r_4 \rangle$ is a valid sequence for the chatbot system that uses the rules defined in Fig. 1. In this case, the user left the conversation without completing the interaction with the bot, considering that the rule $r_7$ defines the endpoint of the system.

Over the years, many sequential pattern mining algorithms have been developed. They produce the same outputs but differ in terms of search strategy and data representation, which impact their computational performance [5]. Thus, we arbitrarily choose the CM-SPAN algorithm [6] that is able to customize the minimum and maximum length of the subsequences; mandatory items; and, the size limit of the gap between events.

## 2.3  Subgroup Discovery

Subgroup discovery algorithms extract interesting relationships between objects considering a particular property or variable [8]. Patterns represent subgroups of the population that have some characteristics in common but differ from the rest when the distribution of a target of interest is observed [4].

For such, it uses quality measures to extract and evaluate the patterns. They can capture the complexity, generality, precision and interest of the subgroups [8]. Specific quality measures can be used to explore particular characteristics of the task, like the chatbot sequences associated with the business criteria, for instance. In this context, business criteria can be defined in terms of the achievement of some rules during the conversation. For instance, hypothetically assuming that in Fig. 1 the rule $r_9$ collects the user's e-mail. By achieving this rule during the conversation, a user meets a business goal.

When the set of rules is complex enough to have many paths of conversation, some paths can be more deterministic to the achievement of the goal than others. An uncommon pattern is a set of events (in this case, a conversation path), whose probability of achieving the goals is notably distinct from the other patterns. Given that, the users' answers determine the path of conversation and the chatbot design may influence the answers, the uncommon patterns are valuable information to the development team. The uncommon patterns can reveal users' behaviors that were not expected when the system was developed, for instance.

## 3  Chatbot Analytic Tools

The scientific literature concerning chatbot analytics is still incipient. A systematic research was conducted using four academic digital libraries: ACM Digital Library, IEEE Explore, Science Direct and Scopus; the query was constructed using the keywords: *chatbot + analytics, chatbot + "log analysis", chatbot + "log visualization"*; only a few and unrelated studies were obtained.

On the other hand, there are commercial tools that provide support for the analysis of chatbots dialogues. Table 1 presents such tools and summarizes their main features. The features include the ability to reproduce past dialogues (C); a dashboard with usage metrics (D); analysis of the flow and dropout (FD); text analysis with natural language processing (NLP); sentiment analysis (SA); and, filtering and query functionalities (Q).

The most common feature is the conversation, which is present in all tools, followed by the dashboard. The flow/dropout and query features are present in three of them. In specific, only the *Chatbase* and *Dashbot* tools use NLP to identify the users' "intent" in the dialogues. Also, only *Dashbot* and *Jani* are able to map the input dialogues in sentiments, offering a qualitative and sensitive information for developers and maintainers.

From the features presented in Table 1, the KnowBots support only the FD analysis. However, by exploring the concept of uncommon pattern for a given user-criteria, it explores a new paradigm when compared with the other analytic tools.

**Table 1.** List of chatbot analytics tools

| Name | Main features | URL |
|------|---------------|-----|
| BotAnalytics | C-D-FD-Q | http://botanalytics.co |
| Botlytics | C-Q | http://www.botlytics.co |
| BotMetrics | C-D | http://bot-metrics.com |
| Chatbase | C-D-FD-NLP | http://chatbase.com |
| Dashbot | C-D-FD-NLP-SA-Q | http://www.dashbot.io |
| Janis | C-D-SA | http://www.janis.ai |

*Features:* C - Conversation; D - Dashboard; FD - Flow/Dropout; NLP - Natural Language Processing; SA - Sentiment Analysis; Q - Query.

## 4  KnowBots

KnowBots provides an easy analysis of the usage logs given user-defined criteria. It allows a simple identification of patterns in conversations that increase or decrease the likelihood of achieving specific goals. Thereby, allowing chatbot maintainers to make decisions about the chatbots rules and oversee the resulting effects.

### 4.1  System Architecture

KnowBots is composed of two components: a batch system and a web interface. The former finds frequent patterns and sorts them by a score of relevance, according to user-defined criteria. The latter presents the results with an interactive user interface to handle the findings. Figure 2 illustrates KnowBots pipeline.

The batch system can be triggered interactively or in a background mode. With regards to the chatbot system, currently only rule-based technologies are supported, however, using the concept of intent and NLP algorithms, the Know-Bots can be extended to support other types of chatbot systems.

As the KnowBots is a standalone application, it does not require a web server or additional environment customization. However, the web module is only a layer of presentation, which restricts the possibilities of interaction with the user. In future versions, by integrating the tool with a web server, it can be extended with new features like supporting multiple chatbot versions simultaneously, dashboards, query and filters dynamically. Nevertheless, we emphasize that the innovative aspect of the KnowBots tool is the discovery of uncommon patterns considering business criteria.

Next subsections detail both components: the batch system and the web interface.

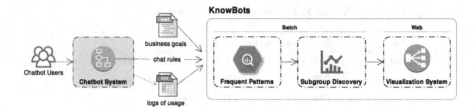

**Fig. 2.** KnowBots pipeline.

## 4.2   Batch System

The batch system receives three input files: a JSON containing the chatbot rules; an XML describing the business goals; and, a CSV containing the usage logs.

The chat rules consist of a directed graph, where the nodes are the interactions and the edges represent the possible paths of the conversation, as presented in Fig. 1. It is defined in a JSON file containing the following structure:

```
{
  "id": "Chatbot example"
  "version" : 1,
  "rules" : [
  {
    "id": "r1",
    "title": "Hello",
    "type": "display",
    "next": [ "r2" ]
  },
  ...
  {
    "id": "r5",
    "type": "decision",
    "next": [ "r6", "r8" ]
  },
  ...]
}
```

A goal can be defined by specific nodes, thus when a user reaches them during a conversation, the respective goal is achieved. Multiple goals can be defined, and each one of them may be associated with a single or multiple nodes. The following structure is used to define the goals:

```
<?xml version="1.0" encoding="UTF-8"?>
<goals>
  <goal>
  <id>Email</id>
  <title>User provides an email</title>
  <rules>
    <rule>r9</rule>
  </rules>
  </goal>
  ...
</goals>
```

The usage logs are the collected dialogues of the users with the chatbot system. Unlike the other systems, KnowBots does not use the dialogue content, but the path followed during a conversation. The CSV file with the usage log contains the following columns:

**Session:** A session identifier. A session begins when a user enters in the chatbot system and ends when the user leaves the chat.
**Timestamp:** Date and time the event occurred.
**Chat_version:** The version of the chatbot system.
**From:** The previous rule identifier.
**To:** The current rule identifier.

To identify the most frequent patterns, the CM-SPAM algorithm [6] provided by the SPMF tool [7] is used. The usage logs are mapped as sequences of items, as described in Subsect. 2.2. We set the size of the subsequences between 2 and 3, and discarded all the frequent patterns that do not have a decision node as part of the subsequence. It will result in patterns that are the smallest possible, containing a fork, which represents users' decision. Such decisions were taken to reduce the number of patterns found by the CM-SPAM, however other hyperparameters' values could be used instead.

The KnowBots tool uses a quality measure to compute a score of interest for each frequent subsequence identified in the previous step. This approach can capture the users' behavior patterns that deviate from the norm, considering the use of a particular chatbot system and the business criteria. The main advantage of using subgroup discovery is that one can score the sequence patterns based on how unusual the patterns are in terms of a particular target (the goal). As the final result, one should expect to obtain unusual (yet interesting) behaviors from the given usage logs of users. The unusualness is both in terms of increasing the chance of reaching the goal and decreasing it.

Let $p(A)$ be the probability of the users to go through the rule $A$; $p(A \mid B)$ be the conditional probability of the users to go through the rule $A$ given that they went through the rule $B$; $r_*$ be the rule of interest in terms of business goals; $\alpha = \langle \alpha_1 \rightarrow \alpha_2 \ldots \rightarrow \alpha_k \rangle$ be a frequent subsequence, where $\alpha_i \in S$ and $\alpha_1$ is the event of the pattern. The quality measure is defined according to Eq. 1:

$$QM = |c_1| * c_2 * p(\alpha),  \qquad (1)$$

where $c_1 = p(r_* \mid \alpha) - p(r_* \mid \alpha_1)$ indicates the improvement (or reduction) that following the pattern represent to achieve the goal, and $c_2 = p(\alpha_1 \mid r_*) - p(\alpha \mid r_*)$ captures how unusual the pattern is for the goal. The third criteria is the support of the pattern.

It is worth highlighting that this framework is extensible concerning the frequent patterns and the subgroup discovery steps. In the former, the patterns could explore other kinds of user-information such as gender, geographic region, operational system and web browser, for instance. In the latter, different quality measures could be used, focusing on distinct characteristics of the chatbot rules.

### 4.3   User Interface

The web interface works as a presentation layer of the results previously computed. By using interactive resources, the chatbot maintainer can explore the relevant patterns as illustrated in Fig. 3. The circle indicates the pattern, while the triangle indicates the node related to the business goal. Text templates and probabilities are used to describe the reasons why each pattern is relevant as illustrated in the *Pattern detail* window.

The tool also has an interface for the analysis of the flow/dropout. It uses a scale of colors to indicate the main paths and the critical dropout points. Additionally, it is possible to visualize the number of session conversations that achieved the goal by going through a specific point.

## 5   Exploratory Analysis

We illustrate the potential of KnowBots by analyzing the usage logs of a real-world chatbot system that provides advice on technical courses for unemployed people. The chat has 87 rule nodes, where 32 of them are decision ones. A single business goal is considered: *"user provides email address"*, which is obtained at the end of the dialogue. The shortest path of conversation to reach this goal will go through 8 decision points, whereas the longest would have more than 15 of them.

After the summarization of the obtained results, a discussion about them including the strengths and limitations of the tool is conducted.

### 5.1   Results

Overall, 24 relevant patterns were identified from the 3266 sessions analyzed. Table 2 presents the top 10 of them. The column *Rules* contains the rules id; columns *c1*, *c2* and *Support* are the components defined in Eq. 1; the quality measure result is presented in column *QM*. Particularly in c1 column, the bold values highlight the negative scores, which indicate patterns that decreased the probability of achieving the goal.

The first pattern is illustrated in Fig. 3. It says: *"users who did not provide their home address rightly showed a lower probability in providing their e-mail*

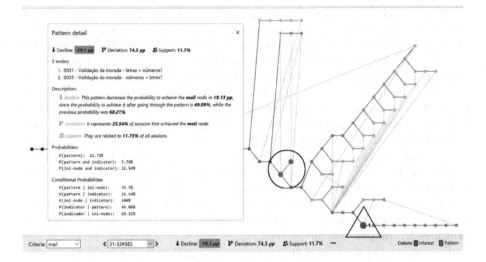

**Fig. 3.** KnowBots interface with details of a relevant pattern.

**Table 2.** Top 10 relevant patterns identified for the KnowBots tool.

| Ranking | Rules | c1 (pp.) | c2 (pp.) | Support (%) | QM |
|---------|-------|----------|----------|-------------|-----|
| 1 | 31-32 | **−19.1** | 74.5 | 11.7 | 0.1665 |
| 2 | 74-75 | **−26.9** | 88.0 | 4.6 | 0.1089 |
| 3 | 29-30 | **−48.1** | 98.2 | 2.1 | 0.0992 |
| 4 | 35-36-48 | 14.6 | 87.8 | 3.0 | 0.0385 |
| 5 | 74-77-78 | 14.0 | 13.0 | 19.6 | 0.0357 |
| 6 | 38-39-50 | 12.2 | 69.4 | 3.5 | 0.0296 |
| 7 | 44-45-46 | **−23.8** | 22.7 | 4.3 | 0.0232 |
| 8 | 39-40-47 | 18.9 | 70.4 | 1.5 | 0.0200 |
| 9 | 41-42-61 | 15.7 | 46.7 | 2.7 | 0.0198 |
| 10 | 29-31-35 | 8.9 | 7.5 | 27.3 | 0.0182 |

*than the ones who did"*. Precisely, the probability of users to give their email decreases in 19.13 pp. when they did not provide their home address correctly. It represents 25.54% of sessions that achieved the goal and it was observed in 11.73% of the sessions. Using this information, the chatbot maintainers can explore alternatives such as the relaxation of the validation rules and/or change the order of the dialogues, for instance.

Some obvious patterns were also found, which helped the validation of the tool and increased the confidence in the results. For instance, before asking for the email, the bot asks if the user desires to receive a newsletter about new courses. It is reasonable to assume that the users who answer "no" for the newsletter would be more resistant in providing their email in the next step (second pattern in Table 2). The probability of providing the email decreased in almost 27 pp. for those who answered "no" to the newsletter. In this case, the KnowBots quantified the impact of this problem making it measurable.

In summary, 4 out of 10 patterns indicate that attaining them reduces the probability to achieve the goal. They may represent possible issues in the chatbot design. The other 6 patterns indicate that the probability of achieving the goal is increased by following them. A positive example is the $4^{th}$ pattern, it is related to 3% of the sessions and increases the probability to achieve the goal in almost 15 pp. when compared with other alternatives. Only users with a background in business domain follow this path, however the main hypothesis to explain it is the order in which the different domains are presented to the user. The fact that it is the first option among several domains may be a relevant factor to the observed users' behavior. Alternatively, different versions could use different sequences to compare the alternatives.

In practice, the analysis of these patterns brought new insights to the developers as well as a better understanding of users' behavior. The other feature available in the KnowBots tool is the flow/dropout analysis. Using it, we realized that 7 rules were not achieved by any of the sessions. They are related to the validation of the zip code address and they are close to the first fork in the path of conversation. Furthermore, iteratively the dropout was analyzed and the following results were observed: 2 rules are related to more than 20% of the sessions' dropout; 7 rules are related to more than 5%; and, 16 rules are related to at least 1%. Some of these rules validate the users' input, which shows that when the bot asks for the same input repeatedly, the users leave the conversation.

## 5.2   Discussion

In this exploratory analysis, the KnowBots was able to identify interesting patterns concerning the business goal investigated. Some results were not expected for the chatbot developers, whereas others were quantified using objective criteria. The analysis of the identified patterns can be used to guide the investigation of the chatbot system, mainly when business goals are taking into account.

The use of language templates to describe the patterns simplified the understanding of the metrics according to the business stakeholders. In comparison with the other tools, they have not employed natural language to present results. Although we did not perform a usability test with the KnowBots users, they were able to use the tool and perform the proposed tasks easily.

Regarding the patterns, other characteristics could be employed to represent the sequences. In addition to the rules, each event can describe details of the user context like operational system and browser; temporal information like the day of week and period of the day; and also, personal information like gender and age, for instance. Such features would increase qualitatively the pattern analysis. Even though they were not used in this study, due to the lack of such information in the available transactional records, the CM-SPAM algorithm supports intrinsically the use of this data. We plan to explore this feature in further studies.

Analogously, the quality measure could also be modified to capture other perspectives of the problem. For instance, the criteria $c_2$ and the support (Eq. 1) are inversely proportional, such that while one increases, the other decreases, and vice-versa. In practice, they are important because the user can choose not to provide the email. In other scenarios, only one of these criteria could be employed, for instance.

In summary, the proposed approach showed to be able to detect possible chatbot problems. The KnowBots is the result of a real-world demand, given the lack of tools that are able to support business criteria in the usage log analysis. To the best of our knowledge, it is the first chatbot analytics tool to address this problem.

# 6 Conclusion

This work presented the KnowBots, a tool for mining and visualization of chatbot usage logs analysis. It finds unusual and relevant patterns by combining sequence mining and subgroup discovery techniques. Specifically, the tool provides useful information concerning users behavior in terms of business goals. It is a descriptive machine learning task that aims to minimize the efforts of chatbot maintainers in the analysis of the chatbot systems.

Despite the rapid growth of chatbot-related technologies, the investigation of analytics tools is still subtly addressed in the literature. In further studies, we plan to explore new attributes in the sequences like temporal data, profile information and user context. Additionally, by supporting multiple versions of the same chatbot system, the KnowBots can be a validation and decision-support tool.

# References

1. Shah, K.B., Shetty, M.S., Shah, D.P., Pamnani, R.: Approaches towards building a banking assistant. Int. J. Comput. Appl. **166**(11), 1–6 (2017). https://doi.org/10.5120/ijca2017914140
2. Chai, J.Y., et al.: The role of a natural language conversational interface in online sales: a case study. Int. J. Speech Technol. **4**(3–4), 285–295 (2001). https://doi.org/10.1023/A:1011316909641
3. Chakrabarti, C., Luger, G.F.: Artificial conversations for customer service chatter bots: architecture, algorithms, and evaluation metrics. Expert Syst. Appl. **42**(20), 6878–6897 (2015). https://doi.org/10.1016/j.eswa.2015.04.067
4. Duivesteijn, W., Feelders, A., Knobbe, A.J.: Exceptional model mining - supervised descriptive local pattern mining with complex target concepts. Data Min. Knowl. Discov. **30**(1), 47–98 (2016). https://doi.org/10.1007/s10618-015-0403-4
5. Fournier-Viger, P., Chun, J., Lin, W., Kiran, R.U., Koh, Y.S., Thomas, R.: A survey of sequential pattern mining. Data Sci. Pattern Recogn. **1**, 54–77 (2017)
6. Fournier-Viger, P., Gomariz, A., Campos, M., Thomas, R.: Fast vertical mining of sequential patterns using co-occurrence information. In: Tseng, V.S., Ho, T.B., Zhou, Z.-H., Chen, A.L.P., Kao, H.-Y. (eds.) PAKDD 2014. LNCS (LNAI), vol. 8443, pp. 40–52. Springer, Cham (2014). https://doi.org/10.1007/978-3-319-06608-0_4

7. Fournier-Viger, P., Gomariz, A., Gueniche, T., Soltani, A., Wu, C.W., Tseng, V.S.: SPMF: a java open-source pattern mining library. J. Mach. Learn. Res. **15**(1), 3389–3393 (2014)
8. Herrera, F., Carmona, C.J., González, P., del Jesús, M.J.: An overview on subgroup discovery: foundations and applications. Knowl. Inf. Syst. **29**(3), 495–525 (2011). https://doi.org/10.1007/s10115-010-0356-2
9. Jia, J.: CSIEC: a computer assisted English learning chatbot based on textual knowledge and reasoning. Knowl.-Based Syst. **22**(4), 249–255 (2009). https://doi.org/10.1016/j.knosys.2008.09.001
10. Jusoh, S., Al-Fawareh, H.M.: Natural language interface for online sales systems. In: 2007 International Conference on Intelligent and Advanced Systems, pp. 224–228. IEEE (2007). https://doi.org/10.1109/ICIAS.2007.4658379
11. Kerly, A., Hall, P., Bull, S.: Bringing chatbots into education: towards natural language negotiation of open learner models. Knowl.-Based Syst. **20**(2), 177–185 (2007). https://doi.org/10.1016/j.knosys.2006.11.014
12. Mooney, C., Roddick, J.F.: Sequential pattern mining - approaches and algorithms. ACM Comput. Surv. **45**(2), 19:1–19:39 (2013). https://doi.org/10.1145/2431211.2431218
13. Pereira, J., Díaz, Ó.: Chatbot dimensions that matter: lessons from the trenches. In: Mikkonen, T., Klamma, R., Hernández, J. (eds.) ICWE 2018. LNCS, vol. 10845, pp. 129–135. Springer, Cham (2018). https://doi.org/10.1007/978-3-319-91662-0_9
14. Shah, H., Warwick, K., Vallverdú, J., Wu, D.: Can machines talk? comparison of eliza with modern dialogue systems. Comput. Hum. Behav. **58**, 278–295 (2016). https://doi.org/10.1016/j.chb.2016.01.004
15. Shawar, B.A., Atwell, E.: Chatbots: are they really useful? LDV Forum **22**(1), 29–49 (2007)
16. Souza, M., Miyagawa, T., Melo, P., Maciel, F.: Wellness programs: wearable technologies supporting healthy habits and corporate costs reduction. In: Stephanidis, C. (ed.) HCI 2017. CCIS, vol. 714, pp. 293–300. Springer, Cham (2017). https://doi.org/10.1007/978-3-319-58753-0_44
17. Toxtli, C., Monroy-Hernández, A., Cranshaw, J.: Understanding chatbot-mediated task management. In: Mandryk, R.L., Hancock, M., Perry, M., Cox, A.L. (eds.) Proceedings of the Conference on Human Factors in Computing Systems, p. 58. ACM (2018). https://doi.org/10.1145/3173574.3173632

# Fast Distance-Based Anomaly Detection in Images Using an Inception-Like Autoencoder

Natasa Sarafijanovic-Djukic[1]([✉]) and Jesse Davis[2][iD]

[1] IRIS Technology Solutions, Barcelona, Spain
natasa.sdj@iris.cat
[2] KU Leuven, Leuven, Belgium
jesse.davis@cs.kuleuven.be

**Abstract.** The goal of anomaly detection is to identify examples that deviate from normal or expected behavior. We tackle this problem for images. We consider a two-phase approach. First, using normal examples, a convolutional autoencoder (CAE) is trained to extract a low-dimensional representation of the images. Here, we propose a novel architectural choice when designing the CAE, an Inception-like CAE. It combines convolutional filters of different kernel sizes and it uses a Global Average Pooling (GAP) operation to extract the representations from the CAE's bottleneck layer. Second, we employ a distanced-based anomaly detector in the low-dimensional space of the learned representation for the images. However, instead of computing the exact distance, we compute an approximate distance using product quantization. This alleviates the high memory and prediction time costs of distance-based anomaly detectors. We compare our proposed approach to a number of baselines and state-of-the-art methods on four image datasets, and we find that our approach resulted in improved predictive performance.

**Keywords:** Anomaly detection · Deep learning · Computer vision

## 1 Introduction

The goal of anomaly detection [5,37] is to identify examples that deviate from what is normal or expected. We tackle this problem for images which is relevant for applications such as visual quality inspection in manufacturing [13], surveillance [32,33], biomedical applications [35,39], self-driving cars [7], or robots [4,22]. This has motivated significant interest in this problem in recent years.

The classic approach to anomaly detection is to treat it as an unsupervised problem (e.g., [2,23]) or one-class problem [25,36]. Recently, there has been a surge of interest in applying deep learning to anomaly detection, particularly in the context of images (e.g., [11,25,26,31]). In this line of work, one strategy is to use (convolutional) autoencoders, which is typically done in one of two ways. First, it is possible to directly use the autoencoder as an anomaly detector. This

© Springer Nature Switzerland AG 2019
P. Kralj Novak et al. (Eds.): DS 2019, LNAI 11828, pp. 493–508, 2019.
https://doi.org/10.1007/978-3-030-33778-0_37

can be done by using an example's reconstruction error as the anomaly score (e.g., [40]). Second, the autoencoder can be used to learn a new low-dimensional representation of the data after which a classical anomaly detection approach is applied on top of this learned representation (e.g., [1, 42]).

In this paper, we take the one-class approach and follow the second strategy for the problem of detecting anomalous images. We begin by training a convolutional autoencoder (CAE) on only normal images. Here our contribution is to propose a novel CAE architecture. It is inspired by the Inception classification model [34] that combines convolutional filters of different kernel sizes. Once the CAE is trained, we use a Global Average Pooling (GAP) operation to extract the low-dimensional representation from the CAE's bottleneck layer. In contrast, existing approaches directly use the bottleneck layer's output. Using the GAP operation is motivated by its successes in reducing overfitting in classification CNN models [21], and extracting image representations from the hidden layers of pretrained classification models for image captioning [38].

At test time, we use a classic nearest-neighbor distanced-based anomaly detector [23] in the learned low-dimensional representation space. Here our contribution is to compute an approximate distance using product quantization [16], which improves the runtime performance and memory footprint of this approach compared to using the exact distance. Empirically, we compare our proposed approach to a number of existing approaches on four standard datasets used for benchmarking anomaly detection models for images. We find that our approach generally achieves better predictive performance.

## 2    Background and Related Work

This work draws on ideas from anomaly detection both in general and for images, deep learning, and fast nearest neighbors. We now review each of these areas.

### 2.1    Anomaly Detection

For a variety of reasons (e.g., what is anomalous changes over time or expense), it is often difficult to obtain labels for examples belonging to the anomaly class. Therefore, anomaly detection is often approached from an unsupervised [2, 23] or one-class perspective [36].

In order to identify anomalies, unsupervised approaches typically assume that anomalous examples are rarer and different in some respect than normal examples. One standard approach is to assume that anomalies are far away from normal examples or that they lie in a low-density region of the instance-space [2, 23]. A common approach [23] that uses this intuition is based on $k$-nearest neighbors. This algorithm produces a ranking of how anomalous each example is by computing an example's distance to its $k^{th}$ nearest neighbor in the data set. Despite its simplicity, this approach seems to work very well empirically [3].

The idea underlying one-class-based anomaly detection is that the training data only contains normal examples. Under this assumption, the training phase attempts to learn a model of what constitutes normal behavior. Then, at test time, examples that do not conform to the model of normal are considered to be anomalous. One way to do this is to use a one-class SVM [36].

## 2.2 Deep Learning for Anomaly Detection

Autoencoders are the most prevalent deep learning methods used for anomaly detection. An autoencoder (AE) is a multi-layer neural network that is trained such that the output layer is able to reproduce its input. An AE has a bottleneck layer with a lower dimension than the input layer and hence allows learning a low-dimensional representation (encoding) of the input data.

AEs are used for anomaly detection in images in two ways. First, an AE can be directly used as an anomaly detector. Here, a typical way to assign an anomaly score for a test example is to apply the AE and calculate the example's reconstruction error (e.g., the mean squared error between the example and the AE's output [28,40]). Second, an AE can be used as part of a two-step process: (1) train an AE on the training data; and (2) learn a standard (shallow) anomaly detector on the transformed training data [1,42]. We follow this strategy.

Deep approaches to anomaly detection for image data often use a convolutional autoencoder (CAE) which include convolutional layers in the AE architecture [24,31]. Another line of work uses Generative Adversarial Networks (GAN) for this task [8,27,29]. This two-step process is also used to make the density estimation task easier by learning low-dimensional representations.

Recently, there have been attempts to design fully end-to-end deep models for anomaly detection. Deep Support Vector Data Description (Deep SVDD) [25] is trained using an anomaly detection based objective that minimizes the volume of a hypersphere enclosing the data representations. Deep Autoencoding Gaussian Mixture Model (DAGMM) [45] uses the representation layer of a deep autoencoder in order to estimate the parameters of a Gaussian mixture model, by jointly optimizing parameters of the autoencoder and the mixture model.

Deep Structured Energy Based Model (DSEBM) [43] belongs to the group of the energy-based models, a powerful tool for density estimation. The energy-based models make a specific parameterization of the negative log probability, which is called the energy, and then compute the density with a proper normalization. In DSEBM, the energy function is a deep neural network.

Another method [11] uses a data augmentation, and generates new training examples by applying a number of geometric transformations to each training example. Then, a multi-class neural network is trained to discriminate among the original images, and all of the geometric transformations applied to the images. Given a test image, the same transformations are applied to it and the prediction is made based on the network's softmax activation statistics.

## 2.3 Fast Nearest Neighbors Search

A notable potential issue with nearest-neighbors-based approaches is that finding the nearest neighbor at test time can be very computationally expensive, particularly for large training sets or high-dimensional examples. Consequently, there has been substantial interest in developing efficient approaches for performing this search either exactly (e.g., using a kd-tree or other index structure) or approximately [10,16].

One prominent recent approach is product quantization [16]. This approach works by compressing the training data by partitioning the features used to describe the training example into $m$ equal width groups. It then learns a code book for each partition. Typically, this is done by running k-means clustering on each partition which only considers the features assigned to that partition. Then the values of all features in the partition are replaced by a single $c$-bit code representing the cluster id that the example is assigned to in the current partition. Hence each example is represented by $m$ $c$-bit code words.

At test time, finding a test example's nearest neighbor using the (squared) L2 distance can be done efficiently by using table look-ups and addition. For a test example, a look up table is constructed for each partition that stores the squared L2 distance to each of the $k = 2^c$ cluster centroids in that partition. Then the approximate distance to each training example is computed using these look up tables and the nearest example is returned.

Locality-sensitive hashing improves efficiency by using hashing to identify a limited number of likely candidate nearest neighbors. Then, a test example is only compared to those examples. Hence, the approximation comes from the fact that not all training examples are considered as the possible nearest neighbor. People have investigated incorporating hashing-based techniques into distance-based anomaly detection systems [12,30,44].

## 3    Our Approach

At a high-level, our approach has two steps: extracting a low dimensional image representation and assigning a distance based anomaly score.

*Extracting a Low-Dimensional Image Representation.* Given a training set of normal image examples, an Inception-like convolutional autoencoder (Inception-CAE) is trained that minimizes the mean squared reconstruction error on the training data. Once the InceptionCAE is trained, the GAP operation is applied on its bottleneck layer to extract a low-dimensional image representation vector.

*Assigning a Distance-Based Anomaly Score.* An distance-based anomaly score is assigned using the learned representation vectors for the images. First, the trained InceptionCAE model is used to convert all training images to the learned low-dimensional representation. Second, it converts a given test image into the same space and assigns an anomaly score by computing the quantized Euclidean distance between the test image and its nearest neighbor in this space.

## 3.1   Inception-Like Convolutional Autoencoder

When using a CAE in a two-step anomaly detection approach, the detector's predictive performance clearly depends on the quality of the learned low-dimensional representation. In supervised image classification, sophisticated deep architectures such as Inception (GoogleNet) [34], Residual Networks [14] or DenseNet [15], have yielded considerable performance gains over a basic CNN architecture. Hence, we expect that adapting these techniques to the CAE setting could improve the quality of the CAE's learned low-dimensional representation.

Inspired by the Inception architecture, we design an Inception-like CAE architecture that combines convolutional filters of different kernel sizes. The main unit of this architecture is an Inception-like layer shown in Fig. 1, where it combines outputs from $1 \times 1$, $3 \times 3$ and $5 \times 5$ convolutions as well as a maximum pooling operation.

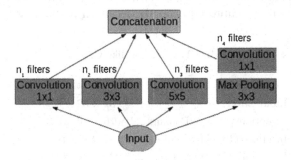

**Fig. 1.** Inception-like layer.

Table 1 outlines the details of our Inception-like CAE architecture. We make our architecture as similar as possible to the baseline CAE architecture [25] and it has the same number of layers and the same number of convolutional filters in each layer to enable a fair comparison. Here, Inception(n) denotes the Inception-like layer with $n_1 = n_2 = n_3 = n_4 = n$. Each convolution operation is followed by Batch Normalization and a Leaky ReLU activation, except the last layer which has a Sigmoid activation.

Beside the architectural change of introducing the Inception-like layer into a CAE, another subtlety in our approach is how we extract the low-dimensional image representation from the CAE. Existing approaches extract a learned image representation simply by using the output of the CAE's bottleneck layer. Consequently, the CAE architecture must be designed such that the bottleneck layer matches the desired dimension of the learned image representation. Our approach extracts the learned image representation by applying a Global Average Pooling (GAP) operation to the output of the CAE's bottleneck layer. The GAP operation on a tensor of the dimension $a \times b \times c$ results in a vector of the dimension $1 \times c$, where each component is an average value over the tensor slice of the dimension $a \times b$ that corresponds to this component. Hence, using the GAP operation as an extractor permits using a wider bottleneck layer than existing CAE's architectures do.

Our intuition behind this architectural choice is that having a wider bottleneck will permit retaining some information that a narrower bottleneck would filter. Thus, the GAP operation on this wider bottleneck would yield a better learned representation. Though the use of GAP is not a novel in deep learning architectures, our contribution is to study its use in conjunction with an CAE for extracting image representations.

The Inception-like CAE is trained on the normal training images, where the objective is to minimize the mean squared error between the input and the output. Applying the GAP operation on the trained network's bottleneck yields a 128-dimensional learned image representation. Note that the GAP operation allows having a wider bottleneck layer in our Inception-like CAE architecture than in the baseline CAE architecture ($4 \times 4 \times 128$ versus $1 \times 128$).

Our experiments show that each of our two architectural choices in designing our CAE contribute to improved anomaly detection performance.

**Table 1.** Inception-like CAE.

| InceptionCAE | Output dimension |
|---|---|
| Input Layer | $32 \times 32 \times n_{channels}$ |
| Inception (8). MaxPooling(2,2) | $16 \times 16 \times 32$ |
| Inception(16). MaxPooling(2,2) | $8 \times 8 \times 64$ |
| Inception(32). MaxPooling(2,2) | $4 \times 4 \times 128$ |
| Inception(16). Upsampling(2,2) | $16 \times 16 \times 64$ |
| Inception (8). Upsampling(2,2) | $32 \times 32 \times 32$ |
| Convolution2D($n_{channels}$) | $32 \times 32 \times n_{channels}$ |

### 3.2  Approximated Distance-Based Anomaly Detection

We assign an anomaly score to a test example by operating on the extracted images representations and not on the raw data itself. Specifically, a test example's anomaly score is the quantized (squared) Euclidean distance in the learned representation space to its nearest neighbor in the training data. The primary advantage of using product quantization instead of the exact distance is that it is substantially faster to compute (at the expense of being an approximation).

Hashing-based solutions have been extensively explored to speed-up distance-based neighbor approaches. While extensively used in nearest-neighbor search, product quantization has received little attention within anomaly detection. One advantage of quantization over a hashing based solution is that it still compares a test example to each training example, it just does so in an approximate manner.

Quantization may provide another benefit when the training data only contains normal examples: its approximation may enforce some regularization on the training data. That is, by mapping each partition of the example to a prototype it may smooth out some variation and make the examples look more "normal," which is beneficial if true. In some cases, we observe empirically that the quantization does indeed improve performance.

## 4    Experimental Results

Our empirical evaluation addresses the following questions:

**Q1** How does our proposed approach compare to existing anomaly detection techniques for images?

**Q2** What is the effect of using product quantization to approximate the distance calculation on performance?

**Q3** What is the effect of using the distance-based nearest neighbors approach to assign an anomaly score compared to using an example's reconstruction error?

**Q4** How sensitive is the performance of our approach to changes in the quantization parameters?

To address these questions, we compare our proposed approach to a number of shallow and deep baselines on four standard benchmark datasets. Next, we describe the approaches, data, methodology and results in greater detail.

### 4.1    Methods Compared

The main empirical comparison considers the following methods.

**Raw NN-QED:** This shallow approach corresponds to applying the classic kNN-based anomaly detection [23] on the raw image data, except that it uses an approximate distance measure. It assigns an anomaly score to a test example as the quantized squared Euclidean distance (QED) to the test example's nearest neighbor in the raw training images.

**DeepSVDD:** This method is a deep extension of the support vector data description method [25]. We use the same baseline CAE architecture for all the datasets as the one used for a CIFAR-10 dataset in the respective paper.

**DSEBM:** This method is a deep extension of energy based models [43], where we adjust a neural network to correspond to the baseline CAE architecture used in Deep SVDD in order to have a fair comparison.

**CAE OCSVM:** This method trains an CAE with the same baseline architecture as in DeepSVDD. Then the learned image data representations obtained from the output of the CAE's bottleneck layer are used as the input to OCSVM.

**CAE NN-QED:** This method trains an CAE with the same baseline architecture as in DeepSVDD. A test example's anomaly score is calculated as the quantized squared Euclidean distance in the CAE's learned representation space to its nearest neighbor in the training set.

**InceptionCAE NN-QED:** This is our approach. It uses our proposed InceptionCAE architecture outlined in Sect. 3.1. A test example's anomaly score is calculated as the quantized squared Euclidean distance in the InceptionCAE's learned representation space to its nearest neighbor in the training set.

## 4.2 Datasets

Our experiments use four common benchmark datasets for both deep learning and anomaly detection approaches. MNIST [20] and Fashion MNIST [41] contains ten classes and have fixed train-test splits with the training set containing 60,000 examples (6,000 examples for each class) and the test set 10,000 examples (1,000 for each class). CIFAR10 has ten class, while in CIFAR100 we consider the 20 super-classes [19]. Both have fixed train-test splits with the training set containing 50,000 examples and the test set 10,000 examples. All the datasets are completely labeled which enables computing standard evaluation metrics.

**Table 2.** Average AUC-ROC and its standard deviation for state-of-the-art deep baselines and our approach on the MNIST and Fashion MNIST datasets.

| Dataset | Normal class | DSEBM | CAE OCSVM | DeepSVDD | InceptionCAE NN-QED |
|---------|--------------|-------|-----------|----------|---------------------|
| MNIST | 0 | 94.9 ± 4.0 | 95.4 ± 0.8 | 99.1 ± 0.1 | 98.7 ± 0.3 |
| | 1 | 98.7 ± 0.1 | 97.4 ± 0.3 | 99.7 ± 0.0 | 99.7 ± 0.0 |
| | 2 | 69.0 ± 11.5 | 77.6 ± 3.3 | 95.4 ± 0.3 | 96.7 ± 0.7 |
| | 3 | 80.2 ± 9.7 | 88.6 ± 1.6 | 95.1 ± 0.5 | 95.2 ± 0.4 |
| | 4 | 83.3 ± 9.1 | 83.6 ± 1.8 | 95.9 ± 0.5 | 95.0 ± 0.5 |
| | 5 | 67.4 ± 6.8 | 71.3 ± 1.8 | 92.1 ± 0.5 | 95.2 ± 0.5 |
| | 6 | 85.6 ± 5.9 | 90.1 ± 1.6 | 98.5 ± 0.1 | 98.3 ± 0.2 |
| | 7 | 90.4 ± 2.1 | 87.2 ± 0.8 | 96.2 ± 0.4 | 97.0 ± 0.3 |
| | 8 | 72.1 ± 7.3 | 86.5 ± 1.6 | 95.7 ± 0.4 | 96.2 ± 0.2 |
| | 9 | 86.8 ± 2.9 | 87.3 ± 1.0 | 97.7 ± 0.1 | 97.0 ± 0.2 |
| | Average | 82.8 ± 12.3 | 86.5 ± 7.5 | 96.6 ± 2.1 | 96.9 ± 1.6 |
| Fashion MNIST | 0 | 89.2 ± 0.1 | 88.0 ± 0.4 | 98.8 ± 0.2 | 92.4 ± 0.4 |
| | 1 | 97.4 ± 0.1 | 97.3 ± 0.2 | 99.7 ± 0.0 | 98.8 ± 0.1 |
| | 2 | 86.0 ± 0.3 | 85.5 ± 0.8 | 93.5 ± 1.4 | 90.0 ± 0.6 |
| | 3 | 90.5 ± 0.1 | 90.0 ± 0.5 | 94.9 ± 0.3 | 95.0 ± 0.3 |
| | 4 | 88.5 ± 0.3 | 88.5 ± 0.5 | 95.1 ± 0.6 | 92.0 ± 0.4 |
| | 5 | 82.4 ± 9.2 | 87.2 ± 0.7 | 90.4 ± 0.8 | 93.4 ± 0.3 |
| | 6 | 77.7 ± 1.5 | 78.8 ± 0.7 | 98.0 ± 0.2 | 85.5 ± 0.4 |
| | 7 | 98.1 ± 0.1 | 97.7 ± 0.1 | 96.0 ± 0.2 | 98.6 ± 0.1 |
| | 8 | 78.8 ± 6.8 | 85.8 ± 1.4 | 95.4 ± 0.4 | 95.1 ± 0.4 |
| | 9 | 96.0 ± 2.7 | 98.0 ± 0.2 | 97.6 ± 0.2 | 97.7 ± 0.2 |
| | Average | 88.5 ± 7.9 | 89.7 ± 6.0 | 95.9 ± 3.8 | 93.9 ± 3.9 |

Following past work on anomaly detection [11,25], we denote the images of one class as normal, while images for all other classes are considered anomalous. The training phase only uses images from the normal class. At test time, test images of all classes are used.

### 4.3 Parameters and Implementations

For all the CAE architectures, we employ the same training procedure as in Ruff et al. [25], with a two-phase learning rate schedule (searching + fine-tuning) with initial learning rate $\nu = 10^{-4}$, and subsequently $\nu = 10^{-5}$. We train $100 + 50$ epochs for MNIST and Fashion MNIST, and $250 + 100$ epochs for CIFAR-10 and CIFAR-100. Leaky ReLU activations use a leakiness of $\alpha = 0.1$. We use a batch size of 200 and set the weight decay hyperparameter $\lambda = 10^{-6}$, and we use an Adam optimization procedure [18]. For CIFAR-10 and CIFAR-100, both the CAE-GAP and InceptionCAE architectures are trained without the GAP layer, but the GAP operation on the bottleneck layer is used at prediction time to extract the image representation. For MNIST and Fashion MNIST, the GAP layer must be included during training to ensure that the bottleneck layer is narrower than the input layer. We implemented the CAEs in the Keras framework [6].

We use the Facebook AI Similarity Search (FAISS) library [17] for computing the quantized Euclidean distance using the parameters $m = 32$ and $c = 4$. We show the effects of these parameters in Subsect. 4.4.

The OCSVM implementation uses the default parameters of Python `sklearn` library, with radial basis function kernel with $\gamma = 1/n_{features}$ and $\nu = 0.5$.

Because we use an identical train-test split, we simply report the AUC-ROCs for prior results for DeepSVDD on CIFAR-10 from the paper. For DeepSVDD, we re-run the experiments for the other datasets using the authors' software in order to use the same CAE baseline architecture. Our code is available online.[1]

### 4.4 Results

We compare the approaches with respect to their predictive performance, where we report the average area under the receiver operator characteristic curve (AUC-ROC) which is a standard performance metric in anomaly detection [9,11,25]. For the methods that use a non-deterministic algorithm (CAE/InceptionCAE NN-QED, CAE OCSVM, DeepSVDD, DSEBM), we train 10 models (with different random seeds) and report the average AUC-ROC and its standard deviation over these 10 models for each considered normal class.

**Results for Q1**

Tables 2 and 3 show detailed AUC-ROC scores for state-of-the-art deep baselines and our method. On average, our approach outperforms the deep baselines for all the considered datasets except on Fashion MNIST. Looking at the 50 individual

---

[1] https://github.com/natasasdj/anomalyDetection.

tasks, our InceptionCAE NN-QED method beats DeepSVDD 32 times, DSEBM 48 times, and CAE OCSVM 48 times. The bigger wins come on the more complex CIFAR datasets.

Table 4 shows how much benefit comes from using our proposed Inception-like CAE architecture with the GAP operation to extract a low-dimensional image representation, compared to using the raw image data or the baseline CAE

**Table 3.** Average AUC-ROC its standard deviation for state-of-the-art deep baselines and our proposed approach on the CIFAR-10 and CIFAR-100 datasets.

| Dataset | Normal Class | DSEBM | CAE OCSVM | DeepSVDD | InceptionCAE NN-QED |
|---|---|---|---|---|---|
| CIFAR 10 | 0 | 64.1 ± 1.5 | 62.4 ± 0.9 | 61.7 ± 4.1 | 66.7 ± 1.3 |
| | 1 | 50.1 ± 5.1 | 44.4 ± 1.0 | 65.9 ± 2.1 | 71.3 ± 1.3 |
| | 2 | 61.5 ± 0.8 | 64.2 ± 0.3 | 50.8 ± 0.8 | 66.8 ± 0.6 |
| | 3 | 51.2 ± 3.0 | 50.7 ± 0.8 | 59.1 ± 1.4 | 64.1 ± 0.9 |
| | 4 | 73.2 ± 0.5 | 74.8 ± 0.2 | 60.9 ± 1.1 | 72.3 ± 0.8 |
| | 5 | 54.6 ± 2.8 | 50.9 ± 0.5 | 65.7 ± 2.5 | 65.3 ± 0.9 |
| | 6 | 68.2 ± 1.1 | 72.4 ± 0.3 | 67.7 ± 2.6 | 76.4 ± 0.8 |
| | 7 | 52.8 ± 1.3 | 51.0 ± 0.7 | 67.3 ± 0.9 | 63.7 ± 0.7 |
| | 8 | 73.7 ± 1.9 | 67.0 ± 1.6 | 75.9 ± 1.2 | 76.9 ± 0.6 |
| | 9 | 63.9 ± 5.9 | 50.8 ± 2.5 | 73.1 ± 1.2 | 72.5 ± 1.0 |
| | Average | 61.3 ± 8.9 | 58.9 ± 10.1 | 64.8 ± 7.2 | 69.6 ± 4.8 |
| CIFAR 100 | 0 | 63.8 ± 0.4 | 63.6 ± 1.2 | 57.4 ± 2.4 | 66.0 ± 1.5 |
| | 1 | 48.4 ± 0.9 | 51.4 ± 0.7 | 63.0 ± 1.2 | 60.1 ± 1.5 |
| | 2 | 63.6 ± 7.6 | 54.5 ± 1.0 | 70.0 ± 3.2 | 59.2 ± 3.1 |
| | 3 | 50.4 ± 3.2 | 48.4 ± 0.8 | 55.8 ± 2.5 | 58.7 ± 0.5 |
| | 4 | 57.3 ± 9.6 | 49.9 ± 1.3 | 69.0 ± 1.9 | 60.9 ± 1.9 |
| | 5 | 44.4 ± 3.3 | 45.3 ± 1.4 | 51.0 ± 2.0 | 54.2 ± 1.3 |
| | 6 | 53.3 ± 5.2 | 53.1 ± 1.6 | 59.9 ± 3.3 | 63.7 ± 1.4 |
| | 7 | 53.4 ± 1.3 | 58.8 ± 0.6 | 53.0 ± 1.2 | 66.1 ± 1.3 |
| | 8 | 66.9 ± 0.3 | 67.8 ± 0.5 | 51.6 ± 3.2 | 74.8 ± 0.4 |
| | 9 | 72.7 ± 4.0 | 70.1 ± 1.2 | 72.9 ± 1.5 | 78.3 ± 0.7 |
| | 10 | 76.2 ± 3.4 | 76.7 ± 0.6 | 81.5 ± 1.9 | 80.4 ± 0.9 |
| | 11 | 62.2 ± 1.2 | 59.7 ± 0.6 | 53.6 ± 0.7 | 68.3 ± 0.6 |
| | 12 | 66.9 ± 0.4 | 68.2 ± 0.3 | 50.6 ± 1.2 | 75.6 ± 0.7 |
| | 13 | 53.1 ± 0.7 | 60.6 ± 0.4 | 44.0 ± 1.2 | 61.0 ± 0.9 |
| | 14 | 44.7 ± 0.7 | 47.1 ± 1.1 | 57.2 ± 1.1 | 64.3 ± 0.7 |
| | 15 | 56.6 ± 0.2 | 59.7 ± 0.3 | 47.7 ± 0.9 | 66.3 ± 0.4 |
| | 16 | 63.1 ± 0.4 | 66.0 ± 0.4 | 54.3 ± 0.8 | 72.0 ± 0.5 |
| | 17 | 73.5 ± 3.6 | 69.4 ± 1.1 | 74.7 ± 2.0 | 75.9 ± 0.7 |
| | 18 | 55.6 ± 2.2 | 54.5 ± 0.8 | 52.1 ± 1.7 | 67.4 ± 0.8 |
| | 19 | 57.3 ± 1.6 | 54.7 ± 1.0 | 57.9 ± 1.8 | 65.8 ± 0.6 |
| | Average | 59.2 ± 9.3 | 59.0 ± 8.6 | 58.9 ± 9.9 | 67.0 ± 7.1 |

**Table 4.** Average AUC-ROC and its standard deviation for Raw NN-QED, CAE NN-QED (the baseline CAE architecture), and InceptionCAE NN-QED (our proposed approach). The AUC-ROC is averaged over treating each of the ten classes as the normal class.

| Dataset | Raw NN-QED | CAE NN-QED | InceptionCAE NN-QED |
|---|---|---|---|
| MNIST | $94.7 \pm 3.8$ | $96.4 \pm 2.4$ | $96.9 \pm 1.6$ |
| Fashion MNIST | $91.4 \pm 4.9$ | $91.6 \pm 4.1$ | $93.9 \pm 3.9$ |
| CIFAR-10 | $59.6 \pm 11.5$ | $60.6 \pm 11.6$ | $69.6 \pm 4.8$ |
| CIFAR-100 | $60.2 \pm 9.2$ | $62.1 \pm 7.9$ | $67.0 \pm 7.1$ |

architecture. On the simpler datasets such as MNIST and Fashion MNIST, both the raw image data and the baseline CAE achieve relatively high AUC-ROCs, but still perform worse than our method. However, on the more complex CIFAR-10 and CIFAR-100 datasets, using our more sophisticated approach to learn a low-dimensional representation of the images results in larger improvements in the average AUC-ROCs.

### Results for Q2

To evaluate the effect of using product quantization on the predictive performance, we consider computing the exact (squared) Euclidean distance instead of computing the approximate quantized Euclidean distance. Again, the quantization is done with the parameters of $m = 32$ and $c = 4$. Table 5 shows the AUC-ROC for our method using the exact Euclidean distance (variants denoted EED) and quantized Euclidean distance (variants denoted QED) for two representative datasets: Fashion MNIST and CIFAR10. Interestingly, in aggregate using the approximate quantized Euclidean distance slightly improves the predictive performance. Depending on which class is considered normal, there are slight differences in performance between EED and QED: sometimes EED results in a higher AUC-ROC and other times QED does. Using QED to assign the anomaly score is about four times faster than using EED.

**Table 5.** The effect of using exact Euclidean distance (variants denoted EED) versus quantized Euclidean distance (variants denoted QED) on predictive performance as measured by AUC-ROC. The AUC-ROC is averaged over treating each of the ten classes as the normal class and the ten models learned for each class.

| Dataset | InceptionCAE NN-EED | InceptionCAE NN-QED |
|---|---|---|
| Fashion MNIST | $93.2 \pm 4.1$ | $93.9 \pm 3.9$ |
| CIFAR-10 | $68.3 \pm 5.7$ | $69.6 \pm 4.8$ |

## Results for Q3

To further investigate where the gains of our approach come from, we explore the effect of the method for assigning an anomaly score on the predictive performance. We compare using the nearest neighbors approach with the quantized Euclidean distance (NN-QED) to using a test image's reconstruction error (RE) as has been done in past work (e.g., [28,40]).

Table 6 shows the results on the Fashion MNIST and CIFAR10 datasets for the baseline CAE and our Inception-like CAE architectures with both methods for computing an anomaly score. We see that using the distance-based approach results in much better performance than using the reconstruction error. Hence, it is probably worth further exploring using distance-based approaches on top of a bottleneck layer.

**Table 6.** Average AUC-ROC when using the reconstruction error (RE) versus the nearest-neighbors approach with quantized Euclidean distance (NN-QED) for assigning the anomaly score. The AUC-ROC is averaged over treating each of the ten classes as the normal class and the ten models learned for each class.

|  |  | RE | NN-QED |
|---|---|---|---|
| Fashion MNIST | CAE | $82.3 \pm 10.9$ | $91.6 \pm 4.2$ |
|  | InceptionCAE | $88.1 \pm 6.5$ | $93.9 \pm 3.9$ |
| FCIFAR-10 | CAE | $56.7 \pm 13.3$ | $60.6 \pm 11.6$ |
|  | InceptionCAE | $55.3 \pm 14.3$ | $69.6 \pm 4.8$ |

## Results for Q4

To explore how the quantization parameters affect predictive and runtime performance, we try all combinations of parameters $m \in \{1, 2, 4, 8, 16, 32, 64, 128\}$ and $c \in \{1, 2, 3, 4, 5, 6, 7, 8\}$. We omit the reduction in memory footprint of using product quantization as the memory tradeoffs are well understood and easily derivable based on the values $m$ and $c$ (see [16]).

Figure 2a shows how the average AUC-ROC (averaged over both treating each class as the normal one and the ten models learned for each class) depends on these parameters for our InceptionCAE NN-QED method on the Fashion MNIST dataset. We see that using values of $c < 3$ has a significant negative effect on the results. The value of $m$ has less of an impact as for a fixed $c$ the average AUC-ROC only varies within a small range regardless of $m$'s value. Until $m = 64$, the AUC-ROC increases with $m$.

Figure 2b shows how the QED search runtime depends on these parameters. We see that for $m \leq 32$ the QED run-time is significantly smaller than the one for the exact distance search, and for these values of $m$ the QED runtime varies only within a small range with the parameters change.

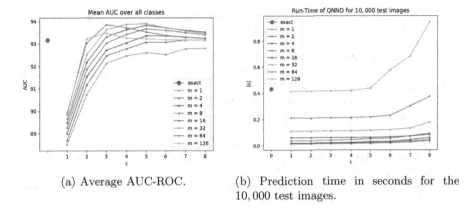

(a) Average AUC-ROC.

(b) Prediction time in seconds for the 10,000 test images.

**Fig. 2.** Effect of the quantization parameters $m$ and $c$ on (a) the average AUC-ROC and (b) the prediction time for the test images in seconds for InceptionCAE NN-QED on the CIFAR-10 dataset. The point "exact" represents computing the exact Euclidean distance (i.e., no product quantization).

## 5 Conclusion

This paper explored anomaly detection in the context of images. We proposed a novel convolutional auto-encoder architecture to learn a low-dimensional representation of the images. Our architecture had two innovations: the use an Inception-like layer and the application of a GAP operation. Then we assigned an anomaly score to images using a nearest neighbors approach in the learned representation space. Our contribution was to use product quantization to improve run time performance of this step. We performed an extensive experimental comparison to both state-of-the-art deep and shallow baselines on four standard datasets. We found that our method resulted in improved predictive performance.

**Acknowledgements.** We thank Lukas Ruff from TU Berlin for help reproducing the results from [25]. This research has been partially funded by the European Union's Horizon 2020 research and innovation program under the Marie Skłodowska-Curie grant agreement No. 752907. JD is partially supported by KU Leuven Research Fund (C14/17/07, C32/17/036), Research Foundation - Flanders (EOS No. 30992574, G0D8819N), VLAIO-SBO grant HYMOP (150033), and the Flanders AI Impulse Program.

## References

1. Andrews, J.T., Morton, E.J., Griffin, L.D.: Detecting anomalous data using auto-encoders. Int. J. Mach. Learn. Comput. **6**(1), 21 (2016)
2. Breunig, M.M., Kriegel, H.P., Ng, R.T., Sander, J.: LOF: identifying density-based local outliers. ACM SIGMOD Rec. **29**(2), 93–104 (2000)

3. Campos, G.O., et al.: On the evaluation of unsupervised outlier detection: measures, datasets, and an empirical study. Data Min. Knowl. Disc. **30**(4), 891–927 (2016)
4. Chakravarty, P., Zhang, A.M., Jarvis, R., Kleeman, L.: Anomaly detection and tracking for a patrolling robot. In: Australasian Conference on Robotics and Automation (ACRA). Citeseer (2007)
5. Chandola, V., Banerjee, A., Kumar, V.: Anomaly detection: a survey. ACM Comput. Surv. (CSUR) **41**(3), 1–72 (2009)
6. Chollet, F., et al.: Keras (2015). https://keras.io
7. Creusot, C., Munawar, A.: Real-time small obstacle detection on highways using compressive RBM road reconstruction. In: 2015 IEEE Intelligent Vehicles Symposium (IV), pp. 162–167. IEEE (2015)
8. Deecke, L., Vandermeulen, R., Ruff, L., Mandt, S., Kloft, M.: Anomaly Detection with Generative Adversarial Networks (2018)
9. Erfani, S.M., Rajasegarar, S., Karunasekera, S., Leckie, C.: High-dimensional and large-scale anomaly detection using a linear one-class SVM with deep learning. Pattern Recognit. **58**, 121–134 (2016)
10. Gionis, A., Indyk, P., Motwani, R.: Similarity search in high dimensions via hashing. In: Proceedings of 25th International Conference on Very Large Data Bases, pp. 518–529 (1999)
11. Golan, I., El-Yaniv, R.: Deep anomaly detection using geometric transformations. In: Advances in Neural Information Processing Systems, vol. 31, pp. 9781–9791 (2018)
12. Hachiya, H., Matsugu, M.: NSH: normality sensitive hashing for anomaly detection. In: IEEE International Conference on Computer Vision Workshops, pp. 795–802 (2013)
13. Haselmann, M., Gruber, D.P., Tabatabai, P.: Anomaly detection using deep learning based image completion. In: Proceedings of 17th IEEE ICMLA, pp. 1237–1242 (2018)
14. He, K., Zhang, X., Ren, S., Sun, J.: Deep residual learning for image recognition. In: Proceedings of the IEEE Conference on Computer Vision and Pattern Recognition, pp. 770–778 (2016)
15. Huang, G., Liu, Z., Van Der Maaten, L., Weinberger, K.Q.: Densely connected convolutional networks. In: Proceedings of the IEEE Conference on Computer Vision and Pattern Recognition, pp. 4700–4708 (2017)
16. Jégou, H., Douze, M., Schmid, C.: Product quantization for nearest neighbor search. IEEE Trans. Pattern Anal. Mach. Intell. **33**(1), 117–128 (2011)
17. Johnson, J., Douze, M., Jégou, H.: Billion-scale similarity search with GPUs. arXiv preprint arXiv:1702.08734 (2017)
18. Kingma, D.P., Ba, J.: Adam: a method for stochastic optimization. arXiv preprint arXiv:1412.6980 (2014)
19. Krizhevsky, A., Hinton, G.: Learning multiple layers of features from tiny images. Technical Report, Citeseer (2009)
20. LeCun, Y., Cortes, C., Burges, C.J.: MNIST handwritten digit database (2010). http://yann.lecun.com/exdb/mnist
21. Lin, M., Chen, Q., Yan, S.: Network in network. arXiv preprint arXiv:1312.4400 (2013)
22. Munawar, A., Vinayavekhin, P., De Magistris, G.: Spatio-temporal anomaly detection for industrial robots through prediction in unsupervised feature space. In: IEEE Winter Conference on Applications of Computer Vision, pp. 1017–1025 (2017)

23. Ramaswamy, S., Rastogi, R., Shim, K.: Efficient algorithms for mining outliers from large data sets. In: ACM Sigmod Record, vol. 29, pp. 427–438. ACM (2000)
24. Richter, C., Roy, N.: Safe Visual Navigation via Deep Learning and Novelty Detection (2017)
25. Ruff, L., et al.: Deep one-class classification. In: International Conference on Machine Learning, pp. 4390–4399 (2018)
26. Sabokrou, M., Fayyaz, M., Fathy, M., Klette, R.: Fully convolutional neural network for fast anomaly detection in crowded scenes. arXiv preprint arXiv:1609.00866 (2016)
27. Sabokrou, M., Khalooei, M., Fathy, M., Adeli, E.: Adversarially learned one-class classifier for novelty detection. arXiv preprint arXiv:1802.09088 (2018)
28. Sakurada, M., Yairi, T.: Anomaly detection using autoencoders with nonlinear dimensionality reduction. In: Proceedings of the 2nd Workshop on Machine Learning for Sensory Data Analysis, p. 4. ACM (2014)
29. Schlegl, T., Seeböck, P., Waldstein, S.M., Schmidt-Erfurth, U., Langs, G.: Unsupervised anomaly detection with generative adversarial networks to guide marker discovery. In: Niethammer, M., et al. (eds.) IPMI 2017. LNCS, vol. 10265, pp. 146–157. Springer, Cham (2017). https://doi.org/10.1007/978-3-319-59050-9_12
30. Schubert, E., Zimek, A., Kriegel, H.-P.: Fast and scalable outlier detection with approximate nearest neighbor ensembles. In: Renz, M., Shahabi, C., Zhou, X., Cheema, M.A. (eds.) DASFAA 2015. LNCS, vol. 9050, pp. 19–36. Springer, Cham (2015). https://doi.org/10.1007/978-3-319-18123-3_2
31. Seeböck, P., et al.: Identifying and categorizing anomalies in retinal imaging data. arXiv preprint arXiv:1612.00686 (2016)
32. Shashikar, S., Upadhyaya, V.: Traffic surveillance and anomaly detection using image processing. In: 2017 Fourth International Conference on Image Information Processing (ICIIP), pp. 1–6. IEEE (2017)
33. Sultani, W., Chen, C., Shah, M.: Real-world anomaly detection in surveillance videos. In: Proceedings of the IEEE Conference on Computer Vision and Pattern Recognition, pp. 6479–6488 (2018)
34. Szegedy, C., et al.: Going deeper with convolutions. In: Proceedings of the IEEE Conference on Computer Vision and Pattern Recognition, pp. 1–9 (2015)
35. Taboada-Crispi, A., Sahli, H., Hernandez-Pacheco, D., Falcon-Ruiz, A.: Anomaly detection in medical image analysis. In: Handbook of Research on Advanced Techniques in Diagnostic Imaging and Biomedical Applications, pp. 426–446 (2009)
36. Tax, D.M.J., Duin, R.P.W.: Support vector data description. Mach. Learn. 54(1), 45–66 (2004)
37. Vercruyssen, V., Meert, W., Verbruggen, G., Maes, K., Baumer, R., Davis, J.: Semi-supervised anomaly detection with an application to water analytics. In: IEEE 2018 International Conference on Data Mining, pp. 527–536 (2018)
38. Vinyals, O., Toshev, A., Bengio, S., Erhan, D.: Show and tell: a neural image caption generator. In: Proceedings of the IEEE Conference on Computer Vision and Pattern Recognition, pp. 3156–3164 (2015)
39. Wei, Q., Ren, Y., Hou, R., Shi, B., Lo, J.Y., Carin, L.: Anomaly detection for medical images based on a one-class classification. In: Medical Imaging 2018: Computer-Aided Diagnosis, vol. 10575, p. 105751M. International Society for Optics and Photonics (2018)
40. Xia, Y., Cao, X., Wen, F., Hua, G., Sun, J.: Learning discriminative reconstructions for unsupervised outlier removal. In: Proceedings of the IEEE International Conference on Computer Vision, pp. 1511–1519 (2015)

41. Xiao, H., Rasul, K., Vollgraf, R.: Fashion-MNIST: a novel image dataset for bench-marking machine learning algorithms. arXiv preprint arXiv:1708.07747 (2017)
42. Xu, D., Ricci, E., Yan, Y., Song, J., Sebe, N.: Learning deep representations of appearance and motion for anomalous event detection. arXiv preprint arXiv:1510.01553 (2015)
43. Zhai, S., Cheng, Y., Lu, W., Zhang, Z.: Deep structured energy based models for anomaly detection. In: Proceedings of the 33rd International Conference on International Conference on Machine Learning, vol. 48, pp. 1100–1109 (2016)
44. Zhanga, Y., Lua, H., Zhanga, L., Ruanb, X., Sakaib, S.: Video anomaly detection based on locality sensitive hashing filters. Pattern Recognit. **59**, 302–311 (2016)
45. Zong, B., et al.: Deep autoencoding Gaussian mixture model for unsupervised anomaly detection. In: International Conference on Learning Representations (2018)

**Time Series**

# Adaptive Long-Term Ensemble Learning from Multiple High-Dimensional Time-Series

Samaneh Khoshrou$^{(\boxtimes)}$ and Mykola Pechenizkiy

Eindhoven University of Technology, Eindhoven, The Netherlands
{s.khoshrou,m.pechenizkiy}@tue.nl

**Abstract.** Learning from multiple time-series over an unbounded time-frame has received less attention despite the key applications (such as video analysis, home-assisted) generating this data. Inspired by never-ending approaches, this paper presents an algorithm to continuously learn from multiple high-dimensional un-regulated time-series, in a framework based on ensembles which with respect to drift level develops over time in order to reflect the latest concepts. Here, we explicitly look into video surveillance problem as one of the main sources of high-dimensional data in daily life and extensive experiments are conducted on multiple datasets, that demonstrate the advantages of the proposed framework in terms of accuracy and complexity over several baseline approaches.

**Keywords:** Long-term learning · Data streams · Ensembles

## 1 Introduction

With the advent of distributed sensor networks in many real world problems such as ambient assisted living, medial diagnosis, and video surveillance making sense of an ever-increasing amount of data is a growing challenge. It is expected that using time-series learning techniques leads to effective and hands-on solutions for such scenarios. In this paper, long-term learning from multiple high-dimensional time-series is approached using an *active ensemble-based strategy*. Inspired by active approaches[1], our ensemble based framework is able to track the environment changes using a change detection test. Exploiting a teacher in the learning loop, which referred as "active learning" in the literature, is one of the new trends for learning from complex environments (e.g. weakly labeled). Inspired by these two groups of methods, we proposed our *active ensemble-based strategy* for long-term learning which posses two main characteristics: (1) perform a

---

[1] The term "active approaches" was first coined in [6] for one of the two main groups of learning methods in non-stationary environments that rely on an explicit detection of the change in the data distribution to activate an adaptation mechanism, as a counterpart of passive approaches which continuously update the model over time without requiring an explicit detection of the change.

© Springer Nature Switzerland AG 2019
P. Kralj Novak et al. (Eds.): DS 2019, LNAI 11828, pp. 511–521, 2019.
https://doi.org/10.1007/978-3-030-33778-0_38

change detection test to activate an adaptation procedure. (2) collaborate with an oracle (teacher) to stay on track. As a use case, we focus on learning from multiple video streams as one of the most common and challenging high-dimensional time-series generated in various real-world problems.

**Running Example.** Consider a network of video cameras employed to monitor a large area. A central issue here is the tracking and recognition of individuals of interest across multiple cameras. These individuals must be recognized when leaving the Field of View (FoV) of one camera and re-identified when entering the FoV of another camera, which is known as *person re-identification (ReID)* in the literature. ReID is a challenging and widely studied problem in video analysis field [13,16] and it underpins many crucial applications such as long-term multi-camera tracking [9], behaviour analysis, and security monitoring. In such environments, the underlying distribution of data changes over time either due to illumination changes, dynamic background, changes in camera angle, and etc. Thus, recognition models need to be continually updated to represent the latest concepts. Moreover, when new objects enter the scene new models need to be trained for the novel classes. The problem gets further complex when the system is faced with *unbounded streams* of data [1]. When entering the scene, the object will enter the coverage area of at least one of the cameras. As objects move around and cross the Field of View (FoV) of multiple cameras, it is more than likely to have multiple unregulated streams, potentially overlapping in time for the same individual object. The surveillance system will have to track that object from the first moment it was captured by a camera and across all cameras whose fields of view overlap the object's path.

**Problem Formulation.** Learning in such scenario can be characterized as follows: Let $v$ be a set of unregulated time-series $v_i$. Streams are potentially subject to concept drift as well as concept evolution. Each observation $x$ within each stream is in a d-dimensional space, $x \in R^d$. Recording is not limited to a bounded period. An effective and appropriate one-pass algorithm to fit in our scenario is required to: (a) learn from multiple unregulated streams; (b) handle multi (possibly high)-dimensional data; (c) handle concept drift; (d) accommodate new classes; (e) deal with massive amount of unlabelled data; (f) be of limited complexity, which is the of the main contributions of this paper.

**Main Contributions.** We propose Intelligent Never Ending Visual Information Learning (InteLL), a novel framework which is designed for long-term learning from multiple un-regulated time-series and here is applied for long-term tracking of previously unseen objects over multiple cameras (Sect. 2). The active approach is based on a change detection strategy that triggers adaptation with respect to the level of drift, by updating or building a classifier. The adaptation process is achieved through the merging of two Gaussian Mixture Models (GMMs) into a single one, which has not been addressed before (See Sect. 2).

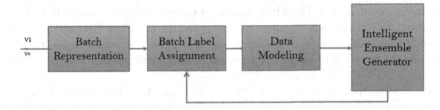

**Fig. 1.** Block diagram of InteLL framework

The codes is made publicly available[2]. The rest of the paper is organized as follows. In Sect. 2 we introduce InteLL our novel framework for Intelligent Never Ending Learning. In Section 3 we discuss the experimental methodology to assess the performance of the proposed approach, and the main results we obtained. An overview of related work on never-ending learning is presented in Sect. 4. Finally, we draw conclusions and discuss directions of future work in Sect. 5.

## 2 InteLL: Intelligent Long-Term Learning from Multiple Time-Series

In this section, the Intelligent Long-term Learning from time-series (InteLL) framework is presented. InteLL is specifically designed for long-term learning from non-stationary environments in which no labelled data is available but the learning algorithm is able to interactively query teachers to label meticulously chosen observations. Figure 1 shows a high-level sketch of INTELL framework.

**Batch Representation.** We adopt the localised average histogram as the appearance feature of a video fragment because of its simplicity and effectiveness.

**Batch Label Assignment.** Once the features of batches of RoIs $(v^{*m}_{t,f})$ at time slot $t$ become available, the framework starts computing the scores $\mathcal{S}(v^{*m}_t|C_k, H_{t-1})$ for every batch. The scores are obtained from the likelihood ratio test of the batch data obtained by the individual class model $C_k$ and the Universal Background Models (as detailed in Sect. 2). The composite model $H_t$ is an ensemble of Micro ensembles $ME^j_t, j = 1, ..., K_t$, where $K_t$ is the number of classes observed until time $t$. Each $ME^j_t$ includes models $h^j_t$ that has been trained on incoming batches of the $j^{th}$ class since it initially appeared until the current time [15]. The prediction output by the composite model $ME^j_t$ for a given batch of RoIs $(v^{*m}_t)$ is

$$S(C_k|v^{*m}_t, ME^j_t) = \sum_{\ell=1}^{t} W^t_\ell \mathcal{S}^j_\ell(C_k|v^{*m}_t) \tag{1}$$

where $\mathcal{S}^j_\ell(.)$ is the score output by $h^j_\ell(.)$ , and $W^t_\ell$ denotes the weight assigned to model $h^j_\ell$, adjusted for time $t$ of the last action to give more credit to

---

[2] https://github.com/SamanehKh77/InteLL.git.

the more recent knowledge. Once batch scores for different classes have been obtained, an uncertainty measure is extracted to evaluate how accurate the ensemble decision is. One of the simplest and most commonly used criterion relies on the probability of the most probable class, defining the confidence level as $\max_{C_k} \mathcal{S}(C_k|v_t^{*m}, H_{t-1})$.

On-line learning from time-evolving environments, where labelled data is scarce, may suffer if labelling errors accumulate. To mitigate this issue, the system is designed to exploit *active learning* strategies. Based on the type of the oracle (teacher) available, these strategies request two groups of teachers to label the data. Strong teacher that is usually but not always is human and assumed to give unambiguous but expensive labels, while weak teachers provide tentative but less expensive labels. Most, but not all weak teachers are assumed to be classification algorithms. InteLL provides the opportunity to take advantage of both groups. The ensemble, $H$, participates in learning as a weak teacher. Additionally, in order to reduce the number of the queries, the label assignment for novel classes is automated. If the scores associated to all observed classes $(\mathcal{S}(C_j|v_t^{*m}, H_{t-1}), j = 1, ..., k)$ are significantly low (below a predetermined threshold), it is very unlikely that this class has been observed before and it is considered novel. A new label $(\ddot{y})$ is automatically assigned to this/these batch(es). Having decided that the batch data belongs to an existing class, one needs to decide which teacher to invoke; either strong teacher (i.e. operator) or weak teacher (i.e. ensemble $H$). If the decision made by $H$ is not reliable enough, i.e. if $\max_{C_k} \mathcal{S}(C_k|v_t^{*m}, H_{t-1}) < T_1$, a strong teacher labelling (in this case by an operator) is requested $(y)$, otherwise we invoke the weak teacher and $\hat{y} = \arg\max_{C_k} \mathcal{S}(v_t^{*m}|C_k, H_{t-1})$ is assigned to the batch.

**Data Modelling.** The batches predicted to belong to the same class are used to generate the class model $h_t^j$ by *tuning the UBM parameters* in a maximum a *posteriori* (MAP) sense at every time slot. The training process of the UBM is performed by fitting a $k$-mixture GMM to the set of feature vectors extracted from a pool of streams of disjoint individuals that is representative of the complete set of potentially observable 'objects'. adaptation process consists of two main estimation steps. First, for each component of the UBM, a set of sufficient statistics is computed from a set of $M$ class specific feature vectors. Each UBM component is then adapted using the newly computed sufficient statistics, and considering diagonal covariance matrices.

**Intelligent Ensemble Generator.** The key contribution of this paper is to propose an incremental learning algorithm for never-ending scenarios where the system can learn continually 24/7 from wide-area surveillance networks. The advantage of ensembles-based algorithms in tackling these problems is the ability to accumulate and aggregate knowledge in the form of learned models [18].

In our framework we consider multiple micro-ensembles that evolve over time either by training new classifiers (in case of abrupt drift) or updating existing classifiers (for gradual drifts) without sacrificing the performance. It is expected that the proposed algorithm has high potential to identify recurrent drift, as it employs change detection strategies at the learner level in order to (re-)identify concept drift thereby provoking a favourable yet less involved response. Once the batch(es), $v_t^{*m}$, in time slot $t$ are classified as class "$m$" (either by invoking teachers or automatically), the framework trains a model $h_t^j$. The similarity scores $d_{tj}^m$ are computed between all the models available in the corresponding micro-ensemble $ME_{(t-1)}^j$ and the newly trained model $h_t^j$. Once similarity scores are obtained, all the models are searched to find the one with minimum distance with the recent one. If $d_{tk}^m = \min d_{tj}^m, j = 1, ..., K$ ($K$ is the number of models inside $m_{th}$ micro-ensemble) is high enough (above a predefined threshold, ($T'$), the model is seen as distinct, and is added to the ensemble, otherwise the closest pair is merged into a single model ($\hat{h}_t^m$). The highest time-based credit will be assigned to either $\hat{h}_t^m$ or $h_t^m$ as the most recent model. Hence, we need a notion of similarity between models as well as a strategy for updating the learners. The Kullback-Leibler divergence is a natural similarity measure between two distributions. Although it cannot be analytically computed for GMMs, an efficient and accurate approximation of KL-divergence for GMMs is proposed in [8]. Assume $h(x) = \sum_{i=1}^n \alpha_i h_i(x)$ and $h'(x) = \sum_{k=1}^m \beta_k h_k'(x)$ are two Gaussian Mixture densities whose KL-divergence we want to compute. Generally, the KL-divergence between two GMMs can be approximated by: $KL(h \parallel h') \approx \sum_{i=1}^n \alpha_i \min_{k=1}^m KL(h_i \parallel h_k')$ The approximation is based on a matching function between each element of $h$ and an element of $h'$ that is the most similar to it. Various methods including the Hungarian algorithm have been employed to find corresponding components.

Since, in GMM-UBM, the GMMs are obtained from a maximum a posteriori adaptation of a universal background model, the both densities have the same number of components and there is a well justified correspondence between components, the KL-divergence can be approximated as:

$$KL(h \parallel h') = \sum_i \alpha_i KL(h_i \parallel h_i') \tag{2}$$

where the KL-divergence between components $h_i(\mu_1, \Sigma_1)$ and $h_i'(\mu_2, \Sigma_2)$ can be formulated as: $KL(h_i \parallel h_i') = \frac{1}{2}(log\frac{|\Sigma_2|}{|\Sigma_1|} + Tr(\Sigma_2^{-1}\Sigma_1) + (\mu_1 - \mu_2)^T \Sigma_2^{-1}(\mu_1 - \mu_2) - d)$

Finally, the distance between two distributions $h$ and $h'$ is computed as:

$$d_{hh'} = \frac{KL(h \parallel h') + KL(h' \parallel h)}{2} \tag{3}$$

**Updating a Concept.** The problem of updating GMM has mostly appeared in situations where: (1) a Gaussian mixture is fitted but the mixture components are not separated enough [11], (2) a non-stationary environment the model is not representative as time passes and the Gaussian mixture model needs to be updated to track environment change [14]. These algorithms mainly focus on updating a GMM, by either merging or splitting the components, while combining two GMMs is a less explored area. Here We propose a method that *incrementally merges GMMs* without the necessity of retaining all data points, which has not been addressed before. Given two GMMs learned from two sets of observations $f(x)$ and $g(x)$ with $n'$ and $n''$ points:[3] $f(x) = \sum_{i=1}^{M} f_i(x'_i, \pi'_i, \mu'_i, \Sigma'_i)$, and $g(x'') = \sum_{j=1}^{M} g_j(x_j'', \pi_j'', \mu_j'', \Sigma_j'')$, where each component is represented by its weight $(\pi)$, mean $(\mu)$, and covariance $(\Sigma)$. First, the corresponding components in different GMMs are found (finding the closest component from two or multiple GMMs). The $i_{th}$ component of $f(x)$ corresponds to $j_{th}$ component of $g(x)$, that can be merged and form the $k_{th}$ component of

$$m(x) = \sum_{j=1}^{M} m_k(x_k, \pi_k, \mu_k, \Sigma_k) \tag{4}$$

Note that the number of points in $j_{th}$ component is expected to be equal to the product of the component weight $(\pi_j)$ and the total number of points in the GMMs $(n)$. Using the definition of mean, variance and prior we derive:

$$\pi_k = \frac{n'\pi'_j + n''\pi''_j}{n' + n''} \tag{5}$$

$$\mu_k = \frac{\sum x_k}{n} = \frac{\sum x'_i + \sum x''_j}{n'\mu'_i + n''\mu''_j} = \frac{n'\pi'\mu'_i + n''\pi''_j\mu''_j}{n'\pi'_i + n''\pi''_j} \tag{6}$$

$$\Sigma_k = E(x_k^2) - E^2(x_k) = \frac{n'\pi'(\Sigma_i + \mu'_i\mu'^T_i) + n''\pi''_j(\Sigma''_j + \mu''_j\mu''^T_j)}{n'\pi'_i + n''\pi''_j} - \mu_k\mu_k^T \tag{7}$$

which are weights, mean, and covariance of the new model, respectively.

## 3    Experimental Methodology and Results

Experiments were conducted on public indoor (seven scenarios of CAVIAR[4], SAIVT-Softbio [3]) as well as outdoor (PETS) datasets. Table 1 presents a qualitative look at the characteristics of the datasets applied in the experiments. Given an image (RoI), the Improved Fisher Vector (IFV) [7] $\upsilon$ is obtained by extracting a dense collection of patches and corresponding local image features (herein, SIFT) from the image at multiple scales. To avoid the curse of dimensionality, Principle Component Analysis (PCA) is applied to the full set of features as a pre-processing step. The number of features in each stream is reduced to 200 dimensions. To evaluate the system, each dataset is divided into 3 disjoint subsets (different individuals). The first subset is used to train UBMs. The second set is used to calibrate all the parameters. The final portion is used to evaluate the performance.

---

[3] For the sake of simplicity, the method is proved by a pair of 2-component GMMs, however the extension for multiple GMMs is quiet trivial.

[4] http://homepages.inf.ed.ac.uk/rbf/CAVIAR/.

**Table 1.** A quantitative look at datasets.

| Datasets | OneLeaveShopReenter1 | OneLeaveShopReenter2 | EnterExitCrossingPaths1 | OneStopEnter2 | OneShopOneWait1 | OneStopMoveEnter1 | PETS2009 | SAIVT-SOFTBIO | SAIVT-NonOver | SAIVT-Recurrent |
|---|---|---|---|---|---|---|---|---|---|---|
| No. of Streams | 15 | 15 | 20 | 19 | 28 | 60 | 29 | 240 | 14 | 28 |
| Range of streams | [85–160] | [63–347] | [34–216] | [51–657] | [36–605] | [10–555] | [85–576] | [21–211] | [21–211] | [63–636] |
| No. of Classes | 15 | 15 | 16 | 16 | 20 | 26 | 22 | 152 | 7 | 7 |

**Evaluation Criteria.** Never-ending learning is a relatively new field, and different problems in the literature are informed by different assumptions with respect to the applications. Thus, there is not a single evaluation metric that everyone agrees is the most reasonable metric for this problem. InteLL is designed to be employed over longer time horizons in an interactive framework, and expected to maintain a reasonable level of performance with controlled complexity. Hence performance as a function of complexity seems to be an informative metric to assess the framework. Never-ending learning are aimed to work over an unlimited time frame. The framework should be able to coverage the knowledge accurately while controlling the complexity.

- Complexity As an ensemble-based framework size of the ensemble in other term the number of the models inside ensemble seems an intuitive notion of complexity.
- Area Under the Learning Curve (ALC) Since, InteLL is an interactive framework, accuracy itself does not seems the most informative measure. Herein, the learning curve is the set of accuracy plotted as a function of their respective annotation effort, $a$. The ALC is obtained by: $ALC = \int_0^1 f(a)da$

**Baseline Methods.** The work closest in spirit to InteLL is [10], that proposed a never-ending framework for one dimensional real value time series. Since, we deal with multiple high-dimensional data streams, the framework is not applicable in our scenario. The InteLL framework is compared with three baseline approaches:

- NEVIL.ubm [15] is an example of algorithms for learning from multiple uneven streams in non-stationary environments where both concept drift and concept evolution are present. The framework adds a new model as new data is received. Thus, the complexity of the ensemble grows over time.
- Incremental UBM learning [17] performs a continuous adaptation of the model once a new observation is received. A single GMM-UBM is incrementally trained on incoming batches of $j_{th}$ class at $t$, $h_t^j$.

– The Active Add is based on change detection mechanisms. Once an abrupt change is detected, a new classifier is built and added to the corresponding micro-ensemble.

**Results.** The optimal batch size varies and is influenced by the characteristics of the streams present in each dataset. However, in our experiments for a frame rate of 25 fps we conducted all the experiments on one second-length TSs (up to 25 frame per batch). In this section, the InteLL is evaluated on the task of long-term learning using the model-level assessment strategy. The system performance is evaluated based on ALC versus the number of classifiers. Figure 2 illustrates the performance of InteLL on various video clips. The ALC is presented as a function of the number of classifiers. We observe that in 7 out of 10 datasets (i.e. SAIVT, SAIVTT Nonover, Enter ExitCrossingPaths1, OneShopOneWait1, OneStop Enter2, WalkBy Shop1front), the framework obtains comparable performance by learning only 20% number of the maximum classifiers possible. The average number of classifiers per person is 1.8 (21 models for 11 classes). With only one exception (OneExitCrossingPath1) InteLL outperforms incremental learning (depicted as red). AA method (plotted as green line) provides the lowest ALC confirming the importance of the updating procedure.

## 4    Related Work

Never-ending learning systems have been one of the latest interest in the field of learning as they are able to learn many concepts "in a cumulative nature". The Never-Ending Language Learning (NELL) [4] research project has been the inspiration of numerous researches to address the never-ending learning problem [2,5,10,12]. NEIL (Never Ending Image Learner) [5], was developed to automatically extract visual knowledge from Internet data. NELTS was proposed in [10] for long-term learning from single time-series. NOEL [19] was designed for never-ending object learning in Robatics. Obviously, the techniques used by research works are informed by different assumptions in respect with the applications and goals. With a few exceptions [19], most of the never-ending literature has focused on coverage of knowledge, while our approach tries to cover knowledge and accuracy as well as efficiency.

To the best of our knowledge, InteLL the first attempt to employ active strategies to control the complexity of a passive method (here an ensemble) to make it applicable for long-term learning of time-series in an online manner. Theoretically there is no upper bound on the number of classes learnt by the system (coverage of the knowledge). Furthermore, results demonstrate high ALC (accuracy) by only generating a quiet reasonable number of models inside each MCE (efficiency). Given these, InteLL is the first step towards a practical constant learner for many challenging real-world problems.

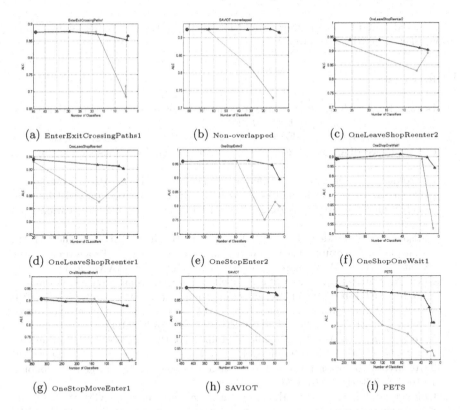

**Fig. 2.** ALC against number of classifiers generated in descending order of InteLL against multiple baseline approaches on our set of real-world datasets. The signs ——•+ denote the results of AA, InteLL, NEVIL.ubm, and Incremental Learning, respectively. (Color figure online)

## 5   Conclusion

We presented a learning setting yet unexplored in the literature but with wide practical relevance, such as in long-term person re-identification over multiple video cameras. Adaptive ensembles are developed over time with respect to drift level, either by updating an existing model or adding a new learner to the ensemble, in order to reflect the latest concepts appearing in the environment as well as bound system complexity. However, long-term learning from multiple video streams is assessed in the experimental part and favourable results on multiple datasets indicate the effectiveness of this method, InteLL can be applied in any multi-stream learning setting.

Utilizing domain adaptation approaches to provide an up-to-date knowledge, as well as employing the framework in other domains including banking, insurance, and home-assisted settings all constitute our future work.

**Acknowledgement.** The authors would like to thank the Dutch Research Council (NWO) for supporting this project.

# References

1. Begum, N., Keogh, E.: Rare time series motif discovery from unbounded streams. Proc. VLDB Endow. **8**(2), 149–160 (2014)
2. Berlin, E., Van Laerhoven, K.: Detecting leisure activities with dense motif discovery. In: Proceedings of the 2012 ACM Conference on Ubiquitous Computing, pp. 250–259 (2012)
3. Bialkowski, A., Denman, S., Sridharan, S., Fookes, C., Lucey, P.: A database for person re-identification in multi-camera surveillance networks. In: 2012 International Conference on Digital Image Computing Techniques and Applications, pp. 1–8 (2012)
4. Carlson, A., Betteridge, J., Kisiel, B., Settles, B., Jr., E.R.H., Mitchell, T.M.: Toward an architecture for never-ending language learning. In: Proceedings of the Twenty-Fourth AAAI Conference on Artificial Intelligence, AAAI (2010)
5. Chen, X., Shrivastava, A., Gupta, A.: Neil: extracting visual knowledge from web data. In: IEEE International Conference on Computer Vision, pp. 1409–1416 (2013)
6. Elwell, R., Polikar, R.: Incremental learning of concept drift in nonstationary environments. IEEE Trans. Neural Netw. **22**(10), 1517–1531 (2011)
7. Perronnin, F., Sánchez, J., Mensink, T.: Improving the fisher kernel for large-scale image classification. In: Daniilidis, K., Maragos, P., Paragios, N. (eds.) ECCV 2010. LNCS, vol. 6314, pp. 143–156. Springer, Heidelberg (2010). https://doi.org/10.1007/978-3-642-15561-1_11
8. Goldberger, J., Aronowitz, H.: A distance measure between GMMS based on the unscented transform and its application to speaker recognition. In: INTERSPEECH 2005 - Eurospeech, 9th European Conference on Speech Communication and Technology, Lisbon, Portugal, 4–8 September 2005, pp. 1985–1988 (2005)
9. Gong, S., Cristani, M., Yan, S., Loy, C.C. (eds.): Person Re-Identification. Advances in Computer Vision and Pattern Recognition. Springer, London (2014). https://doi.org/10.1007/978-1-4471-6296-4
10. Hao, Y., Chen, Y., Zakaria, J., Hu, B., Rakthanmanon, T., Keogh, E.: Towards never-ending learning from time series streams. In: KDD 2013, pp. 874–882 (2013)
11. Hennig, C.: Methods for merging gaussian mixture components. Adv. Data Anal. Classif. **4**(1), 3–34 (2010)
12. Hua, X.S., Li, J.: Prajna: towards recognizing whatever you want from images without image labeling. In: Proceedings of the Twenty-Ninth AAAI Conference on Artificial Intelligence, AAAI 2015, pp. 137–144 (2015)
13. Karanam, S., Li, Y., Radke, R.J.: Person re-identification with discriminatively trained viewpoint invariant dictionaries. In: 2015 IEEE International Conference on Computer Vision, ICCV 2015, Santiago, Chile, pp. 4516–4524 (2015)
14. Khoshrou, A., Aguiar, A.P., Pereira, F.L.: Adaptive sampling using an unsupervised learning of GMMS applied to a fleet of auvs with CTD measurements. In: Second Iberian Robotics Conference, pp. 321–332 (2016)
15. Khoshrou, S., Cardoso, J.S., Teixeira, L.F.: Learning from evolving video streams in a multi-camera scenario. Mach. Learn. **100**(2–3), 609–633 (2015)
16. Wang, T., Gong, S., Zhu, X., Wang, S.: Person re-identification by discriminative selection in video ranking. IEEE Trans. Pattern Anal. Mach. Intell. **38**(12), 2501–2514 (2016)

17. Wu, T.Y., Lu, L., Chen, K., Zhang, H.J.: UBM-based incremental speaker adaptation. In: 2003 International Conference on Multimedia and Expo, ICME 2003, Proceedings, vol. 2, pp. II-721, July 2003
18. Yi, Z.: Constructive and Destructive Optimization Methods for Predictive Ensemble Learning. University of Iowa (2006)
19. Yuyin Sun, D.F.: Neol: toward never-ending object learning for robots. In: IEEE International Conference on Robotics and Automation (ICRA) (2016)

# Fourier-Based Parametrization
# of Convolutional Neural Networks
# for Robust Time Series Forecasting

Sascha Krstanovic$^{(\boxtimes)}$ and Heiko Paulheim$^{(\boxtimes)}$

Data and Web Science Group, University of Mannheim, Mannheim, Germany
{sascha,heiko}@informatik.uni-mannheim.de

**Abstract.** Classical statistical models for time series forecasting most often make a number of assumptions about the data at hand, therewith, requiring intensive manual preprocessing steps prior to modeling. As a consequence, it is very challenging to come up with a more generic forecasting framework. Extensive hyperparameter optimization and ensemble architectures are common strategies to tackle this problem, however, this comes at the cost of high computational complexity. Instead of optimizing hyperparameters by training multiple models, we propose a method to estimate optimal hyperparameters directly from the characteristics of the time series at hand. To that end, we use Convolutional Neural Networks (CNNs) for time series forecasting and determine a part of the network layout based on the time series' Fourier coefficients. Our approach significantly reduces the amount of required model configuration time and shows competitive performance on time series data across various domains. A comparison to popular, state of the art forecasting algorithms reveals further improvements in runtime and practicability.

**Keywords:** Time series forecasting · Neural networks · Fourier analysis

## 1 Introduction

In the age of connected sensors, devices, and services, temporal data is one of the most widespread data types these days. Designing accurate forecasting models typically involves lots of manual work, e.g. data preprocessing, parameter tuning, and model selection. Since time series data comes in different shapes and distributions, these manual steps are usually required for each new dataset.

In this work, we propose a time series forecasting framework based on CNNs that makes no prior assumptions about data distribution and integrates all required preprocessing steps. We demonstrate its predictive power on thirty data series, where the approach outperforms all baselines in two-thirds of the cases without the need to manually adapt any parameter across the different datasets. In addition to this, we show significant improvements in runtime of the training process, therewith, providing a very convenient forecasting method that is fast and robust.

© Springer Nature Switzerland AG 2019
P. Kralj Novak et al. (Eds.): DS 2019, LNAI 11828, pp. 522–532, 2019.
https://doi.org/10.1007/978-3-030-33778-0_39

Our approach combines the predictive power of CNNs with the time series decomposition capabilities of Fourier analysis. Hyperparameter tuning is expensive, because it requires training multiple models. In this paper, we follow a different approach and configure a neural network analytically. Our idea is based on the assumption that the characteristics of a time series – more specifically: its Fourier decomposition – can be used to determine a suitable network layout for a CNN. Hence, we exploit the inherent structure of time series data in order to parametrize the CNN used for forecasting.

Section 2 starts with existing approaches to time series forecasting, discussing their assumptions and strengths. In Sect. 3, we provide a detailed explanation of our contribution and its motivation. These ideas are applied to numerous real world datasets in Sect. 4, demonstrating advantages and limitations. We conclude and discuss future work in Sect. 5.

## 2 Related Work and Time Series Fundamentals

Due to the diverse occurrence of time series data in applications and databases, its analysis has been an active research field for decades. Temporal data has the interesting property that the current value is dependent on a number of past values. In other words, observations are not independent of each other but can be thought of as a function of their past values.

### 2.1 Autoregression and Smoothing

Autoregressive (AR) models are amongst the most popular approaches for time series analysis and forecasting. AR models approximate a time series with a linear combination of the most recent past values and their errors [4]. These models perform particularly well if the assumption is met that the series is generated by a linear process [1], however, this barely holds in practice.

Exponential Smoothing constitutes another relatively simple yet popular approach to forecasting. Here, the series is smoothed by applying an exponential window function. This implies the assignment of weights which decrease over time.

### 2.2 Machine Learning

**Forecasting as a Supervised Regression Problem.** An advantageous property of historical time series data is that transforming it to a supervised machine learning task is easy. Past observations serve as explanatory features to the respective future values that constitute the target variables. Unlike other supervised learning tasks (e.g. image recognition), time series data can be automatically transformed to a supervised problem without the need for manual annotation. This aspect is critical for the success of end-to-end forecasting frameworks such as the one presented in this paper.

Handled this way, the forecasting task follows the same process as any other supervised machine learning challenge, i.e., hyperparameter optimization, evaluation, and model selection. This idea is also implemented in [23] in order to train the base models required for ensemble learning.

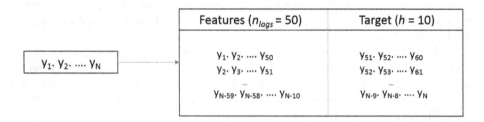

**Fig. 1.** Transforming raw, univariate time series data to a supervised learning task

There are two common strategies of how to design a machine learning system for multi-step time series forecasting, known as the direct and indirect methods. Assuming a forecasting horizon $h > 1$, one can either train a dedicated model for each future point $1, 2, ..., h$ (direct method), or only train a single model and use its forecast as input for the succeeding future point in an iterated fashion (indirect method). Since the direct approach requires the training of $h$ individual models it scales very poorly for longer horizons, leading to a low practicability for actual applications. It was also shown that the performance is inferior to that of the indirect method for AR models [9].

**Artificial Neural Networks.** [11] showed that an ANN with one hidden layer is able to approximate a continuous function arbitrarily well, which makes ANNs highly interesting for regression problems. The spike in popularity of ANNs within the past decade led to significant developments for time series analysis, especially with regard to recurrent neural networks (RNNs). As these models usually make no prior assumptions about data distribution, they have a major advantage over more classical time series models described in the previous section. Intuitively speaking, RNNs can be thought of as standard feed forward networks with loops in them. This sequential architecture enabled RNNs to achieve new state of the art results on a variety of sequential tasks such as machine translation [5,20] and time series forecasting [3,14,17]. Nevertheless, RNNs tend to suffer from vanishing or exploding gradients in case of very long-term dependencies in the data [2,18]. Long short-term memory (LSTM) cells overcome this problem by the introduction of a gating mechanism that regulates the information flow of the network [10]. This allows for more reliable modeling of long-term dependencies, leading to a wide adoption of LSTMs for sequential tasks.

Due to the sequential nature of RNNs, parallelization possibilities are limited and training these models is computationally expensive. As a consequence, the application of convolutional neural networks (CNNs) to sequential problems has been a very active field of study. CNNs rely on filtering mechanisms in order to generate meaningful meta-features out of the raw input data. While primarily used to analyze visual imagery [27] and audio [26], the application of CNNs to sequential regression problems has shown competitive prediction performance at significantly lower computational complexity [7].

### 2.3   Ensembles

An essential requirement for ensemble methods to be effective is that the base learners are heterogeneous and make different errors at prediction time. [22] presents strategies on diversity generation for ensemble models. [23] combines several heterogeneous machine learning base models that are arbitrated by a meta-learner to generate the final forecasts. [12,13] make use of local minima in the LSTM training process by storing model snapshots every time the LSTM converges to a local minimum. While ensemble models boost predictive performance, training of multiple base learners leads to high computational costs.

## 3   Forecasting with CNNs

The general procedure to train a CNN forecasting model follows the steps required for any regression problem, where the lags of the input series serve as features and the respective future values as targets. Figure 1 depicts the splitting logic in the data preprocessing stage. The forecasting horizon $h$ is usually a given parameter that is defined by the problem at hand. More importantly, the number of past lags to include as features must be large enough in order to account for long-term relationships across the series.

### 3.1   A CNN Algorithm for Multi-Step Forecasting

While CNNs are best known for their powerful capabilities within the area of image recognition, they are also widely used for more traditional regression problems. Hence, their structure allows the application to autoregressive tasks such as time series forecasting. Contrary to RNNs, however, CNNs are not naturally built for sequence processing. In order to generate sequence forecasts despite its architecture, we apply the indirect forecasting method, following the logic described before.

### 3.2   Enhancing Parameter Optimization with Fourier Analysis

The major challenge when dealing with ANNs is hyperparameter optimization. As this is a data specific task, a robust forecasting framework greatly benefits from efficient parameter optimization.

For the special case of time series data, we make use of its inherent structure in order to enhance CNN parametrization. Since one of the key tasks of time series models is to correctly identify repeating patterns over time, it is essential to parametrize models in a way that enables them to determine these structures.

Fourier analysis is a powerful and efficient tool that solves the problem of determining autocorrelations [19]. We apply the fast Fourier transform (FFT) [21] to the training data and use the decomposition of the series to extract the strongest autocorrelations. As these frequencies determine the respective sizes of the strongest patterns, we can configure the convolutional layers so that they match those pattern lengths. Therefore, the CNN is directly tailored towards the patterns that dominate the time series under study. More precisely, the proposed method follows these steps to get from data input to forecasting:

1. z-standardize and transform the input data according to Fig. 1
2. Determine Fourier coefficients and top-2 autocorrelations
3. Define a 2-layer CNN and set the lengths of the convolution windows according to the top periodicities inferred from the Fourier coefficients. The CNN uses 96 filters, 50 past lags, a batch size of 16, a dropout rate of 20%, a learning rate of $10^{-4}$, the Adam [8] optimizer, 100 epochs and mean squared error as loss function.
4. Train CNN and evaluate performance on the test set

## 4     Experimental Analysis

### 4.1     Baseline Models and Evaluation

In order to validate the performance of the proposed algorithm, we perform forecasting experiments on 30 real world datasets.

**Methods.** The 10-steps ahead forecasting accuracy, measured in terms of root mean squared error (RMSE), is compared to the following baseline methods (cf. Sect. 2 for details):

- ARIMA, where model selection is based on a parameter grid opting for the AIC ($p = 1, 2, 3$, $q = 1, 2, 3$, $d = 1, 2$)
- Exponential Smoothing (ES)
- Arbitrated ensembles (ArbEns) specified in [23]
- Fourier: continuing reconstructed Fourier signals to generate forecasts
- CNNs parametrized with common filter length selections $\{2, 8\}$ opposed to Fourier based parameter estimation (CNN-Std.)
- Standard 2-layer LSTM architecture as described in [13] (LSTM-Std.)
- LSTM with the same architecture as the previous one, trained using the Snapshot Ensemble approach from [13] (LSTM-Snap)
- Our proposed method, combining CNNs and Fourier analysis (CNN-Fou)

An implementation, written mostly in Python, is available on GitHub[1]. For arbitrated ensembles, the *tsensembler* library written in R is used since the authors released it in that language.

---

[1] https://github.com/saschakrs/CNN-Fou, accessed April 6, 2019.

**Data.** The datasets listed in Table 1 are used for model training and validation. These time series originate from various domains such as air quality measurements and energy loadings. We also report the frequencies, mean values and standard deviations for each series. Furthermore, the length of each series is normalized such that all datasets have between 2.974 and 2.995 observations, making the results comparable.

In this analysis, we focus on the univariate case, i.e., forecasting the target variable based on its own past values. Prior to modeling, all data is standardized according to a z-transformation, i.e., $Y = \frac{Y-\mu}{\sigma}$. Note that mean $\mu$ and standard deviation $\sigma$ are determined based on only the training set as the holdout data points are unknown in a real scenario. For each approach and dataset, the most recent 10% of the series are used as test data in a windowing fashion. Every model is evaluated based on its ability to provide accurate 10-step ahead forecasts.

**Evaluation.** The results are summarized in Table 2. For each method and dataset we report the forecast RMSE for the 10% holdout data sample.

In addition, we provide the runtimes for model training in Fig. 2. The neural networks were trained on a NVIDIA Tesla K80 GPU and an Intel i7-6820HQ CPU was used for all other models as these don't profit from GPU usage.

The key value of our method lies in its robustness across different datasets without the need to manually incorporate domain knowledge. This implies that we leave all training architectures and parameters constant for each series (except for the Fourier coefficients that are learned for each dataset). Therefore, the

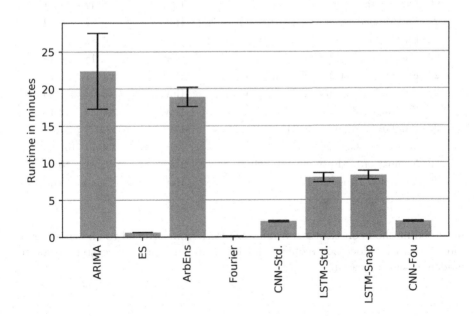

**Fig. 2.** Average model runtime across 30 datasets

**Table 1.** Overview of datasets

| i | Domain | Frequency | $\mu$ | $\sigma$ |
|---|--------|-----------|-------|----------|
| 1 | Water consumption indicators in Porto [23] | 30 min | 1.75 | 1.38 |
| 2 | Water consumption indicators in Porto | 30 min | 1079.03 | 271.90 |
| 3 | Water consumption indicators in Porto | 30 min | 10.08 | 7.03 |
| 4 | Solar radiation [23] | Hour | 930.72 | 289.03 |
| 5 | Solar radiation | Hour | 112.46 | 86.14 |
| 6 | Solar radiation | Hour | 912.74 | 279.47 |
| 7 | Solar radiation | Hour | 84.25 | 45.84 |
| 8 | Various air quality measurements [25] | Hour | 1593.23 | 366.28 |
| 9 | Various air quality measurements | Hour | 948.45 | 380.95 |
| 10 | Various air quality measurements | Hour | 20.29 | 7.89 |
| 11 | Various air quality measurements | Hour | 43.41 | 17.38 |
| 12 | Various air quality measurements | Hour | 1.00 | 0.30 |
| 13 | Various air quality measurements | Hour | 15.96 | 44.31 |
| 14 | Various air quality measurements | Hour | 27.82 | 35.80 |
| 15 | Various air quality measurements | Hour | 202.37 | 287.44 |
| 16 | Various air quality measurements | Hour | 84.96 | 122.25 |
| 17 | Various air quality measurements | Hour | 72.77 | 105.03 |
| 18 | Various air quality measurements | Hour | 58.06 | 29.89 |
| 19 | Various air quality measurements | Hour | 40.35 | 16.90 |
| 20 | Various air quality measurements | Hour | 33.42 | 18.81 |
| 21 | Energy loads such as electricity or gas [23] | Hour | 1.67 | 0.74 |
| 22 | Energy loads such as electricity or gas | Hour | 67.83 | 153.33 |
| 23 | Energy loads such as electricity or gas | Hour | 2.15 | 0.46 |
| 24 | Energy loads such as electricity or gas | Hour | 256.11 | 40.23 |
| 25 | Energy loads such as electricity or gas | Hour | 1021.64 | 203.35 |
| 26 | Exchange rates [24] | Day | 208.32 | 82.10 |
| 27 | Rainfall in Melbourne [24] | Day | 522.93 | 87.65 |
| 28 | Mean river flow [24] | Day | 456.74 | 105.85 |
| 29 | Number of births in Quebec [24] | Day | 23.01 | 10.68 |
| 30 | Mean wave height [24] | Hour | 4.49 | 3.16 |

proposed method constitutes a framework for automated end-to-end time series forecasting that does not require additional, manual processing steps prior to modeling and forecasting.

**Table 2.** RMSE for 30 different datasets, 10-steps ahead forecasting

| i | ARIMA | ES | ArbEns | Fourier | CNN-Std | LSTM-Std | LSTM-Snap | CNN-Fou |
|---|-------|------|--------|---------|---------|----------|-----------|---------|
| 1 | 0.60 | 1.12 | 0.50 | 1.04 | 0.51 | 0.88 | 0.51 | **0.38\*\*** |
| 2 | 0.65 | 0.93 | 0.37 | 0.99 | 0.52 | 0.45 | **0.21\*** | 0.25 |
| 3 | 0.69 | 1.22 | 0.51 | 1.15 | 0.54 | 0.67 | 0.41 | **0.39\*** |
| 4 | 0.60 | 1.09 | 0.43 | 1.09 | 0.41 | 0.53 | 0.36 | **0.31\*\*** |
| 5 | 0.86 | 1.45 | **0.53\*\*** | 1.54 | 0.95 | 0.74 | 0.58 | 0.59 |
| 6 | 0.47 | 0.89 | 0.42 | 0.93 | 0.40 | 0.59 | 0.30 | **0.30** |
| 7 | 0.77 | 1.48 | 0.57 | 1.11 | 0.67 | 0.81 | 0.47 | **0.45** |
| 8 | 0.54 | 1.17 | 0.42 | 0.71 | 0.45 | 0.57 | 0.48 | **0.30\*\*** |
| 9 | 0.95 | 1.90 | 0.67 | 1.34 | **0.40\*\*** | 0.77 | 0.55 | 0.55 |
| 10 | 0.51 | 0.83 | 0.27 | 0.52 | 0.38 | 0.31 | 0.30 | **0.27** |
| 11 | 0.39 | 0.83 | 0.33 | 0.46 | 0.25 | 0.41 | 0.37 | **0.21\*** |
| 12 | 0.73 | 1.34 | 0.62 | 1.35 | **0.48\*** | 0.91 | 0.56 | 0.50 |
| 13 | 1.29 | 1.29 | 1.31 | 1.59 | 1.08 | 1.66 | **1.01\*\*** | 1.19 |
| 14 | 0.56 | 0.41 | 0.29 | 0.69 | 0.29 | 0.37 | 0.28 | **0.27** |
| 15 | 0.46 | 1.87 | 0.40 | 0.50 | 0.56 | 0.42 | 0.45 | **0.40\*** |
| 16 | 0.65 | 1.04 | 0.65 | 0.81 | 0.78 | 0.94 | **0.46\*\*** | 0.53 |
| 17 | 0.63 | 1.01 | 0.63 | 1.09 | 0.72 | 1.13 | **0.50\*** | 0.51 |
| 18 | 0.74 | 1.16 | 0.27 | 1.37 | 0.40 | 0.30 | 0.31 | **0.26\*\*** |
| 19 | 0.94 | 1.06 | 0.44 | 1.41 | 0.51 | 0.69 | 0.43 | **0.32\*\*** |
| 20 | 0.86 | 0.86 | 0.71 | 1.19 | 0.86 | 0.86 | **0.42\*\*** | 0.59 |
| 21 | 1.16 | 1.33 | 0.49 | 1.45 | 0.51 | 0.55 | 0.42 | **0.37\*\*** |
| 22 | 0.11 | 0.02 | 0.03 | 0.12 | 0.03 | 0.05 | 0.03 | **0.03** |
| 23 | 1.05 | 0.27 | 0.07 | 1.46 | 0.06 | 0.13 | 0.06 | **0.05** |
| 24 | 0.76 | 1.16 | 0.57 | 1.19 | 0.37 | 0.78 | **0.44** | 0.45 |
| 25 | 0.74 | 1.06 | 0.23 | 0.90 | 0.34 | 0.38 | 0.21 | **0.18\*** |
| 26 | 0.55 | 1.01 | 0.27 | 0.79 | 0.38 | 0.30 | 0.19 | **0.15\*** |
| 27 | 0.52 | 1.59 | 0.37 | 0.63 | 0.44 | 0.51 | 0.26 | **0.25** |
| 28 | 0.39 | 1.69 | 0.34 | 0.59 | 0.44 | 0.43 | 0.22 | **0.22** |
| 29 | 0.52 | 1.06 | 0.44 | 0.70 | **0.23\*** | 0.80 | 0.33 | 0.32 |
| 30 | 0.72 | 0.61 | 0.63 | 1.19 | 0.62 | 0.91 | 0.73 | **0.55\*** |
| Avg | 0.68 | 1.09 | 0.46 | 1.00 | 0.49 | 0.63 | 0.40 | **0.37** |

## 4.2   Results

The results in Table 2 show that the proposed approach yields superior performance in 20 of 30 cases. We apply the Diebold-Mariano test [15] in order to evaluate whether the top performing model has a significantly different forecasting accuracy than the next best method. The null hypothesis states that the forecasting accuracy of the two methods are not different. One star (*) or two stars (**) indicate a p-value of less than 0.05 or 0.01, respectively. We can observe that:

- In two-thirds of all cases, the Fourier-integrated CNN is superior or equal to all baseline methods with an average performance gain of 10.25% compared to the next best method (Snapshot Ensembles).
- Traditional models such as ARIMA, exponential smoothing and simple Fourier forecasting show poor performance compared to advanced methods.
- In terms of runtime, modern CNN implementations benefit heavily from strong parallelization on GPUs. Here, LSTMs suffer from their sequential nature that makes them harder to train efficiently. Compared to the training of LSTMs, CNNs are faster by a factor of 4.
- While arbitrated ensembles are amongst the top performers for each dataset, they are computationally expensive since a number of base learners must be trained in order for ensembles to be effective.
- Result significance differs depending on the dataset. This is due to varying problem complexity between datasets.

## 5    Conclusions

We presented an end-to-end time series forecasting framework based on CNNs and Fourier analysis which is more computationally efficient and accurate than existing approaches. We made use of the natural structure of time series data in order to capture repeating patterns effectively. It was shown that Fourier analysis can be used to enhance CNN parametrization and improve forecasting performance without the need to adapt the setup for new datasets. The method was compared to various state of the art forecasting methods and generated the most accurate results in the majority of thirty use cases.

While this work focused on the univariate case, its extension to the multivariate scenario will be part of our future research. The basic methodology will be the same, however, the additional amount of features requires more efficient preprocessing and modeling strategies. Apart from that, it is worth investigating the effects when scaling the framework to larger datasets, especially in terms of CNN architecture.

## References

1. Adhikari, R.: A neural network based linear ensemble framework for time series forecasting. Neurocomputing **157**, 231–242 (2015)
2. Bengio, Y., Simard, P., Frasconi, P.: Learning long-term dependencies with gradient descent is difficult. IEEE Trans. Neural Netw. **5**(2), 157–166 (1994)
3. Gers, F.A., Eck, D., Schmidhuber, J.: Neural nets WIRN vietri-01. perspectives in neural computing. In: Tagliaferri, R., Marinaro, M. (eds.) Applying LSTM to Time Series Predictable Through Time-Window Approaches. Perspectives in Neural Computing, pp. 193–200. Springer, London (2002)
4. Hamilton, J.D.: Time Series Analysis, vol. 2. Princeton University Press Princeton, Princeton (1994)

5. He, Z., Gao, S., Xiao, L., Liu, D., He, H., Barber, D.: Wider and deeper, cheaper and faster: tensorized LSTMs for sequence learning. In: Advances in Neural Information Processing Systems, pp. 1–11 (2017)
6. Hipel, K.W., Ian McLeod, A.: Time Series Modelling of Water Resources and Environmental Systems, vol. 45. Elsevier, Amsterdam (1994)
7. Vaswani, A., et al.: Attention is all you need. In: Advances in Neural Information Processing Systems, pp. 5998–6008 (2017)
8. Diederik, P.: Kingma and Jimmy Ba. Adam. A method for stochastic optimization. arXiv preprint arXiv:1412.6980 (2014)
9. Marcellino, M., Stock, J.H., Watson, M.W.: A comparison of direct and iterated multistep AR methods for forecasting macroeconomic time series. J. Econom. 135(1–2), 499–526 (2006)
10. Hochreiter, S., Schmidhuber, J.: Long short-term memory. Neural Comput. 9(8), 1735–1780 (1997)
11. Hornik, K., Stinchcombe, M., White, H.: Multilayer feedforward networks are universal approximators. Neural Netw. 2(5), 359–366 (1989)
12. Huang, G., Li, Y., Pleiss, G., Li, Z., Hopcroft, J., Weinberger, K.: Snapshot ensembles: Train 1 Get M for free. In: Proceedings of the International Conference on Learning Representations (ICLR 2017)
13. Krstanovic, S., Paulheim, H.: Stacked LSTM snapshot ensembles for time series forecasting. In: Proceedings of ITISE 2018, International Conference on Time Series and Forecasting, Godel (2018)
14. Längkvist, M., Karlsson, L., Loutfi, A.: A review of unsupervised feature learning and deep learning for time-series modeling. Pattern Recogn. Lett. 42(2014), 11–24 (2014)
15. Harvey, D., Leybourne, S., Newbold, P.: Testing the equality of prediction mean squared errors. Int. J. Forecast. 13(2), 281–291 (1997)
16. Lichman, M.: 2013. UCI Machine Learning Repository (2013). http://archive.ics.uci.edu/ml
17. Malhotra, P., Vig, L., Shroff, G., Agarwal, P.: Long short term memory networks for anomaly detection in time series. In: Proceedings, vol. 89. Presses universitaires de Louvain (2015)
18. Pascanu, R., Mikolov, T., Bengio, Y.: On the difficulty of training recurrent neural networks. International Conference on Machine Learning, pp. 1310–1318 (2013)
19. Sharma, D., Issac, B., Raghava, G.P.S., Ramaswamy, R.: Spectral Repeat Finder (SRF): identification of repetitive sequences using Fourier transformation. Bioinformatics 20(9), 1405–1412 (2004)
20. Sutskever, I., Vinyals, O., Le, Q.V.: Sequence to sequence learning with neural networks. In: Advances in Neural Information Processing Systems, pp. 3104–3112 (2014)
21. Welch, P.: Network model. Neurocomputing 50(2003), 159–175 (1967). Neurocomputing 50, 159–175 (2003)
22. Oliveira, M., Torgo, L.: Ensembles for time series forecasting. In: JMLR: Workshop and Conference Proceedings, vol. 39, pp. 360–370 (2014)
23. Cerqueira, Vítor, Torgo, Luís, Pinto, Fábio, Soares, Carlos: Arbitrated Ensemble for Time Series Forecasting. In: Ceci, Michelangelo, Hollmén, Jaakko, Todorovski, Ljupčo, Vens, Celine, Džeroski, Sašo (eds.) ECML PKDD 2017. LNCS (LNAI), vol. 10535, pp. 478–494. Springer, Cham (2017). https://doi.org/10.1007/978-3-319-71246-8_29
24. Hyndman, R.: Time series data library. https://datamarket.com/data/list/?q=provider:tsdl. Accessed 6 April 2019

25. Dua, D., Graff, C.: 2019. UUCI Machine Learning Repository. School of Information and Computer Science, University of California, Irvine, CA (2019). http://archive.ics.uci.edu/ml. Accessed 6 April 2019
26. Van Den Oord, A., et al.: WaveNet: a generative model for raw audio. In: SSW (2016)
27. LeCun, Y., Bengio, Y., Hinton, G.: Deep learning. Nature **521**(7553), 436 (2015)

# Integrating LSTMs with Online Density Estimation for the Probabilistic Forecast of Energy Consumption

Julian Vexler$^{(\boxtimes)}$ and Stefan Kramer$^{(\boxtimes)}$

Johannes Gutenberg-Universität Mainz, Staudingerweg 9, 55128 Mainz, Germany
jvexle01@uni-mainz.de, kramer@informatik.uni-mainz.de

**Abstract.** In machine learning applications in the energy sector, it is often necessary to have both highly accurate predictions and information about the probabilities of certain scenarios to occur. We address this challenge by integrating and combining long short-term memory networks (LSTMs) and online density estimation into a real-time data streaming architecture of an energy trader. The online density estimation is done in the MiDEO framework, which estimates joint densities of data streams based on ensembles of chains of Hoeffding trees. One attractive feature of the solution is that queries can be sent to the here-called forecast-based point density estimators (FPDE) to derive information from a compact representation of two data streams, leading to a new perspective to the problem. The experiments indicate promising application possibilities of FPDE, including but not limited to the estimation of uncertainties, early model evaluation and the simulation of alternative scenarios.

**Keywords:** Density estimation · Data stream · Energy consumption

## 1 Introduction

In recent years, Machine Learning (ML) and Data Mining (DM) algorithms have become increasingly popular in the energy sector. The popularity of such algorithms is to a large extent due to the increasing computational power of CPUs and GPUs, making ML and DM algorithms reliable tools to find patterns in data, make highly accurate predictions in real-time settings and optimize processes. Many big companies have embedded those methods into their production systems (e.g. Uber [1], Amazon [2] or Facebook [3]), but there are still many small and medium-sized enterprises (SMEs) in the energy sector, where ML and DM algorithms can simplify processes and deliver more accurate solutions, but are not part of the infrastructure yet.

This paper presents the current status of an on-going project with an energy trader. The main goal is the integration of long short-term memory networks (LSTMs) and MiDEO [4], a framework to estimate joint densities of data streams, into the real-time production system of the energy trader. Our system

© Springer Nature Switzerland AG 2019
P. Kralj Novak et al. (Eds.): DS 2019, LNAI 11828, pp. 533–543, 2019.
https://doi.org/10.1007/978-3-030-33778-0_40

shall not only be capable of automating model handling processes (e.g., hyperparameter optimization, model training, model evaluation and model selection), but we also propose and apply a new way of combining LSTMs with MiDEO.

The main contributions of this paper are: (i) The paper proposes a combination of neural networks for time series forecasting (LSTMs) and *online* density estimators (EDO, i.e., estimation of densities online [4]), to overcome the limitations of either approach. LSTMs give accurate predictions, but lack the possibility to estimate the uncertainty of their output, and density estimators can estimate uncertainty, but are not designed to give accurate predictions. Coupling a density estimator that works in an online manner with an LSTM has the advantage of being able to extract uncertainty information from LSTMs (i.e., LSTM forecasts) quickly. This is a practical option that has not been considered so far. Employing EDO for this task has the advantage that arbitrary inference tasks can be supported on the basis of the density estimate, which is currently not possible for general neural network architectures. (ii) Technically, the integration is achieved by the processing of two input streams, one being the stream of actual values and one being the stream of forecasts. (iii) The approach has been developed for the prediction of energy consumption, where approaches so far have used *either* neural networks *or* density estimation. (iv) We test the framework regarding the sensitivity to hyperparameters, showing the strengths as well as the weaknesses, and compare to the performance of LSTMs alone and of other density estimators. Daily-based scoring functions for conditioned point density estimation are developed as part of the evaluation.

## 2    Related Work

Predicting energy consumption is an important topic with global relevance. A lot of work has been done to address the challenge on different levels, let it be for individual households, buildings, countries or SMEs. Concerning the energy consumption of individual households, Berriel *et al.* [5] studied deep fully connected, convolutional and LSTM networks, Alobaidi *et al.* [6] proposed an ensemble-based artificial neural network (ANN) framework and Kong *et al.* [7] presented another LSTM based framework.

Zhao and Magoulès [8] published a review paper on the prediction of building energy consumption, including engineering, statistical and artificial intelligence methods. In a more recent review, Deb *et al.* [9] analyze different ML forecasting techniques and hybrid models.

Other work addresses the challenge of predicting the energy consumption of certain countries. For example, Kaytez *et al.* [10] analyze least squares support vector machines, multi linear regression models and ANNs to predict the energy consumption of Turkey; or in 2015, Dedinec *et al.* [11] used deep belief networks to forecast the energy load in Macedonia.

While the above approaches mostly focus on various types of neural networks for predicting energy consumption, the latter has also been pursued by applying density estimation methods. Arora and Taylor [12] use conditional kernel density estimation approaches to predict electricity smart meter data, and Hong and Fan [13] provide a review of different probabilistic electric load forecasting methodologies, techniques and evaluation methods.

All the mentioned work mostly focuses on the prediction of energy consumption on distinct levels using different approaches, solely trying to achieve the lowest prediction errors. In contrast, we employ a probabilistic approach to not *only* estimate the distribution of either the forecasts or the energy consumption, but to calculate a probabilistic stream of point densities comprising the information of *both actual consumption and forecast*. Based on this approach, statistical, ML or DM methods can be applied to retrieve more information from the output stream. Moreover, the output is represented in terms of the log-likelihood (LL). This has the advantage that neither the measurement of the true energy consumption nor the unit of the true measurements is revealed. Consequently, the calculated densities can be anonymously shared with other companies or customers, e.g. for data analyses purposes.

# 3 Methodologies

In this section, MiDEO is shortly introduced, which is the framework where EDO is implemented, followed by a formal definition of *online density estimation* to help understanding the FPDE approach. Afterwards, we explain FPDE, give some use cases to point out the diversity of possible applications and conclude with some scoring functions for the evaluation of the point densities.

## 3.1 MiDEO

EDO is an ensemble of online density estimators, which uses classifier chains of Hoeffding trees to model dependencies among features. We choose EDO as our online density estimator because it is competitive with other density estimators [4], can handle mixed types of variables, enables inference tasks, trains fast, and allows to address privacy issues. As EDO is implemented in the framework MiDEO[1], this is the framework to be coupled with LSTMs. In the following, we give a formal definition of online density estimation based on Geilke *et al.* [4], which will be further used to describe FPDE and the scoring functions.

**Online Density Estimation.** Let $X$ be a random variable, *values*$(X)$ a set of possible outcomes of $X$ and $[a, b]$ an interval with $a, b \in values(X)$. We call $X$ a *continuous random variable* if it can take on any value in the range of $[a, b]$. The *joint density* $f$ over random variables $X_1, ..., X_m$ is a non-negative Lebesgue-integrable function $f(X_1, ..., X_m)$, such that

$$Pr(X_1 \in [a_1, b_1], ..., X_m \in [a_m, b_m]) := \int_{a_1}^{b_1} ... \int_{a_m}^{b_m} f(x_1, ..., x_m) dx_1 ... dx_m. \quad (1)$$

---

[1] https://github.com/kramerlab/mideo.

Let $x$ be an *instance* of $f$, denoting an assignment of random variables $X_1 = v_1, ..., X_m = v_m$, such that $v_i \in values(X_i)$ for $1 \le i \le m$. If the joint density is conditioned on some random variables $Y_1, ..., Y_l$, then we write $f(X_1, ..., X_m | Y_1, ... Y_l)$, where we call $X_1, ..., X_m$ the target (random) variables.

**Definition 1.** *Let $F := \{f_i(X_1, ..., X_m | Y_1, ..., Y_l) | 1 \le i \le k \in \mathbb{N}\}$ be a set of joint densities. A data stream of $f \in F$, denoted as $stream(f)$, is a possibly infinite sequence of instances $x_1, x_2, ...$ that are drawn according to the probability distribution induced by $f$, where $stream(f)[1 : N] := \{x_1, x_2, ..., x_N\}$. A data stream over $F$, denoted as $stream(F)$, is a possibly infinite sequence of instances $stream(F) := stream(f_{j_1})[1 : N_{j_1}] \circ stream(f_{j_2})[1 : N_{j_2}] \circ ...$, where $f_{j_i} \in F$ and $j_i, N_{j_i} \in \mathbb{N}$.*

**Definition 2.** *Let $F := \{f_i(X_1, ..., X_m | Y_1, ..., Y_l) | 1 \le i \le k \in \mathbb{N}\}$ be a set of joint densities and $stream(F)$ be a data stream over $F$. An algorithm is called online density estimator, if*

1. *it receives this sequence instance by instance,*
2. *it has a limited amount of memory, and,*
3. *after receiving an instance $x_i$, it produces a density estimate $\hat{f}_i$.*

The estimation of the density of a data stream by an online density estimator is called *online density estimation*.

### 3.2   Forecast-Based Point Density Estimation

Dealing with streaming data incorporates the difficulty of not knowing how the energy consumption forecasts of an LSTM perform in advance. Furthermore, it might be desirable to not only have point forecasts, but to retrieve additional information from the forecasts and the true data. Therefore, we suggest to use a density estimator, e.g. EDO [4], to obtain a condensed representation of the forecast and consumption patterns. In the following, we provide a formal explanation.

Let $y_1, y_2, ...$ be a possibly infinite sequence of instances, where $y_i \in \mathbb{R}, \forall i \in \mathbb{N}$. Having a time window of size $l$, a prediction range $k$ and a prediction interval of size $m$, with $l, k, m \in \mathbb{N}$, let $y_{t_{-l+1}}, y_{t_{-l+2}}, ..., y_{t_0}$ denote the input to a trained LSTM at time point $t_0$. Then the output of the LSTM, formalized as $(\hat{y}_{t_k}, \hat{y}_{t_{k+1}}, ..., \hat{y}_{t_{k+m-1}})$, represents the forecasts for the specified prediction interval. Our focus lies on point density estimates, hence, the conditioned forecast-based density is calculated by EDO for each of the point forecasts:

$$\hat{f}(\hat{y}_j | y_{t_{-d+1}}, y_{t_{-d+2}}, ..., y_{t_0}), \forall j \in [t_k, ..., t_{k+m-1}], \tag{2}$$

where $\hat{y}_j$ and $y_i$ stand for the variable assignment of the target variable $\hat{Y}_j$ and the random variables $Y_i$. Figure 1 shows an example where the FPDE estimates the point densities of the LSTM forecasts based on a time window of the true consumption in the range from $t_{-3}$ to $t_0$. According to Eq. 2 it follows

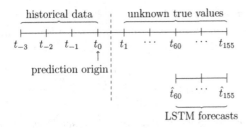

**Fig. 1.** The prediction origin $t_0$ represents the time point until which the energy consumption data is available, where the true consumption starting from $t_1$ is unknown. The prediction interval for the LSTM is from 60 to 155 time steps ahead of $t_0$.

$$\hat{f}(\hat{y}_j | y_{t_{-3}}, y_{t_{-2}}, y_{t_{-1}}, y_{t_0}), \forall j \in [\hat{t}_{60}, ..., \hat{t}_{155}]. \tag{3}$$

Note, the time window $d$ of historical data in Eq. 2 can differ from the time window $l$ of the LSTM, because LSTMs need larger time windows to recognize patterns in the data, whereas density estimators output lower probabilities the more variables are used. Latter can lead to LL-values of $-\infty$ (see Sect. 4).

**Use Cases.** The main purpose of the FPDE-approach is the combination of the LSTM's prediction strength and the inference possibilities with MiDEO.

Alternative scenarios are useful if the energy consumption was influenced by exceptional events or if incidents shall be simulated which are not present in the true data. In case of LSTMs, adapting the training set requires new model training which can be time-consuming based on the complexity of the model and the data. In contrast, MiDEO achieves to train density estimators in approximately 5 s on a dataset with 35000 instances $\times$ 65 features.

Furthermore, FPDE delivers a new condensed data stream incorporating the information of true consumption and LSTM forecasts. This output prevents any conclusion regarding the unit of the data (kWh or MWh). Hence, the new data can be shared for further statistical analyses without revealing the ground truth.

Finally, MiDEO can not only be used to estimate the joint probability of LSTM forecasts over time intervals like hours or days, but also the marginal probability of point forecasts. In the remainder of this paper, we will focus on the latter aspect, as it is rarely considered, although it has its own advantages.

### 3.3    Scoring Functions

To compare the results obtained by varying the hyperparameters of EDO and the time windows of historical data, we introduce scoring functions which represent the relative changes in the point densities, once for relative changes within days and once for relative changes between days. We consider relative changes because absolute values lead to misinterpretations, especially because the densities get lower if the time window of the true consumption is increased.

**Definition 3.** *Let $f$ be a point density, $d \in \mathbb{N}$ a time window, $t_0$ the prediction origin and $a, b \in \mathbb{N}$ a lower and an upper bound for some time interval $[t_a, t_b]$. We define the mean relative change of densities within a day as*

$$mrc_{intra}(f, a, b, d, t_0) := \frac{1}{b-a} \sum_{i=a}^{b-1} \frac{f(\hat{y}_{t_{i+1}}|y_{t-d}, y_{t-d+1}, ..., y_{t_0})}{f(\hat{y}_{t_i}|y_{t-d}, y_{t-d+1}, ..., y_{t_0})}. \quad (4)$$

*The $mrc_{intra}$ averaged over all days of a data set $D$ with a prediction interval of size $m$, representing the measurements of a whole day, is*

$$mrc_{intra}^D(f, m, d, t_0) = \frac{1}{n} \sum_{i=0}^{n-1} mrc_{intra}(f, i * m, (i+1) * m - 1, d, t_{i*m}) - 1, \quad (5)$$

*where $n = \frac{|D|}{m}$. We subtract 1 for an easier interpretation of the result.*

Analogously, if we want to calculate the mean relative change of densities between days, denoted as $mrc_{inter}$, then we use Eq. 4 and set $b = a + 1$, where $a$ stands for the last measurement of a day. It follows

$$mrc_{inter}(f, a, d, t_0) := \frac{f(\hat{y}_{t_{a+1}}|y_{t-d}, y_{t-d+1}, ..., y_{t_0})}{f(\hat{y}_{t_a}|y_{t-d}, y_{t-d+1}, ..., y_{t_0})}. \quad (6)$$

Hence, the average $mrc_{inter}$ between all days of data $D$ is given by

$$mrc_{inter}^D(f, m, d, t_0) = \frac{1}{n-1} \sum_{i=0}^{n-2} mrc_{inter}(f, (i+1) * m - 1, d, t_{i*m}) - 1. \quad (7)$$

## 4 Experimental Setup

The section starts with a short description of the real-world data and the trading type used for the training and forecasting with LSTM and FPDE. Afterwards, experiments with different parameterizations are conducted, followed by an evaluation of the results and a comparison with other methods.

### 4.1 Data Description and Day-Ahead Market

The available real-world data contains information about the energy consumption of a firm location from 2015 to 2016. The instances are measured quarterhourly in kWh, resulting in 96 measurements per day and 70176 records in total.

In the electricity market there are currently two trading types on the exchange EPEX SPOT in Europe, namely the intraday trading and the dayahead (DA) trading, whereas in the latter case the electricity is traded until 12 noon for the consecutive day. Traders have to ensure that their forecasts are timely available, which is a non-trivial challenge due to possible delays of incoming information about energy production and consumption. Therefore, the

LSTM forecasts of the energy consumption for the provided company location are created at 9:00 o'clock for the consecutive day, giving a buffer of 3 h to handle exceptional situations. Hence, the prediction origin $t_0$ is at 9:00 o'clock and the prediction interval ranges from 60 to 155 time steps ahead, representing the consecutive day from 0:00 o'clock to 23:45. This type of forecasting will be denoted as *DA-trade* and is illustrated in Fig. 1.

### 4.2 Experiments

The point probabilistic condensed representation $f(X|Y_1,...Y_l)$ combines information about historical data by the variables $Y_i$ and LSTM forecasts by the target variable $X$. In the following experiments the influence of different time window lengths and hyperparameter values using FPDE with offline training and online forecasting, based on the DA-trade, is analyzed.

**Offline Training with DA-Based Forecasting.** The LSTM network used in this paper was trained for 300 epochs, batch size 1 and a time window of 3 days. Using the stochastic gradient descent as optimizer and the mean absolute error as loss function, experiments have shown that a shallow network structure with one hidden layer and 10 LSTM blocks was sufficient to capture the patterns of the data, achieving a mean absolute percentage error (MAPE) of 2.7% on the test data. Note that hyperparameter optimization and experiments with deeper network structures have been performed, but will not be presented here, as it is not the focus of this paper.

In the current case, FPDE takes a time window of size $d = 4$, where the point densities are estimated for a prediction range $k = 60$ and a prediction interval length $m = 96$ (see Eq. 3). Training was performed on the data from 2015.

The observation of monthly distributions for the year 2016, as illustrated in Fig. 2, is one way to help interpreting the point densities. Figure 2(a) represents the distribution of the forecasts based on the true energy consumption. The scattering during April and August is smaller than in the remaining months which, combined with the higher probabilities, indicates that the distribution is similar to the previous year. March is striking as it has the greatest scattering, explainable by the course of the energy consumption reaching ranges rarely observed before. This is only exceeded by December, where new minima were measured in 2016 such that many FPDEs are marked as outliers.

Taking Fig. 2(b) into consideration, a correlation between estimated densities and errors can be seen, with a negative correlation regarding the medians. Although the scattering of the error bars influences the scattering of the density bars, the latter is also influenced by the true consumption. This explains why distant outliers occur in the densities regarding December although they are not present in the errors. Nevertheless, it is remarkable that the errors can only be computed as soon as the true consumption being predicted becomes available, whereas the estimated point densities of the FPDE are available as soon as the forecasts are made. This correlation indicates that a first evaluation of the forecasts might be performed, without even knowing the true consumption.

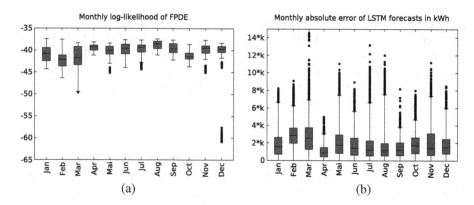

**Fig. 2.** Monthly representation of (a) the estimated point densities of the LSTM forecasts and (b) the absolute errors of the LSTM forecasts. k is an anonymization factor.

To examine the influence of the hyperparameters [4], experiments are conducted with 20, 40 or 100 bins; a maximum of 5000, 10000 or 15000 kernels; and ensembles of 1, 5 or 8 classifier chains. The most noticeable results are summarized in Table 1. First, we consider the $mrc^D_{intra}$. Using 20 bins, 10k–15k kernels, and ensembles with at least 5 classifier chains makes the density estimator more prone to decreases in the forecasts and leads, on average, to larger relative decreases than increases in the density. On the other hand, if only one classifier chain, 5k kernels and 40 or 100 bins are used, then the highest mean relative changes can be observed. The latter can be explained by the vulnerability of a single classifier chain, hence, this result has to be handled with caution.

Concerning the $mrc^D_{inter}$, which is strongly influenced by the true consumption, the bins have the greatest impact on the changes, followed by the ensemble size. The influence of the kernels is comparably small. The more bins and the greater the ensemble size, the larger the mean relative change between days.

In the previous experiments the density was conditioned on four random variables representing an hour of a day, which, based on given conditions, might not be representative. In the next experiment, we condition the densities on larger time windows of the true consumption, i.e. on 4, 8, 16, 32 and 64 random variables. Keeping the number of bins, kernels and ensemble size fixed, we obtain the results shown in Table 2. If the density is conditioned on an increasing number of random variables, then the $mrc^D_{intra}$ has, on average, greater relative decreases than increases and gets smoother, as the changes are smaller than in the case of 4 or 8 random variables. The reason is that larger time windows make the density estimates less prone to strong fluctuations. Furthermore, the $mrc^D_{inter}$ increases until the time window size 16, followed by a decrease for greater sizes. Finally, conditioning the FPDE on too many variables is not a good choice. First, the relative changes get too small, possibly leading to a loss of information, and second, the density estimator calculates the distribution over a great range of possible value combinations. The latter can lead to LLs of $-\infty$ for value assignments not represented in the training set.

**Table 1.** Mean relative density changes for different parameters.

| Bins | Kernels | Ensemble | $mrc_{intra}^{D}$ | $mrc_{inter}^{D}$ |
|------|---------|----------|-------------------|-------------------|
| 20 | 10k | 8 | −2.90e-06 | +1.87e-03 |
| 20 | 15k | 8 | −2.90e-06 | +1.87e-03 |
| 40 | 5k | 1 | +2.62e-05 | +3.33e-03 |
| 100 | 5k | 1 | +3.24e-05 | +3.57e-03 |
| 100 | 10k | 8 | +6.64e-06 | +3.89e-03 |

**Table 2.** Mean relative density changes for different time windows.

| Time window | $mrc_{intra}^{D}$ | $mrc_{inter}^{D}$ |
|-------------|-------------------|-------------------|
| 4 | −1.08e-06 | +1.80e-03 |
| 8 | +1.30e-06 | +2.32e-03 |
| 16 | −7.61e-07 | +3.80e-03 |
| 32 | −4.71e-07 | +2.96e-03 |
| 64 | −5.67e-07 | +0.81e-03 |

**Table 3.** Average log-likelihoods and mean difference between point-wise density estimates of forecasts and true power consumption

| Model | kernel | bandwidth | $\overline{LL}_{prg}$ | $\overline{LL}_{power}$ | $|\hat{f}(\hat{y}) - \hat{f}(y)|$ |
|-------|--------|-----------|-----------------------|-------------------------|-----------------------------------|
| FPDE | | | −40.36 | −40.65 | 0.292 |
| KDE | Gaussian | 10 | −539.54 | −610.01 | 196.857 |
| KDE | Gaussian | 20 | −155.85 | −170.90 | 47.698 |
| KDE | Gaussian | 100 | −39.16 | −40.09 | 2.159 |
| KDE | Gaussian | 200 | −37.19 | −37.58 | 0.888 |
| KDE | exponential | 10 | −47.91 | −49.79 | 4.544 |
| KDE | exponential | 20 | −39.98 | −41.04 | 2.420 |
| KDE | exponential | 100 | −36.83 | −37.18 | 0.768 |
| KDE | exponential | 200 | −37.99 | −38.23 | 0.469 |

**Model Evaluation.** In order to evaluate the results of the FPDE, a comparison with different kernel density estimators (KDEs) is performed, as it is shown in Table 3. The comparison includes the average LLs as well as the mean absolute difference between the density estimates. As we work with point densities, the estimated density curve of the true consumption forms the optimal curve we want to approximate. Hence, the smaller the mean difference between the density estimates, the better. The result of the FPDE is achieved quite fast and the mean difference error was in all of our experiments below 1. In case of the KDEs, an exhausting search for the optimal hyperparameters has to be performed, until the mean difference errors get close to the error 0.292 of the FPDE.

Finally, the advantage of the FPDE lies in the fast approximation of the true density curve, whereas with other methods exhaustive parameter searches have to be conducted until a good approximation can be found.

## 5   Conclusion and Future Work

We proposed a novel way to compactly represent the distributions of two time series, i.e. the energy consumption of a company and the respective forecasts according to the DA-trade. We have shown that it is possible to combine MiDEO with LSTMs and furthermore, these methods can be integrated into real-time streaming architectures of companies. The performed experiments showed promising results and indicate several application areas for the FPDE approach.

In the future, ensembles of LSTMs with MiDEO on top could be developed, in order to create more accurate energy consumption forecasts. Besides the aspect of hyperparameter optimization and a tighter coupling of either approach, data preprocessing is another relevant topic. The application of time-dependent fading factors or the integration of weather forecasts might improve the results and deliver further valuable information. Finally, other trading types and daily segmentations than DA-trade and 15-min intervals can be investigated.

## References

1. Zhu, L., Laptev, N.: Deep and confident prediction for time series at uber. In: 2017 IEEE International Conference on Data Mining Workshops (ICDMW), pp. 103–110 (2017)
2. Dai, H., Kozareva, Z., Dai, B., Smola, A., Song, L.: Learning steady-states of iterative algorithms over graphs. In: International Conference on Machine Learning, pp. 1114–1122 (2018)
3. Lin, F., Beadon, M., Dixit, H.D., Vunnam, G., Desai, A., Sankar, S.: Hardware remediation at scale. In: 2018 48th Annual IEEE/IFIP International Conference on Dependable Systems and Networks Workshops (DSN-W), pp. 14–17. IEEE (2018)
4. Geilke, M., Karwath, A., Frank, E., Kramer, S.: Online estimation of discrete, continuous, and conditional joint densities using classifier chains. Data Min. Knowl. Discov. **32**(3), 561–603 (2018)
5. Berriel, R., Teixeira Lopes, A., Rodrigues, A., Varejao, F., Oliveira-Santos, T.: Monthly energy consumption forecast: a deep learning approach. In: 2017 International Joint Conference on Neural Networks (IJCNN), pp. 4283–4290 (2017)
6. Alobaidi, M.H., Chebana, F., Meguid, M.A.: Robust ensemble learning framework for day-ahead forecasting of household based energy consumption. Appl. Energy **212**, 997–1012 (2018)
7. Kong, W., Dong, Z.Y., Jia, Y., Hill, D.J., Xu, Y., Zhang, Y.: Short-term residential load forecasting based on LSTM recurrent neural network. IEEE Trans. Smart Grid **10**(1), 841–851 (2017)
8. Zhao, Hx, Magoulès, F.: A review on the prediction of building energy consumption. Renew. Sustain. Energy Rev. **16**(6), 3586–3592 (2012)
9. Deb, C., Zhang, F., Yang, J., Lee, S.E., Shah, K.W.: A review on time series forecasting techniques for building energy consumption. Renew. Sustain. Energy Rev. **74**, 902–924 (2017)

10. Kaytez, F., Taplamacioglu, M.C., Cam, E., Hardalac, F.: Forecasting electricity consumption: a comparison of regression analysis, neural networks and least squares support vector machines. Int. J. Electr. Power Energy Syst. **67**, 431–438 (2015)

11. Dedinec, A., Filiposka, S., Dedinec, A., Kocarev, L.: Deep belief network based electricity load forecasting: an analysis of macedonian case. Energy **115**, 1688–1700 (2016)

12. Arora, S., Taylor, J.W.: Forecasting electricity smart meter data using conditional kernel density estimation. Omega **59**, 47–59 (2016)

13. Hong, T., Fan, S.: Probabilistic electric load forecasting: a tutorial review. Int. J. Forecast. **32**(3), 914–938 (2016)

# Author Index

Printed in the United States
by Bookmasters

Printed in the United States
By Bookmasters